Memory Mass Storage

Giovanni Campardo · Federico Tiziani ·
Massimo Iaculo
Editors

Memory Mass Storage

Springer

Editors
Giovanni Campardo
Numonyx
Via C. Olivetti 2
20041 Agrate Brianza Milano
Italy
giovanni.campardo@numonyx.com

Federico Tiziani
Micron
Via C. Olivetti 2
20041 Agrate Brianza Milano
Italy
ftiziani@micron.com

Massimo Iaculo
Micron
Via Remo De Feo 1
80022 Arzano Napoli
Italy
miaculo@micron.com

Front cover:
Giovanni Campardo
Ulivi di Puglia (Puglia's olive trees)
oil on wood, 40 × 30, 2008
[Photographed by Lino Mazzucchi]

ISBN 978-3-642-14751-7 e-ISBN 978-3-642-14752-4
DOI 10.1007/978-3-642-14752-4
Springer Heidelberg Dordrecht London New York

Cover design: WMXDesign GmbH, Heidelberg

Printed on acid-free paper

Springer is part of Springer Science+Business Media (www.springer.com)

Preface

> Creditors have better memories than debtors. (Benjamin Franklin, January 17, 1706–April 17, 1790, one of the Founding Fathers of the United States)

Mnemosyne (Greek Μνημοσύνη), the personification of memory in Greek mythology, was the daughter of Gaia and Uranus and the mother of the Muses by Zeus. Zeus and Mnemosyne lay together for nine consecutive nights during which the nine Muses were conceived. Mnemosyne ruled over a pool in Hades, a counterpart to the river Lethe.[1] One of the five rivers of Hades, the Lethe, which flowed around the cave of Hypnos and through the underworld, was the river from which the souls of the dead drank so they would not remember their past lives when they were reincarnated. Initiates were encouraged to drink from Mnemosyne's pool when they died, rather than from Lethe which induced to oblivion, in order to learn from their past so as to achieve a higher level of wisdom.

This is a book about technologies that allow us to store large quantities of data and provide reliable storage devices. We describe the techniques and methodologies that allow us to store data, large quantities of data, just (or precisely (choose the better word)) called mass memories storage.

However, memory is not an issue that concerns only technicians, engineers, and physicists, but is also very important in the philosophical tradition.

Just think of Plato and Aristotle: when Plato said that all knowledge is reminiscence, his concept was related to memory or, better still, closely linked to memory. That is, he thought that everything that we know is the memory of what we learned in another life, in another world, before falling into this world. On the other hand, in Aristotle's philosophy, the perspective is entirely different. Aristotle carefully distinguishes between memory and reminiscence. Memory is that phenomenon by which we remember things of the past, whereas reminiscence is the attempt to recapture a missing piece of the past. Thus, reminiscence has an aspect of consciousness that is somehow absent in memory. In contemporary culture there is great interest in regard to memory, an interest that is not exclusively the realm of philosophers, but

---

Translated by Marcello Campardo

[1] The term comes from a Greek root, *leth*, "forget," from which derives *alètheia*, "truth" with the alpha privative, thus indicating "that you never forget."

is of interest to neurologists, psychologists, psychiatrists, scholars of the brain in general, and of course technicians.

However, memory is not confined simply to the acts of storing and subsequent recall, but is deeply rooted in the mind of man, who is terrified of being forgotten. Our desire for immortality, whether or not we believe in the immortality of the soul, is still strong. The idea of not being forgotten is so vast that it reflects a peculiar expression: we are full of objects that bring to mind something; limited to the people. Cemeteries are places that bring to mind people who are missing, as well as monuments, steles, tombstones; in short they provide a way to idealize a presence that no longer exists, in a physical image, to recall something.

In the film *Blade Runner*, there are characters called "replicants," artificial beings identical to the humans among whom they live, who do not know that they themselves are not human. And that is their problem. When one of these androids realizes that he or she is not a real human being, but a replicant (a creature who has a memory that has been inserted in his or her brain as in a machine and is not a real memory), there is a crisis. The concern that the memories are false causes terrible anguish because the replicant has no memory or nostalgia for the past. The absence of nostalgia, the absence of memory, is a loss of identity. If we did not have our memories we would not know who we are.

Personal identity is based on memory, on one's autobiography. I know that I am the same person that I was when I was 3 years old, and this is an absolute certainty, Hume would say, even if that knowledge is not confirmed in any way it is nonetheless an absolute certainty that comes from memory and from the use I make of that memory. How my identity is defined by my personal memory. Thus in the same way, within certain limits, I can say that the identity of a group is defined by its memory, so much so that every group, every party, or any human community, e.g., a group of people who gather to play cards, create symbols that that recall the aim or purpose that its members somehow found for themselves.

Modern techniques tend toward retaining all data. Teenagers keep text messages that have been exchanged by phone, pictures, virtually everything possible, and then never look at them again.

However, a person who could remember everything would be in a frightening, pathological, situation. In Borges' story *Funes el memorioso*, Funes is a man who cannot forget anything, and since he cannot forget, he has no memories, but only a huge crowd of things that kill the mind and brain. He not only recalls a glass on a table, but also sees all the clusters of grapes that make up the pergola is over the table and remembers all the fabrics that he has ever seen in the same way as he remembers that specific glass. That is, he remembers individual things, moment by moment. He has no memories because if there were no oblivion, there would be no memory. There would only be a frightening awareness of remembering everything.

In ancient times there were arts of memory and there were people who used these art or claimed to use them, saying that by way of these arts they achieved quite fantastic effects. We find this in Cicero, in Quintilian, in the ancient rhetoric, in Thomas Aquinas, and in the great mnemonists of the fifteenth and sixteenth centuries. The technique was to take a physical place, such as a church or a house with many windows, many columns, in short, a place that could be easily represented geometrically, and store it in the mind in a complete and absolute way, being certain not to miss the "loci," the so-called "places" of memory. Images are located in these places much like images on paper. That is, the places are fixed and I cannot change them, for if I change them, I have built another system—the images are movable, like writing on paper. In this case the art of memory is to place the pictures in their proper places. Then, if the environment is very familiar to me, as I review sites, I again see one image after another. These images are such that by association or contrast I recall the thing that I have to remember.

A rather complicated affair if we think about it, because it is more involved than what we commonly do when we remember. What features those images must have? Pietro di Ravenna, who was one of the most famous theologians of the "*ars memorativa*" of the Renaissance, said they must be images that excite the imagination, so much so that one should hesitate before turning to an audience that is obviously chaste and not sinful, for it must not have sinful images. But his best piece is to think of nude women in particular places, because, he says, men remember the image of a naked girl better than any other image!

> *"And when they ask us what we're doing, you can say, WE'RE REMEMBERING. That's where we'll win out in the long run. And someday we'll remember so much that we'll build the biggest goddamn steamshovel in history and dig the biggest grave of all time and shove war in it and cover it up."* (Ray Bradbury, *Fahrenheit 451*, born on August 22, 1920, American fantasy, horror, science fiction, and mystery writer).

In the painting *The Persistence of Memory* (more commonly known as the *Flabby Watches*, 1931), Salvador Dali deforms watches (the time-measuring instrument par excellence) to invite the viewer to consider the temporal dimension with new eyes. Some vague shapes are included the dream space of the canvas, suggesting that our minds record our memories in an unconventional way. The deformation of the objects corresponds to doubt that what is ordinarily considered the rational is real, and that same doubt is the means by which we are able to elicit one more meaning.

Art critics have debated whether Dali would have imagined the new space-time metric of general relativity. Actually, it almost seems as though the painter was thinking about the softness of Camembert, a typical cheese that he had eaten the evening that he completed the picture by adding the "flabby watches" to the background.

Memory is one of the most complex and important 'worlds' in the life of our consciousness and thought. For St. Augustine memory is the space of subjective interiority, where in addition to images of objects there are the memory of numbers, the first principles of knowledge, the deep tensions that drive us to seek happiness, as we do not find it in the fleeting pleasures of everyday life and so invoke it with all our being. Happiness is real ownership of good, and we call this truth, which means we are happy only when we happen to be in a state that is really good. So we find the answer to the primary question posed in Book X of *Confessions* "I am absolutely certain to love God, but I always have a question about what I really love when I love God. I love something that is in the depths of my memory, where I also remember exploring paths to oblivion, and that (it) structures my throbbing (or pulsating or beating please choose the best) alive (or living) as a great love, and in the same time an acute nostalgia."

According to the Augustinian conception, remnants of an idea create a permanent trace that can be recovered, perhaps through some sort of "involuntary memory," which can, in some irrational way, bring back episodes from the past that we thought were lost forever: a sight, a scent, or a taste might be enough to resurrect a memory of that perception in order to revive "as faithfully as possible what we were at that time; this fragment of time cannot be relived except through our sense of the person we were at that time" (Proust 1913).

On the other hand, according to Bergson, memory is not the faculty by which we classify memories and put them in a drawer or write them on a register. There is no record; there is no drawer. Indeed, strictly speaking, we cannot speak of it as a 'faculty,' as a faculty works intermittently, when it wants to or when it can, whereas the accumulation of the past continues day in and day out. In fact, the past preserves itself automatically. It follows us, whole, all the time: what we have heard, thought, wanted from an early age is there, hovering over the present that it is going to absorb, pressing on the door of consciousness, which would to leave it out."(Bergson 1907).

A man tells his stories so many times that he becomes the stories. They live on after him, and in that way he becomes immortal (Tim Burton, *Big Fish*, 2003). The world is not an external object to us, given and immutable. It does no matter (or it isn't very important) to live things (or events); it is much important to remember them, and if memory can change events by making them more adventurous, so much better, if that can help us to live or die.

"*Once Upon a Time in America*" is a backward journey in memory, an epic effort based on the autobiographical novel of a small-time gangster. The constant swinging between the present and the past, between the present life and the life relived, is the forward and backward path in the memory of the protagonist, Noodles wanted to create an emotional continuum extending over a period of 50 years. The time does not flow; rather it swings in jumps from 1933 to 1968, then again back to the 1920s, and so forth and so on.

It is the 'magic feeling' mentioned by Proust in *Recherche du temps perdu* (1913), a sensation "common to the past and the present, and very essential to both them: [...] it had allowed him to seize, isolate, halt—for the duration of a flash—what he usually does not capture ever: a piece of pure time." It is the feeling (or sensation) that allow to "escape from the present" and "to enjoy the essence of things, that is out of the time."

> Memory. A bag full of junk that rolls out at random and ends up surprising you, as if it had not been you that had collected it, transforming the pieces into precious objects. (Wu Ming, stage name of a group of Italian writers, 1994–1999)

Zen, which developed from Buddhist philosophy, soon spread among the samurai; "In a society dominated by them, death was always present and destroying the fear of death was just one of the tasks of the followers of Zen. Zen awakened the innate aesthetic sense of the Japanese by creating close links with the national characteristics of Japan and gave great impetus not only to architecture, painting, calligraphy, and ceramics, but also to poetry and music.

Zen considers man to be an integral part of the things around him: there is no purpose since there is no victory to be gained and no end to be achieved. According to this principle, then, there is no hurry because the world proceeds to a destination, but does not move toward any goal. We must achieve this purpose without effort; we must be free and detached from our "selves" and ensure that everything flows from the unconscious. However, this condition of unawareness is reached only if one is perfectly free from any technical difficulty and has an absolute mastery of form.

Here the importance of repetition, of veneration for the teacher whose patience to wait and observe the rhythms of the pupil will ensure that the technical ability becomes a spiritual one. Thus Zen requires physical relaxation, which is gained by focusing on breathing, a concentration of all physical and psychological strengths. Hence the importance of memory, which becomes "the instrument through which you can automate and internalize the experience" so that learned art becomes art that is going to be learned.

The experience of a martial arts gymnasium makes the sense of memory concrete. Memory is the means by which we internalize the movement, the breath at the right time, the balance, and the dynamics with an infinite repetition of the action to get to the end, to forget everything and reach a vacuous state.

> "A man's real possession is his memory. In nothing else is he rich, in nothing else is he poor." (Alexander Smith, 1830–1867, Scottish poet)

The chapters of the present book describe various technological approaches.

- Chapter 1, *What Is a Memory, That It May Comprehend Itself?*, by Ernesto Bussola, is an epistemological reflection on the phenomenon of memory, taking into account some formal models related to the potential self-reflexivity of the remembering process.
- Chapter 2, *Mass Storage Memory Market Biography*, D. Caraccio, Nicola Guida, Manuela Scognamiglio, Cristina Tiziani, and Federico Tiziani, traces the history of the development of devices used to remember, a journey through the millennia.
- Chapter 3, *Probe Storage*, Marcellino Gemelli, Leon Abelmann, Johan B. C. Engelen, Mohammed G. Khatib, Wabe W. Koelmans, and Oleg Zaboronski discusses a combination of electronics and micromachines that achieves very promising storage density.
- Chapter 4, *Modern Hard Disk Drive Systems: Fundamentals and Future Trends*, Tong Zhang, George Mathew, Hao Zhong, and Rino Micheloni, describes the techniques and future prospects of what might be today's most common mass storage media.
- Chapter 5, *Introduction to SSD*, Massimo Iaculo, Francesco Falanga, and Ornella Vitale, proposes an alternative to the hard disk for portable applications.

- Chapter 6, *Packaging Trends and Technology in Wireless and SSD Applications*, A. Losavio, D. Codegoni, M. L. Polignano, F. Zanderigo, and L. Zanotti, discusses the technology that enables the assembly devices to keep Moore's law alive.
- Chapter 7, *High Capacity NAND Flash Memories: XLC Storage and Single-Die 3D*, R. Micheloni, L. Crippa, A. Grossi, and P. Tessariol, deals with NAND memories and new technology to increase storage capacity.
- Chapter 8, *Optical Data Storage*, Yang Wang, Yiqun Wu, Haifeng Wang, Mingju Huang, and Yan Wang, is an overview of the latest storage techniques that exploit the properties of light.
- Chapter 9, a nice contamination, *Biological Memory in Animals and in Man*, Raffaele d'Isa, Nicola Solari, and Riccardo Brambilla, is an exposition of the best-known mass storage device—the human brain.
- Chapter 10, *Memories for Everybody* by G. Campardo, is a popular exposition about techniques of solid-state storage.

Agrate Brianza, Italy — Giovanni Campardo
Agrate Brianza, Italy — Federico Tiziani
Arzano Napoli, Italy — Massimo Iaculo
31 March 2010

References

1. Day M (2007) 100 characters from classical mythology: discover the fascinating stories of the Greek and Roman Deities. Barrons Educational Series
2. Cooper JM (1997) Plato. Complete works. Hackett Pub Co.
3. Ross WD, Smith JA (February 2010) The works of Aristotle. Nabu Press
4. Runner B (1982) American science fiction film, directed by Ridley Scott and starring Harrison Ford, Rutger Hauer, and Sean Young. http://en.wikipedia.org/wiki/Blade_Runner
5. Smith NK (2005) The philosophy of david hume: with a new introduction by Don Garrett. Palgrave Macmillan
6. Borges JL (2007) Funes el memorioso in Labyrinths. New Directions
7. Everitt A (2003) Cicero: the life and times of Rome's Greatest Politician. Random House Trade Paperbacks
8. Selman F (2007) Aquinas 101: a basic introduction to the thought of Saint Thomas Aquinas. Christian Classics
9. Yates FA (2001) The art of memory. University of Chicago Press
10. Dalì S. http://www.moma.org/collection/object.php?object_id=79018
11. Saint Augustine of Hippo (1961) Confessions. Penguin Classics
12. Proust M (2003) In search of lost time. Modern Library
13. Bergson H (2007) Matter and memory. Cosimo Classics
14. Big Fish (2003) Fantasy comedy-drama based on the 1998 novel of the same name by Daniel Wallace. The film was directed by Tim Burton and

stars Albert Finney, Ewan McGregor, Billy Crudup, and Jessica Lange. http://en.wikipedia.org/wiki/Big_Fish
15. Once Upon a Time in America is a 1984 epic crime film directed and co-written by Sergio Leone and starring Robert De Niro and James Woods. http://en.wikipedia.org/wiki/Once_Upon_a_Time_in_America
16. Watts AW (2000) What is zen? New World Library
17. Mumon K (2007) The gateless gate: all 48 Koans, with commentary by Ekai, called Mumon. Classic Books Library
18. Tsunetomo Y (2010) Hagakure: book of the Samurai. Spastic Cat Press
19. Musashi M (2007) A book of five rings (Go Rin No Sho: 1645). MSI

Contents

Contributors

Leon Abelmann University of Twente, Enschede, The Netherlands, l.abelmann@utwente.nl

Riccardo Brambilla San Raffaele Scientific Institute and San Raffaele University, Milano, Italy, brambilla.riccardo@hsr.it

Ernesto Bussola Micron, Via Olivetti 2, 20041 Agrate Brianza (MB), Italy, ebussola@micron.com

Giovanni Campardo Numonyx Agrate, Agrate Brianza, Italy, giovanni.campardo@numonyx.com

Danilo Caraccio Micron, WSG Department, Via Olivetti 2, 20041 Agrate Brianza (MB), Italy, dcaracci@micron.com

Davide Codegoni Micron, R&D Department, Via Olivetti 2, 20041 Agrate Brianza (MB), Italy, dcodegon@micron.com

Luca Crippa Forward Insights, luca.crippa@ieee.org

Raffaele d'Isa San Raffaele Scientific Institute and San Raffaele University, Milano, Italy

Johan B.C. Engelen University of Twente, Enschede Area, Netherlands, j.b.c.engelen@ewi.utwente.nl

Francesco Falanga Micron, WSG Department, Via Remo de Feo, 1, 80022 Arzano (NA), Italy, ffalanga@micron.com

Marcellino Gemelli STMicroelectronics, Santa Clara, CA, USA, marcellino.gemelli@st.com

Alessandro Grossi Micron, R&D Department, Via Olivetti 2, 20041 Agrate Brianza (MB), Italy, agrossi@micron.com

Nicola Guida Micron, Software Department, Via Remo de Feo, 1, 80022 Arzano, Italy, nguida@micron.com

Mingju Huang

Massimo Iaculo Micron, WSG Department, Via Remo de Feo, 1, 80022 Arzano (NA), Italy, miaculo@micron.com

Mohammed G. Khatib University of Twente, Enschede Area, Netherlands, mghiathk@gmail.com

Wabe W. Koelmans University of Twente, Enschede Area, Netherlands, w.w.koelmans@alumnus.utwente.nl

Aldo Losavio Micron, Back End Operation, Via Olivetti 2, 20041 Agrate Brianza (MB), Italy, alosavio@micron.com

George Mathew

Rino Micheloni Integrated Device Technology, Agrate Brianza, Italy, rino.micheloni@ieee.org

Maria Luisa Polignano Micron, R&D Department, Via Olivetti 2, 20041 Agrate Brianza (MB), Italy, mpoligna@micron.com

Manuela Scognamiglio Elpida Memory (Italy) S.r.l., Via Colleoni 15, Palazzo Orione, 20041 Agrate Brianza (MB), Italy, Manuela.Scognamiglio@ei.elpida.com

Nicola Solari San Raffaele Scientific Institute and San Raffaele University, Milano, Italy

Paolo Tessariol Micron, R&D Department, Via Olivetti 2, 20041 Agrate Brianza (MB), Italy, ptessari@micron.com

Cristina Tiziani Viale Vittoria 2, 13048 Santhià (VC), Italy, ctiziani@tiscalinet.it

Federico Tiziani Micron, WSG Department, Via Olivetti 2, 20041 Agrate Brianza (MB), Italy, ftiziani@micron.com

Ornella Vitale Micron, WSG Department, Via Remo de Feo, 1, 80022 Arzano (NA), Italy, ovitale@micron.com

Haifeng Wang Data Storage Institute, Singapore Area, WANG_Haifeng@dsi.a-star.edu.sg

Yan Wang

Yang Wang Shanghai Institute of Optics and Fine Mechanics, Chinese Academy of Sciences, P.O. Box 800-211, Shanghai 201800, China, ywang@siom.ac.cn

Yiqun Wu

Oleg Zaboronski Department of Mathematics, University of Warwick, Coventry, UK, olegz@maths.warwick.ac.uk

Federica Zanderigo Micron, R&D Department, Via Olivetti 2, 20041 Agrate Brianza (MB), Italy, fzanderi@micron.com

Luca Zanotti STMicroelectronics s.r.l., MEMS, Sensors and High Performance Analog Device Division, Via Olivetti 2, 20041 Agrate Brianza, Italy, luca.zanotti@st.com

Tong Zhang Electrical, Computer and Systems Engineering Department, Rensselaer Polytechnic Institute, NY, USA, tzhang@ecse.rpi.edu

Hao Zhong Sandforce Inc, San Francisco Bay Area, haozhong@gmail.com

About the Authors

Giovanni Campardo was born in Bergamo, Italy, in 1958. He received the laurea degree in Nuclear Engineering from the Politecnico of Milan in 1984. In 1997 he graduated in Physics from the Universita' Statale di Milano, Milan. After a short experience in the field of laser in 1984, he joined in the VLSI division of SGS (now STMicroelectronics) Milan, where, as a Project Leader, he designed the family of EPROM nMOS devices (512, 256, 128 and 64k) and a Look-up-table-based EPROM FIR in CMOS technology. From 1988 to 1992, after resigning from STMicroelectronics, he worked as an ASIC designer, realizing four devices. In 1992 he joined STMicroelectronics again, concentrating on Flash memory design for the microcontroller division, as a Project leader. Here he has realized a Flash + SRAM memory device for automotive applications and two embedded Flash memories (256k and 1M) for ST10 microcontroller family. Since 1994 he has been responsible for Flash memory design inside the Memory Division of SGS-Thomson Microelectronics where he has realized two double-supply Flash Memories (2 and 4M) and the single supply 8M at 1.8 V. He was the Design Manager for the 64M multilevel Flash project. Up to the end of 2001 he was the Product Development Manager for the Mass Storage Flash Devices in STMicroelectronics Flash Division realizing the 128M multilevel Flash and a test pattern to store more than 2 bit/cell. From 2002 to 2007, inside the ST Wireless Flash Division, he had the responsibility of building-up a team to develop 3D Integration in the direction of System-in-Package solutions. From the 2007 he was the Director of the Card Business Unit inside the Numonyx DATA NAND Flash Group till October 2010. Today he is the Design Director of the Technoprobe, Italy. He is author/co-author of more than 100 patents (68 issued in USA) and some publications and co-author of the books: "Flash Memories", Kluwer Academic Publishers, 1999, and the book "Floating Gate Devices: Operation and Compact Modeling", Kluwer Academic Publishers, January 2004. Author of the book "*Design of Non Volatile Memory*", Franco Angeli, 2000, and "*VLSI-Design of Non-Volatile Memories*", Springer Series in ADVANCED MICROELECTRONICS, 2005. "Memorie in Sistemi Wireless", Franco Angeli Editore, collana scientifica, serie di Informatica, 2005. "*Memories in Wireless Systems*", in CIRCUITS AND SYSTEMS, Springer Verlag, 2008. He was the Co-Chairs for the "System-In-Package-Technologies" Panel discussion for the

IEEE 2003 Non-Volatile Semiconductor Memory Workshop, 19th IEEE NVSMW, Monterey; Ca. Mr. Campardo was the co-Guest Editor for the Proceeding of the IEEE, April 2003, Special issue on Flash Memory and the co-Guest Editor for the Proceeding of the IEEE, Special issue on 3D Integration Technology to be published in 2008. He was lecturer in the "Electronic Lab" course at the University Statale of Milan from 1996 to 1998. In 2003, 2004 and 2005 he was the recipient for the "ST Exceptional Patent Award".

Federico Tiziani was born in Rome, Italy, in 1969 and graduated in electronic engineering at the University of Vercelli in 1997, having written his thesis on microelectronics devices. In 1999, he joined the Memory Product Group in STMicroelectronics as firmware design engineer for Flashcard device. In 2001 he moved to the System & Application Group and by 2004 he was promoted to System & Application manager for Flashcard market. In Numonyx, now he is in charge as the Managed Memory Product Concept and Market Enablement Manager inside the eMMC Development Group. He is co-authors of chapter 4 in "Memories in Wireless System", Springer 2008 and has been co-inventor of some patents.

Massimo Iaculo obtained a bachelor's degree in electronic engineering at Federico II University (Naples – Italy) with a thesis in engineering and technology of control systems. He has worked in IT field and then in software for industrial automation since 1990. Since 2001 he has been working, for ST Microelectronics before and for Numonyx now, as micro controller team leader for NAND based storage media devices for which he is co-inventor of patents on algorithms.

Ernesto Bussola Ernesto Bussola after studies in Physics and Music (classical guitar), has taken the degree in Philosophy (Università Statale di Milano) with a dissertation in formal logic and epistemology. Currently he works in Numonyx Italy (Micron group), managing public funding R&D programs.

Danilo Caraccio was born in Benevento on May the 9th 1975. After Scientific Lyceum High School, he attended University of Naples 'Federico II' and graduated in Electronic Engineering on 2001. On October 2003 earned a post-laurea Master Degree in Software Engineering at University of Sannio College of Engineering. In November 2003 he started working in STMicroelectronics as Embedded Software Engineer working on micro-controller firmware for Secure Digital and MultiMedia Flash Cards. In October 2007 he began being in charge of following Technical Marketing activities for embedded MultiMediaCard devices in Wireless Applications and became one of STMicroelectronics representatives in JEDEC Committees for e.MMC and Universal Flash Storage standardization activities. On April 2008 he started to work in Numonyx covering the Managed Memory Product Concept and Market Enablement Engineer position keeping his role in JEDEC. During his working activity, he was co-author of several patents and articles related to Managed Memory applications. Moreover he promoted new features in e.MMC 4.4/4.41 and Universal Flash Storage standardization in JEDEC.

Nicola Guida is born in Torre del Greco, on November, the 14th. He studied at "Pitagora" high school in Torre Annunziata (NA), then he continued his studies at the "Universita' degli Studi di Napoli – Federico II", where he obtained Computer Science Engineering Degree – Automation and Industrial Systems at the, (marks 110/110) in 2001. He discussed a thesis (Industrial Robotics) named: "Image Processing in horizontal traffic signs recognition for automotive applications and cruise control". He's been working as Firmware Engineer since November 2001 for ST Microelectronics starting from the Agrate Brianza (MI) site. Currently he works as software engineer in Arzano (NA) with Numonyx, and belongs to the CTO division.

Manuela Scognamiglio was born in Cercola (NA) in 1981. She got a diploma in the Technical Industrial Institute "E. Fermi" in San Giorgio a Cremano (NA) in 2000 with title of Capotecnico Perito Informatico. In December of the same year, she was selected to participate to one course of "logical programming" based on language Cobol CICS/DB2 in DQS (Data Quality System) in Rome. In 2001 she proceeded her working activity into the same company in the bank software management, before in Rome in Banca di Roma, in 2002 in Florence in Banca Toscana and finally in Turin in San Paolo IMI. In 2003 she had the possibility to work in STMicroelectronics in Naples as Software/Firmware Designer into design team of flash card. Since 2007 she belongs to the Managed Memory Product Concept and Market Enablement group for the eMMC Development group for Numonyx Company. In her Product Concept experience she has been co-inventor of four patents.

Cristina Tiziani was born in Rome, Italy, in 1972 and graduated in Contemporary Literature a Faculty of Literature and Philosophy "A. Avogadro" of Vercelli in 1999. She started working in "A. Avogadro" Western Piedmont University for Press Office in charge of internal and external public relations; in 2006 she achieves specialization in teaching vocational training for secondary school in Interateneo Specialization Institute of Torino. Starting from 2004 she teaches in secondary state school in Vercelli County.

Marcellino Gemelli received the 'Laurea' degree in Electronic Engineering at the University of Pavia, Italy in 1994, while in the Italian Army and an MBA from MIP, the Milano (Italy) Polytechnic business school. He joined STMicroelectronics in 1995 at the "Castelletto" Research and Development center close to Milano, Italy, where he developed and supported tools for electronic design automation for integrated circuit designs requiring high current and/or high voltage. From 2000 to 2005 he has been project manager for research involving MEMS (Micro Electrical Mechanical Systems) and mixed-signal integrated circuits. He currently is technical marketing manager for power control products for data storage applications at the Santa Clara (CA) USA STMicroelectronics site. He was contract professor for the Microelectronics course at the Milano (Italy) Polytechnic from 2000 to 2002.

Leon Abelmann Leon Abelmann is associate professor in Electrical Engineering at the University of Twente. His research is part of the MESA+ Research Institute for Nanotechnology, where he specialises in nanomagnetism, MEMS and probe based data storage. He received the best teacher award for Electrical Engineering in 2005, and for Advanced Technology in 2010. He was selected junior research fellow for the Royal Academy of Sciences in 1999, and received a VIDI innovation grant in 2002. Leon Abelmann is co-chairing the International Probe Storage Workshops since 1999, member of the International Mass Storage Technology conference which he chaired in 2007 in Enschede, co-chair of the Micro Machining Europe conference 2010 and member of the IEEE Magnetics Society technical committee. He is co-author of over 60 international publications.

Johan B.C. Engelen Johan Engelen received his MSc degree in Electrical Engineering cum laude from the University of Twente, Enschede, the Netherlands, in 2006, studying the reversal mechanism of magnetic nanodots. Currently, he is pursuing the PhD degree in the Transducers Science and Technology group of the MESA+ Research Institute for Nanotechnology at the University of Twente, studying the application of electrostatic actuators in probe data storage systems.

Mohammed G. Khatib Mohammed Khatib received his MSc degree in Computer Science from the Technical University of Braunschweig, Germany, in 2004 and the PhD degree in Computer Science from the University of Twente, the Netherlands, in 2009. He was on leave at Storage Systems Research Centre (SSRC) at the University of California at Santa Cruz from January 2008 through May 2008. Since July 2009 he is a postdoctoral researcher at the Department of Electrical Engineering, University of Twente. Dr. Khatib received several awards of academic excellence. He is a member of IEEE. His research interests include computer architecture, storage systems, and green ICT systems.

Wabe W. Koelmans Wabe Koelmans received his MSc degree in Electrical Engineering from the University of Twente, the Netherlands, in 2006, where his work involved the fabrication and characterization of tips for spin polarized scanning tunneling microscopy. Currently, he is pursuing the PhD degree in the Transducers Science and Technology group of the MESA+ Research Institute for Nanotechnology at the University of Twente, studying parallel readout technologies for probe based data storage systems.

Oleg Zaboronski received the MSc degree in theoretical physics at the Moscow Engineering Physics Institute in 1993, the PhD degree in mathematical physics at the University of California at Davis in 1997 and was a postdoctoral member at the Institute for Advanced Studies, Princeton, New Jersey from 1997 to 1998. Since 1998 he is an Associate Professor (Reader) at the Department of Mathematics, University of Warwick, Coventry, UK. Dr. Zaboronski was a Royal Society Industry Fellow from 2004 to 2008 where he worked on data detection and decoding algorithms for data storage, in particular magnetic storage and probe storage. Technology developed during the tenure of the fellowship was acquired by a Japanese read channel start-up Siglead Inc (Yokohama, Japan). Since October 2009 he is Fellow of Siglead Inc on research leave from Warwick University and

President and CEO of Siglead Europe – European research and development facility of Siglead Inc. He is work package leader for the FP6 EU-funded PROTEM project since 2006.

Tong Zhang is an associate professor in Electrical, Computer, and Systems Engineering department at Rensselaer Polytechnic Institute. He co-authored over 90 refereed papers in the areas of memory circuits and systems, VLSI signal processing, and computer architecture. He currently serves as an Associate Editor for the IEEE Transactions on Circuits and Systems - II and the IEEE Transactions on Signal Processing. He is a Senior Member of IEEE.

George Mathew is with the Read Channel Architecture division of LSI Corporation, Milpitas, California. His main research area is signal processing for data storage systems. His focus includes channel modeling and characterization, equalization, synchronization, detection, reduced complexity algorithms, and development of drive-friendly channel features for characterization and optimization. He has published about 90 papers in journals and conferences, and has about 15 pending patent applications.

Hao Zhong has extensive experience of research and development on algorithms and VLSI architecture for data storage, digital communication and signal processing. He is currently a principle architect at Sandforce Inc, working on high performance controller for Solid State Drive (SSD). Prior to SandForce, he was an architect at LSI corporation, where he worked on LDPC, SOVA/BCJR and Turbo equalization. Hao Zhong earned his PhD in Electrical Engineering from Rensselaer Polytechnic Institute at Troy, NY. He has more than 20 US patents and pending applications.

Rino Micheloni (rino.micheloni@ieee.org) is Lead Flash Technologist at IDT (Integrated Device Technology). He has 15 years experience in NAND/NOR Flash memory design, architecture and algorithms as well as the related intellectual property. Before IDT, he was Senior Principal for Flash and Director of Qimonda's design center in Italy, developing 36 nm and 48 nm NAND memories. From 2001 to 2006 he managed the Napoli design center of STMicroelectronics focusing on the development of 90 nm and 60 nm MLC NAND Flash. Before that, he led the development of MLC NOR Flash. He is co-author of 95 U.S. patents and four Springer books on NOR/NAND/ECC. He is IEEE Senior Member and received the STMicroelectronics Exceptional Patent in 2003 and 2004, and the Qimonda IP Award in 2007.

Ornella Vitale was born in Naples (Italy) in 1982. She obtained a degree in Electronic Engineering at Federico II University (Naples) in 2006, with a thesis about the co-simulation of SystemC Transaction Level Models and the VHDL Register Transfer Level Models, developed as a result of collaboration between the University Electronic Department and STMicroelectronics. She began working as design engineer for ST Microelectronics in April 2006. Currently she works in Arzano (Naples) for Numonyx in Managed NAND Microcontroller Group.

Francesco Falanga was born in Pozzuoli in 1975 and graduated from the Politecnico di Napoli in Naples, Italy in 2003 with a degree in Electronic Engineering. Falanga began working on Multi Media Card/Secure Digital micro controller for ST Microelectronics (and for Numonyx since 2007) as firmware designer in January 2004. He was, further on, also involved in system architectural optimization and in definition of standard evolutions as well. He is currently responsible for eMMC product development.

Aldo Losavio graduated in physics degree from the University of Milan in 1989 and started his career at STMicroelectronics 1990. He has worked initially in Central R&D on process technology development for Non Volatile Memory CMOS process by covering most of the technology process steps. In 1993, he was appointed task leader high energy ion implantation process development and integration. In 1996, he became project leader for silicide process integration into CMOS process. In 1998, he was appointed Intellectual Property and Licensing Country Manager for all the ST Microelectronics Italian Sites. In 2003, he became Technology and Program Manager for 3D Vertical Integration into Wireless & High Density Memory Group. He is author/co-author of tens of international technical papers. In 2008 he joined Numonyx as Package Design Manager into R&D organization. He is also author/co-author of patents granted in the field of memory architecture and process technology.

Davide Codegoni received his degree in physics in 2002 from the University of Milan Italy. In the same year Davide began to work in a European project between INFN (National Institute of Nuclear Physics) and ST Microelectronics regarding the radiation effect on bipolar transistors. In 2004 Davide joined ST Microelectronics in Agrate Brianza as material characterization engineer in the Physics and Material Characterization Group and is involved in contamination and gettering problems. In the year 2008 he moved to Numonyx, still involved in the same subjects. In this activity he published about 10 papers and participated in one EC financed cooperation program.

Maria Luisa Polignano received her degree in physics in 1978 from the Genoa University. In 1979 she joined ST Microelectronics in Agrate Brianza as a process engineer. In 1985 she moved to the Physics and Material Characterization Group and since then she is involved in problems related to mechanical stress and crystal defects in devices, metal contamination and gettering techniques. In 2008 she joined Numonyx, still involved in the same subjects. In her activity she published more than 100 papers, among which 2 invited papers, filed three patents and was involved in various European financed cooperation projects. In the years 1999–2000, she was a contract professor of the post-degree School in Material Science at the Parma University.

Luca Zanotti received a degree in Physics, Solid State address, from Milan State University in 1988, when he joined STMicroelectronics (formerly SGS-THOMSON). After covering a role as responsible for the dielectric materials by CVD and Wafer Finishing for non volatile memory applications, in Agrate Brianza STM site, he joined the MEMS, Sensor and High Performance Analog Division for process integration development. He is co-author of several papers concerning Microelectronics technology and international patents.

Federica Zanderigo She received a degree in Solid State Physics from Padova University in 1997 and in the same year joined STMicroelectronics Central R&D Department. She worked on fundamental physical characterization and on process integration of non volatile flash memories (NOR, NAND and PCM technology platforms). She is working in the parent company Numonyx, where is involved also in 3D integration technology. She is author of several technical papers.

Luca Crippa is Senior Technical Analyst for Design Architecture for Forward-Insights (www.forward-insights.com) with more than 10 years of experience in MLC flash memory design. Previously, he was Senior Designer for 48 nm floating gate and 36 nm floating gate NAND flash memories at Qimonda AG as well as 90 and 60 nm MLC NAND flash products at STMicroelectronics. He was instrumental in the development of 64, 128 and 256 Mb MLC NOR Flash products at STMicroelectronics and is the author/co-author of 20 U.S. patents and the book Memories in Wireless Systems (Springer-Verlag ed., 2008). Luca received his Masters degree in Electronic Engineering at the Politecnico of Milan in 1999.

Alessandro Grossi received the Laurea degree (*cum laude*) in Physics from the University of Milan in 1991. In 1993 he joined STMicroelectronics R&D working in Flash technology development starting from 0.6μm node. From 1999 he has been Project Leader for NOR technologies and from 2007 he is involved in NAND working on both floating gate and charge trap cell architectures. Since April 2008, he has been with Numonyx R&D as Flash Memory Manager and his current activities are mainly related to new architectures for Flash devices. He is author of various publications on international journals and conference proceedings and holds several patents. He has been lecturer in Electron Device Physics at the University of Milan and University of Parma and in Non-Volatile Memory Devices at the University of Udine.

Paolo Tessariol was born in Montebelluna, Treviso, Italy in 1972. He received his doctoral degree in Physics from University of Padova (Padua), Italy. From 1998 to 1999 he has been with Italian National Institute for Material Physics working with MEMC Electronic Materials, Meran, Italy on electrical characterization of Czochralsky single silicon crystal. From 1999 to 2008 he has been with ST Microelectronics, Agrate Brianza, Italy working in Non Volatile Memory Process Development Group of Central R&D. He work has been focused in NOR and NAND Flash memory process development. Since 2008 he is with Numonyx appointed NAND Flash memory Manager. He is also Technical Coordinator and Member of Strategic Board of EC FP7 "GOSSAMER" project. He is author of several patents in the field of Flash memory process architecture.

Yang Wang received his Ph.D degree of material sciences in 2001 from Shanghai Institute of Optics and Fine Mechanics (SIOM), Chinese Academy of Sciences (CAS). At present, he is the research professor and project leader of Laboratory for

High Density Optical Storage Technology, SIOM, CAS. His research fields include super-resolution materials and technology for optical memory, phase change optical materials and multilayer structure design for optical disk applications etc. He is co-author of more than 50 scientific papers and 20 technical patents on optical storage materials and technology.

Yiqun Wu received his Ph.D degree of chemistry in 1999 from Fuzhou University, China. At present, she is the research professor of SIOM, CAS, and the professor of Heilongjiang University. She is the director of Laboratory for High Density Optical Storage Technology, SIOM, CAS. Her research fields include organic materials for optical disk applications, two-photon absorption materials and multilayer optical storage, recordable optical disk technology and industrial dye technology etc. She is co-author of more than 100 scientific papers and 30 patents on opto-electronic materials and optical disk technology.

Haifeng Wang received his PhD. in Optics in 2001 from Shanghai Institute of Optics & Fine Mechanics, Chinese Academy of Sciences and then did his post-doctoral research in Delft University of Technology (TU Delft) and Free University Amsterdam (VUA). He joined Data Storage Institute (DSI) A-STAR Singapore in 2005. He has over 40 publications, covering Nature photonics, Applied Physics Letters and OSA Journals. He has experience in theoretical optics, theoretical physics, theoretical photonics, surface Plasmon optics, modeling and design of optical antennas, optical system design and experimental optics.

Mingju Huang received his Ph.D degree of Optics in 2002 from Shanghai Institute of Optics and Fine Mechanics (SIOM), Chinese Academy of Sciences (CAS). At present, he is Professor and vice president of School of Physics and Electronics, He'nan University, China. He is served as member of Optical Material Committee, Chinese Optics Society. He has engaged in the research of holographic recording material and technology and has already published more than 40 scientific papers on this field.

Yan Wang graduated student of School of Physics and Electronics, He'nan University, China. Her main research field is high-density holographic storage technology.

Raffaele d'Isa was born in London in 1983. He obtained a bachelor's degree with honours in Psychological Sciences in 2005 and graduated with honours in Cognitive Neuroscience in 2007 at San Raffaele University of Milan, where he is now PhD student of the Molecular Medicine Phd program in the section of Experimental Neurology and performs his research activity in the Molecular Genetics of Behaviour Unit, Institute of Experimental Neurology and Division of Neuroscience. He has been a member of the Italian Society for Neuroscience (SINS) since 2009.

Nicola Solari was born in 1984 in Genoa (Italy): he achieved his Bachelor Degree cum laude in Philosophy at the Vita-Salute San Raffaele University, Faculty of

Philosophy, and obtained the Master Degree in Cognitive Neuroscience at the Vita-Salute San Raffaele University, Faculty of Psychology. He is currently enrolled at the PhD course in Molecular Medicine program, section of Experimental Neurology at the San Raffaele Scientific Institute, conducting both behavioural and molecular biology researches in the Unit of Molecular Genetic of the Behaviour, Institute of Experimental Neurology and Division of Neuroscience.

Riccardo Brambilla is Head of the Unit of Molecular Genetic of the Behaviour, Institute of Experimental Neurology and Division of Neuroscience, at the San Raffaele Scientific Institute and Professor of Psychobiology at the San Raffaele University. His research focus on the role of synaptic signalling in behavioural plasticity. He is Member of the Society for Neuroscience (since 1997), Council Member of the Molecular and Cellular Cognition Society (MCCS, since 2003) and President of the European branch of MCCS (European MCCS, since 2006). He served in the past as reviewer and rapporteur for both the NEST Programme and the 6FP of the European Commission as well as reviewer for a number of European Granting Agencies (The French, Spanish, Dutch Ministries of Research; the British MRC; the US-Israel Binational Science Foundation, The Israel Science Foundation) and Peer Review Journals in the field of Neuroscience (Journal of Neuroscience, Molecular and Cellular Neuroscience, Brain Research, TINS, Neurobiology of Disease, Biological Psychiatry, Neuroscience Methods). He is Review Editor of Frontiers in Neuroscience, since 2007.

Chapter 1
What Is a Memory, That It May Comprehend Itself?

Self-Referential Implications of the Phenomenology of Remembering

Ernesto Bussola

Abstract The phenomenon of memory, thought as a general property of some natural or artificial systems, can be analyzed only within the frame of a process-approach to reality. In this view, some epistemological paradigms, related to the notions of autopoiesis, self-organization, and self-reference, seem to bear a profound relevance for the essence of memory, as function ensuring temporal coherence to complex systems. On the other hand, the meta-mnemonic power of memory systems, that can self-represent themselves, potentially gives rise to self-referential paradoxes. These antinomies involve some critical logical issues: the problem of self-encoding within a formal untyped language, the threat of infinite regress, and the need for non-isomorphic models of memory. Many formal and/or computational models and notions have been developed to cope with such issues, in the last few decades, by logicians and mathematicians – like indication calculus, hyper-set theory, co-algebras – all aimed at explaining self-reference as a constitutive and non-paradoxical process in complex systems, like living organisms or concurrent information systems. Most of these ideas may suggest new metaphors and models to think about memory and may induce further reflection toward a definition of the essential systemic nature of the phenomenology of remembering involving and connecting together different notions, as self-description, reversibility, entropy, and life itself. Moreover, the application of these models to the problem of memory could bring to consider some general epistemological issues, related to the structural coupling between the notions of time and consciousness.

E. Bussola (✉)
Micron, Via Olivetti 2, 20041 Agrate Brianza (MB), Italy
e-mail: ebussola@micron.com

G. Campardo et al. (eds.), *Memory Mass Storage*,
DOI 10.1007/978-3-642-14752-4_1,

Introduction: Memory as Autopoietic Phenomenon

> BORDEU – You can be sure that there is only one center of consciousness. [. . .]
> MADEMOISELLE DE L'ESPINASSE – But what if my finger had a memory?
> BORDEU – Then your finger would be able to think.
> MADEMOISELLE DE L'ESPINASSE – And *what, precisely, is memory?*
> BORDEU – It is a property of the center, the specific sense to the hub of the network, just as sight is the specific property of the eye. It is no reason for astonishment that the eye has no memory, any more than that the ear has no sense of vision.
>
> Denis Diderot, *D'Alembert's dream*[1]

When we ask a question, we are looking for a center.

Versailles, August 1769: in an imaginary dialogue, the physician Théophile de Bordeu explains to Mademoiselle de l'Espinasse, pupil of him and d'Alembert's companion, that consciousness is not a scattered entity: it has one center "to which all sensations are transmitted, where the memory functions, where comparisons are made". That center – consciousnesses, which constitutes us as selves – is in turn constituted as center by virtue of a converging force, the centripetal property that connects the entangled and dispersed network of experience.

This force will be, along the following pages, our center.

The Trouble in Defining Memory

The left part of this chapter's title[2] recalls this very basic and profound question: "What, precisely, is *memory*?" We may answer it in different general ways. For instance

Memory is the faculty to retain information.

Such a statement doesn't ultimately answer our question, of course: as every definition does, it merely shifts the definitory charge to three new questions, equally deep and basic: "What do 'faculty', 'retain', and 'information' mean?" We suddenly suspect that other wide concepts are subsumed in the rising of the relevant second-level definitions: the notion of "faculty" unveils those of *cognition, system,* and *cognitive system*; "retain," those of *time, duration,* and *invariant*; "information," those of *sign* and *communication.* And all of these third-level concepts seem to involve, among several others, that of . . . *memory.*

---

[1] Barzun and Bowen (2001) Hackett (trans).

[2] The title has been initially inspired, periphrastically, by McCulloch [50] and von Foerster [23]. After the completion of this contribution, I discovered that Ranulph Glanville wrote a beautiful paper with a more resembling title – see Glanville [29]. The content being quite different, in spite of the similar titles, I decided to keep it (*non habitus monachum reddit*).

Like in the infant game of the infinitely chained questions,[3] the task of defining a general concept leads us to the vertiginous perspective of an endless process; or, what is more discouraging and frustrating, to come back, in some way, upon our starting step – what we might call "Sisyphus syndrome."

Certainly, memory can be defined in several other ways. For example we may refer to memory (unsystematically)

- as device for the continuous re-categorization of perceptive data or cognitive and action schemes;
- as attractor in the semantic space of a cognitive system, or parallely in the phase space of its supposedly immanent neuronal activity;
- as inner (simplified) modeling of the external world to make an organism able to anticipate possible changes in it;
- as cognitive organ for the subject's perception of time or as synchronization device to link the subject to the environment;
- as temporal dimension of information or its hypostatization to the condition of symbol;
- as information loop in the four-dimensional manifold of space-time;
- as structural and purposive inertia of a learning system (hysteresis);
- as process aimed at the local containment of entropy increase;
- (and so forth).

The difficulty would then lie in understanding how to harmonize such different views in a coherent and more general conceptual frame.

So, let us step back once. The scope of this short essay, indeed, is not to definitively establish what memory should, would, or even could be. We rather try to explore some general aspects of the phenomenology of remembering processes, which can suggest ideas and further reflections on a possible epistemological status of the concept of memory. Our starting – and ending – point for this brief excursion, as promised by the second half of the title, is the notion of *self-reference*, or circularity, as arisen even in the primal step of defining memory.[4] More specifically, we intend to refer, as a leitmotiv or methodological equipment in this brief inquiry, to a set of metaphors and models of self-referential phenomena. In this sense, memory can fulfill its flair for being a center.

---

[3] "Daddy, why *X*?" – "Because *Y*." – "And why *Y*?" – . . . This attitude underlies any research activity.

[4] For the sake of honesty, we should add that there are no terms that can be defined without finally falling into a circular path. To be convinced, one may think of a complete collection of definitions given within a language, which is in fact a dictionary of that language; this list of definitions is made by chains that connect together the words by the threads of the definitions themselves; the list of the defined terms is forcefully finite; *ergo*, no matter is the term we start from, following the branched definitory chains derived from it, we shall exactly end, early or later, up the same initial term. This is not strange: it's an almost trivial theorem in finite graph theory.

Memory and Self-Reference

There are several different areas where circularity – in the shape of functional loops or recursive patterns – emerges while reasoning about memory:

a. The underlying structure of most memory phenomena is based on some form of *self-reinforcing process*, which characterizes the continuity of the remembered data throughout time in a modifying environment.
b. The main function of memory for self-organizing autonomous systems, like living agents or populations, is to give them structural continuity in time: system biologists taught us that the *autonomy* of an organism is fundamentally based on some recursion principle.
c. The faculty of remembering structurally includes the ability of the system, which holds memory, to be in some way aware, consciously or procedurally, of this same function, so that it can use it: this kind of intentionality is often called *meta-memory*, that is memory-of-memory or self-reference memory, and is a (proper) part of the meta-cognition apparatus of a cognitive system.
d. Recursive phenomena bear a constructive role in the subjective and objective *experience of time*, typically through the synchronization of multiple oscillating resonant systems, while memory is involved as necessary means to perceive the continuous flowing of events from past to future, including the ability of a system to anticipate external events through a continuous model making activity, which is the essence of learning or memory. An inquiry on the models of circularity will show us that this parallel between memory and circular patterns could unveil their unexpected and profound relations to the notion of time.
e. Thinking of memory, at a pure physical level, as any form of embanking toward the diffused increasing of *entropy* – *id est* as the faculty of a system to preserve its internal form of organization or "in-formation" – we may suspect that the self-organizing structure of a system, typically based on dynamical entangled loops, could be involved in this process of "information retention."

Our aim is then to sketch the first steps of a possible reflection around such cues, hoping that this could contribute to a wider characterization of the notion of memory.

While our discourse unfolds, it will become also clear that the theme of self-reference is a deep and fertile approach for the convergence of different notions related to memory – like those of attractor, resonance, periodicity, stability, coherence, categorization, anticipation, modeling, computable function, information, negative entropy, reversibility, time, life, consciousness – which look lying at first sight far from each other, but that finally turn out to be in some ways linked, at some deep and quite uninvestigated levels.

Following Wittgenstein, we may recall that the meaning of a word is the collection of all the possible conditions of its use; in one term: its phenomenology. So, any reasoning about the general notion of memory should start from a phenomenological approach, which can take into account any differentiated possible ways this specific notion is pointing to, shaping a semantic landscape that can give meaning to the abstract idea of memory.

Memory as Object vs. Memory as Process

We may look at the world in two opposite and complementary ways: considering it as fundamentally made by (even abstract) objects or by processes. These views are structurally dual: from the first standpoint, processes are interpreted as temporal sequences of configurations of objects, while, from the second, objects are viewed as hypostatization of processes.[5] In the case of memory, to bear one or another view leads to two different ways of modeling this phenomenon.

If we think about memory as an (even abstract) object, then the metaphor of the wax tablet, engraved with the information to be kept, is the natural way to model its function. This is what we could refer to as the *trace metaphor* of memory, and its first appearance in the history of ideas goes back to Plato's dialogue *Theaetetus*[6]:

> [*Socrates*]: ... whenever we want to remember something we've seen or heard or conceived on our own, we subject the block to the perception or the idea and stamp the impression into it, as if we were making marks with signet rings.

This is the source of every simulation of our inner faculty of memory by means of external physical stable supports. Instantiations of this kind of objectification of memory are petroglyphs, molds, contracts, portraits, portolans, labyrinths, temples, knotted hankies, continuos, films, museums, DNA banks, this book and so on, up to microelectronics devices and databases containing encoded traces of all these kinds of things plus software programs: the history of the technology of memory, that is the progressive externalization of this faculty, is the history of our culture.

The trace-memory paradigm, where every memory is viewed as an *engram*, can be thought as the first step in the process of symbolization, which coincides with the evolution of the very human faculties of speech and abstract thinking.

The prominent feature of the trace-like memory is that its reliability statically grounds only on the endurance of its physical support throughout time, and consequently that obliviscence is irreversible, being only related to the mechanical, chemical, or electrical deterioration of the traces stamped on the material support.

This way to conceive memory also embeds the notion of *isomorphism* between the content of memory and its image inside the memory device: the paradigmatic example of trace-like memory is that of a mold, that stamps its (negative) shape in a

---

[5]There is an important tradition, in Western philosophy, that supports the process approaches to ontology, metaphysics, epistemology, and so on, since Heraclitus up to Leibniz and Whitehead [82]. Contemporary physics often refers to it.

[6]For a well-documented history of the cultural and scientific models of memory, see Draaisma [17]. To the "trace" model is dedicated the second chapter, "*Memoria*: Memory as Writing", and particularly the section "Like a Seal in Wax". For a deeper and more specific philosophical inquiry on this topic, see Sutton [73].

pliable material, so that memory is the memory of a form. For this reason the trace metaphor is at the heart of all forms of conceptual *spatialization* of memory, including the mental techniques of mnemonics, as codified since Cicero and Quintilian, up to Ramon Lull, Giordano Bruno, and Leibniz.[7] This flattening of the memory function to the spatial dimension, while essential to implement it by physical devices, turns out to essentially limit the ways we may think about memory.

On the other side, the idea of *memory as process* encompasses a dynamical view of it and opens room for more complex models of its possible ways of working and functions.

Since Aristotle, who dedicated a short treatise *On memory and Recollection,*[8] it has been pointed out that the phenomenon of memory is composed and that it encompasses two distinguishable phases: memory and recollection. In this sense, *memory* is the pure process of retaining collected information, while *recollection*, or reminiscence, is the process of recalling the stored information to mind's presence.

What is clear from this model is that the very phenomenon of memory arises at the crossing of *two subjective intentional stances*: the first, pointing from present to future, consisting of the subject's will to preserve some information for a possible use to come of it; and the second, in the opposite sense, from present to past, consisting in the (same or another) subject's decision to evoke the preserved information. We may call the first *forward memory process* (or *calling*) and the second *backward memory process* (or *recalling*). This coupling of intentionality stances constitutes the dynamic of *retention* and *protention* of the phenomenology of the consciousness of internal time, as depicted by St Augustine and Husserl, and explains why the cognitive ability of *anticipation* is grounded on memory.

No memory process is possible without this convergence of these two mutually constitutive intentions,[9] as no information comes to existence without somebody's intention to communicate something to somebody else. To give a plain example, simple states of things, like a white stone placed in a field or a notch carved in a wood stick, could be accidental facts or purposive messages passed to fix the knowledge of some meaningful facts: what makes the difference is the coupling of the intentionalities of the acts of setting up and interpreting these states of things. In other words, memory is always living memory.

Actually, a parallelism between the communication process – as formalized by Claude E. Shannon in his *Mathematical Theory of Communication* (1949) – and the memory process is useful to visualize the articulation of the latter (see Fig. 1.1),

[7]See Yates [86] and Rossi [66], and Carruthers [8]. It is worth to remember that these techniques were based on the mental encapsulation of structured spaces (*topoi*): memory has born as imaginary topography or virtual architecture.

[8]See Aristotle [3].

[9]Even if several examples can be done of unintentional memory episodes (from everyday life to literature), it should be clear that the necessary condition for a memory process is the existence *in principle* of these two intentional stances, which in their turn entail the presence of the relevant cognitive subjects experiencing them.

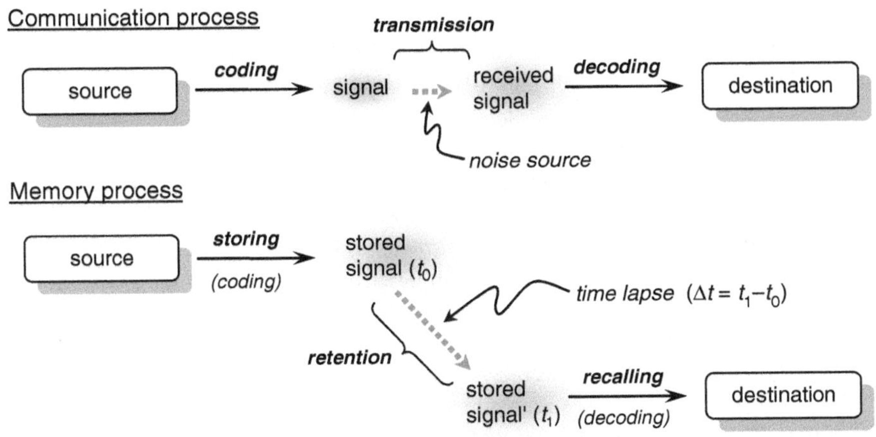

Fig. 1.1 Parallelism between communication and memory process

where transmission (typically almost synchronic) is replaced by retention (strictly diachronic).

The communication process is based on the sequence *coding – transmission – decoding*, while the sequence in the memory process is *storing – retention – recalling*.

We notice that, in the memory process, the source and the destination can typically be the same subject, so that it consists of a delayed circular self-communication process.

This scheme, even being a simplification of the process, highlights some essential features of memory:

- The coding–decoding process, like in communication, presumes that storing information is based on its *translation* into another expressive mean. In human and animal memory this process is related to the way the nervous system works, through association[10]: the perceived contents take a different form when stored, not necessarily isomorphic to the original one.
- Consequently, while not explicitly mentioned in the diagram, in both communication and memory processes the notion of *information* plays a central role: the memorized content undergoes a metamorphic process that turns it into a virtual form, the recalling if which requires the reverse decoding process.
- The intervening step, while being affected by noise in the transmission of communication, is essentially subject to a *temporal delay* in memory: time elapsing is the defining element in the retention process; so, the effect of noise in the communication process corresponds to the progressive deterioration of the stored signal,

[10]James [38], p. 653: "The cause both of retention and of recollection is the law of habit in the nervous system, working as it does in the "association of ideas"."

depending on the length of the time interval Δt and on the endurance properties of the storing system (it is for this that the second occurrence of the stored signal is marked differently in the diagram).

Figure 1.1 shows us that "trace metaphor" of memory mirrors only a limited part of the process, notably the static or passive element of it, that is, the pure retention phase. The entire process of memory, like that of communication, is part of a pair, or even a *network of intentional stances*, related to the semantic aspects of communication, which give meaning to it. This fact is fundamental to understand the possibility of second-order memory, or meta-memory, which is in turn the basis for the arising of memory process itself. In this sense, calling "memory" a physical device, like a microelectronic circuit, is a metonymy (substitution of the part for the whole): a memory device owes its name to the fact that it is enclosed in a wider flux of dynamical relations determining its function.

The limitations of the trace view of memory phenomena have been emphasized by several authors, both from the scientific and from the philosophical account. For the first we may mention the cognitive psychologist James J. Gibson, holder of the systemic echological approach to perception, and his followers; for the second, we may quote as example Ludwig Wittgenstein:

> An event leaves a trace in the memory: one sometimes imagines this as if it consisted in the event's having left a trace, an impression, a consequence, in the nervous system. As if one could say: even the nerves have a memory. But then when someone remembered an event, he would have to *infer* it from this impression, this trace. Whatever the event does leave behind, it isn't the memory. [11]

The following passage from the cybernetician Ross W. Ashby better expresses how the process view on memory is involved by the relational cognitive context where it appears, and connects its appearing to the invisible presence of past in the present state of things:

> Thus the possession of "memory" is not a wholly objective property of a system – it is a relation between a system and an observer; and the property will alter with variations in the channel of communication between them. [. . .] Clearly, "memory" is not an objective something that a system either does or does not possess; it is a concept that the observer invokes to fill in the gap caused when part of the system is unobservable. [12]

On the other side, the trace metaphor conveys the idea that memory shares some basic elements with the nature of *form*, going as far as to be confused with it. Actually, every form can be thought as the visible concretion of the dynamical process that shaped it. So, a smooth curve on a blackboard is the sensible memory of the gesture of the arm that traced it. In a broader sense, looking at living organisms,

---

[11] Wittgenstein [84] § 220. For a comprehensive epistemological discussion, see again Sutton, where the quotation is taken from.

[12] Ashby [4], pp. 116–117.

a hundred years ago the Scottish naturalist Sir D'Arcy W. Thompson[13] based his study of the physiology of plants and animals on the basic idea that every form is the outcome of an evolutionary process, where the intervening external and internal forces concur in forging and preserving the essential type or individual's appearance: the form is the diagram of a flux of forces, the equilibrium phase of their concurrent actions in time, their crystallized history, or, to say it in a word, their memory.[14]

The idea of memory as form, maintained also in some different scientific contexts,[15] allows connecting together the object and processing perspectives. On the other hand, it introduces the paradigmatic model of memory as image. The classical mnemonic techniques, taught to young Greek and Roman students in rhetoric courses, were entirely based on the visuo-spatial approach to cognition, and each memorized concept was associated with an *imago* or image.[16]

There are also some models of memory that in some way bridge the trace metaphor to the process account. Among them we underline the importance of the connectionistic paradigm, based on the mathematical formalization of neural networks. On the one hand, these models record information as traces in the "weights" that label the links, or "synapses", in the network; on the other hand, from the behavior of the network a subsymbolic level that can be interpreted as holistic phenomenon emerges.

Continuity as the Natural Function of Memory

Most definitions in natural sciences seem to be teleological, starting from their final point: the right way to define something that works as a function is to say what it actually does when it works properly or for what purpose it appears to be designed. It is like to see a fact from the reverse perspective of its potential effects, rather than of its actual causes.[17] So, the analysis of the natural functions of memory is helpful to understand its essential, or constitutive, features.

In the following passages, we explicitly consider the role of memory for biological systems, leaving in the background its function in cognitive systems and

---

[13]Thompson [77]. We may also recall the idea, developed by the German ethologist Konrad Lorenz, that the innate behavior of an animal is the invisible part of its comprehensive morphology: in this way, the application of the notion of memory as form goes up from the phylogenetic to the ontogenetic level.

[14]The same biological evolution can be seen as a learning process; Edelman's *neuronal Darwinism* would seem a logical consequence of this vision (see below, section "Feeding Back Processes Underlying Memory Systems").

[15]See, e.g., Leyton [44] where the entire notion of *geometry* is subsumed under that of memory storage. In a more structured mathematical discourse, Thom [74] equates *form* and *information*, being the latter a scalar measure of the topological complexity of the form(er).

[16]See Yates [86].

[17]We are coping with the Aristotelian distinction between "final cause" and "efficient cause".

technological artifacts. This emphasis is tacitly justified by the fact that cognitive and artificial systems are substantially a prolongation, by means of other tools, of the same finalities of natural ones.

In natural living systems – like cells, organs, organisms, or populations – the most essential features of their organization are related to the inner scope of keeping and saving the integrity of the system itself, or its *continuity*, in two senses: *passively*, along with the flowing of time; but mostly *actively*, against threats or toward opportunities from the external world (and sometimes also against the peril of inner disorders).

The main aim of an organism is to survive, while the aim of its organized parts (cells, tissues, organs) is to contribute to the survival of the whole organism. The notion of aim, applied to living systems, has been a central and over-discussed issue since the beginning of the reflections about life, from the above-mentioned Aristotelian idea of *final cause* to that of *vitalism* by Driesch, passing throughout Kant's *regulative idea* and Bergson's *élan vital.* Most thinkers are now convinced that this notion has found an ultimate arrangement on the basis of self-organizing structures, as defined by the cybernetics school of thought, in the mid. 1950s.[18] The idea is that the finality of a system is embedded within its own structure, based on a circular causal process, or feedback. Actually, every system regulated by a negative feedback loop is intrinsically self-controlled or, said in other terms, is built in such a way that its behavior turns out to be naturally oriented to a target status, or attractor, which a biologist can interpret as the finality of the system itself. Conversely, every controlling system is based on the principle of negative feedback.

One of the deepest characterizations of this idea has been worked out by the Chilean biologists Humberto Maturana and Francisco Varela, since 1970, through the notions of autopoiesis, organizational closure, structural coupling, and autonomy.[19]

An *autopoietic system* is defined as

> ... a machine organized (defined as a unity) as a network of processes of production (transformation and destruction) of components which
> (i) through their interactions and transformations continuously regenerate and realize the network of processes (relations) that produced them and
> (ii) constitute it (the machine) as a concrete unity in space in which they (the components) exist by specifying the topological domain of its realization as such a network.[20]

To sum up, we may say that circular causal processes are a necessary and sufficient condition for the self-sustainment, or autonomy, of life as such. And, applied to the

[18]The seminal paper for this idea is Rosenblueth et al. [65].

[19]See Maturana and Varela [49], and Varela [80]. The term "autopoiesis" etymologically means "self-making" or "self-building." This notion has been recognized in the last 40 years as one of the founding pillars to a renewed vision in several different fields, other than biology and neuroscience, like epistemology (von Glasersfeld), computer science (Winograd and Flores), sociology (Luhmann), philosophy of law (Teubner), semiotics (Brier), linguistics.

[20]Maturana and Varela [49], pp. 78–79.

cognitive systems, autopoiesis could be the explicative ground for the highest level unity of the system, which is one of the most complex phenomena to be understood: consciousness.

Now, coming back to the memory process, we point out that there is a mutual entailment between it and the notion of continuity, both related to the properties of autopoietic systems:

- *continuity* is the essential feature of a memory system, being the preservation of a status over time flowing its most fundamental defining element,

  while

- *memory* is the function that ensures continuity to a system, because the system can react to external changes only by referring to past experience ("backward memory process") or by learning a new association of a cause–effect phenomenon that it might turn out to be useful in the future ("forward memory process");

The last point, that recalls that every memory process is essentially a matter of *learning*, is better explained in these excerpts by the same authors of the notion of autopoiesis:

> Learning as a process consists in the transformation through experience of the behavior of an organism in a manner that is directly or indirectly subservient to the maintenance of its basic circularity. [...] Otherwise the organism disintegrates.[21]

Now, the tight relationship between memory and continuity being clearer, we may ask how the stored contents could be preserved facing the pressure of a continuously changing environment.

Remembering the distinction between passive and active continuity, the autopoietic paradigm suggests that the process of preserving information in a living system should be of active type. This means that the phenomenon of maintenance of the state of affairs for a system, meant as reactive attitude toward external changes, needs some underlying circular or feedback mechanism, namely the possibility to effectively refresh the content of memory. This constraint, as we will see later, involves the necessity to have a circular process which could self-feed the content of memory.

The following modification (Fig. 1.2) of the diagram shown in Fig. 1.1 is a graphical highlight of the circular way the retention phase could work.

Conversely, if we think of a pure passive form of retention, as explicit in the memory-trace models, the recorded trace is inevitably subject to a progressive

---

[21] Maturana and Varela [49], pp. 35–36. This point also suggests that memory systems are the functional basis for the biological and cognitive faculty of *anticipation*. See also von Foerster [23], explicit since its same title.

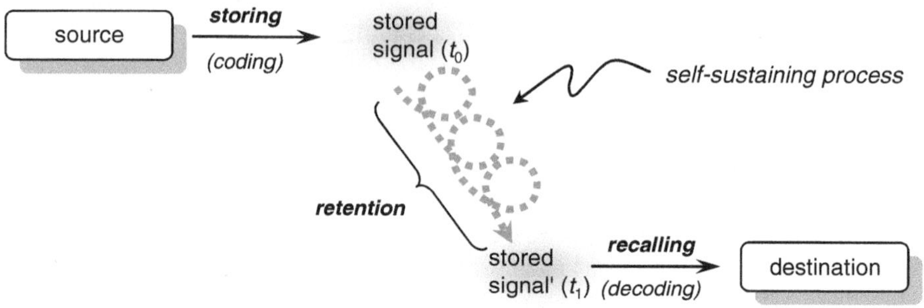

Fig. 1.2 Circular self-sustaining retention process

destructive decay, the duration of it depending only on the physical reliability of the memory device. [22]

As we shall see hereunder, the self-reinforcing behavior of a system is always caused by some form of feedback in its functional structure. For this very reason, we propose to connect the essential features of every general model of memory – intended as dynamical and reactive process, rather than as static and passive object – to the self-referential shape of the hidden functional patterns that constitute a memory system as such.

In this sense, taking autopoiesis as paradigmatic form of circularity applied to dynamical systems, we claim that

$$\alpha_{\to\mu} \equiv \textit{Memory can be considered, in its deepest essence, as an autopoietic phenomenon.}$$

In the next chapters we try to better define these rough ideas, on the one hand recalling the ways memory systems actually work, on the other hand comparing these mechanisms with formal models of self-referential processes. Our approach, albeit this is not explicitly claimed, owes its grounding epistemological attitudes to some ideas developed within the cybernetics and second-order cybernetics traditions over the last 50 years.

At the end of this swift excursus, we will return to the point $\alpha_{\to\mu}$, to discover if also the converse is tenable i.e., if the deep structure of self-constructive systems has something of essential of its own to share with the nature of memory processes.

[22]Having this fact in mind, Heinz von Foerster addressed one of his first scientific papers to the idea that natural memory decay process, or obliviscence, would follow the same decay law of the population of atoms of a radioactive material – see von Forester [22].

Remembering Process as Attractor

> Memory is the form of the being's change, – progressive, relative, addictive. [...]
> Memory does not record things but their connections ... – their conditions – what is needed so that they would obtain .. answers. [...]
> It seems to be necessary that all the elements of mental activity are closed circuits (of time) always trodden in the same direction. [...]
> Memory is the shortest path – a geodetic of implexus ... *What has been, it is a minimum.* Passage to order. – Its essential trait is not *restitution*, but its role as *reducer* and *simplifier*.
>
> Paul Valéry, *Cahiers [Notebooks]*

In the Introduction we have seen that there would be some kind of *isomorphism* between the function of memory processes and their internal mechanisms. An early deep intuition about this isomorphism – other than more generic ideas within the *Gestalt* theoretic approach to psychology – is due to the Austrian economist and epistemologist Friedrich A. Hayek in his precursive essay on *The sensory order*:

> In discussing the relationships between the network of connexions which will thus be formed, and the structure of external events which it can be said to reproduce, it will be useful sometimes to employ the simile of the *map* which in a somewhat analogous manner reproduces some of the relations which exist in certain parts of the physical world. [23]

For living organisms memory is primarily a learning system, aimed at embedding collated models of the external world within the organism.[24]

So, if the external world acts on the organism through circular patterns, like those elicited by stimulus-answer processes, the memory systems are supposed to mirror in their circuitry such self-referential schemes. This operative circularity, which we have assimilated to the self-reinforcing processes to keep the integrity of the stored information, appears in the form of a cause–effect feedback circuit.

Feeding Back Processes Underlying Memory Mechanisms

Circular causality, or *feedback*, is the basic principle of every regulatory or control system. This fact, recognized by the first cyberneticians (Wiener, Ashby, McCulloch, etc.) as a founding element of their modeling activity, occurs no matter if the control apparatus is part of a natural or artificial system. Actually, the only way to stabilize in time the output of some measurable parameter of a system (temperature, position, neurotransmitter flux, pH, electric charge, and so forth) is to feed it back into the system together with some form of comparison of its value to a target one or a fixed threshold. The reinjection of the output value into the system can be

---

[23] Hayek [32], par. 5.25.

[24] Actually, Hayek distinguishes the "map" and the "model" as two kinds of memories, acting at different logical levels (Hayek [32], parr. 5.40–42). For the purpose of our argument, we may leave out this distinction. More widely, Whitehead thought of memory as a form of perception: see Whitehead [82], p. 120.

directed to reinforce or, conversely, to reduce its previous level: what is respectively called "positive feedback" and "negative feedback."

The following picture (Fig. 1.3) illustrates the recursive structure of a feedback system.

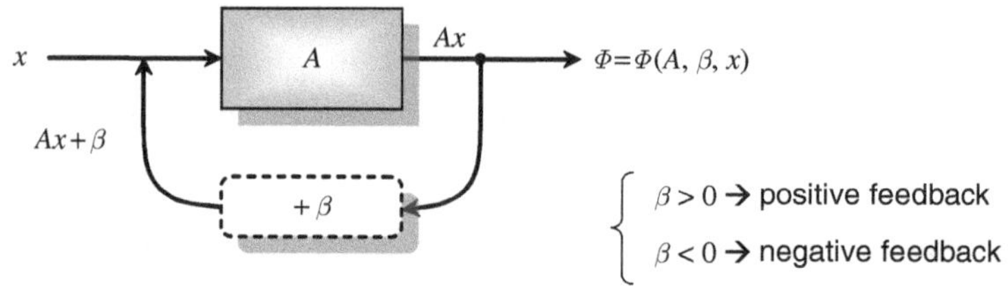

Fig. 1.3 Scheme of a positive/negative feedback system

It is immediately clear that the transfer function (or behavior) Φ of the system critically depends – other than on the input x and on the amplification function A – on the role of the feedback module β, which can act either as adder or as subtractor.[25]

The two main effects of feedback on the system's response[26] are

- positive feedback → *multistability*: the system is forced to occupy a few limited and disjoint regions in its phase space;
- negative feedback → *homeostatic regulation*: the system is attracted by a point or a closed periodic trajectory in its phase space (i.e., with or without oscillation).

A positive feedback triggers a centrifugal drift of the system's state, after exogenous or endogenous perturbations, and its outcome is a categorization of the possible states of the system.

The typical processes so elicited are of the kind of *hysteresis*, *differentiation*, or *memory*.[27] The following two figures illustrate the typical behavior of a bistable positive feedback system.

Figure 1.4 is a phase space diagram, where a continuous application U describes the potential status of a system in function of the spatial position x of it. The application shows two minimum states, related to two different positions, that in some sense compete each other to "attract" the evolutionary trajectory of the system (i.e. the temporal sequence of its states)[28], as shown in Fig. 1.5, depending on initial conditions[29].

---

[25] The degenerate case where $\beta = 0$ is here clearly worthless, because the feedback effect disappears.

[26] For a detailed characterization, we refer to Demongeot et al. [16].

[27] See, e.g., Thomas and Kaufman [76].

[28] The model visibly "mimics" the gravity effect of attraction.

[29] This is a case of finite behavior. Often a positive feedback can cause a divergent trajectory, forcing the system until it breaks down.

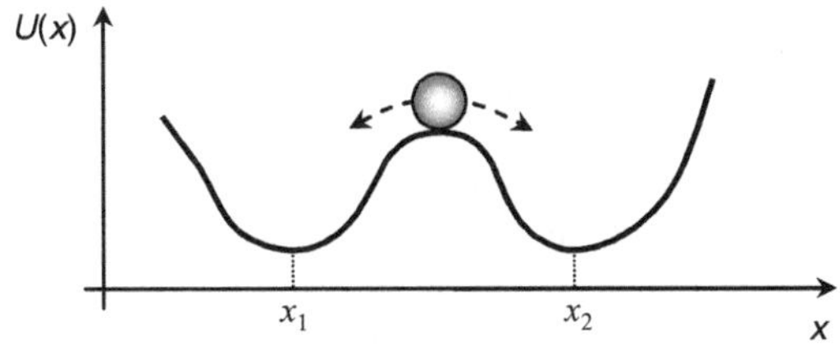

Fig. 1.4 Phase space diagram of a bi-stable system

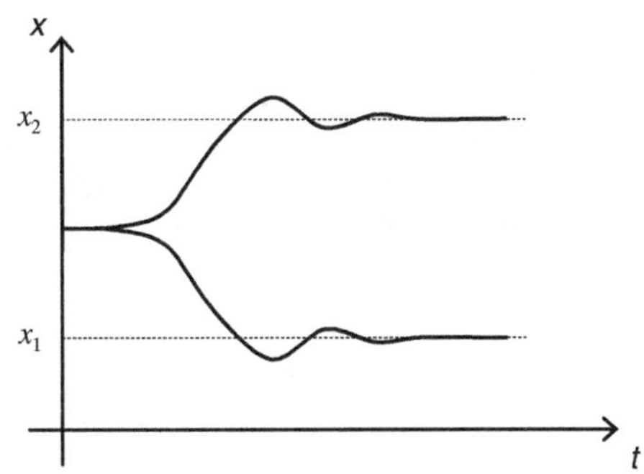

Fig. 1.5 Possible temporal trajectories of the bi-stable system of Fig. 1.4

Conversely, a negative feedback triggers a centripetal draw of the system's state toward a limit point or a periodic oscillatory regime. This is a *fixed point* (or region) of the transfer function, that is, its steady or dynamical *equilibrium*. We can express this fact by a *fixed point equation*:

$$U(x) = x. \tag{1.1}$$

An example of behavioral stability in a system induced by negative feedback is the homeostat, an equipment invented by Ashby to prove that circular causal paths bring to convergence trajectories in the phase space of the system. The diagram of Fig. 1.6 illustrates the dynamics of a homeostat toward its fixed point.

In both these specific cases, the final statuses of the systems are an *attractors* in their phase spaces, using the language of dynamical systems science. In some

Fig. 1.6 Convergent temporal trajectory of a homeostat

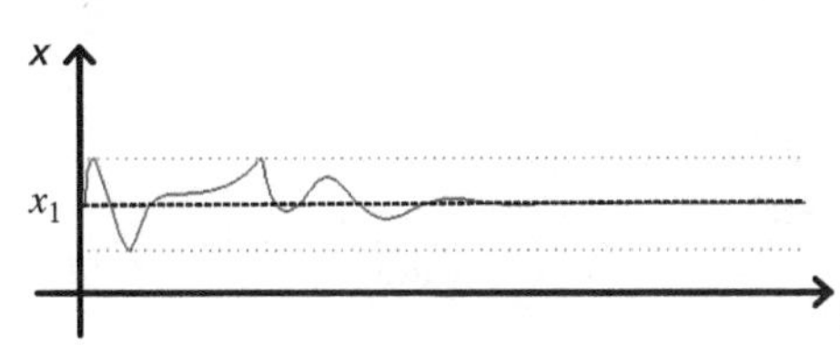

cases, the properties of a system and the study of its possible trajectories (temporal sequences of its states) can be fully characterized by its Jacobian matrix, whose elements are the partial derivatives of the transfer function.[30]

Most complex systems resort to combinations of both the two sorts of feedback, showing sophisticated behaviors in responding to environment solicitations.

While positive feedback, involving self-reinforcing processes, is the natural candidate to implement memory functions, also negative feedback plays a role in active storage systems. The effect of such a structure is a continuous refreshment of the equilibrium status or the information content, which ensures its continuity over time.

We can exhibit some examples of feedback structures implementing memory functions. The first example is a basic electronic realization of a memory cell circuit: the so-called *flip-flop* or bistable multivibrator.[31] The system consists of a pair of NAND logic gates[32] coupled through mutual feedback connections (see Fig. 1.7). The output x of the flip-flop depends on both the input signals a and b, which have different functions: a represents the datum to be stored ("set" variable), while b is the variable that can switch the system from the status of storing to idle state ("reset" variable).

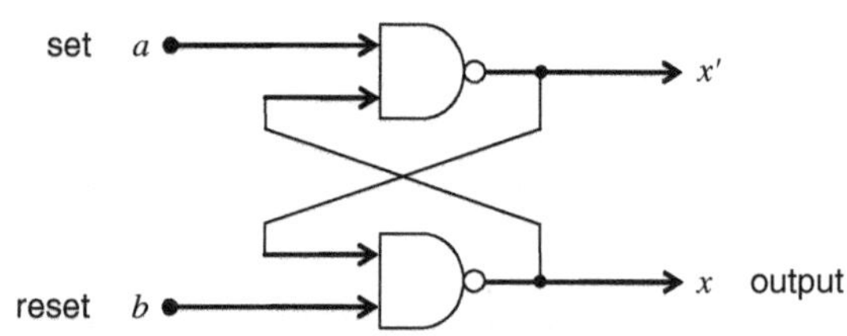

Fig. 1.7 Circuit scheme of a "flip-flop"

The logical expression of the flip-flop, considering the intermediate variable x' (the value of which is the complement to the output x), is given by the coupled equalities:

$$\begin{cases} x' = \neg(a \wedge x) \\ x = \neg(b \wedge x') \end{cases} . \tag{1.2}$$

This circuital model is typically implemented as an elementary memory bit in microelectronic SRAM (static random access memory) devices.

An example of natural system based on feedback is the first model of a *mnemon*, or memory unit, designed by John Z. Young after his studies on the *Octopus* neurophysiology.[33] We reproduce here, in Fig. 1.8, the simplified version of the mnemon proposed by Heinz von Forester.

[30] Cf. Thomas [75].

[31] See, e.g., Feynman [21], section 2.2.

[32] NAND is the *not-and* logic operator, i.e., NAND(a, b) $\equiv \neg\,(a \wedge b)$.

[33] Young [87]

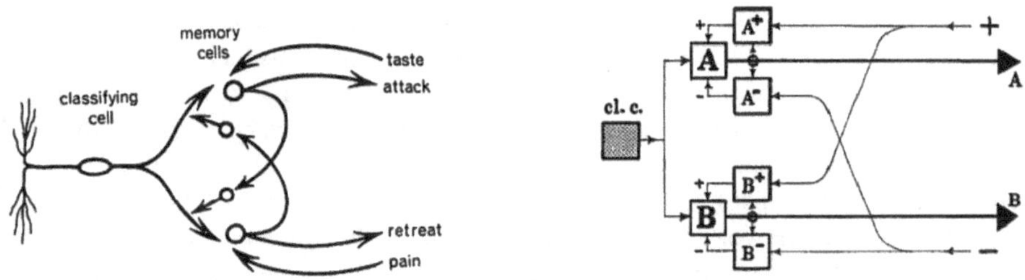

Fig. 1.8 Schematic structure of a "mnemon," or memory unit, of an Octopus, and its logic arrangement (from von Foerster [23])

This is a more complex circuit, in which positive and negative feedbacks concur to the overall learning function, as crossing of excitatory and inhibitory signals: taste vs. pain, which, respectively, trigger off reactions of attack vs. retreat.

For more recent models based on feedback mechanisms see, e.g., Rolls [64]. As shown in Fig. 1.9, the storing and the recalling processes are reconstructed too in terms of mixed positive and negative feedbacks in the nervous systems.

In this neural network model the vector of inputs $\{e_j\}$ produces the intermediate outputs $\{y_k\}$ that circularly retroact towards the network changing the matrix of weights[34] $\{w_{jk}\}$ of the "synapses", which in turn influences the transfer vector function producing the final output.

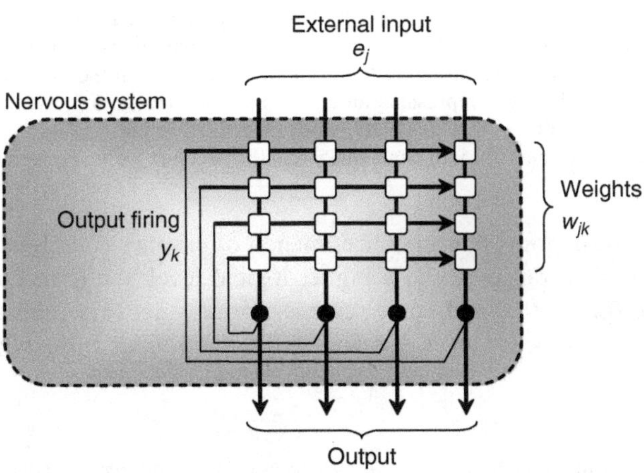

Fig. 1.9 The architecture of an autoassociative or attractor neural network (adapted from Rolls [64])

[34] This learning mechanism through selective strengthenings of the connections has been proposed by the neuropsychologist D. O. Hebb in the '40s.

This kind of models belong to the category of the artificial *neural networks*, which application to simulate the behavior of learning systems has shown the emergence of fundamental properties of memory, like the ability to robustly respond overcoming second order errors, the collective and evolutionary nature of the process, and so forth (see, e.g., the seminal paper by Hopfield [37], but the literature on this topic is immense). Hopfield, a Nobel laureate physicist, had the idea to compare the neuronal network to a physical system described by an energy function, so that memorized patterns correspond to the minima of this function, and the recalling process to the reaching of these states of equilibrium.

Complex models with circular paths are often used to explain higher level functions of human memory. We may still mention just two more significant examples from neurophysiology[35]: the reentrant maps and the resonant cell assemblies.

Another Nobel laureate, Gerald M. Edelman, has proposed a complex model of consciousness, based on the idea that neuronal cells are subject through experience to a proliferation-selection process, homologous to the Darwinian evolution, which he called *neural darwinism*[36]. The central mechanism of learning is based on the formation in the brain web of *"reentrant maps"*, that are subnetworks of mutual links among neuronal groups (from different and often distant areas of the brain) concurrently (and competitively) activated during experiences, so connecting perceptual data to action schemes. In this model the function of memory emerges as a dynamic *process of continuous recategorization* of the sensory inputs, that is of the external (and the internal) world. This procedural view on memory chimes in with the autopoietic definition of this function:

> Learning is not a process of accumulation of representations of the environment; it is a continuous process of transformation of behavior through continuous change in the capacity of the nervous system to synthesize it. Recall does not depend on the indefinite retention of a structural invariant that represents an entity (an idea, image, or symbol), but on the functional ability of the system to create, when certain recurrent conditions are given, a behavior that satisfies the recurrent demands or that the observer would class as a reenacting of a previous one.[37]

Edelman also supposes that the phenomenon of primary consciousness is based on the same model, transposed at a higher logical level: the (conceptual) categorization of the (perceptual and value) categorizations gives rise, through reentrant connections, to a process of *self-categorization*, or *bootstrap*, which leads to

---

[35]To say nothing of the models in cognitive psychology. We could just mention, as an example, that Alan Baddley proposed the mechanism of the *"articolatory loop"*, or "phonological loop", within a comprehensive model of working memory, to explain the ability to prevent the decaying of verbal information during a learning process.

[36]Edelman [18] and [19].

[37]Maturana, Varela [49], p. 45. The idea of memory as recategorization of the external world would bring us to consider the complex notion of *apprehension*, that puts the cognitive process on the axis of the complementarity between predicativity and indicativity of speech acts. For a deeper insight on the subject, see Seiler [71].

the construction of complex mental scenarios, which constitute our conscious experience:

> Perceptual (phenomenal) experience arises from the correlation by a conceptual memory of a set of ongoing perceptual categorizations. Primary consciousness is a kind of *"remembered present"*.[38]

This idea of the "remembered present" strongly interlinks the phenomenon of memory to that of consciousness in a fascinating sense.

In the last years of his too short life the already mentioned polyhedric scientist Francisco Varela studied, *inter alia*, the collective behavior of resonant cell assemblies[39]. The main assumption underlying this paper is that some global states of the brain depends on the dynamical large-scale synchronization of neuronal cell assemblies. The resonance phenomena are supposed to emerge thanks to the reciprocal stimulations within the neuronal network.

Actually, various forms of self-organization are the ground for memory properties also for non living systems, as Hermann Haken's *synergetics* recalls:

> ... in opern systems, even in the inanimate world, specific spatial or temporal structures can be generated in a self-organized fashion. Examples are provided by the laser which produces coherent light, by fluids which can form specific spatial or temporal patterns, or by chemical reactions which can show continuous oscillations, or special spirals, or concentric waves. Even at this level we can speak to some extent of creation or storage of information.[40]

What we can learn from all these models is that, other the central role of circular patterns for the emergence of complex behavioral schemes, and namely of memory, the phenomenon of recalling for living organisms seems to be always based on a process of real *reconstruction* of the past data, rather that a passive retention of them. Moreover they reaffirm that memory plays a central role in a cognitive system, being instrumental in other basic functions like perception, time experience and, finally, consciousness, as Diderot, through the words of doctor de Bordeu, suspected.

The Paradox of Meta-Memory: Self-Encompassing Maps and Impredicativity

The pervasive occurrence of circular patterns in the hidden functional structure of most memory systems seems to mirror the higher level circularity of *memory self-reflexivity*. Actually, the conscious use of memory reveals the necessity of an integrated process that embeds the same property of self-awareness in the very process of memory itself, often called *meta-memory*.

---

[38]Edelman [20], p. 120 (emphasis added). This theory recalls some profound intuitions of St Augustine about the mutual and dynamical relationships between memory, perception and thought (see *Confessiones*, *De musica*, etc.).

[39]Cf. Varela et al. [81].

[40]Haken [31], p. 24. There is here an implicit reference to entropy: this topic will be considered later, in section "Time as Memory: Information and Entropy".

Since Aristotle the superposition of these two phenomena is clear:

> ... when a person actualises his memory for the fact that he has seen, heard or learned something, he senses *in addition* that he did this earlier. [41]

The problem of the superposition of meta-memory to memory is parallel to the problem of self-consciousness that involves the cognition/meta-cognition relationship. This parallelism allows transferring to memory the typical paradox of infinite regress that arises when we try to depict how the mind can represent itself within itself.

The typical image of a self-containing thinking system is the ancient image of a *homunculus* ("little man") inside our mind, to whom the problem to explain mental functions, including perception, dismantling the underlying mechanisms, is shifted.

The homunculus hypothesis presupposes the memory-trace viewpoint: this inner subject is an observer of the images kept in the warehouse of memories.

The two following figures (Figs. 1.10 and 1.11) present the trace-memory isomorphical effect (an arrow is projected through the eyes inside the brain, as if the brain could have another internal eye to look at this image).

The obvious risk of this conception is to fall in infinite regress, which recalls the story of the *homunculus*: the subject looks at an object, but the projected image needs another inner subject that looks at it, ... , and so on, in a never-ending recursive process.

As summarized by Sutton:

> Encoded traces, it is said, require an internal interpreter or reader to recognise a new input as matching an existing trace, or to know in advance which trace to search for and recall for a given purpose. But who is this inner subject behind the engram? Such an intelligent *homunculus* merely [shifts] the problems of retrieval one step deeper inside ... [42]

The *homunculus* is often used as an argument against the memory-trace model.

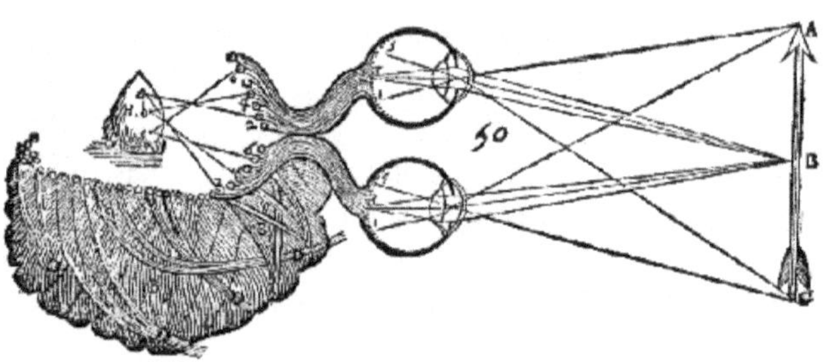

Fig. 1.10 Isomorphic projection of an object inside the brain (from Descartes, Traité de l'homme, 1664)

---

[41] Aristotle [3], p. 29.

[42] Sutton [73], p. 310.

Fig. 1.11 Infinite regress of self-embedding observing brains or *homunculi* (from Fig. 1.10)

On the other hand, what is of interest for our discourse is that it implies paradoxical infinite regress, in a manner that recalls the literary figure of the *mise-en-abîme*.

The Hegelian British philosopher Josiah Royce, in the appendix to his masterpiece *The World and the Individual*, poses the problem of self-representative systems within the human mind and illustrates the ensuing paradox by means of a metaphor: let us imagine that we want to draw a perfect map of England, where no details should be omitted, including the map itself.[43] So, this perfect map shall contain infinite images of itself, more and more miniaturized, but ever showing the same quantity of particulars (Fig. 1.12).

A potential issue of the infinite regress arises along with all forms of self-referential definitions, including feedback systems. Taking the logical definition

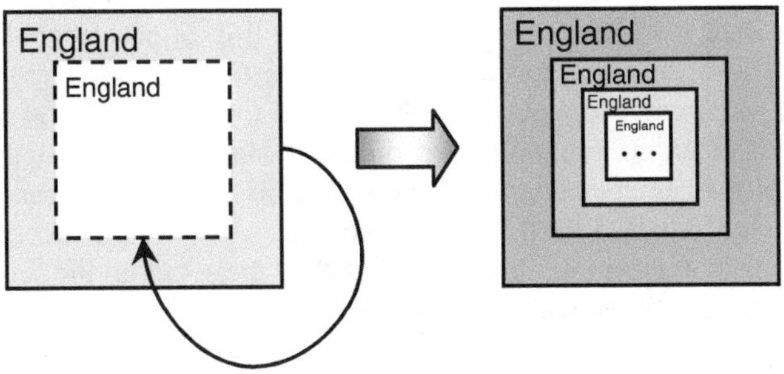

Fig. 1.12 Royce's infinitely nested maps of England

[43]Royce [67], pp. 473–588: "The One, the Many, and the Infinite." This metaphor has been successfully reutilized by Alfred Korzybski and Gregory Bateson.

of a flip-flop device (1.2), we can unfold the pair of recursive formulas through progressive alternating replacements of the unknown variables x and x':

$$\begin{aligned} x &= \neg(a \wedge x') \\ &= \neg(b \wedge \neg(a \wedge x)) \\ &= \dots \\ &= \neg(b \wedge \neg(a \wedge \neg(b \wedge \neg(a \wedge \neg(b \wedge \neg(a \wedge \neg(b \wedge \neg(a \wedge \dots)))))))) \\ &= \dots \end{aligned} \tag{1.2a}$$

The sequence appears to be a never-ending list of increasingly imbricated formulas, where the unknown variables x and x' disappear only beyond the horizon of infinity. This evolution at first glance gives no hope to solve the original equations and so to determine the actual behavior of the circuit. A different notation immediately unveils the implicit circularity of the system:

$$x = \neg(\neg(x \wedge a) \wedge b). \tag{1.2b}$$

This is the consequence of (1.2), where it is said that the variable x depends on x', while x' in turn depends on x.[44]

Our theme is, How is it possible that a memory system can perfectly remember itself or that it can contain a complete self-image? Or, put in another way, Can a system embed a complete description of itself?

Certainly we are landing ourselves in the hazardous territory of *impredicativity*: this term, coined by Jules-Henry Poincaré, denotes when a definition of a set is based on a property that invokes the same elements we are defining.

A fundamental historic example of the devastating effects impredicative definitions may cause on a theoretical consistent construct is done by the *Russell's paradox*. This antinomy, communicated by Russell to Frege in a short letter written in 1902, provoked an insoluble impasse to Frege's attempt to consistently build the foundations of mathematics on the ground of formal logic. The semantic antinomy found by Russell is based on the simple recursive definition of irreflexivity with respect to the set–membership relation $\in$. Let R be the set of all the sets that do not contain themselves (that is, which are irreflexive):

$$R = \{x | \neg(x \in x)\}. \tag{1.3}$$

[44]Let us notice that (1.2b) is a fixed point equation, like (1.1). Notice that the seeming lack of a solution is due to the neglecting of the temporal dimension of the phenomenon.

Now, we can infer by this same definition that both

$$R \in R \quad \text{and} \quad \neg(R \in R),$$

simultaneously hold, which leads to a patent contradiction.[45]

All the formal accounts of self-referential relationships risk to fall in the webs of inconsistency or infinite regress, and the idea of a self-encoding memory a priori should not escape this fate. But we think that in spite of the intuitive paradoxes related to infinite regress, in principle this would not be impossible.

We may consider a suggestive abstract example. The mathematician John H. Conway invented an easy way to write a number sequence that describes another number sequence: as when we dictate a long phone number, every cipher could be preceded by the quantity of its consecutive occurrences.

For instance, the sequence

$$r = 3321111355$$

can be first divided by grouping repeated numbers

$$33\text{-}2\text{-}1111\text{-}3\text{-}55$$

and finally described with this sequence[46]

$$\text{two``3''} - \text{one``2''} - \text{four``1''} - \text{one``3''} - \text{two``5''}$$

that is,

$$s = 2312411325.$$

Assuming this passage as a *describing function* δ, we say that "s describes r" and write

$$s = \delta(r).$$

Now, let us consider the following triple of infinite sequences a, b, and c:

$$\begin{aligned} a &= 11131221131211132221\ldots \\ b &= 3113112221131112311332\ldots \\ c &= 1321132132211331121321 23\ldots \end{aligned}$$

[45]Mathematicians and logicians of the early 20th century strove for proposing consistent solutions to this kind of contradictions, typically through some interdictory axiom that could limit the field of self-referential or circular constructs. The *vicious circle principle* proposed by Russell and Whitehead is a paradigmatic instance of these solutions.

[46]This makes intuitive why this procedure has been called *audio-active sequence*.

We suddenly realize that they are describing each other in a circular way[47]:

$$b = \delta(a); \quad c = \delta(b); \quad a = \delta(c). \tag{1.4}$$

So, we encountered a closed system $a+b+c$ which, in some sense, contains a complete self-description; and more self-imaging constitutes its proper essence.

Actually, this example does not fully solve our problem, because it deals with infinite chains of symbols, while every memory system we can treat is supposedly finite. However, it opens the perspective that self-description could be a sound non-empty concept.

Another possible way to put the question of whether an information system can contain an encoded image of itself is to refer to some ideas from the *algorithmic information theory*, developed by the mathematician Gregory J. Chaitin. Here the fundamental notion is that of *algorithmic information content*, or "complexity", $H(X)$ of an object (a string of numbers) X, defined as

> the smallest possible number of bits in a program for a general-purpose computer to print out X. In other words, $H(X)$ is the amount of information necessary to describe X sufficiently precisely for it to be constructed.[48]

What Chaitin's approach suggests is that the minimal dimension of $H(X)$ is a critical datum about the intrinsic structure of X: if its value is comparable to the same dimension of X, then X is a random entity, while if $H(X)$ is sensibly smaller than X it means that X can be "compressed", that is described in a shortest way or, which is the same, "comprehended". In this sense, 'casual' (or 'random') is synonymous with 'algorithmically uncompressible', while the role of a natural law, e.g., is that to compress the information about the possible states and evolution of a physical system.

Now the question could be expressed in these terms: is it possible to have a memory system M such that it is a fixed point to the application H, or better, such that a part or a member m of it corresponds to a self-description of the whole:

$$m \in M \quad \text{and} \quad m = H\,(M)\,. \tag{1.5}$$

This way to face the problem presupposes some tacit assumptions about the possible computable nature of memory as such: H is at last the code of an algorithm, which is supposed to be reducible to a Turing machine. So, to go forward with this hypothesis, we should build a model of memory compatible with a plausible algorithmic description, involving the notions of *mutual information content* $H(X : Y)$ between X and Y – which is defined as "the extent to which knowing X helps one to calculate

---

[47] We owe this intriguing example to Kauffman [42]. It is possible to prove, by induction, that there are no proper fixed points of the function δ, i.e. $\forall p \neg (p = \delta(p))$.

[48] Chaitin [10], p. 93. Elsewhere (e.g. at p. 46) $H(X)$ is defined as "algorithmic entropy". See also Chaitin [11]

Y"[49] –, and that of *conditional information content of* X (given Y) $H(X \mid Y)$ – which is defined to be "the size of the smallest program to calculate X from a minimal program for Y"[50]. So, our request on M (memory system) and m (self-description of it) could be expressed, for instance, by an equivalence like this:

$$H(M \mid m) = H(M) \quad - H(M : m)\,. \tag{1.6}$$

Anyway, beyond these tentative and raw sketches about the possibility for a system to be self-descriptive, we would conclude this section reminding how profoundly critical could turn out to be the notion of self-reference. The recognition of the presence of self-referential aspects, in analyzing abstract concepts, is often interdicted, or underestimated, due to an ancestral suspect of epistemological unreliability aprioristically charged to circularity.

This negative fame of supposed unreliability basically rests on two paradigmatic idiosyncratic forms: we may refer to them as *autology* and *antilogy*. The first is a self-statement, like "This sentence is true": nothing really paradoxical, but a sentence that is logically empty, because it can be either true or false; the second is exemplified by the famous *liar paradox*, based on a statement such as: "This sentence is false," which mirrors Russell's paradox, again being simultaneously true and false.

Eigen-Behavior and Dynamical Stability

What should be clear, after the above bird's eye view over some examples of memory patterns based on feedback mechanisms (See "Feeding Back Processes Underlying Memory Mechanisms), is that the issue of giving a formal account to self-referential processes could have a remarkable role in describing memory. The notions of fixed point and attractor, from function theory and dynamics, are much more than metaphors: they have the descriptive power allowing capturing some essential features of remembering systems behavior.

A synthetic and incisive account of this idea is given by the notion of *eigen-behavior*, proposed by Heinz von Forester in his cybernetic approach to ontology.[51] Referring to Piaget's cognitive psychology, von Forester recalls that the organization of sensory–motor interactions is primarily based on circular causal patterns between the observing subject and the observed object. Summarily, the operator of inferential coordination (COORD) applied to an observational act (obs_0) gives rise to a new observation (obs_1):

$$\text{obs}_1 = \text{COORD}(\text{obs}_0). \tag{1.7a}$$

[49] Chaitin [10], p. 65.: the mutual information measures the dependence of X and Y, being $H(X : Y) = 0$ when they are independent variables, and is symmetric.

[50] Ibid. This function is not symmetric.

[51] Von Foerster [24].

The recursive application of the same operator to obs_1 and so forth, which represents the act of including the same act of observing within the observation process, may define a convergent trajectory, whose limit point is

$$\text{obs}_\infty = \lim_{n\to\infty} \textsc{coord}^{(n)}(\text{obs}_0) = \textsc{coord}(\textsc{coord}(\textsc{coord}(\ldots))). \tag{1.7b}$$

Another way to symbolize the infinite nesting of the operator into itself is the following:

$$\text{obs}_\infty = \text{COORD}(\ \lrcorner\ ). \tag{1.7c}$$

The last equality shows that obs_∞ is a fixed point of the operator COORD[52], or its *eigenvalue*, which represents the convergence of the system's sensory–motor behavior: whence the coinage of *eigen-behavior*.[53]

Like the eigenvalues of functions, the eigen-behaviors of an observing system show the properties to be discrete, stable, separable, and composable and correspond to equilibria determining themselves through circular processes. Von Foerster claims that *objects* are ontologically indiscernible from the related fixed points of the observer's behavior, so that they can be defined as "tokens for eigen-behaviors."[54] This definition can be applied also to the subject himself: "I am the observed link between myself and observing myself." [55]

According to the previous argument, the notion of eigenvalue could be used also to define memories. Actually, the fundamental property of stability, which is intrinsically embedded within the same definition of eigenvalue or eigen-behavior, fits with the very nature of memory. Francisco Varela, when illustrating the notion of biological autonomy as eigen-behavior, resorted to the example of a flip-flop, it being a bistable device capable of storing a bit of information.[56]

The following diagram depicts a typical attractor in the phase space of an oscillating system ($U(x)$ being a secondary variable of the system, like potential, energy or speed, depending on its position x): every trajectory, wherever it starts, is forced by the system to land on the attractor closed curve. (Fig. 1.13).

In this perspective, all memories can be viewed as attractors in the phase space of all the possible continue configuration of our nervous system, as suggested by

[52]That is, $\text{obs}_\infty = \textsc{coord}\,(\text{obs}_\infty)$.

[53]For example, the iterated application of the function cosine converges to a fixed point $\gamma = \lim_{n\to\infty} \cos^{(n)}(x)$ through an infinitely oscillating damped trajectory, whatever is the initial value of x. The word *eigen-behavior* is shaped on the cast of the mathematical terms "eigen-value", "eigen-function," etc.; in German "eigen-" corresponds to "self-."

[54]Cf. Kauffman [42]: "An object, in itself, is a symbolic entity, participating in a network of interactions, taking on its apparent solidity and stability from these interactions."

[55]Von Foerster, "Notes on an epistemology for living things," in von Foerster [25].

[56]Varela [80]. Varela's more formal approach leads to define these notions in terms of Σ-algebras.

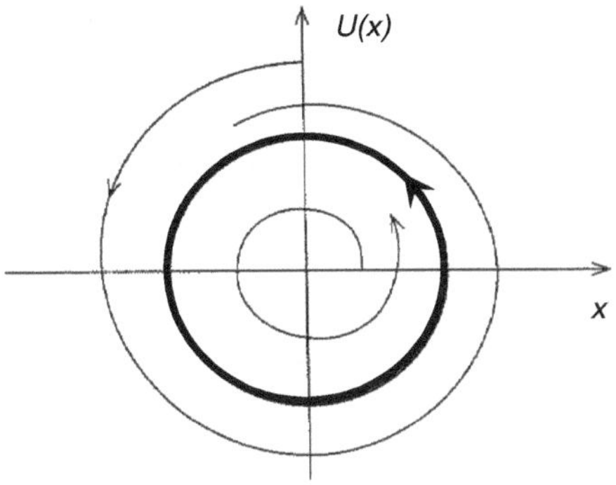

Fig. 1.13 Typical shape of an attracting periodic orbit in the phase space of a system

some models described in section "Feeding Back Processes Underlying Memory Mechanisms"[57]. Properly speaking, an attractor like that in Fig. 1.13 is closer to the dynamical notion of *homeorhesis*, or *chreod*, rather than that of fixed point. In any case, the notion of eigen-behavior seems to be sufficiently comprehensive to catch some aspects of memories as attractors for a dynamical system. This way, the notion of geodetic path, invoked by Valéry to catch the way memory works, finds a natural visual and phenomenological meaning.

Finally, when speaking about memory in terms of system theory, it could be worth to mention that Haken extends this vision subsuming the same field of semantics within the study of dynamical systems, proposing to consider the meaning of an exchange of information as assimilable to the response of the receiving systems, that is the evolving landscape of its attractors[58]. In a similar approach, we would also cite a sort of paradox of recalling illustrated by the French mathematician René Thom[59]. Recalling is a process that depends on two systems: the remembering subject A and the memory M, defined as a set of stable states. Since A is in turn a system, how could be possible that the interaction between A and M is so asymmetric that A does not turn out to be affected by it?[60] Thom's solution to the paradox is based on the notion of "free interaction", that is possible among "metabolic forms", that is systems of attractors with topological properties such that they can resonate with similar forms. This is again close to the dynamical vision of memory.

---

[57]Mostly with reference to the Hopfield neural network model. For a useful summary of the theory of dynamical systems applied to cognitive issues, see Port [60].

[58]Haken [31], p. 21. He also says: «It will be an interesting problem to determine the minimum number of bits required to realize a given attractor (or to realize a given value of "relative importance")», *ibid.*, p. 22. The idea to measure the "cost of memory" connects the system vision to the information-computational one (see below, section "Time as Memory: Information and Entropy").

[59]Thom [74]. René Thom is known as a prominent author of the topological catastrophe theory (or singularity theory), which has been applied to semantics too.

[60]This problem could be equivalently restated in terms of information algorithmic theory, recalling that the mutual information $H(X : H)$ is a symmetric function: $H(M : A) = H(A : M)$.

Algebraic and Set-Theoretic Models of Self-Referential Processes

> A world that was simple enough to be fully known would be too simple to contain conscious observers who might know it.
>
> John Barrow, *Impossibility*

Having seen the relevance of self-reference for a general definition of the memory process, we illustrate in this section some formal approaches to model circularity. The first, the calculus of indications, is a sort of algebraic language for logical expressions; the second, the hyperset theory, is a recent branch of axiomatic set theory, expanded also to the area of the theory of categories, with interesting applications in computer science and other fields. Both are presented succinctly, to give some suggestions of them as possible formal tools to think about the self-referential implications of memory.

Spencer Brown's Calculus of Indications

In 1969, the British logician George Spencer Brown published an atypical essay with an intriguing title: *Laws of Form.* Spencer Brown proposes a new kind of algebraic calculus, which stimulated in the following years a lot of discussions and expectations among several scientists and logicians. The core idea is to take as primitive notion the act of *distinction* or *indication*, so systematically explained:

> The theme of this book is that a universe comes into being when a space is severed or taken apart. The skin of a living organism cuts off an outside from an inside. So does the circumference of a circle in a plane. By tracing the way we represent such a severance, we can begin to reconstruct [. . .] the basic forms underlying linguistic, mathematical, physical, and biological science, and can begin to see how the familiar laws of our own experience follow inexorably from the original act of severance. The act is itself already remembered, even if unconsciously, as our first attempt to distinguish different things in a world where, in the first place, the boundaries can be drawn anywhere we please.[61]

From this idea Spencer Brown builds his algebra on the basis of the symbol he takes as the *distinction operator*:

$$\overline{\quad}|$$

which can dually act both as function and as argument in logical expressions. This primary fact introduces the idea that this would be a calculus about events, not objects, where expressions represent processes rather than things (or at least both).

The calculus is grounded on two elementary axioms or substitution rules:

---

[61] Spencer Brown [72], p. v.

$$\overline{\ \ }|\ \overline{\ \ }| = \overline{\ \ }| \qquad \text{condensation,} \tag{1.8a}$$

$$\overline{\overline{\ \ }|}| = \qquad \text{cancellation.} \tag{1.8b}$$

The first axiom is equivalent to *idempotence* and the second to *bivalence*[62]. Starting from these simple elements, with the introduction of letters as variables, it is possible to compound expressions and to pass from one to another through substitutions or applications of the axioms: it is a logical calculus which can be proved being analogous to a boolean algebra. For example, the following equalities hold (i.e., are theorems[63]):

$$\overline{\overline{a}|\ a}| = \qquad \text{position,} \tag{1.9a}$$

$$\overline{\overline{ac}|\ \overline{bc}|}| = \overline{\overline{a}|\ \overline{b}|}|\ c \qquad \text{transposition.} \tag{1.9b}$$

An interpretation of the terms as classical logical expressions is possible turning the distinction mark into negation ("not"), and the juxtaposition into disjunction ("or"), as follows:

$$\overline{a}| \quad \leftrightarrow \quad \neg\, a,$$

$$a\, b \quad \leftrightarrow \quad a \vee b,$$

In this interpretation, (1.9a) and (1.9b), respectively, become the excluded third:

$$(a \wedge \neg a) = \mathbf{F} \quad (=\text{false}), \tag{1.10a}$$

and distributivity[64]:

$$(a \wedge c) \wedge (b \wedge c) = (a \wedge b) \wedge c. \tag{1.10b}$$

---

[62] Spencer Brown [72], pp. 1–2. In summarizing the elements of Spencer Brown's calculus, we also follow further elaborations from Varela [80].

[63] For the proofs, see Varela [80], pp. 131–132.

[64] For the sake of simplicity, in (1.10b) every occurrence of a, b, and c is replaced by its negation.

En passant, it is worth to remark some curious aspects, which have no correspondence in traditional logical calculi:

- in (1.8b) "appears" the blank space as an actual symbol of the formal calculus (called *unmarked state*);
- the same symbol of distinction is visually evocative of the idea of *boundary*, the crossing of which corresponds to the idea of distinction or indication;
- like in Church's *lambda calculus*, operators and their arguments are at the same level and so interchangeable as per their roles.

All these features of Spencer Brown's algebra are to some extent related to the possibility to introduce *self-reference* in the calculus. Let us consider, for example, the infinite sequence of equalities[65]:

$$\overline{\overline{a\,|}\;b\,|} = \overline{\overline{\overline{\overline{a\,|}\;b\,|}\;a\,|}\;b\,|} = \cdots = \overline{\overline{\overline{\overline{\cdots\; a\,|}\;b\,|}\;a\,|}\;b\,|}\,. \tag{1.11}$$

This sequence can be expressed in a finite form by the "second-order" equation:

$$x = \overline{\overline{x\,a\,|}\;b\,|}\,.$$

The variable x being in both the two sides of the equal sign, to mark this co-occurrence, Spencer Brown introduces the notion and notation of *re-entry*, that is, the self-embedding function with respect to the indication mark[66]:

$$x = \underline{\overline{\overline{\,a\,|}\;b\,|}}\,, \tag{1.12}$$

where the ascending trait of the embedding symbol of distinction indicates the "self-occurrence" of the variable x in the left expression. An example of definition made possible by the use of re-entry is the paradox of the self-negating element, analogous to the above mentioned *liar statement*, the most celebrated semantic paradox in logic[67]:

$$x = \overline{x\,|} \qquad x = \underline{\overline{\quad|}}\,. \tag{1.13}$$

The function of re-entry turns out to be one of the most interesting among the innovative features of this calculus. Spencer Brown noticed that it is possible to

[65] The proof is in Spencer Brown [72], p. 55.

[66] The relevant symbol is often called "curl," for an apparent reason. The term "re-entry", amazingly, recalls the "reentrant maps" of Edelman.

[67] Remember that the distinction mark is equivalent to negation; so (1.13) expresses the self-negation of a term.

solve equation (1.13) only if we can project the function's behavior over the variable time, or if we interpret the re-entry symbol as an *imaginary Boolean value*, like the imaginary unit for complex numbers $i = \sqrt{-1}$:

> And since ⏋ and   represent the only states of the form hitherto envisaged, if we wish to pretend that [1.13] has a solution, we must allow it to have a solution representing an *imaginary state*, not hitherto envisaged, of the form. [...] Since we do not wish, if we can avoid it, to leave the form, *the state* we envisage is *not in space but in time*. [...] One way of imaging this is to suppose that the transmission of a *change of value* through the space in which it is represented *takes time* to cover distance.[68]

The diagram of Fig. 1.14 is a representation of the behavior of x in (1.13):

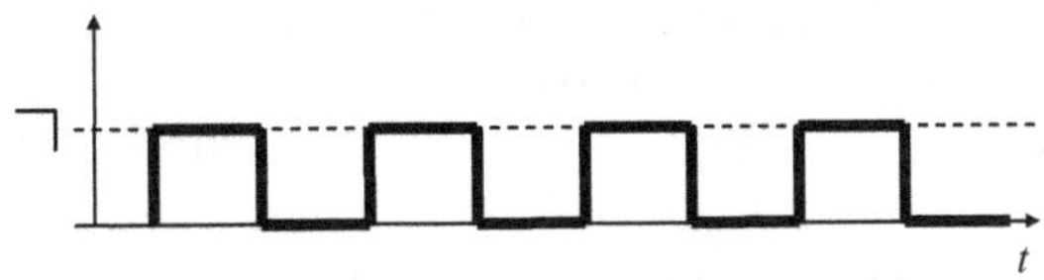

Fig. 1.14 Oscillating dynamic solution to the self-correlated equation (1.13)

The function x is oscillating because it is imagined that the passage of x from a value to the opposite, instead of producing a synchronic antinomy, is unfolded over time, being considered a process.[69] A more complete calculus, that includes specific axioms for re-entry, has been introduced by Varela [79, 80].

Louis H. Kauffman [41] developed on the basis of these ideas a matric *"wave-form algebra"*, useful to model the behavior of undulatory systems. The profound connection we may find between the notion of *re-entry* (i.e. self-reference) and that of *time* flow has been clearly highlighted by Varela, who involves in these relations also the idea of *infinity*:

> ... we find a peculiar equivalence of self-reference and time, insofar as self-reference cannot be conceived outside time, and time comes in whenever self-reference is allowed. [...]
>
> We should not be surprised by the connection between infinity and time since the nature of a re-entering expression is precisely that of an infinite recursion in time of a closed system. [...]
>
> We may interpret a self-cross [...] as an alternation of the other values in time. Conversely we may take the states, marked and unmarked, as timeless constituents of a self-cross occurring in an oscillation in time. Either point of view reattaches time to our dealing with self-referential forms.[70]

---

[68] Spencer Brown [72], pp. 58–59. See also Kauffman [40].

[69] An analogous periodic solution has been given to the semantic liar paradox by Herberger, Gupta, and Belnap in the frame of the "revision theory of truth".

[70] Varela [79], pp. 20–21. To give a more philosophical flavor to the subject, let us quote the great phenomenologist Maurice Merleau-Ponty, who dedicated a lot of intellectual energies to understand the phenomena of perception and cognition: «Time is the affecting of self by self», Merleau-Ponty [51], engl. transl. p. 494.

We might not be surprised, at this point, to uncover that Spencer Brown calls *memory function* the expression (1.12). After a short reflection we may realize that the value of the variable x depends on the past values of a and b: it "remembers" the status of "marked" or "unmarked" of their own. Using the correspondence rules (1.10a) and (1.10b), we may prove that (1.12) is actually equivalent[71] to the flip-flop boolean equation system (1.2), or better to (1.2b).

So, even after such a succinct recapitulation of Spencer's Brown work, we may argue that this calculus

- has the flexibility to catch in a natural way a characterization of a memory element, equivalent to a flip-flop circuit, based on a circular correlation between two logic gates;
- introduces the temporal dimension that emerges as intrinsic property of self-referential expressions;
- allows the equal treatment of functions and values, like in untyped algebras, where there are no hierarchy levels: what is also called an *untyped language*, because Russell and Whitehead called *types* the levels of stratification of a language made to avoid circularities.

We recalled here Spencer Brown's calculus as a curiosity, but it unveils the behavioral aspects hidden in logical notations and, moreover, emphasizes the tight links between self-referentiality and temporality.

The Anti-foundation Axiom and Hypersets

Classical axiomatic set theories, like ZFC (by Zermelo and Fraenkel) or NGB (by von Neumann, Gödel and Bernays), include an axiom called "foundation axiom" (FA) or "restriction axiom," aimed at preventing the system from allowing the existence of infinite chains of decomposable elements. These infinitely descending sequences, using the reciprocal of the membership relation $\in$, are like

$$a_0 \ni a_1 \ni a_2 \ni \cdots \text{ (ad infinitum)}. \tag{1.14}$$

The resemblance of the sequence (1.14) with (1.2a), (1.7b), (1.11), and with those of Fig. 1.11 and 1.12 is not accidental: they all represent the same paradoxical idea.

The sets like a_i are said to be *non-well-founded* sets, or *hypersets*. The *foundation axiom* can be expressed as follows[72]:

$$\textsf{FA} \qquad \forall x(x \neq \varnothing \Rightarrow \exists a\,(a \in x \wedge (a \cap x = \varnothing)))$$

and prohibits also the availability of circular objects in the universe of sets.

[71] Again up to some equivalent replacement of variables.

[72] Where $\varnothing$ is the symbol for the empty set.

In the mid-1980s, Forti, Honsell [26], and Aczel [1] proposed to drop this axiom, replacing it with the opposite one, typically called *anti-foundation axiom*, or AFA, and developed the consequent hyperset theory.[73]

There are several equivalent forms of AFA. We choose the more intuitive and elegant, based on interpreting *sets as (oriented) graphs*, rendering their underlying structure visible. If we represent the membership relation $a \in b$ as a directed arrow

then every set, considered into its atomic elements, can be depicted as an oriented graph. For instance, the set $x = \{\{a, b\}, c\}$ is equivalent to the following tree-like graph:

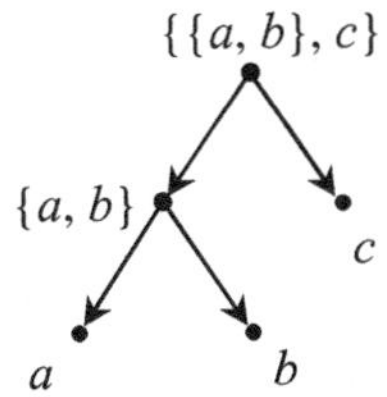

Here every node represents a set or (if terminal) an atom. The bijective function labeling the nodes with the elements of the underneath subsets or atoms is called *decoration*.

The *anti-foundation axiom*, in this context, can be expressed like this[74]:

AFA *Every graph has a unique decoration.*

The new axiomatic set theory, with AFA instead of FA, has been proven to be consistent,[75] and allows existence of circular sets, or hypersets.

For instance, the set $x = \{\{x, a\}, b\}$, which is circular because it contains itself as a proper element, is depicted by the following graph:

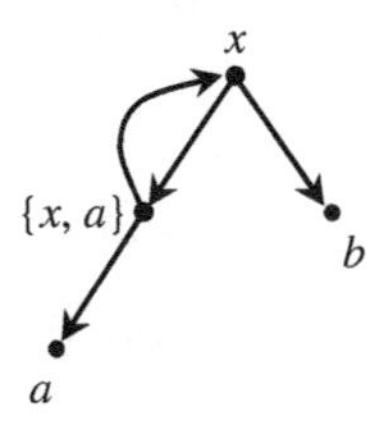

[73]The mathematician who first realized the possible existence of such sets has been the Russian D. Mirimanoff, in 1917, having called them "extraordinary sets." Barwise and Moss [5] is the fundamental reference to dig into the matter in the broadest way.

[74]There are at least eleven variants of AFA.

[75]The first proof, which is based on the existence of an actual model of the theory, is the "main theorem" in Forti and Honsell [26]. The uniqueness constraint has been imposed in AFA to save the categoricity of the theory, i.e. the equivalence of its models up to isomorphisms.

(Needless to say that this is another representation of the logical structure (1.2b) or (1.12) of a flip-flop memory element.)

The simplest hyperset is the so-called *self-singleton*: $\Omega = \{\Omega\}$, which may be represented as follows[76]:

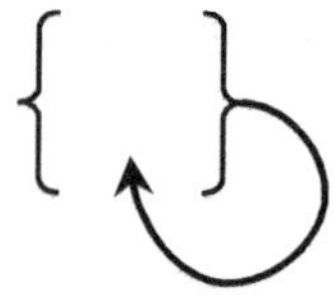

(1.15)

This intuitive notation recalls the self-referential re-entry form and illustrates in a natural way the property of self-membership. This property is inconceivable with standard sets: in a naive interpretation of sets as objects, we should figure a thing that contains itself.

Hypersets like Ω are very strange objects. The classical equivalence relation of extensionality, which ensures a natural and reliable identity principle for sets, does no longer hold in the universe "founded" on AFA: if you try to compare, element by element, two hypersets to discover whether they are equal or not, you will fall in a never-ending sequence, like (1.2a) or (1.11). So, you cannot "capture" by the standard extensionality axiom the true essence of an hyperset. For these reasons they have been banished from set theory for about 70 years.

The solution to this problem comes from the idea of *bisimulation*: told in simple and rough words, two relational structures are bisimilar if they "show" the same behavior, that is, if we can prove that there is equivalence between the classes of all their respective "unfoldings".[77] Bisimulation is an equivalence relation (i.e., it is reflexive, symmetric, and transitive) which is embodied in the *strong extensionality* axiom (SEA) and, like simple extensionality, is an expression of Leibniz's principle of the identity of indiscernibles[78]: if two hypersets are bisimilar, then they are the same hyperset:

SEA $\quad \forall a, b \, \exists R (R \subseteq a \times b \Leftrightarrow a \sim_R b) \quad \Rightarrow \quad a = b$

---

[76]We owe this good visual idea to the mathematician Louis Kauffman. Let us notice its resemblance to Fig. 1.12 and to (1.13).

[77]The definition of bisimulation is more complex, but we do not need to go deeper here. Barwise and Moss [5] is a wonderful guide for all these topics. The first idea of bisimulation comes from the French logician Roland Fraïssé, who in the 1950s studied the "local isomorphisms" between relational structures.

[78]The *identity of indiscernibles* is a second order fundamental principle of identity, saying that if two entities A and B share all their apparent properties ϕ they are the same entity:

$$\forall \phi \, (\phi\,(A) \Leftrightarrow \phi\,(B)) \Rightarrow A = B.$$

where a and b are two (hyper-)sets, $a \times b$ is their cartesian product, R is a bisimulation, and $\sim_R$ is the symbol signifying that a and b are bisimilar.

To give the simplest example, the following two graphs α and β are bisimilar:

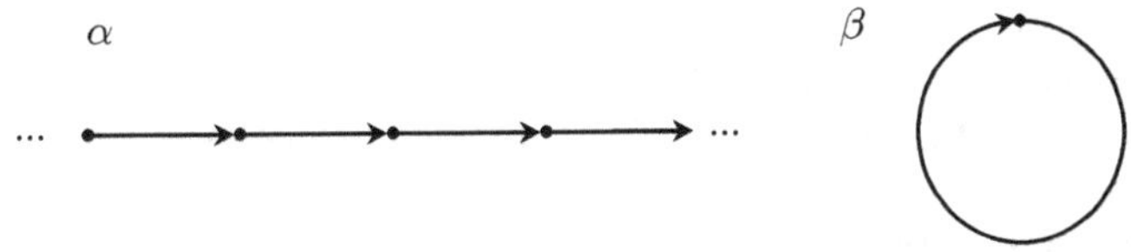

The system α is infinite, whilst β is finite but non-well-founded. α can be thought of as the "unfolding" of β.

The notion of bisimulation has known important applications in computer science, where it makes possible a meaningful modeling of the behaviors of programs (*computational semantics*). Hypersets and bisimulation are useful formal tools where other formalisms (like those based on standard set theory) risk failing due to the fact that the computational systems we want to model often encompass self-referential processes. A typical example is concurrent communication processes, where information fluxes can experience circular paths.[79]

Systems as Co-algebras

A more recent and fertile development of hyperset theory comes from its interpretation within the mathematical *theory of categories*.[80] Here we have the category ***Set*** whose objects are of all the sets, equipped as every category with its set of morphisms and the subset of identity morphisms. The applications between categories with certain properties (identity preservations, etc.) are called *functors*.

Given a set S and an endofunctor F in ***Set***, and a function f, we say that a pair $\alpha = \langle S, f \rangle$ is an *algebra* or a *co-algebra* if, respectively,

$$f : F(S) \to S \quad \text{(algebra)}$$

or

$$f : S \to F(S) \quad \text{(co - algebra)}$$

Now, without giving here more details, we may summarize the matter recalling that the anti-foundation axiom AFA can be restated in the language of categories

[79] See the seminal study of Milner [52].

[80] The classical introduction to this field is: Mac Lane [48].

in an elegant and synthetic way.[81] Moreover, the duality between algebras and co-algebras can be perfectly mirrored in the duality between ordinary sets and hypersets.[82] Co-algebras have turned out to be a powerful tool to formalize the behavior of theoretical machines, like *labeled transition systems* (which are a generalization of formal automata). On the mentioned categorial duality is based on the development of further interesting notions, like *co-induction* and *co-recursion* as definition or proof principles, which are the dual, respectively, of induction and recursion.

Moreover, the hyperset universe introduces the notion of *deconstruction*, which is assimilable to that of observation. This fact can be intuitively caught in the infinite sequence (1.14): at every step, we "extract" (observe) an element from the previous set, triggering off a potentially infinite process of analysis of the system.

The following concise synopsis sums up the reciprocal relations among all these concepts (Fig. 1.15).

sets universe	$\longleftrightarrow$ *(duality)*	hypersets universe
foundation axiom (FA)	$\leftrightarrow$	anti-foundation axiom (AFA)
hierarchical structures (trees)	$\leftrightarrow$	circular systems (networks)
identity (extensionality)	$\leftrightarrow$	bisimulation (strong extensionality)
algebras	$\leftrightarrow$	co-algebras
induction/recursion	$\leftrightarrow$	co-induction/co-recursion
construction	$\leftrightarrow$	deconstruction (observation)
finite behavior unfolding	$\leftrightarrow$	infinite behavior unfolding

Fig. 1.15 Synoptic summary of the dual notions involved in set vs. hyperset theory

The lesson we can learn from hyperset theory for memory systems is that in this context we may find plenty of formal concepts and tools to describe both recursive and self-representational aspects of memory processes. We will give some more hints in the next section.

About the possible use of hyperset theory to model consciousness phenomena, it is interesting to see that some authors are recently finding in this area new inspirations for their research. For instance, Miranker and Zuckerman [55] are proposing an inquiry on self-referential qualities of consciousness based on hyperset theory. Here

[81]The categorial version of AFA sounds this way: "The universe of all the sets is the (minimal) fixed point of the endofunctor of the category of classes that associates to every class the class of all its subsets."

[82]Exhaustive treatments of these topics can be found in Rutten [68, 69].

consciousness is defined as an operator K applied to the domain of experiences: the awareness of an experience x is symbolized by $K(x)$. Then, the first axiom of this theory is[83]

$$\forall x K(x) \subseteq x,$$

which means that "experience generates its own awareness," that is, *self-awareness*. It is clear that such a formal system needs to apply in a hyperset-theoretic frame, where self-reference is allowed. Now, interpreting the consciousness operator K as a memory operator, we could start to formally design the main features of an alleged *meta-memory*. This is only one example on how the hyperset theory could find applicative new openings.[84] More generally, as we have seen that the objectification of sets is no longer plausible with hypersets, they rather induce to naturally interpret them as *processes*.[85]

As a consequence, it is possible to model, through the notion of *bisimulation*, the fact that memory processes typically do not retain the mere images of the remembered things or events, but *preserve their structures*. This fact has two main consequences:

- bisimulation mirrors a deeper form of isomorphism, which turns out to be more robust against minor changes, neglecting transient details and keeping the essential relational features of the object, or its intrinsic organization, also allowing the emergence of symbolic generalization: a cake is not isomorphic to the recipe you used to make it, but to its "idea", because this late reflects the nature and scope of the cake better than its physical instantiation;
- bisimulation treats as equivalent a process and its temporal unfolding, while the first could be finite and recursive (circular) and the second linear and infinite: the description of an oscillating system can be compressed in a single differential equation, while its actual behavior may last as long as the system doesn't stop for some reason[86].

These observations suggest that the fundamental *duality* between a *process* and its temporal *unfolding* – or a dynamical structure and the relevant possible behavior – could impact the algorithmic description size of a system, as defined by algorithmic information theory[87]. What we would mean is that the critical issue of the informational size of a system that would encompass itself could be regarded

[83] Cf. with (1.5).

[84] See, e.g., Williford [83], Goertzel [30].

[85] See Rutten [68].

[86] Friction combined with the second law of thermodynamics, typically.

[87] See above, section "The Paradox of Meta-Memory". Cf. also Varela's reflections on the connections between self-reference, time and infinity quoted in section "Spencer Brown's Calculus of Indications". About the structure/unfolding duality, let us keep in mind the simple example of the bisimilar graphs α and β reported in the previous section.

under a different light if we consider that the process of self-encompassing can be done with respect to its structure rather that to its actual unfolding: this would allow, in principle, for a memory to *embed a "compressed version"* (whatever we mean with this) *of itself*. For the same reason we do not memorize all the passages of a theorem proof, but only the few rules and hints necessary to reconstruct them in details. Memory is smart, and notions like bisimulation open the way to think of it in terms of smart processes, which proceed by dynamic synthesis, not by creating rigid and static copies of the reality inside memories. Furthermore, this approach recalls again the idea of *memory as reconstruction*, rather that replication, as always evidenced above, when mentioning autopoietic approach to cognitive systems[88].

Beyond the mentioned studies, there is a quite broad literature of contributions that connect, with different approaches and perspectives, self-reference to consciousness[89], where we might imagine that hyperset theoretical machinery could bring new ideas. The same problematic importance of self-reference in the study of cognition is testified by several authors, like in the last page of the monumental essay of Philip N. Johnson-Laird on *Mental models*:

> It is a proposition, however, that may lead us in time to revise our concept of computation. At the moment that I am writing this sentence, I know that I am thinking, and that the topic of my thoughts is precisely the ability to think about how the mind makes possible this self-reflective knowledge. Such thoughts begin to make the recursive structure of consciousness almost manifest, like the reflections of a mirror within a mirror, until the recursive loop can no longer been sustained.[90]

Life, Computation, Time: Toward a Natural History of Memory

> The power of the memory is prodigious, my God. It is a vast, immeasurable sanctuary. Who can plumb its depths? And yet it is a faculty of my soul. Although it is part of my nature, I cannot understand all that I am. This means, then, that *the mind is too narrow to contain itself entirely*. But where is the part of it which it does not itself contain? Is it somewhere outside itself and not within it? How, then, can it be part of it, if it is not contained in it?
> I am lost in wonder when I consider this problem. It bewilders me.
>
> St Augustine of Hippo, *Confessions*[91]

In the previous section we have seen that the self-referential aspects of memory can be described through different formal systems that encompass and overcome the difficulties that may arise from paradoxical circularity. Moreover, we have seen that

[88] Pribram [62] recalls that the fundamental features of remembering processes are the "four R's": *representation*, *reconstruction*, *registration* and *rearrangement*. The neurophysiologist Karl H. Pribram is known for his holographic model of memory. For an ecological perspective on memory, see also Turvey and Shaw [78].

[89] See, e.g., Perlis [59], Hofstadter [35], and the classics Hofstadter [34] and Hofstadter and Dennett [36]. Cordeschi, Tamburrini and Trautteur [14] is a subtle inquiry on the role of different kind of loops (or hierarchies of loops) in self-consciousness, confronting MacKay's cybernetic approach to the more recent Gray's neurophysiological account.

[90] Johnson-Laird [39], p. 477.

[91] Pine-Coffin, R. S. Penguin classics, (1961) (emphasis added).

such models supply structural features that naturally introduce a process approach to the described phenomena, taking into account the dynamical implications of the systems functioning.

Now we try to enlarge our view by encompassing more abstract properties or notions related to the phenomena of remembering, to understand if they can say something more about memory, or vice versa. These elements could be thought of as a possible ground to the development of a *natural history of memory*, where its properties and features have seen as actual evolutive contributions to the same existence or right functioning of wider aspects of reality.

In light of these ideas, we approach the relationship between memory and life, computation and time, to discover that the other side of memory, which is information, is always involved.

Life as Memory: Self-Description and Self-Reproduction

As mentioned before, a potential criticality in modeling memory could be the need to have a self-encoding process, such that the memory can remember itself.

The necessity to think about self-representing memories lies mainly in the fact that any memory is useless if it is not immersed in a process flow that includes a subject's intentionality, which requires that something in the memory process acts knowing that it is a memory: a certain degree of self-awareness is the necessary condition to make the memory actually working. The segregation of this function, i.e., the external reification of memory implemented via the trace-like approach, does not solve the problem: it simply shifts its impact to a higher level, like in the *homunculus'* regress.

At first sight, the idea of a completely self-containing memory recalls Royce's paradox of the self-encompassing map, or that of the self-belonging set Ω depicted in (1.15).

What we are asking is that, given a memory function M that extracts the content x of memory, this function could be supposed to have a *fixed point*, that is,[92]

$$\exists x\ \delta(x) = x. \tag{1.16}$$

The theme is studied by prominent scientists and philosophers,[93] from different viewpoints: neurophysiological, system-theoretic, ontological, etc. We are more interested here in a possible computational approach.

As a matter of fact, we know that there are real systems that contain, in some way, a description of themselves: living organisms, which are capable of self-reproduction, "by construction." Since the discovery of the DNA molecular structure in 1953 it is universally clear that memory is a fundamental ingredient

---

[92]Cf. with the similar formulas (1.4), (1.6) and (1.7b).

[93]Cf., e.g., Churchland [12].

of life. Most studies are being carried out to discover the logical, informational, and computational implications of this fact for life.[94] The most prominent historical contribution to this issue is the famous essay on *self-reproducing automata* by von Neumann[95]: a first proposal for a purely formal account that, besides the new born artificial intelligence, inspired the more recent studies in artificial life. Von Neumann's writing was purposed to understand

> what kind of organization is sufficient for an automaton to be able to reproduce itself. [...] He wished to abstract from the natural self-reproduction problem its logical form.[96]

As summarized by Langton,

> Von Neumann was able to exhibit a universal Turing machine embedded in a cellular array using 29-states per cell and the 5-cell neighborhood. His Turing machine is suitably modified so that, as output, it can "construct" in the array any configuration which can be described on its input tape. Such a machine is called a *universal constructor*. His machine will construct any machine described on its input tape and, in addition, will also construct a copy of the input tape and attach it to the machine it has constructed. Now, self-reproduction follows as the special case where the machine described on the tape is the universal constructor itself. The result of the construction process is a copy of the universal constructor together with an input tape *which contains its own description*, which can then go on to construct a copy of itself, together with a copy of its own description, and so on indefinitely.[97]

So, the central idea of such an abstract model of the main function of life is based on two modules: a constructor and a memory. While this model has been successively improved and simplified by other authors,[98] the main features have been preserved. As stressed by Arbib,[99] in designing the automaton memory function von Neumann faced the risk of a possible infinite regress. In fact, let us suppose that the automaton A is a universal constructor that, having received as input the description[100] $\delta(X)$ of the automaton X, it can build X itself; indicating by $\rightarrow_R$ (or by R) the function of reproduction, we have

$$A + \delta(X) \rightarrow_R X \qquad \text{that is,} \qquad R(A + \delta(X)) = X.$$

Now, A is not capable of self-reproducing, because when provided with its own description $\delta(A)$ it can build a copy of itself, but without a copy of its own

---

[94] Even if with completely different approaches, we might mention several scholars and research fields, from Erwin Schrödinger to Robert Rosen, from system biology to biosemiotics.

[95] Von Neumann [58]. Similar to this approach is that of the *self-reproducing programs*.

[96] Burks [7] p. xv; quoted in Langton [43].

[97] Langton [43]; emphases added. Christopher Langton has been one of the first promoters of artificial life studies. Good syntheses of this topic could also be found in Johnson-Laird [39] and Poundstone [61].

[98] Codd, Minsky, Arbib, Langton, among others.

[99] Arbib [2].

[100] Here we use intentionally the same notation as in (1.4) and (1.16). The standard symbol used for the descriptive instructions of a system X is I_X.

description, which should be added again, but together with a description of the description:

$$A + \delta(A) + \delta(\delta(A)) \to_R A + \delta(A).$$

To avoid infinite regress, von Neumann added two more automata: *B* that can make a copy of the input description and *C* that inserts this copy in the automaton under construction. Now we have a new composed automaton

$$U = A + B + C + \delta(A + B + C), \tag{1.17}$$

which finally turns out to be a fixed point of the reproduction function, i.e., a faithful self-reproducer:

$$U \to_R U \quad \text{that is,} \quad R(U) = U.$$

The definition (1.17) clearly depicts a machine that contains a self-description, but we see that it is a partial one: it does not contain the description of the description. This is the price paid for avoiding an unmanageable circularity, while on the other side the problem of finding a fixed point for *R* is brilliantly solved.

So, the result we can draw from this example is that if we consider a purely extensional account of memory, even though it is possible to use it for defining a self-reproducing machine, when we try to have a full self-encompassing of the memory device we always fall in the paradox of infinite regress.[101]

A more subtle approach, with respect to our issue, has been formulated by Marvin Minsky in a theoretical essay on the relationship between mind and modeling activity.[102] In Minsky's notation, *M* is a man, *W* is the world, and a superscript star denotes that W^* is a model of *W* for *M*, i.e., an object such that "*M* can use W^* to answer questions that interest him about *W*." Initially Minsky wisely splits the man *M* into two parts:

$$M = (M - W^*) + W^*,$$

[101] As a matter of fact, it seems that von Neumann was worried about the possible implication of circularity in his account for self-reproduction, heaving left some allusions in his uncompleted manuscript: "I am a little twisting a logical theorem, but it's a perfectly good logical theorem. It's a theorem of Gödel that the next logical step, the description of an object, is one class type higher than the object and is therefore asymptotically infinitely longer to describe." To cope with this problem, after von Neumann's death, Burke wrote directly to Kurt Gödel, asking for his opinion. Gödel answered that his theorem mentioned by von Neumann should be "the fact that a complete epistemological description of a language *A* cannot be given in the same language *A*." (Cf. von Neumann [58], pp. 47, 51–56.)

[102] Minsky [53].

where

> W^* really contains the knowledge and M–W^* contains only general-purpose machinery for coding questions, decoding answers, and general administrative work.[103]

We may paraphrase saying that W^* represents the memory part of M's consciousness, while (M–W^*) represents all his other cognitive functions.

Now the problem is, again, how to consider the self-modeling activity of the man M:

> If W^* contains a model M^* of M, then W^{**} may contain a model M^{**} of M^*.[104]

Infinite regress is still just round the corner. But here Minsky is ready to recognize the intrinsic limits of the extensional account of the traditional spatial modelization of memory as container:

> I think we must envision W^* as including an interpretative mechanism that can make reference to W^* – using a sort of computer-program subroutine – to a certain depth of recursion. In this sense W^{**} must contain W,** but in another more straightforward sense W^* can contain W^{**}. This suggests (1) that the notion *contained in* is not sufficiently sophisticated to describe the kinds of relations between parts of program-like processes and (2) the intuitive notion of *model* used herein is likewise too unsophisticated too unsophisticated to support developing the theory in technical detail. It is clear that in this area one cannot describe intermodel relationships in terms of models as simple physical substructures.[105]

In a few statements Minsky poses a lot of capital questions related to the epistemological status of the models. For our discourse, it is sufficient to remark that the hyperset and the co-algebraic approaches to modeling would seem to be a sound way to answer the questions left open by Minsky. In fact, he invokes the availability of formal structures that, overcoming the intuitive and "unsophisticated" notion of model, could supply a different type of relationship among models, where the limits of the spatial metaphor of the container/content disappear. In hyperset theory, as we have seen, a set can be a member of itself, so that a reciprocal containing relationship between W^* and W^{**} is also acceptable.

Actually, the system

$$W^* \in W^{**} \quad \text{and} \quad W^{**} \in W^*$$

is inconceivable within standard set theories, while perfectly acceptable with AFA, which precisely asserts that it has a (unique) solution.

Also John Case poses the problem of a logical account of self-reproduction, and focuses its essay in adapting Kleene recursion theorem, exposed in terms of the computability of a self model:

---

[103] Ibidem.

[104] Ibidem.

[105] Ibidem.

> For any algorithmic thing one would want to compute about a complete low level self model, one can algorithmically find a program *e* which creates an *external* complete low level self model and subsequently computes that thing about that self model.[106]

This approach suggests new hints to cope with infinitary self-reference.

Anyway, the embedding of a self model is a fundamental need for all the self-regulative systems, as described in Conant, Ashby [13].

Computation as Memory: Recursion and Reversibility

Another way to look at memory is to consider its connections to *computation.*

On the one hand, the same processes of categorizing, selecting, encoding, storing, retrieving, and decoding information, whatever their logic and material substrates, are reducible to computational processes; so, remembering is a form of calculation.

On the other hand, we may notice that any computational process entails some form of information retention to be performed or even conceived. An immediate but particular evidence of this fact is that, when we perform even an elementary arithmetic calculus or a trivial logical inference that requires more than one step, we must keep the intermediate results to execute the next steps: the scribbled draft of a paper-and-pencil calculus is a plain example. The same Herman H. Goldstine and John von Neumann, when designing the general architecture of the first computing machines[107], claimed that the principal requirements for a memory functions are related to the need to perform operations and store intermediate results.

The most general (but not immediate) evidence is that the set of all possible *computable functions* is coextensive to the set of all *Turing machines*, and every Turing machine has inside a memory, by definition.[108] This memory, being an ideal object, is supposed to be unlimited.[109] Actually, being computable functions fully identifiable with *partial recursive functions*, we may point out that every recursive process also needs to keep in a temporary storage the interim steps, when the sequence of the recursion reaches some depth and then gets back to the initial level.

106Case [9].

107Goldstine, von Neumann, "On the principles of large scale computing machines" (1946), in: von Neumann [57], pp. 1–34.

108A Turing machine could be roughly described as consisting of a (potentially) infinite tape divided in discrete slots (memory) coupled with a machinery that can print/erase symbols of a finite alphabet on the tape, following (finite) instructions that determine step by step its internal status (from a finite set of possible statuses, including an initial one) and a possible single-step shift to right or left on the tape.

109Often a Turing machine's memory is said to be "infinite," but we think that this term could induce the misleading idea of an actual infinity. It is an interesting exercise to define the subset of computable functions corresponding to the set of Turing machines with strictly finite memory. Curiously, Kurt Gödel was firmly convinced that human memory is actually infinite: for him the infinity of the tape of a Turing machine could have sound not so an exotic idea.

Let us think, for example, of a cycle of nested loops in a computer algorithm (like an IF-THEN-ELSE block), which behaves like a LIFO stack or a palindromic string:

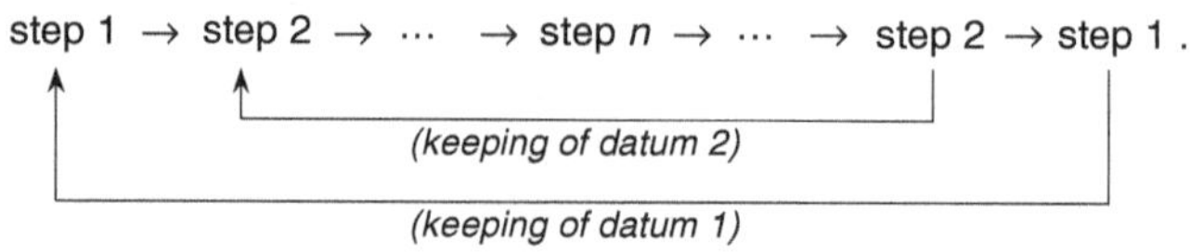

For these reasons, inside a computer – other than the "official" memories, like the hard disc – memory functions are also implemented within the computational core, as temporary local registers and registers which keep for some fractions of seconds a myriad of ever-changing variables. Memory is pervasive to all processes assimilable to computation.

At a deeper level, we may suppose that the same notion of circular process underlies both the general idea of recursion and that of dynamical storage, so that at the computation level this correspondence emerges in the form of the essential co-occurrence of computability and memory.

The ordinary formal notation systems for mathematical and logical calculus typically do not highlight the stratification depth entailed by the memory processes associated with them. An exception is represented by *linear logic*, the formal language devised by the French mathematician Jean-Yves Girard,[110] in which to each variable is associated an initially available resource quantity that is progressively "consumed" at every step of an inference. This calculus perfectly embeds in itself and manifest the function of memory.

In all the abstract "machinery" that implements the different ideas of calculus – like automata, transition systems, Petri nets, neural networks–, *ça va sans dire*, the function of memory is implicitly embodied in the permanence of the temporary statuses of the machine, during the computation process.

To sum up, we may say that, while computation is not reducible to memory, memory is absolutely necessary to it, to such an extent that memory is an essential part of the same definition of universal computability.

We may also mention that another interesting topic that can be related to the computational aspects of memory is that of *reversible computing*. This area, developed from some basic work by Charles H. Bennett, Edward Fredkin, and Tommaso Toffoli,[111] is based on considering that most logic connectives, like the AND and OR boolean gates, are not reversible, because it is impossible to retrace their inputs after the output is performed.[112] This topic is not only a theoretical field of

[110]Cf. Girard [28].

[111]Cf. Bennett [6], and Fredkin and Toffoli [27].

[112]For example, if we know that x OR y gives 1, we cannot know what are the values of x and y. There are three possibilities: (0, 1), (1, 0), and (1, 1): we are not able to "come back" from the output 1 to the initial inputs. The NOT function is just an exception.

speculation on computing but is also related to the problem of energy consumption in computational processes:

> ... conservative logic shows that it is ideally possible to build sequential circuits with zero internal power dissipation.[113]

Some authors proposed new kinds of logical gates, like Fredkin gate, that are reversible and that can be combined to simulate the functioning of standard boolean gates. Typically reversibility is obtained using more than three input ports, rather than two, and an equal number of output ports.

This is, for instance, the structure of Fredkin gate (Fig. 1.16):

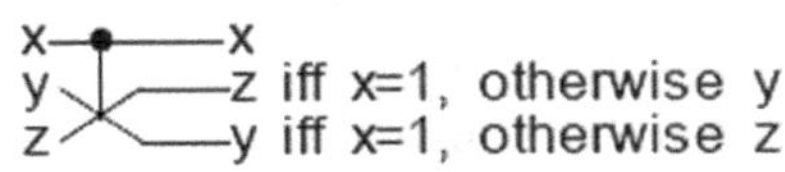

Fig. 1.16 Fredkin reversible logic gate

These logical gates have been used also to design reversible memory elements (Fig. 1.17), like different kinds of flip-flop, like this[114]:

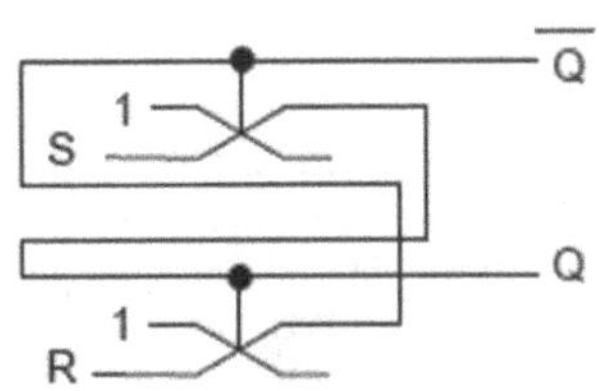

Fig. 1.17 Reversible memory element

This draws the notion of memory up to its essential function of conservative or preserving principle, in an ideal perspective where entropy slows down and time tends to be frozen: the essence of memory thought as crystallization of the present instant.

There is no room here to go deeper with this topic, that would lead us into the deep debates about the physical Church-Turing thesis, the Landauer principle, and so on.[115] It is enough, for our purposes, to notice that it lies on the boundary between physics and computation, where the convergence of the notions of entropy and information takes on a full significance. It could be the starting point to a deeper

---

[113]Fredkin and Toffoli [27]. See also Feynman [21], chapter 5.

[114]Cf. Rice, JE. [63] A new look at reversible memory elements. In: 2006 IEEE International Symposium on Circuits and Systems, IEEE 2006. In the diagram S and R represent the "set" and "reset" input signals, while Q and its negation are the outputs.

[115]See, e.g., Bennett [6] and Lloyd [46], but the literature is vast.

investigation about the physical roots of memory, involving the more general ideas and discussions about the possible computational nature of mind.

Anyway, somebody could think that a reversible or bidirectional memory would be a better and more useful one, as claims the White Queen:

> [*Alice*:] « . . . I can't remember things before they happen.»
> «It's a poor sort of memory that only works backwards,» the Queen remarked. [116]

Time as Memory: Information and Entropy

The observations made in the previous section about the physical basis of information and memory processes suggest that some deep relationships could be found at this level. Without giving a formal treatment to the discussion, we would sketch some rough ideas that could be followed up elsewhere.

The convergence of the notions of *thermodynamic entropy* and *information* is subject since some decades to long-lasting debates, to such an extent that the definition of information is often overlapped with that of entropy:

> The concept of information is too broad to be captured completely by a single definition. However, for every probability distribution we define a quantity called *entropy*, which has many properties that agree with the intuitive notion of what a measure of information should be. This notion is extended to define *mutual information*, which is a measure of the amount of information one random variable contains about another. Entropy then becomes the self-information of a random variable. Mutual information is a special case of a more general quantity called *relative entropy*, which is a measure of the distance between two probability distributions.[117]

The exact definition of the information/entropy H of a discrete random variable X over an alphabet A is given (up to a change of sign) by the equation

$$H(X) = -\sum_{x \in A} [p_X(x) \log p_X(x)],$$

where $p_X(x)$ is the probability mass function, i.e., the probability that $x = X$.

More concisely, when N is the total number of (equi-)possible states of the system, the total entropy of the system is:

$$H = -k \ln N, \tag{1.18}$$

[116]L. Carroll, *Through the looking glass and what Alice found there*, London 1871 (in: *The annotated Alice*, ed. by M. Gardner, Penguin, London 2001, p. 206).
[117]Cover and Thomas [15], p. 12.

where k is the Boltzmann's constant[118].

What is clear is that the information carried by a message within a system is in some sense a measure of the order/disorder associated with it, depending on the probability of the related state with respect to all the possible configurations of the system.[119] We should also recall that the quantity of total entropy of an isolated system not in equilibrium is subject, according to the second law of thermodynamics, to increase. A local lowering of entropy within a confined system is always compensated by an increase of it outside the system: it is likely to pump heat outside a box to keep the inner temperature low (a refrigerator). Lastly, it has been said[120] that the asymmetric direction of the arrow of time could be a consequence of the second law of thermodynamics: when we experience an increase of the disorder of a system we perceive this fact as the flowing of time.[121]

Now, in a few informal statements, we may arrange these issues to encompass memory as follows:

- information is the measure of the order within a system, which is its entropy; the more the event related to the information is improbable, the more is its informational content high;
- memory is a process aimed at keeping (freezing) information over time;
- so, memory is a process aimed at keeping lower the value of entropy (disorder) within a system: it is the function that gives continuity to the level of order in the system;
- so, memory is a process that "runs" in the opposite direction with respect to time, having the apparent function to suspend the effect of its flux: it is the way the past emerges in the present.

This reasoning can be summarized giving a characterization of *memories* as *processes* (or devices implementing them) *aimed at local and temporal confinements of entropy*, like dams or embankments erected against the flooding disorder coming from the open environment.[122] In this sense, the thermodynamical decay of the content of a memory device is the physical equivalent to oblivion, which is always a loss of order.

---

118 It is significant that von Neumann, when discussing about how to measure memory capacity of automata, «suggested using the logarithm (to the base two) of the configuration number (i.e., the number of alternatives)», as refers Burks: cf. Von Neumann [58], p. 40.

119 Analogously, in Thom [74] entropy is geometrically characterized as "topological complexity".

120 Sir Arthur S. Eddington, I presume.

121 As useless excuse to the oversimplified presentation of these complex physical and epistemological issues, we invoke the fact that this is here a border argument with respect to our scope.

122 Cf. Haken [31], pp. 23–24: "... a system in thermal equilibrium cannot even *store* information. [...] Any memory consisting of a closed system is out of thermal equilibrium and it is always necessary to ask *how long* the information can be stored in each specific case."

More accurately, every bit has a measurable *"entropic cost"*, which can be evaluated[123] with (1.18) as a variation in entropy equal to

$$H_{bit} = -k \ln 2. \tag{1.19}$$

Consequently, the minimal amount of energy required to erase a bit of memory is

$$E_{bit} = kT \ln 2, \tag{1.20}$$

where T is the temperature (or noise in the information channel). Conversely, the *Bekenstein bound* defines the upper limit of the information that can be stored in a system, depending on the volume of the system:

$$I \leq H(E) \;/\; (k \ln 2) = r\,E/\,(hc \ln 2)\,, \tag{1.21}$$

being E the average energy of the system, r the radius of a sphere that exactly contains the system, h the Planck's constant and c the speed of light.[124]

These reflections can also be associated with the idea that the typical phenomenon opposing entropy is *life*; quoting an exquisite classic:

> How would we express in terms of the statistical theory the marvellous faculty of a living organism, by which it delays the decay into thermodynamical equilibrium (death)? We said before: "It feeds upon negative entropy," attracting, as it were, a stream of negative entropy upon itself, to compensate the entropy increase it produces by living and thus to maintain itself on a stationary and fairly low entropy level. [...]
>
> Thus the device by which an organism maintains itself stationary at a fairly high level of he orderliness (= fairly low level of entropy) really consists continually sucking orderliness from its environment.[125]

So, as a further step, we are again to consider the proximity of the phenomenon of memory to life itself. In a certain sense, the common link to the notion of (negative) entropy could be a confirmation of the supposed overlapping of memory processes and biological organization we mentioned reasoning about the notion of autopoiesis.

We are aware that the previous statements are only rough and imprecise ideas, connecting together distant and complex concepts, sometimes not well defined at all.[126] However, what is worth to mention here is that circularity comes again back on the stage, because just through the notion of autopoiesis and self-organization we have been able to connect memory to life. The following Fig. 1.18, although quite naive, summarizes the previous reflections.

---

[123] See Lloyd [45] and Lloyd [46] for details.

[124] Ibid.

[125] Schrödinger [70].

[126] About the use of the term "entropy" to denote the average information, Feynman [21], p. 123, refers that: "Shannon adopted this term on the advise of the mathematician John von Neumann,

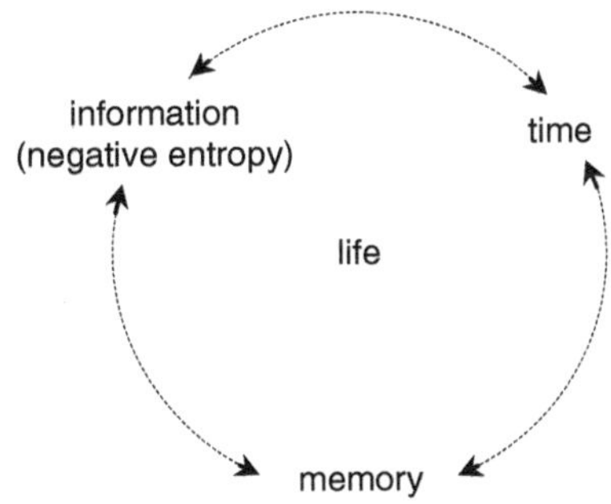

Fig. 1.18 Mutual definitory dependencies of some basic notions related to life itself

Perspectives on Artificial Memory Systems

> The biggest advances will come not from doing more and bigger and faster of what we are already doing, but from finding new metaphors, new starting points.
>
> Terry Winograd, *The design of interaction*

In the previous pages we mostly dealt with the features proper of natural memory, and specially the human faculty. Our aim, however, is not to forget a more general account that could comprehend all kinds of memory.

So, after some reflections starting from the process perspective on memory and the self-referential properties so entailed, we can sketch now some ideas about how these notions could fit a purely extensional objectification of the remembering process in memory technologies.

The evolution of memory technologies has followed two routes: internal mental techniques (mnemonics) and external implementations (from writing to nanoelectronic systems). As seen, both are based on the trace-memory metaphor, which forgets the intensional and dynamical side of memory, enhancing the extensional and spatial dimension of it.

Leaving on shelf the mnemotechnics, we recognize that the evolution of artificial devices to support memory has known an impressive acceleration in the last decades. The very basic idea for recording a bit of information by means of a stable state of a physical device is still unchanged, as Fig. 1.19 seems to suggest.[127] Nevertheless, the jump is gigantic: we inexorably fell from the domain of visible things to that of micro- and nano-metric objects, far below our capability of direct observation. The road of this technological evolution is one of the most astonishing story in

---

who declared that it would give him " ... a great edge in debates because nobody really knows what entropy is anyway"."

[127] The idea is to categorize the states of things such that they can be expressed as sequences of "yes/not" answers. (Curiously, the section of a microelectronic memory cell visually looks like a knot.)

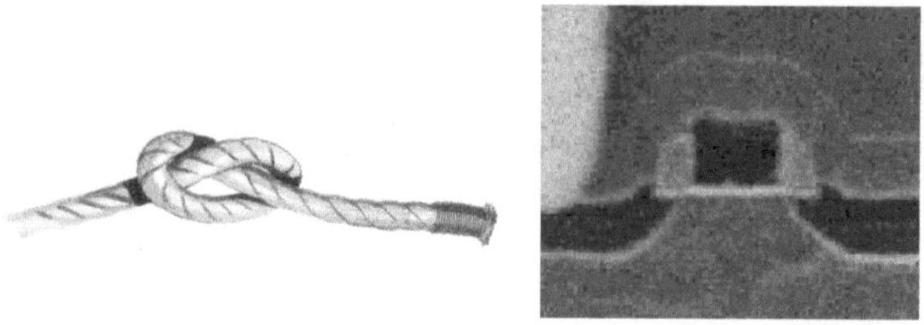

A B

Fig. 1.19 Examples of artificial memories. (**A**) Rudimentary digital device to store a bit of information. (**B**) Advanced digital device to store a bit of information

contemporary technology. The main chapters of this same book tell this story with plenty of details. This evolution is expressible only with progressions of figures that require logarithmic scales to be shown in a readable way. The most celebrated evolutionary trajectory is the so-called Moore's law,[128] but its shape is isomorphic to the other technology parameters' evolution, as depicted in Fig. 1.20.

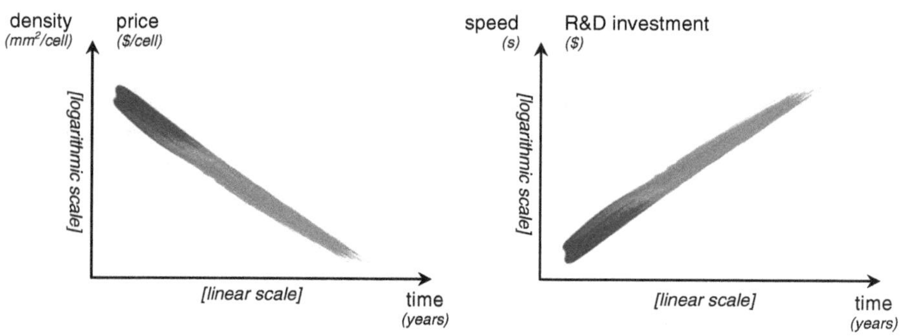

A B

Fig. 1.20 Evolutive trends of memory microelectronic technology – (**A**) Downscaling parameters. (**B**) Raising parameters

This astonishing evolution, that has requested the tremendous effort of some millions of highly qualified person-years of technology research and development, is based on some fixed paradigms that lead to asymptotically approach, as close as possible, the bottom physical limits of matter and computation. We don't know

[128] While we know that it is not a law (in the sense a physicist intends the term), but more a "prophecy that fulfill itself" (which is more likely to a law in the sense an anthropologist could mean).

where and when such limits will exactly be reached. What we can imagine is that the market demand will absorb (or accept) the saturation of the physical constraints.

At that point, we suppose, researchers would address again natural memory to understand what we can learn more from it (and in fact this reference has never been discontinued), in order to find new paths that could turn out to be "orthogonal" with respect to the lines traced in the preceding diagrams. What they can do, at this point, as the computer scientist Terry Winograd suggests, is to find "new metaphors, new starting points." Our brief excursus would point out that the process and self-referential accounts to memory are potential sources of (not necessarily new) interesting metaphors for further ideas. To conclude this section we may recall the most significative:

- *Intensionality* – While the extensional and spatial metaphor is reflected on the same term of "capacity" (ability to contain), an intensional account should explore the dynamic dimension of information, which is not a fixed entity, but can be at every moment compressed or re-created following associative relationships based on semanticity, i.e., on the content of the recorded information. In such a way, every recalling turns out to be in fact a "re-membering" (giving again the members/limbs, i.e., a body) process, where the information, as negative entropy, plays as the dematerialized face of reality.
- *Embodiment* – The typical way to design a system is to segregate the memory function in a definite box, with fixed boundaries. In natural systems such a separation is not so neat, while it depends on the dynamic evolution of the process, every pattern storage being a really learning activity, a structural modification of the internal network of relations that makes the system able to recognize at first sight any new occurrence of the experienced action/reaction or visuo-motor schemes. Some theoretical work is already in place in the sense of "confusion of memory",[129] that is, computational models which do not separate input from working memory. This kind of possible memory would also be able to embody in some way a model of the external environment, turning out to anticipate what we are requesting to it: here the boundary between cognitive (or "intelligent") and storage functions tends to disappear.
- *Self-referentiality* – For the reason mentioned above, the confusion between different logical levels, even if source of possible paradoxes, induces in real systems the creation of new categorizations and higher level schemes to manage information and to model the same interactions with it. Self-representation features could suggest that the memories could not only learn (store information) but also learn-to-learn (second-level learning or meta-memory). Moreover, the ability of self-reproduction, as proposed by von Neumann, could be a base to imagine memories that clone their own contents and use such self-images to other high-level purposes, like error detecting and history tracking.

[129]Cf. Moss [56]. A conspicuous computational example are cellular automata, where all the system *is* its own memory; cf. Wolfram [85].

- *Reversibility* – A closer attention to reversible features of the logic immanent to the circuits could help in coping with the limits related to energy dissipation and power consumption of the physical devices. Even if this would turn out to be a purely speculative account, without measurable benefits, coming back to the theoretical basis of computation and to the correlations between the physical and the symbolic-informational levels may enhance more reflections on the structural limits of implementability of computational paradigms.
- *Resonance* – The process and self-organizational approach to memory suggest also the importance of collective behaviors, based on mutual resonance phenomena among groups or assemblies of communicating cells, where the same assemblies are not fixed, but continuously regroup with variable geometries, following new exogenous stimuli or the same internal evolution of their own behavior.[130] In resonant cell assemblies there is also a new parameter that regulates the same resonance: rhythm. Such parameter may play a regulating role at higher levels.

We do not mention here other likely esoteric fields of research, like non-von-Neumann architectures, cellular automata, Gandy machines, quantum computing and supertasks, because they shall bring us far from our more modest scope.[131]

Conclusion: Autopoiesis as Memory Phenomenon

> D'ALEMBERT – If there were no memory, there would be no awareness of self, because if a creature were aware of its existence only during the instant of that awareness, it would have no history of its life. Its life would be only an interrupted series of sensations without anything to bind them together.
> DIDEROT – Very good. Now, tell me *what memory is*, and explain where it comes from.
> D'ALEMBERT – It comes from a certain organization of matter – an organization that grows or disintegrates, and sometimes disappears altogether.
>
> Denis Diderot, *D'Alembert's dream*[132]

To conclude, we come back to Diderot and to our initial question about the essence of memory.

If memory "comes from a certain organization of matter," we may say that this organization is the guarantor of the very unity of the living organism and, if it is a thinking one, of its cognitive sphere too.

Memory, in a general way, can be thought of as the *function that gives continuity to systems*, or diachronic coherence – what vests them with functional solidity across the liquid essence of the fading present –, at different levels:

---

[130] See Varela et al. [81]

[131] We acknowledge that the language used to sketch such metaphors is vague and consequently far from real technology life. However, metaphors are metaphors and have the privilege that they can be nobly neglected without shame for their authors.

[132] As per footnote 1.

- *ontogenetic level*, in the form of the biological autonomy of the organism, necessary condition to its same survival, and realized by means of learning processes, that is, the ability of the system to categorize perceptions, formulate values, and embody models of the external and internal dynamic environment (→individual memory);
- *phylogenetic level*, which is the same, but at population level, aimed at the survival of the species, through the hereditary genetic code transmission (→ biological or genetic memory);
- *noogenetic level*, i.e., cognitive, symbolic and cultural level, where the autonomy is reflected to the spiritual and social behavior of individuals and the ideas and pieces of culture (often called "memes") they keep in memory are mutually connected in a network of intersubjective self-replicating patterns that form what we call humanity (→ cultural or "memetic" memory, collective and individual).

This function of continuity, which is the temporal synoptic *analogon* to the logical consistency and the spatial solidarity of system, is implemented through processes, and this dynamical dimension is not negligible if we want to catch the very essence of memory.

Memory, to realize its scope, uses some abstract relational means: positive and negative feedbacks, autopoiesis, entropy containment. All these means can be subsumed under the category of self-reference and tend to confuse themselves with the same nature of memory, so that we could define in terms of these elements.

Finally, we remark that the phenomenon of memory is intrinsically related to that of consciousness. It is an asymmetric relation, because there can be memory without consciousness (like in immune systems), but not consciousness without memory. The tight link between consciousness and self-consciousness (fully recognized since Kant) suggests by analogy that also the self-memory function would be of importance for the memory as such, at least as ability to use the faculty itself. The study of self-reference for memory would methodologically be a first step to approach the study of the self-reference of consciousness.

A further step is the recognition that memory is never a private fact, but always a collective phenomenon, like social conventions, songs, theorems, and languages.[133] This fact may be thought in terms of a multi-agent characterization of cognition,[134] considering the ideas we believe as our property emerge and live in a scattered way, beyond individuals. Here, in the sphere of intersubjectivity comes into existence the source of our ability to think and communicate, that is, symbolism. Its biological roots have to be looked for right in our neuronal ability to recognize and categorize the co-occurrences, transforming them to abstract schemes (no matter if sensorial, motorial, or conceptual), that in turn become gestures, rites, words, thoughts; in a word, memories. For these reasons the history of mnemonic techniques is like a loom where we can read a map of our knowledge and culture.

---

[133] Wittgenstein maintained that a language is like a village.

[134] Cf. e.g., Minsky [54].

Finally, we may recall that several scientists are nowadays oriented to think at the universe as a gigantic computational machine, where physical events and physical laws are thought of as the continuous unfolding of elementary algorithms.[135] Within the frame of this quite dematerializing and pan-computationalist vision we can look at memory as the ultimate ingredient of the same universe, under the form of programs and variables. The shape of a leaf of grass, the hunting project of an animal, the trajectory of a river, the gesture of a child bringing an object to the mouth, everything carries inside itself a part or the whole process that led it there. All is memory, and the universe is the memory of itself.

Leaving to philosophers further speculations around such abstract ideas, we may summarize the last reflections by reverting to the initial statement $\alpha_{\to\alpha}$ and turning it over to its converse:

$$\mu_{\to\alpha} \equiv \textit{Autopoiesis can be considered, in its deepest essence, as a memory phenomenon.}$$

So, all our discourse recalled us that memory, this organization of matter to which we owe life and consciousness, seems strongly related to the circular fluxes of forces, and as such it justifies its own definition as a property of our center, as said doctor de Bordeu. But remembering that this organization is precious, because it "grows or disintegrates, and sometimes disappears altogether.": the other side of memory is oblivion, the entropic destiny of disintegration, to which memory resists.

Let me end with a thought from the autobiographical memoirs of a great filmmaker, Luis Buñuel, who dedicated his life to an art which seems to shadow the virtues of memory. (Cinematography is in itself a form of memory, and furthermore it mirrors the mental mobility of our mind back and forth through time, as memory does, giving depth to life.)

> You have to begin to lose your memory, if only in bits and pieces, to realize that memory is what makes our lives. Life without memory is no life at all, just as intelligence without the possibility of expression is not really intelligence. Our memory is our coherence, our reason, our feeling, even our action. Without it, we are nothing.[136]

Acknowledgments The writing of this contribution owes its same existence to the kind invitation of Gianni Campardo. I'm grateful to him, also for fruitful discussions.

I dedicate this essay to the fond memory of my brother-in-law Conrad Klemm, great musician, unforgettable personality, and missing human presence.

The domain of memory is vast, and it encompasses a huge amount of the human experience. However, the deepest feeling of its functioning is when somebody is missing you: here memory discloses the majestic potency if its work, giving you back what seems lost in the darkness of past, and restoring in the palace of your mind the true essence of a presence, in a strange and profound ontology, which every moment enlarges your visible world. Memory is our life – without it, ...

---

[135] This is the other face of the so called "physical Church-Turing thesis", often referred as "digital physics", and held by different scholars with different perspectives, like John A. Wheeler, Stephen Wolfram, Seth Lloyd, etc. A good manifesto of this view is Lloyd [47].

[136] Buñuel, L (1982) My last breath. Paris.

References

1. Aczel P (1988) Non-well-founded sets. CLSI Lecture Notes, 14, CLSI Publications, Stanford
2. Arbib MA (1988) From universal Turing machines to self-reproduction. In: Herken R (ed) [33], pp. 177–190
3. Aristotle (ca 350 B.C.) (2007) Peri mnēmēs kai anamnēseōs /De memoria et reminiscentia [On memory and recollection]. In: Bloch D (ed) Aristotle on memory and recollection: text, translation, interpretation, and reception in western scholasticism. Brill, Leiden
4. Ashby WR (1956) An introduction to cybernetics. Chapman & Hall, London
5. Barwise J, Moss L (1996) Vicious circles. On the mathematics of non-well-founded phenomena. CLSI Lecture Notes, 60, CLSI Publications, Stanford, CA
6. Bennett ChH (1973) Logical reversibility of computation. IBM J Res Dev 17:525
7. Burks AW (ed) (1968) Essays on cellular automata. University of Illinois Press, Illinois
8. Carruthers M (1990) The book of memory: a study of memory in medieval culture. Cambridge University Press, Cambridge
9. Case J (1996) Infinitary self-reference in learning theory. J Exp Theor Artif Intell 6:3–16
10. Chaitin GJ (1987) Information, randomness & incompleteness. Papers on algorithmic information theory, World Scientific, Singapore
11. Chaitin GJ (2005) Meta math! The quest for Omega. Knopf, New York, NY
12. Churchland P (2002) Self-representation in nervous systems. Science 296:308–310
13. Conant RC, Ashby WR (1970) Every good regulator of a system must be a model of that system. Int J Syst Sci 1(2):89–97
14. Cordeschi R, Tamburrini G, Trautteur G (1999) The notion of loop in the study of consciousness. In: Taddei-Ferretti C, Musio C (eds) Neuronal bases and psychological aspects of consciousness. Proceedings of the international school of biocybernetics. World Scientific, Singapore, pp 524–540
15. Cover ThM, Thomas JA (2006) Elements of information theory. Wiley, New York, NY
16. Demongeot J, Kaufman M, Thomas R (2000) Positive feedback circuits and memory. C.R. Académie des Sciences de Paris, Sciences de la vie/Life Sciences 323:69–79
17. Draaisma D (2000) Metaphors of memory: a history of ideas about the mind. Cambridge University Press, Cambridge
18. Edelman GM (1987) Neural Darwinism: the theory of neuronal group selection. Basic Books, New York, NY
19. Edelman GM (1989) The remembered present. A biological theory of consciousness. Basic Books, New York, NY
20. Edelman GM (1992) Bright air, brilliant fire. On the matter of the mind. Basic Books, New York, NY
21. Feynman RP (1996) Feynman lectures on computation. (edited by AJG Hey, RW Allen) Addison-Wesley, Reading
22. von Foerster H (1949) Quantum mechanical theory of memory. In: von Foerster H (ed) Cybernetics: circular causal, and feedback mechanisms in biological and social systems. Josiah Macy Jr. Foundation, New York, NY, pp 112–145
23. von Foerster H (1969) What is memory that it may have hindsight and foresight as well?. In: Bogoch S (ed) The future of the brain sciences: proceedings of the 3rd International Conference. Plenum Press, New York, NY, pp 19–64 (reprinted in: von Foerster H (2003) Understanding understanding. Springer, Berlin, pp 101–131)
24. von Foerster H (1976) Objects: tokens for (eigen-)behaviors. ASC Cybern Forum 8(3–4):91–96 (reprinted in: von Foerster (1981))
25. von Foerster H (1981) Observing systems. Selected papers, Intersystems Publications, Seaside
26. Forti M, Honsell F (1983) Set theory with free construction principles. Annali della Scuola Normale Superiore di Pisa, Classe di Scienze 4, 10, 3:493–522
27. Fredkin E, Toffoli T (1982) Conservative logic. Int J Theor Phys 21:219–253
28. Girard JY (1987) Linear logic. North Holland, Amsterdam

29. Glanville R (1978) What is memory, that it can remember what it is?. In: Trappl R, Pask G (eds) Progress in cybernetics and systems research. Hemisphere, Washington, pp 27–37
30. Goertzel B (2010) Hyperset models of self, will and reflective consciousness. URL: http://goertzel.org/consciousness/consciousness_paper.pdf
31. Haken H (1988) Information and self-organization. A macroscopic approach to complex systems. Springer, Berlin
32. Hayek FA (1952) The sensory order. An inquiry into the foundations of theoretical psychology. Routledge, London
33. Herken R (ed) (1988) The universal Turing machine: a half-century survey. Oxford University Press, Oxford
34. Hofstadter DR (1978) Gödel, Escher, Bach. An eternal golden braid. Basic Books, New York, NY
35. Hofstadter D (2007) I am a strange loop. Basic Books, New York, NY
36. Hofstadter DR, Dennett DC (1981) The mind's I: fantasies and reflections on self and soul. Basic Books, New York, NY
37. Hopfield JJ (1982) Neural networks and physical systems with emergent collective computational properties. Proc Natl Acad Sci 79:2554–2558
38. James H (1890) The principles of psychology. Henry Holt & Co., New York, NY (reprinted by Harvard University Press, Cambridge 1981)
39. Johnson-Laird Ph N (1983) Mental models. Towards a cognitive science of language, inference, and consciousness. Cambridge University Press, Cambridge
40. Kauffman LH (1987) Imaginary values in mathematical logic. Proceedings of the 17th international symposium on multiple-valued logic, IEEE, Boston
41. Kauffman LH (1998) Space and time in computation, topology and discrete physics. Int J Gen Syst 27(1–3):249–273
42. Kauffman LH (2005) EigenForm. Kybernetes 34(1–2):129–150
43. Langton ChG (1984) Self-reproduction in cellular automata. Physica 10D:135–144
44. Leyton M (2001) A generative theory of shapes. Springer, Berlin
45. Lloyd S (1989) Use of mutual information to decrease entropy: implications for the second law of thermodynamics. Phys Rev A 39(10):5378–5386
46. Lloyd S (2000) Ultimate physical limits to computation. Nature 406:1047–1054
47. Lloyd S (2006) Programming the universe. A quantum computer scientist takes on the cosmos. Knopf, New York, NY
48. Mac Lane S (1971) Categories for the working mathematician. Springer, Berlin
49. Maturana H, Varela F (1980) Autopoiesis and cognition: the realization of the living. Reidel Publishing Company, Dordrecht
50. McCulloch WS (1960) What is a number, that a man may know it, and a man, that he may know a number?. Gen Semantics Bull 26–27:7–18
51. Merleau-Ponty M (1945) Phénomenologie de la perception. Gallimard, Paris. English edition: Merleau-Ponty M (2002) Phenomenology of perception (trans: Smith C). Routledge, London
52. Milner R (1989) Communication and concurrency. Prentice Hall, New York, NY
53. Minsky M (1965) Matter, mind and models. In: Proceedings of IFIP congress 1965, I, Spartan Books, pp 45–49
54. Minsky M (1985) The society of mind. Simon & Schuster, New York, NY
55. Miranker WL, Zuckerman GJ (2009) Mathematical foundations of consciousness. J Appl Logic 7(4):421–440
56. Moss LS (2008) Confusion of memory. Inf Process Lett 107(3–4):114–119
57. von Neumann J (1963) Collected works. Volume 5: Design of computers, theory of automata and numerical analysis (ed: Taub AH). Pergamon, Oxford, pp 81–82
58. von Neumann J (1966) The theory of self-reproducing automata. (edited and completed by AW Burks) University of Illinois Press, Urbana, IL
59. Perlis D (1997) Consciousness as self-function. J Consciousness Stud 4(5–6):509–25
60. Port RF, van Gelder T (eds) (1995) Mind as motion. Explorations in the dynamics of cognition. MIT Press, Cambridge

61. Poundstone W (1985) The recursive universe: cosmic complexity and the limits of scientific knowledge. Contemporary Books, Chicago, IL
62. Pribram KH (1969) The four R's of remembering. In: Pribram KH (ed) On the biology of learning. Harcourt Brace & World, New York, NY
63. Rice JE (2006) A new look at reversible memory elements. In: 2006 IEEE international symposium on circuits and systems, IEEE, Grenoble
64. Rolls ET (2010) Attractor networks. WIREs Cogn Sci 1:119–134
65. Rosenblueth A, Wiener N, Bigelow J (1943) Behavior, purpose and teleology. Philos Sci 10:18–24
66. Rossi P (1983) Clavis universalis. Arti della memoria e logica combinatoria da Lullo a Leibniz, Il Mulino, Bologna. English edition: Rossi P (2000) Logic and the art of memory (trans: Clucas S). University of Chicago Press, Chicago, IL.
67. Royce J (1912) The world and the individual. First series, Macmillan, New York, NY
68. Rutten J (1992) Processes as terms: non-well-founded models for bisimulation. Math Struct Comput Sci 2(3):257–275
69. Rutten J (1996) Universal coalgebra: a theory of systems. CWI report CS-R9652, Centrum voor Wiskunde en Informatica, Amsterdam
70. Schrödinger E (1944) What is life?. Cambridge University Press, Cambridge
71. Seiler H (2002) Object, language, and communication. Stud Commun Sci 2(2):83–108
72. Spencer Brown G (1969) Laws of form. Allen & Unwin, London
73. Sutton J (1998) Philosophy and memory traces. Descartes to connectionism. Cambridge University Press, Cambridge
74. Thom R (1972) Stabilité structurelle et morphogénèse. Essai d'une théorie générale des modèles, Inter Editions, Paris. English edition: Thom R (1976) Structural stability and morphogenesis (trans: Fowler DH). Benjamin, Reading, MA
75. Thomas R (2006) Circular causality. IEE Proc Syst Biol 153(4):140–153
76. Thomas R, Kaufman M (2001) Multistationarity, the basis of cell differentiation and memory – I. Structural conditions of multistationarity and other nontrivial behavior. Chaos 11:165–179
77. Thompson D'AW (1917) On growth and form. Cambridge University Press, Cambridge
78. Turvey MT, Shaw R (1979) The primacy of perceiving: An ecological reformulation of perception for understanding memory. In: Nilsson L-G (ed) Perspectives on memory research: essays in honor of Uppsala University's 500th anniversary. Lawrence Erlbaum, Hillsdale, pp 167–222
79. Varela F (1975) A calculus for self-reference. Int J Gen Syst 2:5–24
80. Varela F (1979) Principles of biological autonomy. Elsevier North Holland, New York, NY
81. Varela F, Lachaux J-P, Rodriguez E, Martinerie J (2001) The brainweb: phase synchronization and large-scale integration. Nat Rev Neurosci 2:229–239
82. Whitehead AN (1929) Process and reality: an essay in cosmology. Macmillan, New York, NY
83. Williford K (2006) Self-representational structures of consciousness. In: Kriegel U, Williford K (eds) Self-representational approaches to consciousness. MIT Press, Cambridge, pp 111–142
84. Wittgenstein L (1980) Remarks on the philosophy of psychology. Blackwell, Oxford
85. Wolfram S (2002) A new kind of science. Wolfram Media, Champaign
86. Yates FA (1966) The art of memory. Routledge & Kehan Paul, London
87. Young JZ (1964) A model of the brain. Oxford University Press, Oxford

Chapter 2
Mass Storage Memory Market Biography

Federico Tiziani, Danilo Caraccio, Nicola Guida, Manuela Scognamiglio, and Cristina Tiziani

Abstract This chapter describes the basic storage technologies such as traditional paper, magnetic, optical, semiconductor, and uncommon memories, providing an outline of the history and a brief technical description of each device in terms of its volatility, accessibility, addressability, capacity, and performance.

Keywords Devices · Storing · Data

Introduction

Right from the beginning, communication activities among the human societies required the capability of storing and passing on information acquired through life experiences through various channels: vocalization or gestures, or backing the case of ancient man, cave paintings, and drawn maps and the written word.

Writing was and probably still is the most common method for storing data all over the world, but the past centuries have seen many kinds of new energies that move from the manual muscle power needed for writing by hand to acoustic vibrations in phonographic recording, to electromagnetic energy that modulates magnetic tape, to digital content written with laser light inside optical disks, and finally data electronically stored in a semiconductor device or molecular materials.

Paper and Book

Ancient History of Paper and Older Memory Methods

From the time that human society first realized the need for written expression, there has been a demand for proper support to store ideas.

F. Tiziani (✉)
Micron, WSG Department, Via Olivetti 2, 20041 Agrate Brianza (MB), Italy
e-mail: ftiziani@micron.com

G. Campardo et al. (eds.), *Memory Mass Storage*,
DOI 10.1007/978-3-642-14752-4_2, © Springer-Verlag Berlin Heidelberg 2011

For many centuries paper was the principal means for transmitting and storing all kinds of thought and all manifestations of human inspiration, but it is important to briefly recall the earlier tools that were adopted and used [1–5].

Prior to Paper

We cannot talk about memory support at all without introducing the concept of the written word; in fact the popularity of writing generated a real "revolution" in social, political, and cultural realms.

Writing was invented during the second half of fourth millennium BC, Uruk, the oldest Sumerian town. Its development was due to the need to enlarge communication capabilities so that information would be acquired not only through oral communication, which is basically ephemeral, but also through methods that would widely transmit and preserve all types of data and collected know-how to other people.

Bars of Clay

People in the ancient world did not know that the original primitive implements for transmitting and preserving information were simple bars of clay. At the end of fourth millennium BC, the Sumerians in Mesopotamia became the first people to attempt this extraordinary method of communication by using a sharpened tip to write on bars of clay. At first they realized a kind of logogram, that is a simple object outline, then gradually moved through a process of reproducing simplified figures and wedges until they developed a large index of syllabic signs (around 600), characterized, depending by different cases, of meaning of a real word and so also used to write abstract terms that are not typically expressible using symbols or schematic drawings.

After the bars were written on, they were baked, which turned the clay into a ceramic material that could be preserved for several millennia. Thousands of bars kept valuable information on the historical and political events of the time, the mythology, the religion, and the social and economic organization of the inhabitant of Mesopotamia, western Iran, Turkey, and Syria (Figs. 2.1 and 2.2).

Wax Bars and Papyrus

As in Mesopotamia, inscription on clay bars was the first form of writing in Egypt,. However, it seems that as early as around 3000–3500 BC the Egyptians began to write on foils of papyrus (*Cyprus papyrus*), which was the product of a complex process using material extracted from marsh reeds, which grew plentifully in the Nile Delta. The reeds were cut into strips around 40 cm long and then squeezed. The resulting weft was beaten with a wooden club and then rolled up inside flax, until the plant's sticky marrow spread across the surface to create a foil. This process produced long rolls that could be written on and decorated with pictures. The rolls

Fig. 2.1 Example of on a Sumerian bar

Fig. 2.2 Fragment of terra-cotta of Gilgamesh Poem, Londra, British Museum

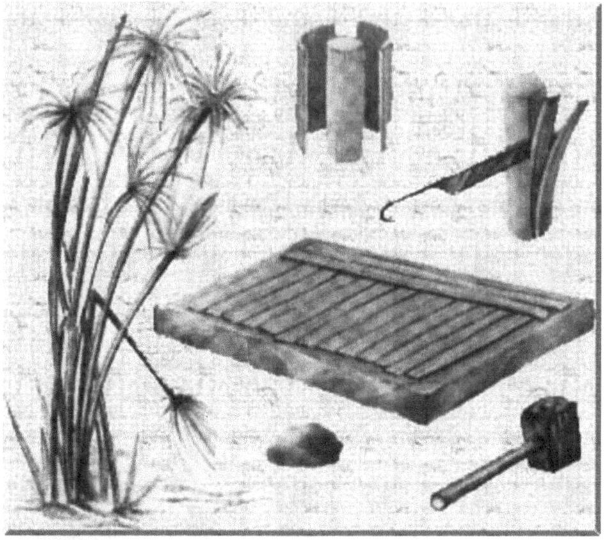

Fig. 2.3 Manufacturing papyrus foils

Fig. 2.4 Papyrus scroll from the famous *Book of the Dead*

were then stored inside sealed amphorae, with an inscription of the content within it (Figs. 2.3 and 2.4).

Papyrus arrived in Greece in the sixth century BC and then found its way to the Roman Empire, where other kind of plants, maple, plane tree, and lime were already being used for writing.

Wax Bars

Papyrus was used in Rome, as was parchment (animal skin), ivory, earthenware, and bars of wax. Egyptian papyrus had an uneven texture and could only be used on one side, but the Romans improved on it and were able to create a perfectly smooth surface using a pressing tool or a hammer. Rome had many papyrus factories (*horrea chartaria, officinae*), the most important located in Fannio.

The paper *fanniana* became famous for its lightness and smoothness respect the Egyptian *amphitheatrica* rough paper, so called because prepared closed to Alessandria amphitheatre.

Several foils of papyrus were glued together to make one long strip that was furled into a roll (*scapus*), which was kept rolled up by a few small sticks (*umbilicus*) on the top and the bottom so that that last foil would not get dirty by trailing on the ground; a label with the title of the book was located on the top side of the roll.

A shape similar to what we know of today was created during the Imperial Age by binding some foils of parchment together (*quaterniones*) to make a kind of exercise book with a cover (*codice membranei*), but the process proved to be too expensive and was not developed.

It was just at this point that the first schools were founded in which pupils made use instead of wooden tables spread by wax (*cerae*), to be able to erase and to correct in easily way.

Were wooden tables with relevant edges and inside was spread the wax and written on it bearing characters with a wooden or metal stick (*stilus*), sharpened on one side and flatten on the other side, to erase. Some holes on the borders allow binding two or more tables together in order to create an exercise book, a set of tables has been called *caudexes* or *codex* (Figs. 2.5 and 2.6).

Parchment

Parchment (a writing material made from animal's skins (lamb, colt, ram, donkey, calf and pig) was introduced in Rome in the first half of the second century BC. The parchment (also called *cartapecora or carta pecudina*) made from untanned of animal skins was very popular until the fourteenth century. From Latin *membrane*

Fig. 2.5 Wax table with stylus

Fig. 2.6 Round portrait made of Roman fresco, ca. 50 AD, woman with book and stick came from Pompei Napoli, National Archeological Museum

Fig. 2.7 Parchment manufacturing process

o vellum was so called due to the city *Pergamo* (in Asia Minor) where, based on traditions referred to "Pliny the Elder" (Gaio Plinio Second, known as Pliny the Elder), was invented around the second century BC, as a replacement for papyrus (Fig. 2.7).

It has stayed some medieval prescription to the parchment manufactory process and the oldest one is stored inside *Compositions* a manuscript of 490 located in Capital Lucca's library (VIII century).

The animal skin was placed in a *calcinatory* (a solution of water and quicklime), over a "back of donkey" tripod and a sharp blade was used firm to removed the fur. The denuded skin was then placed tautly on a frame and left to dry. It was important to eliminate residual flesh and this was done with a special knife. After the skin was completely dried, the parchment could be removed from the frame and finally used. Further steps to refine this material were carried out using a pumice stone to reduce the differences between rough side, which was hard to the touch, and meat side, which was soft and smooth.

. . . And finally paper.

Etymology of Paper

The meaning of the word "paper" is quite uncertain. Someone suppose that it derives, from the Latin *charta*, from the Greek *charassò*, which means *to etch, to carve.* The terminology corresponding to the Anglo-Saxon word *paper, papel* Spanish and *papier* French comes from the papyrus plant, used in ancient Egypt until 3000 BC and, subsequently by the Greeks and the Romans.

Based on Chinese documents, paper was invented in 105 AD towing to Ts'ai Lun, a civil servant of Ho-ti from the Han dynasty, emperor's court. As paper was discovered in territories owned by the Chinese empire, with little evidence of its use earlier than 105 AD, is likely that Ts'ai Lun introduced it in the vicinity of the imperial factory, and perhaps improved on a technique that had been in use for many years. The Chinese process made use of tree bark, probably from the mulberry tree (*Brussonetia papyrifera*), and old fish net, which was treated and filtered in a bamboo mold.

The oldest well-known paper extant today was made around 150 AD using rags. For the next 500 years the art of paper making was confined to China but it was introduced in Japan in 610 AD and in the Central Asia around 750 AD.

One of the first pieces of evidence in Italian concerning the Chinese ability to produce paper was received from Marco Polo inside a *Milione* passage in which he mentions the material used to fabricate the stamp paper, referring in particular to the quality of the bamboo and waste hemp.

Writing about the Emperor of China, he noted "they bring bark of the tree called Mulberry, – whose leaves are eaten by silk worms,—and take the thin bark that is between the thick bark and the core of wood, and from this bark extract paper similar to cotton wool."

The Chinese emperors long kept the secrets of these techniques, which were only known elsewhere beginning in the seventh century, first in Korea, then in Japan, and finally in Central Asia in Samarkand city, where the Arabs learned directly from Chinese people and subsequently introduced them in the Middle East and elsewhere in the Mediterranean area.

The Arabs spread the knowledge of paper making in Europe in the twelfth century, particularly in Spain at Jativa (where they built the first European paper mill) and in Italy in Palermo. During thirteenth century the city of Fabriano became the most important center of paper manufacturing, using local workforce; from Italy production moved across all of Europe, primarily to France (fourteenth century), the Netherlands (fifteenth century) and to England (fifteenth century), where, in the

Fig. 2.8 Paper's ancient route around Asia

sixteenth century, development benefited from the availability of cash assets and low-cost combustibles for manufacturing (Fig. 2.8).

History of Paper in Italy

The first paper mill in Italy was founded in Fabriano (the Marches, Province of Ancona) in 1276. The first official document on Fabriano paper dated 1283, was a legal document recording the purchase of a building from a "carthaio," and six other paper-making masters served as witnesses of the event.

In brief, the Italian paper manufacturing process consisted of few steps.

A hydraulic pile based on multiple power hammers, invented by a paper-making master from Fabriano, was made up of a large wooden container, in which a large nailed pestles or mullers, powered by a big water mill, threshed some vegetable fiber rags until they formed a mash. This mash was transferred to large wooden vats full of water and stirred with a stick. A sieve was then inserted into the vat and moved around until the mash formed a layer of uniform paste that could be spread all over the surface of a cloth. The foils so obtained were placed on wool felts and squeezed in a mechanical press to remove all the water inside. The foils were then hung up to dry (Figs. 2.9, 2.10, and 2.11).

Other Italian cities had paper-making factories, including the sea-faring cities of Amalfi, Venice, Genoa, and Palermo as well as university towns such as Padova and Bologna.

The year 1456, which saw Johann Gutenberg's invention of the printing press, a method of printing using movable type, marked a fundamental change in the history of paper production and the evolution of writing. Very people could afford to purchase expensive parchment, but with the availability of relatively inexpensive paper and the new method of printing, general knowledge rapidly spread worldwide.

Fig. 2.9 Hydraulic pile based on multiple power hammers

Fig. 2.10 Soaking

Fig. 2.11 Drying the foils

Punch Cards

Basile Bouchon was the inventor of the first perforated paper loop used in couture arte to store some outlines on a textile. But the first use as data storage is dated back September 23rd, 1884 by Herman Hollerith who patented his invention and subsequently it was adopted for ten years until the mid 1970s.

Figure 2.12 shows an example of a classic punch card. Composed of 90 columns, it held a limited amount of stored data was and was typical used to configure parameters in special devices.

In 1846 Alexander Bain created the first paper tape; he was also the inventor of the electric printing telegraph and the fax machine. Characters were symbolized by a punched rows and owing to the simplicity of the method it was possible to create a fan-folded length of paper that could hold more data than a punch cards (Fig. 2.13).

Fig. 2.12 A classic punch card

Fig. 2.13 Example of a punched tape

Magnetic Memories

History

The first description of an audio recording method based on a magnetic storage medium was released in September 1878 by Oberlin Smith, who hadfirst started using magnetic recording 10 years earlier [6–10].

However, it was Valdemar Poulsen, the owner of a patent filed in 1989, who, during the Paris Exposition of 1900, first demonstrated publicly a device for recording a signal on a wire wrapped around a drum. The first magnetic tape, made up of metal strips on paper was patented by Fritz Pfleumer (Figs. 2.14 and 2.15).

Progress in magnetic recording was slow and it was only in 1932 that the first magnetic recording devices appeared on the market.

Fig. 2.14 Oberlin Smith

Fig. 2.15 Valdemar Poulsen

During the second half of the twentieth century most of common audio and video contents were based on magnetic recording devices such as the audio and video tapes used professionally or in homes to distribute movies and music.

Several years ago there was a move to replace the tape recordings with hard disks and other kinds of memories that were not volatile, but there are thousands of people still working in the magnetic storage market for companies located around the world.

Historically initial development focused on magnetic tape for audio recording. In 1945 the era of video recording began. From the early 1950s data recording was accomplished using primarily the floppy disk and hard disk drives. Today the latter is becoming the main device used not only for data content but also for audio and video recording. The main four elements in the progress of this technology were as follows:

- The Magnetophon audio recorder was developed in Germany between 1930s and 1940s. The mechanism consists of a magnetically coated plastic tape and a stationary head to read/write and erase information stored in analog way (Fig. 2.16).
- The quadruplex video recorder was based on four heads rotating and connected on a wheel and a magnetic tape, used mainly for video applications, which replaced the earlier photographic method (Fig. 2.17).
- The floppy disk drive developed by IBM in the 1960s introduced data storage in a portable support based on a flexible disk and heads with minimal movement capability but with a semirandom access method, which was an improvement on any radial position of the disk. This was the best cheap device for storing data in a removable medium a few years ago (Fig. 2.18).
- The RAMAC disk file, also known as hard disk drive, developed by IBM in 1956, uses a rigid disk and a head with a random access capability. The growing interest in the electronic computer required storing large amounts of data in a nonvolatile device with a fast access technique and large storage density (Fig. 2.19).

Fig. 2.16 The Magnetophon

Fig. 2.17 The quadruplex

Fig. 2.18 Different floppy disk sizes

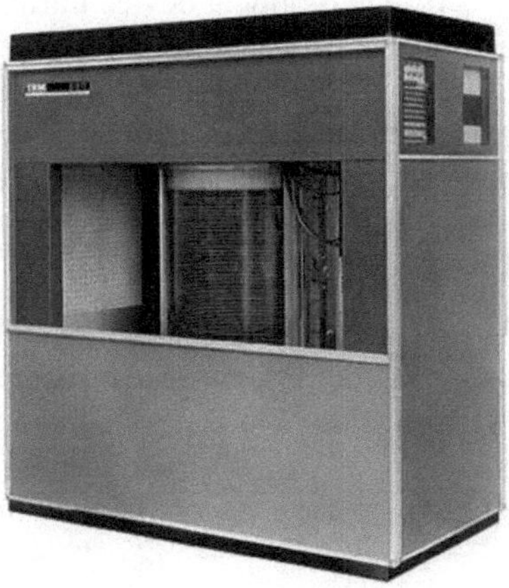

Fig. 2.19 The RAMAC system

Magnetic Recording Classification

The first two developments described in the previous paragraph are based on analog method to record the audio or video information on the medium, whereas the last two are based on a digital system, but there are also two other ways to store information using magnetic effects: optical-magnetic and domain propagation memory.

Analog Recording

Analog recording schemes come from the property of some materials to maintain the magnetic configuration depending by magnetic field applied. The devices are made up of a tape initially not magnetized; during the recording process the tape runs under the charged fixed head to magnetize the support in accordance with the applied magnetic field. The magnetic tape is composed of iron oxide or chrome oxide particles assembled on a thin plastic film. The analog recording was very popular for audio and video recording, but has now been replaced by digital recording systems.

Digital Recording

Unlike the analog method, digital recording requires only two stable conditions, identified by the opposing conditions of a hysteresis loop. The most important examples are the floppy disk drive (FDD) and the hard disk drive (HDD); the latter will eventually replace all other recording magnetic media.

Magneto-Optical Recording

In a magneto-optical device during the writing operation the medium is heated by a laser that brings about a rapid decrease in the coercive field. Subsequently using a small magnetic field it will be possible to modify the magnetic polarization. The reading process is based on the Kerr effect. There are not yet many devices based on magneto-optical behavior, the most well known being Minidisc developed by Sony.

Domain Propagation Memory

The method is based on the capability of controlling domain wall motion in a magnetic medium free of microstructure. It is also called Bubble memory because of its cylindrical domain. The data are recorded depending the presence/absence of bubble domain. Due to the high insensitivity to vibration and shock resistance these devices are used especially in aeronautics.

Access Method

Magnetic storage devices can be classified mainly into two different categories depending on the kind of access: sequential access or random access.

Typically the magnetic tape used for audio and video recording is based on a sequential access method, in which the end user has to wait an average of tens of seconds before accessing the sought for information. On the other hand the HDD and FDD are based primarily on a random access method system, which employs a cylinder and sector structure; when the head is positioned on the right cylinder and sector it may take some time to read the information needed, but the typical access time is tens of milliseconds, several times less than with the sequential system. A ferrite-core memory, not a common magnetic device in the market, can be considered a pure random access based device.

Owing to the huge growth of the personal computer market the HDD is the most important magnetic device on the market today.

Hard Disk and Floppy Disk Drives

Description

The hard disk is composed of one or more aluminum or glass disks covered by iron-magnetic particles; two heads, one for each side of the disk, rotating rapidly disk spinning tens of nanometers from the surface, reading/writing digital data. The heads are kept raised from the air flow itself and the speed rotation can exceed 15.000 revolutions per minute (rpm); currently the standard rotation values are 5.200, 5.400, 7.200, 10.000, and 15.000 rpm.

A floppy disk is a portable and removable storage support system composed of a thin, flexible disk within a plastic wrapping that protects data from the dirt in the environment.

Hard Disk Drive Biography

Hard disk drive commercial use began in 1956 with the first production of IBM RAMAC 305, including the memory system disk IBM 350, the original fixed disk drive. The term hard disk, as opposed to the term used for the removable drives, floppy disks, was only introduced after the 1970s disks.

The IBM 350 disk was invented by Reynold Johnson and was made up of 50 disks, 24 in. in size, with a read/write head made with two components with a typical access time around 1 s and a maximum capacity of 5 MB.

In 1961 IBM introduced the first hard disk drive with spinning heads that skim over an air bearing and named the system IBM 1301. The first portable disk version 1311 was able to store up to 2 million characters.

These first hard disk prototypes were so large and cumbersome that they could only be used in labs or industrial environments; also owing to their high power consumption and fragility they not suitable for home use.

In 1973 IBM introduced the 3340 Winchester version, so-called for a comparison to the popular shotgun "0.30–30 Winchester" since was equipped with two 30 MB disks; this name was widely used and became the synonymous with HDD because it was the predecessor of every current modern hard disks.

Fig. 2.20 Seagate ST-506

Prior to the 1980s hard-disk drive dimensions were between 8 and 14 in. and three-phase power supply was needed to provide enough energy for the large engines used to rotate the disk. This was the reason hard disks were not used for microcomputers until the introduction of first $5\frac{1}{4}$ HDD manufactured by Seagate with a 5-MB formatting capacity. A step-by-step engine to manage the read/write head (voice coil control was introduced in the market only few years later). This Seagate ST-506 model equipped AT&T PCs with 286 microprocessors assembled at Olivetti sites located in Scarmagno near Ivrea in the north of Italy in a collaboration between the American and Italian companies. At the same time, OPE (Olivetti Peripheral Equipment), a Olivetti's partner, provided HDD for the M24 personal computer. Historically this company was the only one in Europe to engage in manufacturing for these types of peripherals (Fig. 2.20).

From this point HDD capacity has grown exponentially during the last three decades, starting from the first portable PC with a 20 MB disk inside.

From the middle to the end of the 1990s, stored content included not only text files but also images and videos, reaching the current capacity of hundreds of GB. Today portable hard disk drives can store up to 350 GB and the HDD in a desktop PC can reach up to 2 TB.

Floppy Disk Drive Biography

The floppy disk drive (FDD) was invented by IBM in 1967 as a simple and low-cost system for loading microcode into mainframe System/370, the first result was a read-only disk with an 8 in. diameter called a "memory disk."

Floppy disk development subsequently underwent many changes during its lifetime, assuming many different form factors, standards, and densities. Olivetti was the first company to produce a personal computer system able to manage the FDD and during the Hannover trade fair in April 1975 unveiled its first P6060 system (Fig. 2.21).

Fig. 2.21 Olivetti P6060 system

The most well-known standards adopted by industry were the 5¼ inches used mainly by IBM for its first personal computer and the latest 3½, introduced by Sony on its MSX platform, subsequently used by Apple and finally used in any kind of laptop or desktop personal computer.

At the moment there are no FDDs on the market because optical media (CD, DVD) and devices such as the USB key and flash card based on NAND solid-state technology have made FDDs obsolete.

Optical Memory

Optical memories are storage devices in which the information is recorded and read using optical methods, for example, the light produced by a laser diode which impacts a material with variable reflection features [6–10].

These types of memory devices provide powerful storage capacity, together with a very low cost per recorded bit. They have longer access times (a few hundred milliseconds) than magnetic memories, but a very high recording density (owing to their ability to focus laser light to a very small point) and a strong resistance to weathering. Optical memory is not yet practical for use in computer processing, but can be an ideal solution for storing large quantities of data very inexpensively. The development of optical memory technology is due to the commercial success of the digital audio Compact Disk, introduced in 1983.

Optical Disk

Data are recorded on an optical disk by means of a laser beam that impacts a reflective surface, resulting in small areas (pits) with a reflectivity different from areas that have not been impacted (lands). A very-low-power laser beam is used by the reading unit to read the sequence of pits and convert it into electrical signals with a scanning photo detector; the laser beam is scattered by the pits and reflected by the lands.

The start and the end of a bit represent a "1," while a continuous surface represents a "0." Unlike with magnetic disks, data on optical disks are written sequentially, in a continuous spiral track, from the inner track to the outer one, with a constant recording density. If compared with concentric tracks, the access to the stored information is slower, but spiral tracks are better for reading long strings of data. Furthermore they increase storage capacity.

With this storage technique all sectors have the same length. They must all be read at the same linear velocity so, the disk angular velocity must be gradually reduced when the laser beam goes from the inner area to the outer one.

Optical disks can be divided into three groups:

- Compact Disks
- Digital Versatile Disks
- Others

Compact Disks

A Compact Disk (CD) is a digital storage medium that consists of a disk of transparent thermoplastic resin (polycarbonate) that stores the information on a thin sheet of metal (aluminum or gold). The reflective surface is protected from dust and scratches by an acrylic cover that can also be used to imprint a label.

CDs of 1.2 mm thick, with a 120-mm diameter and a 15-mm central hole; they read and write with a 780-nm infrared laser beam. This technology was introduced in the early 1980s and initially used for audio CDs, offering a significant improvement in audio quality. But apart from this primary application, it also represented an enormous leap from traditional data storage media because of its 650-MB storage capacity at a very low cost.

CDs can be divided into the following types:

- *CD-Read Only Memory* (CD-ROM), which is a prerecorded, read-only device based on a reflective layer made of aluminum.
- *CD-Recordable (CD-R)* which is a one-time recordable device based on a gold sheet covered by a special transparent paint that turns dark if impacted by a laser beam.
- *CD-Rewritable (CD-RW)*, an erasable and re-recordable device in which erasure is accomplished by writing with a beam at a different wavelength.

Write operations are similar on all three CD types but not identical.

A CD-ROM writing process is performed using a glass master disk, written with a high-intensity, finely focused laser beam that etches the tracks into the disk surface. A liquid polycarbonate is injected into the glass matrix in order to reproduce the tracks on the CD. Later the written surface is covered first with a highly reflective layer and then with a protective one (Fig. 2.22).

In a CD-R the pit and land reflectivity is simulated by using a paint layer in between the transparent and the reflective layers. Initially the paint layer is

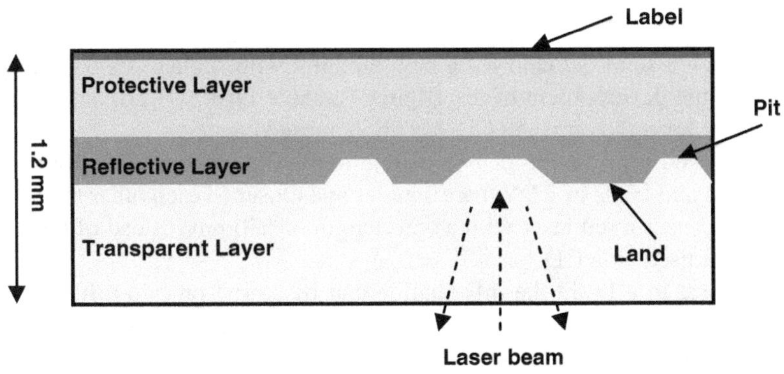

Fig. 2.22 Structure of a CD-R

transparent, but when impacted by a high-intensity laser beam it becomes dark and simulates a pit (Fig. 2.23).

In a CD-RW the reflective layer has amorphous and crystalline states that have different reflectivities: it is slightly reflective in the amorphous state and highly reflective in the crystalline state. When impacted by a high-intensity laser beam, the reflective substance becomes amorphous and simulates a pit. If impacted by a medium-intensity laser beam the mean becomes crystalline again and simulates a land. A low-intensity laser beam can be used to read the information that has been written without changing the mean phase.

For all CD types the reader/writer performances (i.e., write, rewrite, and read performances) are evaluated in comparison with the "standard" data transfer rate of 150 kB/s. So, a throughput of "*nx*" means a transfer rate *n* times greater than the standard throughput.

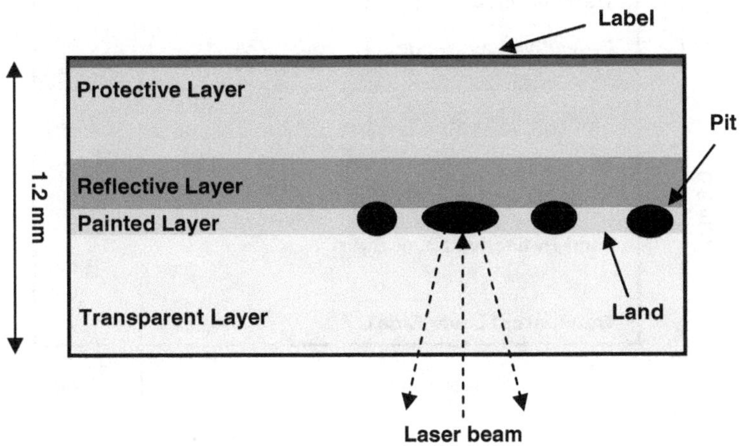

Fig. 2.23 Structure of a CD-RW

Digital Versatile Disk

The 1990s saw a high demand for a new medium with higher storage capacities, which led to the development of the Digital Versatile Disk (DVD), an optical disk with the same form factor as the CD, but an increased capacity.

This increased information density is due to the dimensions of the pits and the lands: the pits and lands of a DVD are smaller and closer to each other than those of a CD. A DVD uses a red laser with a wavelength of 650 nm instead of the 780-nm wavelength of used by a CD.

Furthermore, in a DVD the information can be stored on two different layers. Data contained in the second layer can be read by changing the focus of the laser beam to make it penetrate the first layer, which is semitransparent. This technique doubles the disk capacity. As this procedure can be applied to both sides of the disk, it is possible to quadruple the basic disk capacity.

The laser beam can read only one side at a time, so the disk has to be turned in order to read the second side.

DVDs can be classified into four groups:

- DVD-5: 4.7 GB, single side and single layer.
- DVD-9: 8.5 GB, single side and double layer.
- DVD-10: 9.4 GB, double side and single layer.
- DVD-18: 17 GB, double side and double layer (Fig. 2.24).

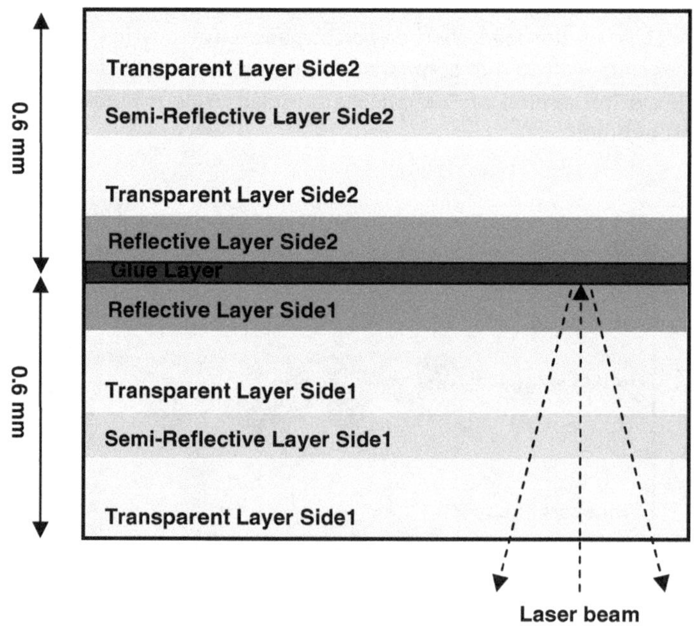

Fig. 2.24 Structure of a DVD

A DVD is also faster than a CD: the standard data rate from a DVD (1350 kB/s) is ninefold greater than that of a CD (150 kB/s).

Other Optical Disks

At the beginning of twenty-first century the development of optical memory technology led to a new generation of high-capacity optical disks that were used primarily for high-definition movies and video games: the High Definition Digital Versatile Disk (HD DVD) and the Blue-Ray Disk (BD).

Both of these devices are optical media with the same form factor as the CD and DVD, and they both use a blue laser beam, which has a wavelength of 405 nm, for writing and reading operations. The reduction in the laser wavelength allows for an increase in disk capacity: HD DVDs have a fixed capacity of 15 GB per layer; BDs can use three different pit sizes, so they can have three different capacities per layer (23.3, 25, and 27 GB). These basic capacities can be increased by using multiple-layer disks.

The data density of the BDs is higher than that of HD DVDs because of a thicker transparent layer (0.1 mm for the BD vs. 0.6 mm for the DVD), which reduces the laser diffusion and the pit dimension and minimum distance between tracks. The higher capacity of BD makes it the winner in the competition with HD DVD (Fig. 2.25).

All the optical disks described up to now are based on reflective materials. The maximum number of layers on these disks is limited by the effects of scatter,

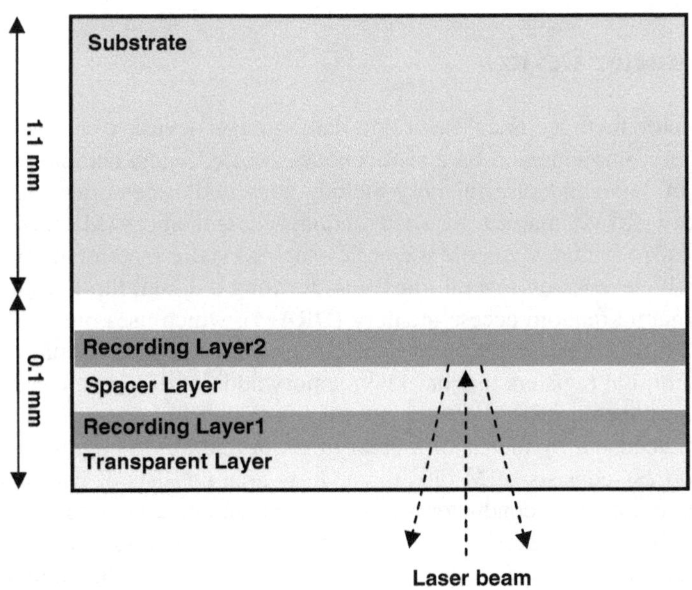

Fig. 2.25 HD DVD structure

interference, and cross-talk. However, this limitation can be overcome by the use of fluorescent materials.

The *Fluorescent Multilayer Disk (FMD)* is an optical disk in which the pits are filled with a fluorescent material. When impacted by a laser beam, the pit emits an incoherent light with a different wavelength, which is decoded by the reader.

An FMD can have up to 100 layers, a limitation that is mainly due to the disk thickness. FMDs that use red laser beams (640 nm) have a maximum capacity of 140 GB, whereas FMDs that use blue laser beams (405 nm) have a maximum capacity of 1 TB.

Magneto-Optical Disk

In magneto-optical memory devices data recording and reading is performed with a laser beam on a magnetic medium.

The magnetic disk is made of special materials that vary their capacity to retain the magnetization induced by an external magnet (coercivity) with temperature: at room temperature, they have high coercivity and do not change their magnetization even in the presence of a magnetic field; when heated by an intense enough laser beam, they can change their magnetization, even with a field that is not very intense. Thus during the writing phase, a well-focused laser beam heats the points where one has to write information bits, while an applied magnetic field alters the direction of the magnetization. The read operation is performed with a laser beam whose polarization direction is changed by the action of the magnetic field of the memory on which it is dropped.

Semiconductor Devices

Semiconductor memory is an electronic data storage device, often used as computer memory, implemented on a semiconductor-based integrated circuit [11–17]. Examples of semiconductor memory include nonvolatile memories such as read-only memory (ROM), magnetoresistive random access memory (MRAM), and flash memory. It also includes volatile memories such as static random access memory (SRAM), which relies on several transistors forming a digital flip-flop to store one bit, and dynamic random access memory (DRAM), which uses one capacitor and one transistor to store each bit. Shift registers, processor registers, data buffers, and other small digital registers that have no memory address decoding mechanism are not considered memories.

Data are accessed by means of a binary memory address to the memory. If the memory address consists of M bits, the address area consists of two raised by M addresses per chip. Semiconductor memories are manufactured with a certain word length (number of 1-bit cells sharing the same memory address) that is a power of two, typically $M = 1, 2, 4$, or 8 bits per chip. Consequently, the amount of data stored in each chip is $M \times N \times 2$ bits. Possible figures are 1, 2, 4, 8, 16, 32, 64, 128, 256 and 512 bits, kbits, Mbits, Gbits, and Tbits, defined here by binary prefixes. By combining several integrated circuits, memory can be arranged for a longer word length and/or larger address space than what is offered by each chip, often but not necessarily a power of two.

As noted before semiconductor memories can be divided in *volatile* and *non-volatile* memories. Typically the volatile memories are SRAM, DRAM, MRAM, and so on. *Nonvolatile* memories, which are also called *storage devices*, include SSD, memory cards, USB drives, and so on.

Flash memory is becoming the primary technology used in storage devices. A particularly important form of semiconductor memory,t is now widely used and is possibly one of the most important forms of medium-term storage.

Flash memory has become increasingly popular in recent years and can be seen in many forms today including flash memory USB memory sticks, digital camera memory cards in the form of compact flash or secure digital, SD, memory. Flash memory storage is also used in many other items from MP3 players to mobile phones, and in many other applications such as SSD.

Flash memory storage is a form of nonvolatile memory: the data held within the flash memory do not disappear when the power is turned off and it can be re-written as required. Each flash memory cell is made up of a single field effect transistor. One of the main advantages of flash memory is the fact that it can be erased electrically. However, it is not possible to erase each cell in a flash memory individually unless a large amount of additional circuitry is added into the chip.

The flash memory is structured in m blocks, typically a power of two (the number of blocks depends on the memory capacity). Each block is divided into k pages; a page is the minimum programmable dimension, whereas the minimum erasable area is a whole block. Each page is divided into several bits physically built by a transistor in floating gate technology (Fig. 2.26).

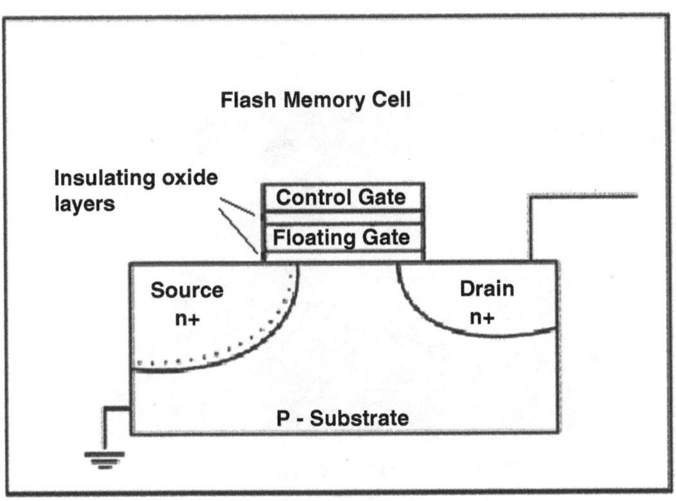

Fig. 2.26 A floating gate cell – 1

Each flash memory cell consists of source and drain electrodes separated by a channel about 1 μm long. Above the channel in the flash memory cell is a floating gate that is separated from the channel by an exceedingly thin oxide layer. The quality of this layer is crucial for the reliable operation of the memory. A control gate located above the floating gate is used to charge up the gate capacitance during the write cycle. The flash memory cell functions by storing charge on the floating gate. The presence of charge then determines whether the channel will conduct or not. During the read cycle a "1" at the output corresponds to the channel being in its low resistance or ON state (Fig. 2.27).

Programming the flash memory cell is a little more complicated and involves a process known as hot-electron injection. When programming, the control gate is connected to a "programming voltage." The drain then sees a voltage of around half this value while the source is at ground. The voltage on the control gate is coupled to the floating gate through the dielectric, raising the floating gate to the programming voltage and inverting the channel underneath. The erase process generally only lasts a few milliseconds. When completed each flash memory cell in the block is checked to ensure it has been completely erased. If not a second erase cycle is initiated.

In the early days of one of the limiting features of flash memories was in their programming, as they had a limited number of erase program cycles, owing to the destructive breakdown of the thin gate oxide layer. Some of the early examples of flash memories only had a few hundred cycles. Now flash memory technology is vastly improved and manufacturers quote figures that indicate that the flash memory life is no longer of concern.

Most of the improvement in flash memory has been brought about by enhancing the quality of the oxide layer. When samples of flash memory chips are found to

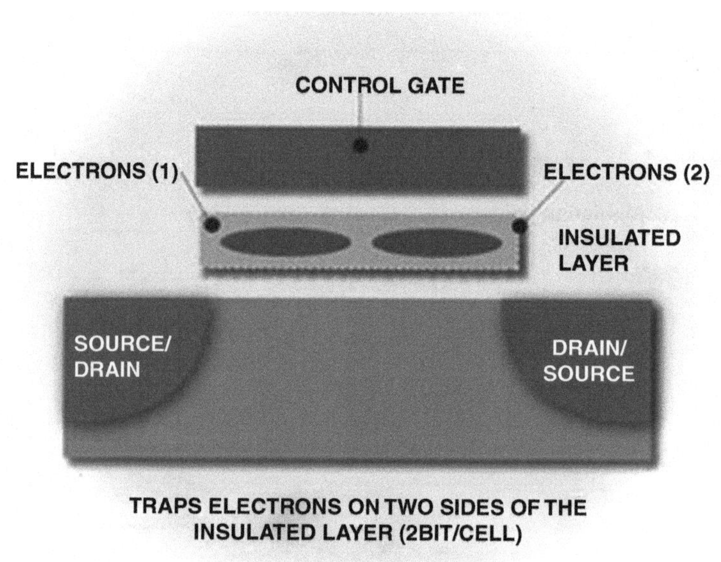

Fig. 2.27 Floating gate cell 2

have a shorter lifetime, it is usually a result of the manufacturing process not being optimized for the oxide growth. Now programming flash memory is not a problem and when using flash memory the chips are, within reason, not treated as items with a limited life.

Flash memory is different from most other types of electronic memories in that while reading data can be performed on individual addresses on certain types of flash memory, erase and write activities can only be performed on a block of a flash memory. A typical block size will be 128, 256, or 512 kB. The low-level control software used to drive flash memories has to take account of this if the read and write operations are to be performed correctly.

There are two basic types of flash memory. Although they use the same basic technology, the ways they are addressed for reading and writing are slightly different.

Over time flash technology has evolved following two different paths originating two different families: NOR and NAND flash. For both, the atomic information is stored in memory cells based on floating gate technology evolving through a progressive lithographic shrink that determines continuous area reduction and cost effectiveness. Each cell is capable of retaining the information represented by a single bit (single-level cell – SLC) or combinations of bits (multilevel cell – MLC). Cells are grouped and organized in arrays whose structures differ in NOR and NAND flash memory:

- NOR flash memory is able to read individual flash memory cells, and as such it behaves like a traditional ROM in this mode. For the erase and write functions, commands are written to the first page of the mapped memory (Fig. 2.28).
- NAND flash memories are accessed much like block devices such as hard disks. When NAND flash memories are to be read, the contents must first be paged into memory-mapped RAM. This makes the presence of a memory management unit essential (Fig. 2.29).

The most common applications are flash memory cards *(Compact flash, SD, miniSD, and microSD)* and USB flash drives, while one of the most recent applications is the SSD *(solid-state drives)*.

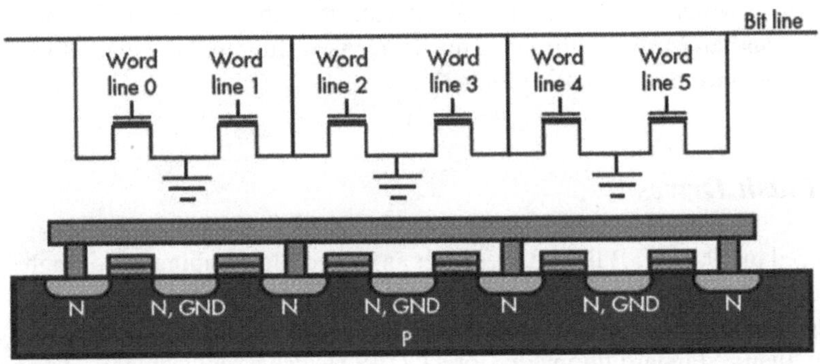

Fig. 2.28 NOR structure

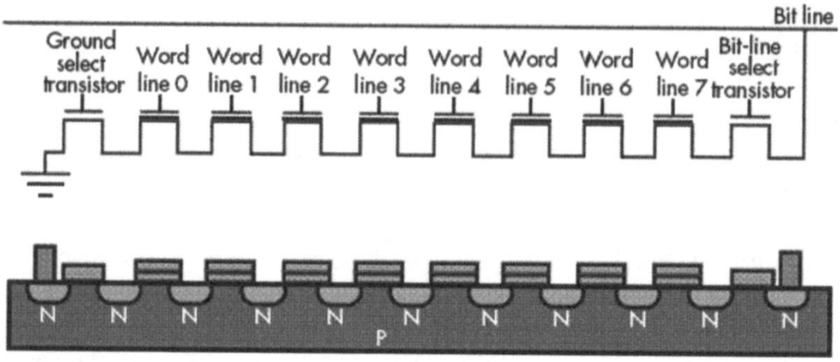

Fig. 2.29 NAND structure

Flash Memory Card

Flashcards are solid-state storage devices, used mostly in small mobile devices such as MP3 players, game consoles, digital and video cameras, mobile phones, and in other portable devices such as PDAs. Through external card readers, PCMCIA-adapters or proprietary slots, the flashcard data can be read in notebooks/PCs. The typical size of a flashcard is that of a large postage stamp. Storage capacity is typically 512 MB up to 32 or 64 GB. Basic types are Compactflash, Smart Media, Multi Media, Secure Digital, and Memory Stick.

The performances offered by these devices may be very high. In fact they can reach up to 40 MB/s in read and 20 MB/s in write. There are lots of different performance values among the various devices depending on which flash memory is used: NAND or NOR. Typically NOR-based flash memory cards are used in applications that require very high-read performances; otherwise NAND-based flash memory cards are used.

Using a flash card, instead of the flash memory stand-alone, guarantees a simplified application. This is due to a series of algorithms implemented in the flash card that gives it a high performance, high reliability, and ensures long life of the device itself.

One of the brand new MMC (multi media card) evolutions is the e·MMC (embedded MMC), which is a nonremovable device that integrates one or more Flash memories and an MMC controller. This device is dominating the market for a lot of wireless and automotive applications.

USB Flash Drives

Introduced in 2002, USB flash drives offer an incredible combination of high storage capacity, fast data transfer rates, and great flexibility, all in the palm of one's hand. Heralded as a floppy or CD drive alternative, USB flash drives have far more storage capacity than a standard floppy disk or a CD-ROM drive replacement. They provide an easy method for quick downloads and transferring digital files to and from a computer or device. USB flash drives incorporate NAND flash and a controller in a capsulated case and work with the vast majority of computers and devices that incorporate the Universal Serial Bus interface, including most PCs, PDAs, and MP3 players.

SSD (Solid State Drives)

A solid-state drive (SSD) is a data storage device that uses solid-state memory to store persistent data. An SSD emulates a hard disk drive interface, and thus easily replaces it in most applications. An SSD using SRAM or DRAM (instead of flash memory) is often called a RAM-drive, not to be confused with a RAM disk.

The original use of the term "solid-state" (from solid-state physics) refers to semiconductor devices rather than electron tubes but, in the present context, has been adopted to distinguish solid-state electronics from electromechanical devices. With no moving parts, solid-state drives are less fragile than hard disks and are also silent (unless a cooling fan is used); as there are no mechanical delays, they usually enjoy low access time and latency (Fig. 2.30).

Owing to these features, SSDs are starting to be used for more and more applications rather than HDDs and are already being employed more often than HDDs for industrial/embedded and phone products. The trend indicates that SSD will completely replace HDD in applications such as ultra low-cost Netbook, Notebook, Smartbook, and Tablet devices by 2014.

Fig. 2.30 Example of an SSD

Uncommon Storage Media

Current predominant nonvolatile memory technologies for mass storage applications rely upon flash memories (used, for instance, on solid-state disk drives and other embedded memory modules) and magnetic-based solutions [18–24].

A general trend is to extend the use of NAND flash technology in solid-state disk applications or for hybrid solutions (with traditional HDD) to create better products in terms of mechanical specifications, robustness, and performance.

Moore's law will continue to drive memory technology scaling but technological complexity will increasing address the fundamental limits of physics. In respect of Moore law's predictions, there is constant research activity in industrial and academic contexts leveraging not only on the future possibilities of current transistor-based solutions but also on the development of material and structural innovations. This research will lead to the identification of scalable nonvolatile technologies for applications in various market segments in the next 5 years and beyond. Such technologies will be used in mass storage solutions or as supportive memory modules in hybrid architectures.

The driving factors for innovative solutions are:

- High performance
- Low power consumption
- Long term scalability
- Cost
- Technological complexity

The best compromise among those parameters will determine the success of a technology with respect to the others.

The following, most of which are based on the use of new materials in the elementary structure to store the information, can be considered uncommon storage media :

- FeRAM or FRAM (ferroelectric RAM)
- PCM (phase change memory)
- PMC (programmable metallization cell RRAM)
- MRAM (magnetoresistive RAM)
- Others (molecular memories, probe storage, carbon nanotube)

The rest of this chapter focuses on the primary alternatives that will be introduced in the memory market.

FeRAM or FRAM (Ferroelectric RAM)

Ferroelectric RAM is a random access memory similar in construction to DRAM but uses a ferroelectric layer instead of a dielectric layer to achieve nonvolatility.

Development of FeRAM began in the late 1980s. Much of the current FeRAM technology was developed by Ramtron, a fabless semiconductor company. One of its major licensees is Fujitsu, which operates what is probably the world's largest semiconductor foundry production line with FeRAM capability. Since 1999 they have been using this line to produce stand-alone FeRAMs, as well as specialized chips (e.g., chips for smart cards) with embedded FeRAMs. Fujitsu produces devices for Ramtron. Texas Instruments has collaborated with Ramtron since 2001 to develop FeRAM test chips in a modified 130-nm process. In the fall of 2005 Ramtron reported that they were evaluating prototype samples of an 8-Mbit FeRAM manufactured using the Texas Instruments' FeRAM process. In 2005 Fujitsu and Seiko-Epson collaborated in the development of a 180-nm FeRAM process.

Storage Mechanism

The physical principle underlying the storing mechanism is the permanent polarization of a ferroelectric dielectric. In a ferroelectric material there is a characteristic nonlinear relationship between the applied electric field and the apparent stored charge, which has the shape of a hysteresis loop (see Fig. 2.31).

Over some range of temperature, ferroelectric materials exhibit a spontaneous electric polarization that can be oriented by application of an electric field. When

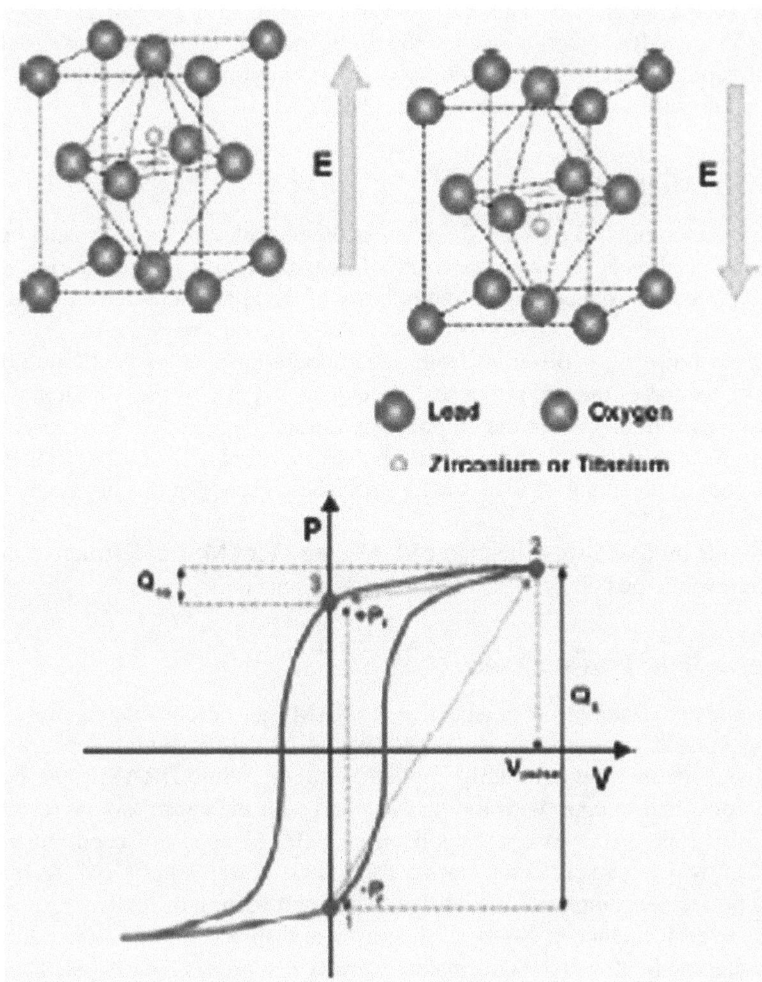

Fig. 2.31 Ferroelectric material polarization and hysteresis loop

an external electric field is applied across a dielectric, the internal semipermanent dipoles in the crystal structure of the material tend to align themselves with the field direction, produced by small shifts in the positions of atoms and shifts in the distributions of electronic charge in the crystal structure. After the charge is removed, the dipoles retain their polarized state. Typically binary "0's" and "1's" are stored as one of two possible electric polarizations in each data storage cell.

Storage Cell Architecture

Each cell consists of one capacitor and one transistor, a so-called "1T-1C" device. The 1T-1C storage cell design in an FeRAM is similar in construction to the storage cell in the widely used DRAM in that both cell types include one capacitor and one access transistor. A linear dielectric is used in a DRAM cell capacitor, whereas in an FeRAM cell capacitor the dielectric structure includes ferroelectric material, typically lead zirconate titanate (PZT), strontium-bismuth-tantalate (SBT), or lanthanum substituted-bismuth-titanate (BLT).

Read/Write Mechanisms

Writing is accomplished by via a field across the ferroelectric layer created by voltage applied to the capacitor plates, which forces the atoms inside into an "up" or "down" orientation (depending on the polarity of the charge), thereby storing a "1" or "0."

Reading, however, is different from a read operation in a conventional DRAM. The transistor forces the cell into a particular state, say "0." If the cell already holds a "0," nothing will happen in the output lines. But if it holds a "1," the reorientation of the atoms in the film will cause a brief pulse of current in the output as they push electrons out of the metal on the "down" side. The presence of this pulse means that the cell held a "1."

Since this process overwrites the cell, reading FeRAM is a destructive process and requires that the cell be re-written if it was changed.

Evaluation of the Pros and Cons

From a power consumption perspective, FeRAM can be considered more effective than DRAM for both the low-power nature of the elementary operations on the cell and for the low externally applied voltages. While DRAM must be periodically refreshed to cope with the loss of charge on the capacitors increasing the refresh rate done by an external application, FeRAM does not need any refresh mechanism which actually determines a continuous power supply. This technology requires power consumption only while reading and writing the cell.

Intrinsic performance is based on the physical displacement of atoms due to an external electric field, which is intrinsically a very fast process (<100 ps). The external circuitry to control the mechanism and to execute read and write operations adds some delay in the overall performances, but write and read operations (<100 ns) can be considered fast when compared to flash technology. The technology is also characterized by a high write endurance (>10^{12}), but the read endurance is limited by the destructive read-out mechanism.

PCM (Phase Change Memory)

Phase change memory (PCM) is a term used to describe a class of nonvolatile memory devices that store information by utilizing a reversible phase change in materials. Materials can exist in various phases—solid, liquid, gas, condensate, and plasma. PCM exploits the differences in the electrical resistivity of a material in different phases.

PCM technology uses a class of materials known as chalcogenides, which are alloys that contain an element in the oxygen/sulfur family of the Periodic Table. Most companies performing research and development in PCM today are using GST (germanium, antimony, and tellurium) or closely related alloys (Fig. 2.32).

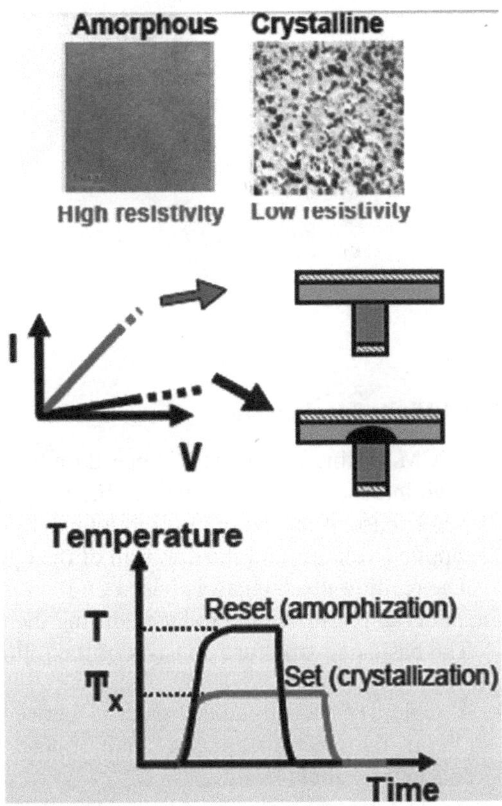

Fig. 2.32 Chalcogenide phase change behavior

Storage Mechanism

Phase change chalcogenides exhibit a reversible phase change phenomenon (see Fig. 2.33) when changed from the amorphous to the crystalline phase. In the amorphous phase, the material is highly disordered–there is an absence of regular order to the crystalline lattice. In this phase, the material demonstrates high resistivity and high reflectivity. In contrast, in the polycrystalline phase, the material has a regular crystalline structure and exhibits low reflectivity and low resistivity. The difference in resistivity between the two phases of the material is what associates the information to be stored in a nonvolatile way.

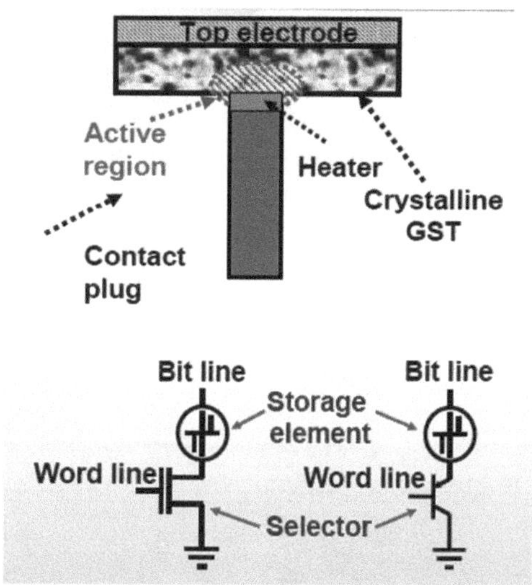

Fig. 2.33 Structure of a PCM cell

Read/Write Mechanism

The PCM writing mechanism is based on the phase change induced in the material through intense localized Joule heating caused by current injection (see Fig. 2.33). The end phase of the material is modulated by the magnitude of the injected current, the applied voltage, and the duration of the operation.

The reading mechanism is based on the resistance measured after the change in the material phase change induced during the write operation.

The basic cell structure consists of the following:

- A region of chalcogenide material between two electrodes (storage element). One of the electrodes has a small contact area and a high-resistivity region to concentrate Joule heating
- A selector represented by a transistor (BJT or MOSFET)

Pro and Con Evaluation

The particular mechanism behind PCM technology offers a series of functionalities capable of enabling possible new models of a nonvolatile memory peripheral for different applications:

- Bit alterability. Unlike flash technology, no intermediate erase steps are required to write a portion of the memory array, which can be directly written with a smaller granularity.
- Read speed. Fast random access can be achieved through the very low latency in reading a PCM cell making easier the adoption of this technology in the code execution host usage model.
- Write speed. Write speed is faster than NOR flash technology. As with RAMs no separate erase operation is required.
- Retention/endurance. The physical characteristics and the circuitry used do not impact massively on the endurance of the device. In particular the read endurance is especially high.
- Scaling. Scaling is an important factor for cost reduction and with PCM, as the memory cell shrinks, the volume of GST material shrinks as well.

Temperature sensitivity is perhaps its most notable drawback, one that may require changes in the production process of manufacturers incorporating the technology.

MRAM (Magneto Resistive RAM)

Storage Mechanism

For this type of nonvolatile memory, the storage principle is not based on an electronic mechanism but rather on magnetic storage elements. In particular, the basic storage element is made up of two ferromagnetic plates, capable of holding a magnetic field, separated by a thin insulating layer (see Fig. 2.34). One of the two plates is a permanent magnet with its own particular polarity permanently set; the second varies its field to match an external field. This is the basic structure of a cell, and a memory device is built from a grid of such "cells."

Read/Write Mechanisms

Reading consists of measuring the electrical resistance of the cell. An associated transistor is used to select the cell; it switches current from a supply line through the cell to ground. The electrical resistance of the cell changes owing to the orientations of the fields in the two plates. The new value of the resistance is obtained by measuring the resulting current. Typically if the two plates have the same polarity it is conventionally considered as "0," whereas if the two plates have opposite polarity, the resistance will be higher and the associated information is "1."

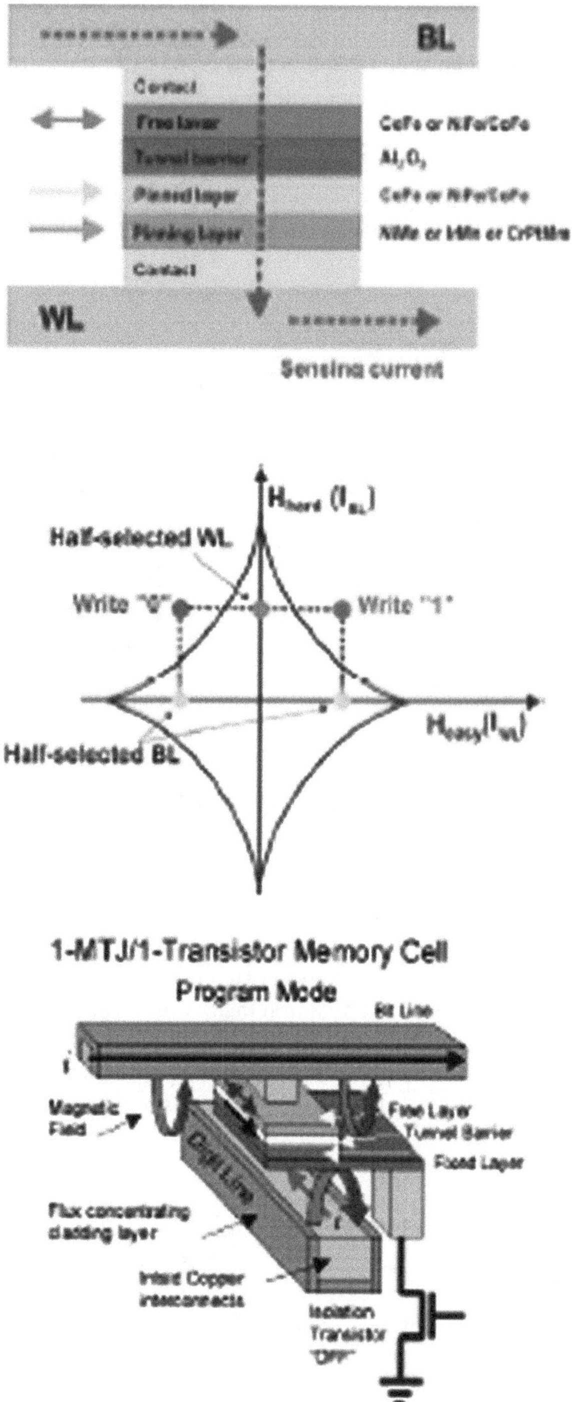

Fig. 2.34 MRAM transistor memory structure

Over the years the writing mechanism has evolved using different methods. The basic technique makes use of the current flowing in the bit and digit lines in the array containing the cell to be written on: the induced external magnetic field modifies the polarity of the plate whose field can be changed. This technique has two immediate implications: power consumption (due to the current flowing in the array) and scaling issues (when the cell size is reduced, the inducted field can cause spurious write operations). The current research effort is to minimize the relevance of these two implications

MRAM does not require a refresh mechanism. However, the write process requires more power in order to overcome the existing field stored in the junction, varying from three to eight times the power required during reading.

Evaluation of the Pros and Cons

MRAMs have some advantages such as fast write operations, high write endurance, and low-voltage writes but, at the same time, there are issues that must be solved (process integration difficulties, large cell size compared to flash and DRAM, high write currents, scaling limitation).

PMC (Programmable Metallization Cell RRAM)

The programmable metallization cell, or PMC, is a new form of nonvolatile memory being studied and developed at Arizona State University and its spin-off, Axon Technologies.

Storage Mechanism

PMC nonvolatile technology is based on the physical relocation of ions within a solid electrolyte. A PMC memory cell is made up of two solid metal electrodes, one relatively inert (e.g., tungsten) the other electrochemically active (e.g., silver or copper), with a thin film of the electrolyte between them. A control transistor can also be included in each cell.

Applying a negative bias to the inert electrode causes a migration of the metal ions in the electrolyte, as well as some originating from the now-positive active electrode, toward the inert electrode, where they are reduced. After a short period of time the ions flowing into the filament form a small metallic "nanowire" between the two electrodes. The "nanowire" reduces the resistance along that path indicating that the "writing" process is complete.

Read/Write Mechanisms

Reading the cell simply requires the control transistor to be switched on and a small voltage applied across the cell. If the nanowire is in place in that cell, the resistance will be low, leading to higher current, which is read as a "1." If there is no nanowire in the cell, the resistance is higher, leading to low current, which is read as a "0."

Erasing the cell is identical to writing, but uses a positive bias on the inert electrode. The metal ions migrate away from the filament, back into the electrolyte, and eventually to the negatively charged active electrode. This breaks the nanowire and increases the resistance again.

Others

Molecular Memories

Molecular memory is a term for data storage technologies that use molecular species as the data storage element, rather than, e.g., circuits, magnetic, inorganic materials, or physical shapes. The molecular component can be described as a molecular switch and can perform this function by any of several mechanisms, including charge storage, photochromism, or changes in capacitance. In a perfect molecular memory device, each individual molecule contains a bit of data, leading to massive data capacity. However, practical devices are more likely to use large numbers of molecules for each bit, in the manner of 3D optical data storage (many examples of which can be considered molecular memory devices). The term "molecular memory" is most often used to indicate very fast, electronically addressed solid-state data storage, as is the term computer memory. At present, molecular memories are still found only in laboratories (see Fig. 2.35).

One approach to molecular memories is based on special compounds, such as porphyry-based polymers, that are capable of storing electric charge. Once a certain voltage threshold is reached the material oxidizes, releasing an electric charge. The process is reversible, in effect creating an electric capacitor. The properties of the material allow for a much greater capacitance per unit area than with conventional DRAM memory, thus potentially leading to smaller and cheaper integrated circuits.

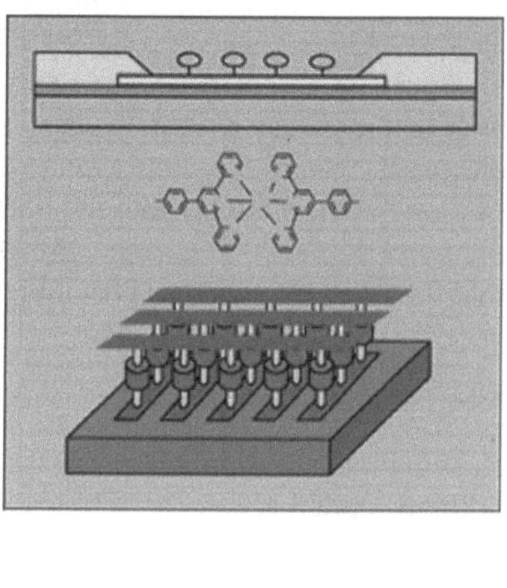

Fig. 2.35 Molecular memories

Millipede

Millipede is a nonvolatile memory stored on nanoscopic pits burned into the surface of a thin polymer layer, read and written on by a MEMS-based probe.

A millipede stores data in a "dumb" medium that is simpler and smaller than any cell used in an electronic medium and accesses the data by moving the medium under a "head" (probe). However, millipede uses many nanoscopic probes that can read and write in parallel, thereby dramatically increasing the throughput to the point where it can compete with some forms of electronic memory. Additionally, millipede's physical medium stores a bit in a very small area, leading to densities even higher than current hard drives. Each probe in the cantilever array stores and reads data thermo-mechanically, handling one bit at a time.

To accomplish a read, the probe tip is heated to around 300°C and moved in proximity to the data sled. If the probe is located over a pit the cantilever will push it into the hole, increasing the surface area in contact with the sled, and in turn increasing the cooling as heat leaks into the sled from the probe. In the case where there is no pit at that location, only the very tip of the probe remains in contact with the sled and the heat leaks away more slowly. The electrical resistance of the probe is a function of its temperature, rising with increasing temperature. Thus when the probe drops into a pit and cools it registers as a drop in resistance. A low resistance is translated to a "1" bit, or a "0" bit otherwise. While reading an entire storage field, the tip is dragged over the entire surface and the resistance changes are constantly monitored.

To write a bit, the tip of the probe is heated to a temperature above the glass transition temperature of the polymer used to manufacture the data sled, which is generally acrylic glass. In this case the transition temperature is around 400 K. To write a "1," the polymer in proximity to the tip is softened and the tip is then gently touched to it, causing a dent. To erase the bit and return it to the zero state, the tip is pulled up from the surface, allowing the surface tension to pull the surface flat again.

Conclusions

This chapter has presented an overview of the evolution of mass storage devices during the past centuries and provided some information on the newer memories currently available in the mass memory market. One important segment that is now interested in mass storage memories is the wireless market: in the last 2 years a lot of mobile phones have begun using big mass storage devices based on NAND technology, especially the Smartphone platforms focusing on multimedia content (audio, video, maps, etc.).

Today there are many mobile phones with, for example, 16/32-GB devices using the widely adopted e·MMC standard interface promoted and described by the JEDEC organization. It appears that this solution and the next evolution, called UFS, will be very familiar item in the future Smartphone market.

In the consumer and personal computer segment HDD is still the predominant choice. An attempt to introduce SDD in the netbook segment met with only limited and temporary success owing to the high cost of the NAND technology compared to the magnetic disk and mechanics of the HDD device.

Currently SSD devices are employed primarily in some dedicated market segments like web servers, banking, military and medical applications.

In the video segment DVD and HD-DVD are still robust in the market but there are a lot of initiatives to distribute these kinds of contents through DVB (Digital Video Broadcasting) and the internet. On the other hand, audio contents using CD devices are not so widespread because they are also distributed via the internet; unfortunately audio magnetic tapes have definitely become extinct.

Finally the most important device used to transmit the greatest amount of information and knowledge during the past centuries is the paper book. There are a few tentative challenges to replace this fundamental tool using solid-state memory-based products but at the moment is not yet clear as to what will be the official new-book replacement for the third millennium.

References

Paper

1. Castagnari CG, Lipparoni N, Manucci U (January 1, 1991) *L'arte della carta a Fabriano*. Meseo della carta a dellla filigrana Fabriano. ASIN: B0012TXTI8
2. Pedemonte E, Princi E, Vicini S *Storia della produzione della carta* (2005). In La chimica e l'industria, anno 87, ottobre 2005, Treccani, Enciclopedia Italiana di Scienze, Lettere ed Arti
3. http://www.archeoempoli.it/istruzione.htm. Accessed 1 Feb 2010
4. http://www.fmboschetto.it/religione/Genesi/Diluv_2.htm. Accessed 1 Feb 2010
5. http://www.museodellacarta.com/default.asp?lang=eng. Accessed 1 Feb 2010

Magnetic Devices

6. Mee C, Daniel E (August 1, 1996). Magnetic storage handbook, 2nd edn. McGraw-Hill Professional. ISBN-10: 0070412758, ISBN-13: 978-0070412750
7. Daniel ED, Denis Mee C, Clark MH (August 17, 1998). Magnetic recording: the first 100 years. Wiley-IEEE Press. ISBN-10: 0780347099, ISBN-13: 978-0780347090
8. Jorgensen F (January 1970). Handbook of magnetic recording. McGraw-Hill/TAB Electronics. ISBN-10: 0830605290, ISBN-13: 978-0830605293
9. RAMAC History (May 2005) http://www.magneticdiskheritagecenter.org/MDHC/MILESTONE/Emerson%20Pugh%20talk.pdf. Accessed 1 Feb 2010
10. http://en.wikipedia.org/wiki/Hard_disk_drive#History. Accessed 1 Feb 2010

Optical Devices

Semiconductor Devices

11. http://www.radio-electronics.com. Accessed 1 Feb 2010
12. http://www.eu-energystar.org/index.html. Accessed 1 Feb 2010
13. http://www.kingston.com.fash. Accessed 1 Feb 2010
14. http://www.arstechnica.com. Accessed 1 Feb 2010
15. http://www.lascon.co.uk/dh00300.htm. Accessed 1 Feb 2010
16. http://www.spansion.com/.../MirrorBit-cell.jpg. Accessed 1 Feb 2010
17. http://www.archive.electronicdesign.com/.../figure_01.gif. Accessed 1 Feb 2010

Uncommon Memory Devices

18. http://www.fujitsu.com/global/services/microelectronics/technical/fram/. Accessed 1 Feb 2010
19. http://www.numonyx.com/en-US/MemoryProducts/PCM/Pages/PCM.aspx. Accessed 1 Feb 2010
20. Huai Y (December 2008) Spin-transfer Torque MRAM (STT-MRAM): challenges and prospects. AAPPS Bull 18(6):33
21. Balakrishnan M, Thermadam SCP, Mitkova M, Kozicki MN (2006) A Low Power Non-Volatile Memory Element Based on Copper in Deposited Silicon Oxide. In: Proceedings of the 7th Non-Volatile Memory Technology Symposium (NVMTS), pp 104–110
22. http://www.zurich.ibm.com/sto/probe/storage.html. Accessed 1 Feb 2010
23. http://www.zurich.ibm.com/news/05/millipede.html. Accessed 1 Feb 2010
24. http://www.nasa.gov/centers/ames/research/technology-onepagers/nonvolatile_memory.html. Accessed 1 Feb 2010

Chapter 3
Probe Storage

Marcellino Gemelli, Leon Abelmann, Johan B.C. Engelen, Mohammed G. Khatib, Wabe W. Koelmans, and Oleg Zaboronski

Abstract This chapter gives an overview of probe-based data storage research over the last three decades, encompassing all aspects of a probe recording system. Following the division found in all mechanically addressed storage systems, the different subsystems (media, read/write heads, positioning, data channel, and file system) will be discussed. In the media subsection various recording media will be treated, such as polymer based, phase change, ferroelectric and magnetic. In the probe array subsection various generations of thermal probes will be discussed, as well as examples of alternative write probes using currents, electric or magnetic fields, and different principles for data detection. Special attention is paid to parallel readout of probe arrays. In the positioning section, examples will be shown of electric and magnetic scanners, either precision engineered or realized by micromachining technologies, or combinations thereof. In the systems subsection the data channel will be discussed, including the read/write electronics circuitry, data detection, and coding algorithms. Special attention is paid to the writing strategy and considerations for probe-based storage file systems.

Keywords MEMS · Probe storage · Nanomechanical storage · Nano-tips · Thermomechanical recording · Homogeneous recording media

Introduction

The area of probe-based data storage leverages two important developments of the last decades. On the one side there is the invention of the Scanning Tunneling Microscope by Binnig and Rohrer [18] in 1985, which triggered the very fruitful field of Scanning Probe Microscopy (SPM). On the other side, the industrial maturity of silicon-based MEMS (Micro Electrical Mechanical System). It is now

M. Gemelli (✉)
STMicroelectronics, Santa Clara, CA, USA
e-mail: marcellino.gemelli@st.com

G. Campardo et al. (eds.), *Memory Mass Storage*,
DOI 10.1007/978-3-642-14752-4_3, © Springer-Verlag Berlin Heidelberg 2011

practicable, for example, to use SPM technology to modify the surfaces of materials on the nano-scale, rather than just for microscopic imaging. Such surface modification might comprise the writing and reading of data, so providing a storage system with, ultimately, atomic resolution. Indeed, such atomic resolution was demonstrated in 1990 by Eigler and Schweizer [43] who placed individual Xe atoms on single crystal nickel substrate to spell out the IBM logo. However, this approach was exceedingly slow and, from a system perspective, offered an impracticably low data rate. A much higher data rate was achieved at IBM Almaden by using a heated AFM probe mounted over a rotating disk with polymer film. Data were written by creating small indents, at impressive data rates over 10 Mbps [111, 112], but still too slow for most applications.

Since the active part of AFM probes has dimensions in the order of 100 μm, the logical step forward is to use arrays of probes to multiply the data rate. Dense stacking of probes was made possible by micromachining technology [21], which developed from integrated circuit technology following Feynman's [Genitive] famous 1959 lecture "There's Plenty of Room at the Bottom." The manufacturing technologies had matured by the end of the last century, entering the VLSI age of MEMS (as Binnig calls it) with the development of arrays of AFM probes [119] and integrated nano-positioning systems [24].

The chapter provides a description of the architectures that make up storage devices based on probe technology that was first proposed by IBM [105] (Fig. 3.1).

The chapter details recording medium, probe arrays, and positioning systems in sections "Recording Medium," "Probe Arrays and Parallel Readout," and "Positioning Systems." Probe storage electronics and data coding are discussed in sections "Probe Storage Electronics and System Considerations" and "Coding for Probe Storage". The chapter will close with an overview of the implications of this new type of mechanically addressed storage device for the computer architecture, specifically the file system.

Why Probe Storage?

Probe storage concepts are being studied and developed alongside the three existing main mass storage technologies: tape drives, flash memories, and hard disk drives.

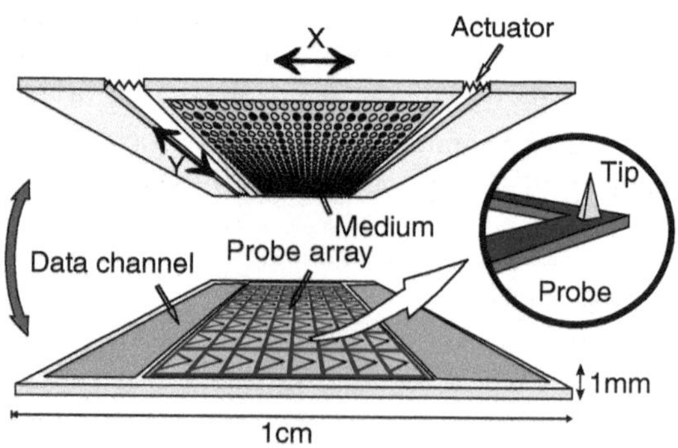

Fig. 3.1 Overview of a probe-based recording system

Considering just the cost of the cartridges, tape drives have the lowest cost per bit but have slow access times due to their sequential storage. Their use is therefore limited to backup. The evolution of this storage technology depends on the ability to increase track density, through improvements in track-following and reel-to-reel servomechanisms, and the ability to increase linear density, through head-tape interaction control and the advances in magnetic particle size control.

On the other hand the hard disk drive roadmap largely depends on the super-paramagnetic limit of its media: as bit volumes decrease, the energy stored in one bit becomes comparable to its thermal energy and the magnetization state becomes unstable. Moreover, the decrease in access time has not kept the pace with the increase in areal density. The access time is limited by the fact that a single head is addressing an increasing amount of data as areal density (and therefore disk capacity) increases. Furthermore, the access time is limited by the disk rotational speed.

The density of flash memories depends on the resolution of the electrical interconnections used to address the individual bit (or groups of bit levels, as in multilevel recording) through a planar silicon manufacturing process. Flash memory's evolutionary roadmap is therefore driven by the ability to decrease the minimum lithographic feature size, which in turn is fueled by continuous investments in new costly semiconductor fabs. At the time of publication a NAND flash manufacturing facility requires an investment of $3.4 billion. Such an investment would generate less than 1% of the yearly shipped capacity of hard disk drives. Due to such huge capitalization it is unlikely that lithographically defined patterned media or SSDs will replace a substantial amount of hard disk drive production [53].

A storage pyramid is used to map the hierarchy of storage products against total industry capacity. The higher levels of the pyramid are the highest performing and most expensive storage products. Thus the lower level is occupied by tape storage,

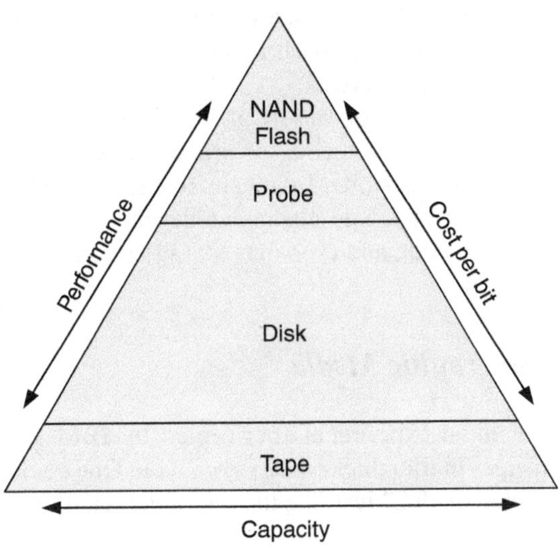

Fig. 3.2 The mass storage pyramid

due to its lowest cost (per bit) and higher capacity. NAND flash, on the other hand, is currently mapped as the tip of the storage pyramid but due to its cost its use will be very selective, using its scalability to target lower capacity applications where flash level performance is required. The majority of data for the majority of the time will not require such performance and will be mapped in the greatest area of the pyramid hierarchy occupied by the magnetic disk.

Probe storage overcomes the limitations of both of these technologies since the ultimate bit size is neither lithographically defined nor bound by the superparamagnetic limit. Unlike in hard disk drives, access time in probe storage benefits from the smaller movements of the read/write heads relative to the media. Parallel probe operation allows data rates to be competitive with hard disks and NAND flash. When represented on the storage pyramid, the economics and performance of probe storage place it between the magnetic disk and the NAND flash (see Fig. 3.2).

Recording Medium

Requirements

A mass storage, non-volatile recording medium has to meet stringent requirements in order to be successful. The most important are bit size (for density), endurance, (the number of read–write cycles the memory can sustain), and the retention. The combination of probe and medium characteristics determines the bit density and therefore the final capacity of the device. Not only should the medium be able to support the small bit diameters that can be written by the probe tip, it also should be able to keep the data for over a sufficient length of time. This means that the energy barrier for erasure should be sufficiently high (over 1 eV for 10 years storage at room temperature). Next to this, the medium should be capable of data overwrite, with sufficient suppression of previously written data. And moreover, the medium should be resilient against wear caused by multiple (read) passes. A minimum target is the endurance of flash memory chips that are typically guaranteed to withstand 10^5 write–erase cycles.

There are a wide variety of recording media for probe storage under investigation. The most studied media for probe data storage are polymer-based media, in which information is stored by means of indents. Other storage principles that have been investigated include chalcogenide phase change media, ferromagnetic and ferroelectric media, and even data storage into single molecules and atoms.

Topographic Media

The initial experiments performed by IBM groups in Almaden and Zürich used changes in film thickness to store data. One could characterize these media as “topographic media,” just like the old punch card or CD.

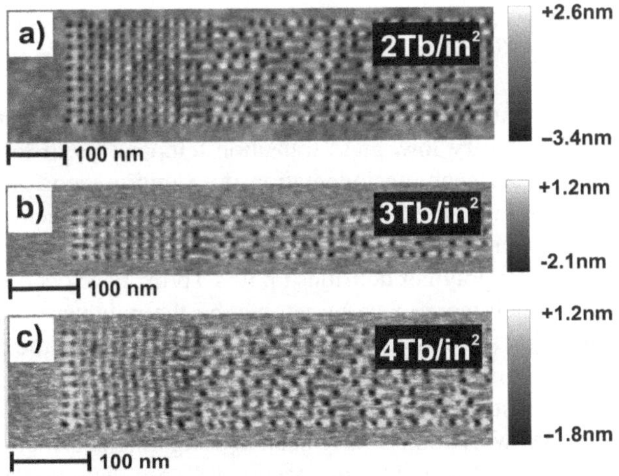

Fig. 3.3 AFM images of a random pattern of indentations recorded in a PEAK polymer. A density of 2 Tb/in^2 was obtained on a normal spin-coated sample. Densities up to 4 Tb/in^2 were achieved on a templated sample. Experimental data by IBM [185]

The first media used for probe-based data storage were simply thick PMMA (perspex) disks of 1.2 mm [110]. Using a single cantilever heated by a laser through the PMMA disk, Mamin et al. were able to write bits with a radius below 100 nm and a depth of 10 nm, allowing for data densities up to 30 Gb/in^2.

In the following work, the bulk PMMA or polycarbonate (compact disk material) disks were replaced by silicon wafers on top of which a 40 nm PMMA recording layer on top of a 70 nm cross-linked hard-baked photoresist was deposited [19]. This allowed for small bit dimensions down to 40 nm, and data densities up to 400 Gb/in^2 were shown.

Next to PMMA, other polymers were studied, such as polystyrene and polysulfone [183]. The method of trial and error was taken out of the research by the development of a write model by Dürig [183]. He discovered that a balance needs to be found between stability and wear resistance of the medium on the one side, requiring highly cross-linked polymers [58], and wear of the tip on the other side, for which a soft medium is necessary. Based on this knowledge, a so-called Diels–Adler (DA) polymer has been introduced [57]. These DA polymers are in a highly cross-linked, high molecular weight state at low temperature, but dissociate at high temperature into short strains of low molecular weight. This reaction is thermally reversible: rather than a glass transition temperature, these polymers have a dissociation temperature. Below the transition temperature, the polymer is thermally stable and has a high wear resistance, above the transition temperature the polymer becomes easily deformable and is gentle on the tip. Using the new DA polymer, densities up to 1.1 Tb/in^2 were demonstrated.

The work was continued with a polyaryletherketone (PAEK) polymer, which incorportes diresorcinol units in the backbone for control of the glass transition temperature and phenyl ethynyl groups in the backbone and as endgroups for cross-linking functionality [185]. The results are illustrated in Fig. 3.3. As with the DA polymer, this polymer is highly cross-linked to suppress media wear during reading and to enable repeated erasing. In contrast to the DA polymer, however, it has a conventional, but very low, glass transition temperature of less than 160°C in the cross-linked state, enabling indentation on a microsecond timescale using heater temperatures of less than 500°C. It exhibits exceptional thermal stability up to 450°C, which is crucial for minimizing thermal degradation during indentation with a hot tip. Using this polymer densities up to 4 Tb/in^2 have been achieved, onto ultra-flat polymers which were made by templating the polymer on a claved mica surface [143]. Modeling shows that in this type of polymer media the density is limited to 9 Tb/in^2 [185]. Further improvements could be made by evaporating material rather than indenting, but of course rewritability is sacrificed.

Apart from the IBM work, others have been investigating polymer media as well. Kim et al. from LG demonstrated bit diameters of 40 nm diameter [85] in PMMA films. Bao et al. of the Chinese Academy of Sciences investigated friction of tips with varying diameters on PMMA and concluded that blunt tips can be used to determine the glass transition temperature, whereas 30 nm diameter tips can be used to detect local (β) transitions [8].

Next to deformation of polymers, topographic changes can also be induced in other materials. The density of phase change materials decreases, for instance, when a transition from the amorphous to crystalline phase occurs, thus changing the film thickness [17, 55, 62] by a few percent. Densities up to 3.3 Tb/in^2 have been obtained at IBM Watson laboratories using an AFM tip heated by a laser. Shaw et al. have investigated shape memory alloys for probe-based data storage, using thermally heated probes [169], and have calculated a maximum data density of 0.5 Tb/in^2.

It is also possible to induce topographic changes by simply removing material with excessive heat pulses, such as can be applied by STM tips. Eagle et al. have written bits as small as 3.5 nm in amorphous carbon films [42]. Of course the method is write only, since material is removed.

Conductive Media

Rather than writing indentations, one can use changes in electrical conductivity caused by phase transformation in thin films, such as in phase change solid-state memories. From a generic point of view one can define a category "electrical probe storage," which might be viewed as using an electrical potential applied to a probe that is in contact (or quasi-contact) with a medium whose properties are altered in some way by the resulting flow of electrical current through the medium toward a counter electrode. The change in medium properties should be electrically detectable, e.g., by a change in electrical resistance.

Major work has been done on probe recording on phase change media at Matsushita [78, 182], Hokkaido University [56], CEA Grenoble and the University

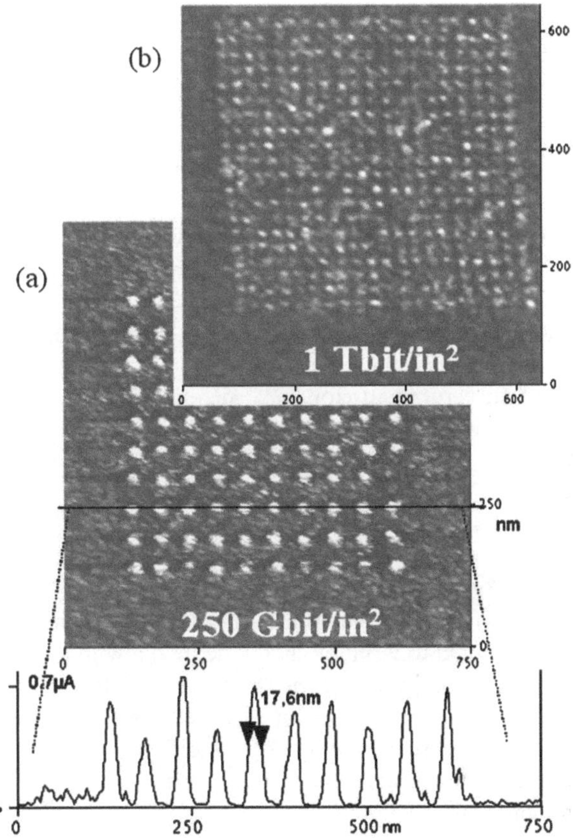

Fig. 3.4 Conductance AFM images of bits written in a $Ge_2Sb_2Te_5$ layer [55]

of Exeter [6, 35, 55, 186] (Fig. 3.4), Hewlett-Packard [120], and NanoChip. By passing a current from the tip, the medium can be locally heated. At sufficiently high temperatures, a phase transition from amorphous to crystalline is induced, which increases the conductivity by several orders of magnitude. The write process is self-focusing, resulting in very high bit densities over 1 Tbit/in^2 [55]. The large change in conductivity can then be used for readback, by measuring the conductance of the medium by conductive AFM in contact [17] or in non-contact modes by changes in field emitter currents [120] or tip/sample capacitance by Kelvin probe force microscopy [124], which measures the work function of the surface.

Rather than heating the phase change material by passing a current from tip to sample, one can use heated tips. At Tohoku University, Lee et al. used dedicated heater tips to write in to AgInSbTe films [101]. Readout was achieved again by measuring the conductance.

As an alternative to the phase change media based on inorganic compounds such as GeSbTe alloys, one can use polymers which become conductive under application of a voltage [178]. The change in conductivity can be due to a change in phase, for instance, caused by polymerization [170] or by electrochemical reactions [50, 190, 191]. The latter method is especially interesting, since it is reversible. The exact nature of the reaction is unknown: it could be either an oxidation–reduction or protonation–deprotonation reaction. Polymer media are softer than alloys, and tip wear is expected to be less of an issue. Rewritability, however, could be a problem since the polymer tends to polymerize.

Electrical probe recording as a generic approach has several attractions, in particular that the power consumption for the writing process is low with respect to other technologies (≈0.1 nJ per written bit). This is because only the dot memory volume, as opposed to the entire tip volume, is heated. Also, since the electrical current only passes through the Hertzian contact area between tip and media, the tip/media contact area could be very small (for hard materials) even if the tips themselves are not necessarily sharp (and a "smoother" tip should alleviate tribology and wear issues).

Magnetic Media

Magnetic recording is one of the oldest data storage technologies, and many researchers have attempted to write on magnetic media with probes. The stray field of magnetic force microscopy probes is, however, too low to write into modern recording media, so some type of assist is necessary. There are essentially two methods: applying a uniform external background field or applying heat.

An external field can be easily applied by means of a small coil mounted below the medium. As early as 1991, Ohkubo et al. of NTT have used permalloy tips on a CoCr film used for perpendicular recording [125–127]. By applying the field in opposite directions, the magnetization of the tip could be reversed and higher bit densities could be obtained by partially overwriting previously written bits. Bit sizes down to 150 nm could be obtained [128], and overwriting data was possible. Similar bit sizes were obtained by Manalis [113] at Digital Instruments, using a CoCr alloy and CoCr-or NiFe-coated tips.

The bit sizes are relatively large, limiting data densities to somewhere on the order of 30 Gb/in^2. This is either due to the media used or by the limited resolution of the MFM tip. Detailed analysis at CMU by El-Sayed shows, however, that densities up to 1.2 Tb/in^2 should be possible with a rather conventional 30 nm tip radius [44, 45].

Onoue et al. at the University of Twente showed that care must be taken when applying high voltages to coils below the medium. In the case that the medium is not grounded, a large capacitive charging current will flow from the tip into the sample, unintentionally heating the medium [131] and resulting in relatively large bits. Without grounding, no bits could be written due to the high switching field distribution in the Co/Pt multilayer used.

Rather than applying background fields, with the risk of erasing previous information, one can heat the medium to reduce its switching field. This is a method also suggested for future hard disk systems with extremely high anisotropy media [151, 166].

In contrast to hard disk recording, in probe storage delivering heat to the medium is surprisingly easy. The most straightforward method for heating is to pass a current from the tip to the medium. Watanuki et al. of IBM Japan [184] used an STM tip made from an amorphous magnetic material, around which a coil was wound. The tip–sample distance was controlled by the tunneling current. A bit size on the order of 800 nm was achieved.

For testing purposes, one does not even have to use a magnetic tip or apply a background field. When starting from a perpendicularly magnetized film, the demagnetization field of the surrounding film will reverse the magnetization in the heated area. Of course this only allows for write-once experiments. Hosaka et al. at Hitachi have experimented with writing bits into magnetic films by passing a current from an STM tip into a Co/Pt multilayer with perpendicular anisotropy. The minimum domain size, observed by optical microscopy, was 250 nm [72, 121], but smaller domains might have been present. In a joint cooperation between Carnegie Mellon University and Twente University, the experiment was repeated but now the bits were imaged by Magnetic Force Microscopy (MFM) [194–200]. Bits sizes down to 170 nm were observed in Co/Pt multilayers and weakly coupled CoNi/Pt granular media at the University of Electronic Science and Technology of China [202].

The big disadvantage of using STM tips is that direct imaging of written bits is only possible by spin polarized tunneling, which is a difficult technique. By using MFM tips, imaging can be done immediately after writing. Hosaka et al. used an MFM tip in field emission mode [72] and wrote bits as small as 60×250 nm. Onoue et al. combined this method with applying a pulsed background field [129, 130], so that bits could be erased [131] (Fig. 3.5). The minimum bit size obtained was 80 nm, which is close to the bubble collapse diameter for the Co/Pt films used in these experiments [131].

Rather than using currents, one can also use heated tips, similar to those used on the polymer media described above. Algre et al. from Spintec proposed to write by means of a heated AFM tip [4]. They start from a Co/Pt multilayer patterned medium prepared by sputtering on pillars of 90 nm diameter, spaced by 100 nm. The pillars are etched into nano-porous silicon to achieve good thermal insulation. The authors show that the media are suitable for heat-assisted magnetic probe recording. The readback method is not clear, however, and the authors do not demonstrate actual recording experiments.

Readback of magnetic information can be done by techniques based on Magnetic Force Microscopy. Being a non-contact mode however, the operation is complex and resolution is strongly determined by tip–sample contact. One can also imagine integrating a magneto-resistive sensor at the end of the probe, similar as in hard disk recording. An initial step in this direction has been taken by Craus et al. at the University of Twente using scanning magnetoresistance microscopy. The magnetic layer in the probe can be used as a flux focusing structure, so that the same probe can be used for writing [33].

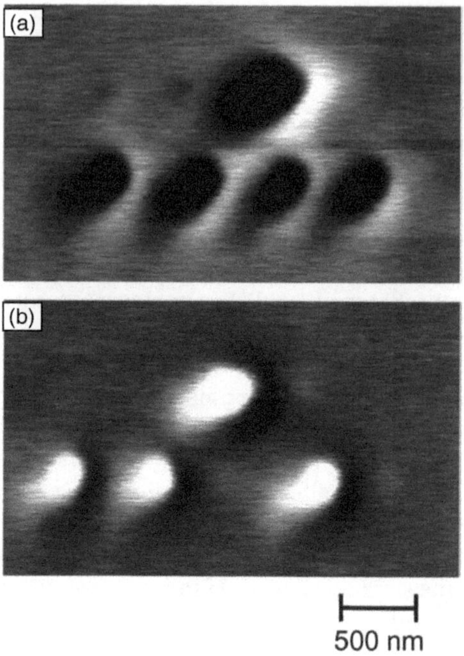

Fig. 3.5 Magnetic Force Microscopy images demonstrating the erasure of individual bits in a Co/Pt multilayer [131]

Ferroelectric Media

The electric counterpart of magnetic recording, ferroelectric storage, has been investigated for decades. In ferroelectric media, the domain walls are extremely thin, indicating a very high anisotropy. A promising piezoelectric material such as PZT has a typical coercive electric field of 10–30 MV/m [142] and a polarization of 0.5 C/m^2 [203]. The energy densities therefore appear to be in the order of 5–15 MJ/m^3, which is a factor of 2 above the highest ever reported energy densities for magnetic materials. More important, however, is the fact that the write head field is not material limited, as is the case with the yoke in the magnetic recording head.

Franke et al. at IFW Dresden [54] were the first to demonstrate the modification of ferroelectric domains by conductive AFM probes. In their case the probe was in contact with the surface. Writing was achieved simply by applying a tip–sample voltage up to 30 V. Readout was achieved by monitoring the response of the probe to a small AC tip/sample voltage at a frequency of about 1 kHz. The sample thickness varies with this frequency due to the piezoelectric effect and with double the frequency due to electrostriction. Soon after Hidaka, Maruyama et al. at Hewlett-Packard in Japan obtained storage densities up to 1 TB/in^2 [64, 114].

Readout can also be performed in non-contact mode. In the early 1990s, Saurenbach and Terris at IBM Almaden induced and imaged charges in polymer disks, using tungsten probes [157, 158]. Imaging was done in non-contact mode by measuring the electric field generated by the polarization charges. Saurenbach measured in dynamic mode, monitoring the changes in resonance frequency of the cantilever caused by changes in the force derivative. It should be noted, however,

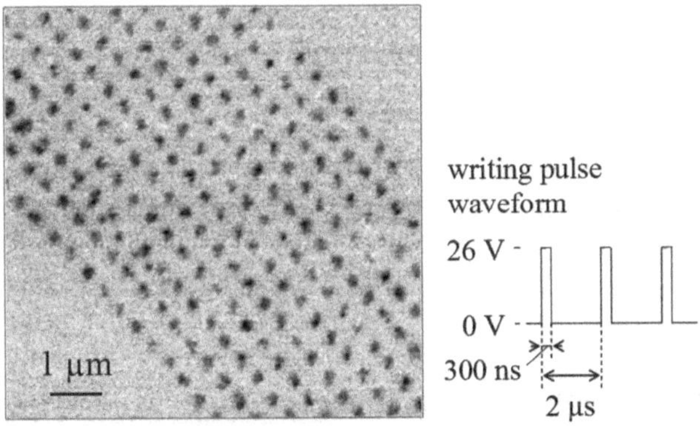

Fig. 3.6 Scanning Nonlinear Dielectric Microscopy images of bits written in a $LiTaO_2$ film at a bit rate of 1 Mb/s. The bit spacing is 206 nm [68]

that on application of an electric field, also the permittivity changes, which gives rise to second harmonics as well [54].

In 2000, Shin et al. at KAIST experimented with AFM data storage on sol–gel deposited PZT. Bits were written at 14 V. Data were read back by measuring electric forces either in non-contact or contact mode [171]. Dot diameters on the order of 60–100 nm were written. Data retention appeared to be a problem, because either free charges accumulated on the medium surface or polarization was lost. Later work in cooperation with Samsung revealed that the polycrystalline nature of sol–gel-deposited PZT films [84] limits the data density, similar as in hard disk storage. The authors conclude that the grain size needs to be decreased.

At Tohoku University, ferroelectric probe storage research started in the same period with experiments on PZT by Lee et al. [101] and $LiTaO_3$ by Cho et al. [25]. A frequency modulation technique was used for data readout. The method is based on the fact that the storage medium's capacitance changes slightly on reversal of the ferroelectric polarization due to the nonlinear terms in the permittivity tensor. This minute change in capacitance causes tiny changes in the resonance conditions, which can, for instance, be detected by monitoring the cantilever vibration if excited with a fixed AC voltage, preferably using a lock-in technique.

Experiments at Tohoku University were continued on $LiTaO_3$ [27, 28, 66, 67], which has superior stability. Since epitaxial films were used, pinning sites are needed for thermal stability [26]. By using thin (35 nm) $LiTaO_3$ single crystal films and a background field, arrays of domains could be written at a density of 13 Tb/in^2 [179]. A realistic data storage demonstration was given at 1.5 Tb/in^2. A raw bit error rate below 1×10^{-4} could be achieved at a density of 258 Gb/in^2 at data rates of 12 kb/s for reading and 50 kb/s for writing [68]. An illustration is shown in Fig. 3.6. The data retention was measured by investigating readback signals at elevated temperatures, and an activation energy of 0.8 eV at an attempt frequency of 200 kHz was found, which is sufficient for a data retention of 10 years [179]. A spin-off of this activity has started at Pioneer, mainly in the production of probes [174, 176, 177].

Rather than using probes, Zhao et al. at Seagate realized a read/write head similar to hard disk heads, where bits are defined at the trailing edge of the head

[201]. Readout was done by Piezelectric Force Microscopy (measuring resonance frequency shifts) or by a destructive readout technique where the displacement currents are measured while the bits are erased by a high electric field emanating from the head. Using this novel type of head, densities up to 1 Tb/in^2 were demonstrated.

Rumors were that also nanochip was attempting probe storage on ferroelectrics. Shaffhauser [159] claims the company licensed a medium based on chalcogenide glass from Ovonix. In 2009, however, Nanochip ceased to exist.

Molecular and Atomic Storage

With ever shrinking bit dimensions, it is inevitable that mechanically addressed data storage will become impossible in continuous thin films, whether they are polymer based, ferroelectric, magnetic, or phase change. We will finally end up at the single molecular or atomic level. That this is not mere science fiction is elegantly proven both in molecular and atomic systems.

Cuberes, Schlitter, and Gimzewski at IBM Zürich demonstrated as early as in 1996 that C_{60} molelcules can be manipulated by STM and positioned on single atomic Cu steps [34]. The experiments were performed at room temperature, and molecules remained stable during imaging. If the binding energy of the molecules is above 1 eV, indeed this method could be used for long-term data storage. Instead of fullerenes, which bind by van der Waals forces, Nicolau et al. therefore suggest to use ionic and chelation bonds between the molecules and the metal surface [123].

Storage of data into single atoms has been beautifully demonstrated by Bennewitz et al. in 2002 [11], who deposited Si atoms from an STM tip onto a 5×2 reconstructed silicon–gold surface (Fig. 3.7). Due to the nature of the reconstructed surface, every bit is stored into an area of 20 surface atoms, resulting in a density of 250 Tbit/in^2. Of course, the method used by Bennewitz is a write-once

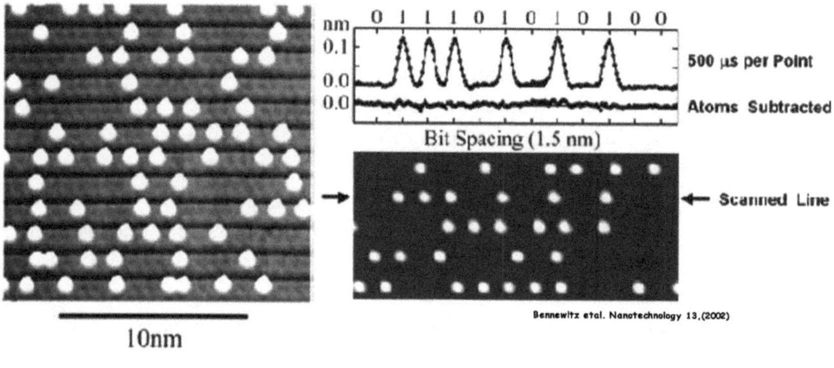

Fig. 3.7 Atomic storage at room temperature: silicon atoms are positioned on top of a reconstructed silicon surface, leading to a density of 250 Tbit/in^2 [11]

technique, but one can also envision deposition of atoms from the gas phase, using, for instance, H or Cl [9, 148].

Even higher storage densities can be achieved, if one does not use the position of molecules or atoms, but modifies their state. For molecules one could use conformal changes or change the charge state of single atoms [147].

Probe Arrays and Parallel Readout

Although the probes used in probe-based data storage are originally derived from standard AFM and STM probes, with time they have become very complex. Not only do the probes require electrical actuation and readout, they also have to be extremely wear resistant (see section "Tip Wear") and have to fit within a restricted area. Another challenging task is to realize probe arrays with thousands of tips in parallel, which will be discussed in the section "Parallel Readout", followed by an analysis of the integration challenges in the section "Integration Challenges". Readout of these arrays is far from trivial, especially when the number of probes is going to increase.

Probe Technology and Arrays

The most advanced probe arrays have been realized by the IBM Zürich probe storage team. Already in 1999, Lutwyche et al. realized a 5×5 array of probes with tip heaters and piezoresistive deflection readout [105]. Ultrasharp tips were obtained by oxidation sharpening of isotropically etched tips. The tips are located at the end of cantilevers that are bent toward the medium by purposely introducing stress gradients, in order to clear the lever anchors from the medium. Boron implantation of specific regions of silicon cantilevers was used to define piezoresistors and tip heaters. In order to increase the sensitivity up to a $\Delta R/R$ of 4×10^{-5}/nm, constrictions were introduced at the base of the cantilever. These constrictions, however, lead to higher resistance, increasing the $1/f$ noise.

Most likely by accident, the team discovered that the resistance of the heater platform dropped with cantilever/medium separation, caused by an increase in heat transfer from the platform to the medium. The effect could be used for readout by heating the platform to about 300°C, well below the temperature needed for writing, and a sensitivity of 1×10^{-5}/nm was obtained [40]. In the next 32×32 array therefore piezoresistive heating was abandoned [36]. In this array the cell size was reduced from 1000 to 92 μm, while keeping the cantilever spring constant at 1 N/m with a resonance frequency of 200 kHz. The array size was 3×3 mm, and thermal expansion deteriorating the tip alignment became an issue. Integrated heaters were positioned in the array to keep temperature variations within 1°C over the chip. The array worked remarkably well, 80% of the cantilevers worked [106], and a density of 200 Gb/in^2 at 1 Mb/s net data rate was shown [40].

The thermal readout was investigated in more detail by King et al., who showed that the fraction of heat transferred through the tip/medium interface is very small and most of the heat flow passes across the cantilever-sample air gap [89]. This observation opened the possibility to heat a section of the cantilever, and avoid reading with heated tips, which causes unwanted erasure and increased medium wear. Simulations were performed to optimize the probe design. By decreasing the tip heater dimensions, the heating time could be decreased down to 0.2 μs [88], and a shorter tip increased the sensitivity to 4×10^{-4}/nm. Of course, these values are power dependent.

The design by King et al. was realized using a mix of conventional and e-beam lithography [38]. The cantilevers in this design are 50 μm long and only 100 nm thick, yielding extremely low spring constants of 0.01 N/m. The heater platform size was reduced down to 180 nm, resulting in time constants on the order of 10 μs. The writing energy was less than 10 nJ per bit, mainly caused by parasitic effects and an inappropriate measurement setup, so there is potential for improvement. In order to guide and speed up the design of more sensitive probes and assist in the readback data analysis, Dürig developed a closed form analytical calculation for the response of the height sensor [39].

The storage density is limited by the medium properties, but more importantly by the probe tip dimensions. Lantz et al. tried to achieve higher densities by applying multiwalled CNT tips with a tip radius down to 9 nm. The advantage of the carbon nanotube tips is that the tip radius does not increase by wear, instead the tip just shortens. Densities up to 250 Gb/in^2 were reached [96], which was disappointing since at that time densities up to 1 Tb/in^2 were already attained with ultrasharp silicon tips. However, power efficiency was improved due to better heat transfer through the nanotube. Data could be written at heater temperatures of 100 K lower than comparable silicon tips.

Since tip wear is reduced by applying less force to the tip, the probe design was modified so that the spring constant reduced to 0.05 N/m. As a result, during read actions the probe applies very little force to the tip. During write actions this force can be electrostatically increased up to 1 μN by means of a capacitive platform at a potential of 20 V. With the new polymer media developed at that time, densities up to 1 Tb/in^2 could be reached. Using this array a read/write demonstration at 641 Gb/in^2 was given, following the stringent rules of the hard disk industry [145]. Figure 3.8 displays a SEM image of the thermo-mechanical probe used in the demonstration.

An impressively tight integration of the probe array with CMOS was demonstrated by Despont et al. [37]. In this method only the integrated cantilevers are transferred to the CMOS chip, and the MEMS carrier wafer is first ground and then etched away. As many as 300 high electrical copper interconnects of 5 μm were realized per square millimeter. An array of 4096 probes with outer dimensions of 6.4×6.4 mm was realized (Fig. 3.9), and the interconnects had a yield of 99.9%.

The work of IBM triggered the interest of other companies. For heated tip writing on piezoelectric and phase change media, researchers at LG Electronics in Korea realized a small array of thermal probes [101]. Heater platforms were integrated in boron-doped silicon by realizing a constriction at the cantilever end and covering

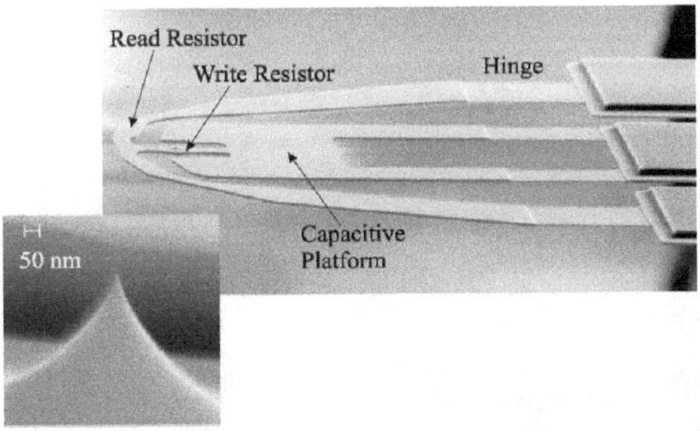

Fig. 3.8 SEM image of the three-terminal thermo-mechanical probe which was used to perform the read/write demonstration at 641 Gb/in^2. During read operation the read resistor is heated to 200–250°C, during write operation the write resistor is heated to approximately 400°C. The inset shows an enlarged view of the sharp silicon tip which is located on the write resistor [145]

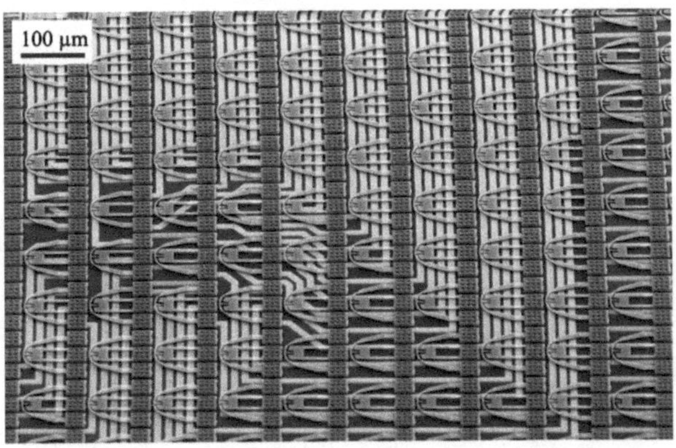

Fig. 3.9 Scanning Electron Microscopy image of part of a 4096 cantilever array, interconnected to a wiring wafer [37]

the cantilever legs with gold. Conductive tungsten tips were grown by focused ion beam deposition.

In the next generation, readout was added by integrating piezoelectric PZT layers on the cantilever legs [100]. Feature heights of 30 nm could easily be distinguished. The array was extended to a size of 128×128 probes and sensitivity improved to 20 nm [122]. A wafer transfer method was developed for a 34×34 array [86, 87], very much along the lines of the IBM process. Rather than silicon, 300 nm thick silicon nitride probes were used with polysilicon integrated heaters. The spring constant was still relatively high (1 N/m). Sharp tips were realized by KOH etching of pits into the silicon wafer and subsequent filling with silicon nitride, enabling bit dimensions of 65 nm.

Researchers at the Shanghai Institute of Microsystems and Information Technology have realized a small cantilever array, with integrated heater tips and piezoelectric deflection detection [188]. These arrays have been used to characterize wear of polymer recording media as a function of tip temperature and radius [8].

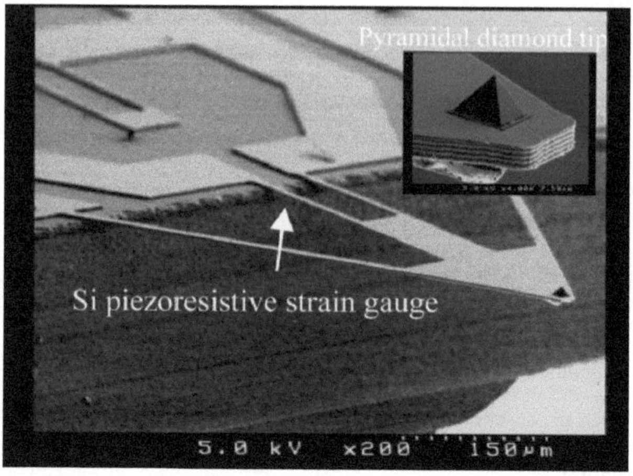

Fig. 3.10 Diamond probe with silicon-based piezoresistive strain gauge [175]

At the Data Storage Institute in Singapore, Chong et al. realized 20×1 and 15×2 arrays, using a fabrication technique along the lines of the early IBM work [31]. The scanning concept is, however, different, featuring a large stroke actuator in one direction. The total storage area can thus be much larger than the size of the array [187]. The 90 μm long and 1 μm thick cantilevers of the array had a spring constant of 1 N/m and resonance frequency of 164 kHz. The tip radius was rather large (220 nm), resulting in bit dimensions in the order of 600 nm. An interesting experiment was performed where the temperature of the heater platform was monitored by means of an infrared camera. Cooling times of 2 μs were measured this way.

Researchers at Pioneer and Tohoku University in Japan investigated probes with diamond tips and integrated piezoresistive sensors for ferro-electric data storage [175, 176], see Fig. 3.10. The boron-implanted silicon piezoresistive Wheatstone bridge was very sensitive (1×10^{-7}/nm). In contrast, the diamond probes had relatively poor radius of curvature. Attempts were made to replace the diamond tip by metal versions, and it was demonstrated that ruthenium tips perform relatively well [177].

Tip Wear

Tip wear poses one of the largest problems for a probe storage system. To get a feel for the extent of the problem a description of the wear issue in thermo-mechanical storage is taken from [99]. Consider a system operating at 1 Tb/in^2 and a data rate of 100 kbps per probe. With the data storage industry lifetime standard of 10 years and continuous operation of the device each probe slides a distance of 10–100 km. This translates to a maximum tolerable wear rate on the order of 1 atom per 10 m sliding distance in order to maintain the 1 Tb/in^2 density. When operated in normal contact mode on a polymer medium, a silicon tip loaded at 5 nN wears down in 750 m, sliding to a bluntness that corresponds to data densities of 100 Gb/in^2.

A first estimate of the tip–sample force threshold at which wear starts to become an issue is reported by Mamin et al. [111]. A loading force of 100 nN is mentioned to maintain reliable operation for the relatively large-sized indentations (100 nm) described in this early work. Such a force is detrimental for any reported probe–medium combination when densities above 1 Tb/in^2 are targeted. Strategies to reduce tip wear include hardening of the tip, softening of the medium, and modulation of the tip–sample forces. One of the first attempts to reduce tip wear is the inclusion of a photoresist layer of 70 nm in between the silicon substrate and the storage medium (PMMA) [19]. Several more measures to reduce tip wear from the medium side have been taken, see section "Recording Medium" for details. Another approach to reduce tip wear is hardening of the probe. Coating the tip with a hard material or molding a tip leads, however, to larger tip radii. For thermo-mechanical storage silicon is therefore preferred [99]. Usage of carbon nanotube tips has been reported for Tb/in^2 storage. In this case an AFM is used to store data by anodically oxidizing titanium [32]. Tips for phase change recording have been successfully made more wear resistant by changing the fabrication material. By first depositing platinum on a silicon tip and then performing an anneal step a hard layer of PtSi is obtained. A second measure to strengthen the tip is encapsulating the conductive PtSi tip with siliconoxide. The loading pressure on the tip apex is now decreased due to the increase of the tip area. The resolution of storage is, however, still determined by the conductive PtSi core [13–15].

A third way of reducing the tip wear is the modulation of the tip–sample contact. It is known from AFM that the intermittent-contact mode of operation reduces tip wear and this is one of the foremost reasons that intermittent contact is preferred over contact mode in many microscopy environments. Application of intermittent-contact mode is not very straight forward for probe storage. There are many requirements on the probes and some of them conflict with the requirements for intermittent contact, e.g. a high stiffness cantilever required for TM-AFM conflicts with the feeble cantilever used in thermo-mechanical storage to allow easy electrostatic actuation. In [153] a solution is presented by using an amplitude modulation of the cantilever by electrostatic actuation, despite high nonlinearity in the cantilever response. The authors show successful read and write operation at 1 Tb/in^2 densities after having scanned 140 m. An impressive reduction of tip wear is demonstrated by Lantz et al. [99] by application of a 500 kHz sinusoidal voltage between the sample and the substrate. The wear over a sliding distance of 750 m is reduced to below the detection limit of the used setup.

Parallel Readout

The massive parallelism of the probe arrays, which is requisite for obtaining data rates comparable to magnetic hard disk heads, poses a large challenge in probe-based data storage. Several thousands of probes have to be simultaneously addressed. The main functions of each probe are positioning, reading, and writing. Positioning, as described in the section "Positioning Systems", is in most cases done by moving the complete array or the storage medium in plane and parallel to the probe array, simplifying the task of each probe. Scanning the medium has two distinct advantages over scanning the probes. To obtain desired read and write speeds arrays have to scan at considerable rate, thereby inducing high-frequency vibrations that create unwanted cantilever movement. Preventing the occurrence of these vibrations is a major extra challenge for any control loop. Second, electrical connections to the probes can more easily be realized, because the probes are not moving with respect to the read-channel electronics. Researchers at the Data Storage Institute show a solution where the coarse positioning has a flexible wire to the readout electronics, though also in this design the fine positioning is directly connected to the cantilever array [187].

Movement in the z-direction, where z is defined as the direction normal to the medium, can be done on a per-array basis instead of a per-probe basis. This hugely simplifies the control required to operate an array of thousands of probes. On the other hand the fabrication tolerances of the array and medium have to be such that every probe in the array is in the appropriate tip-medium distance range. A too large tip–sample separation results in a failure when an attempt to write or read a bit is done. The other extreme leads to a probe that is pushed into the medium with considerable force (depending on the spring constant of the cantilever) leading to excessive tip wear. Without independent z motion these demands on the medium and probe array increase tremendously. The technically most mature probe storage system, described in [135], is based on thermo-mechanical storage and features a 64×64 array where 32 levers are active. By determining the electrostatic pull-in voltage for each cantilever the initial tip–sample separation is calculated to have a standard deviation of 180 nm. With a cantilever spring constant of 1 N/m this would lead to a maximum additional loading force of 180 nN.

The read and write operation requires the independent addressing of each probe. Traditionally AFM probes are monitored by an optical readout system of which the optical beam deflection technique [116, 117] is the most widely used. Although optical readout has been demonstrated for arrays of cantilevers [5, 95, 173] none of the readout schemes has been implemented in probe storage. Optical readout alone would, however, not suffice. The probes have to be actuated for the write operation. Wired schemes are implemented to achieve this. Wireless schemes and passing signals through the storage medium have also been proposed [1] but are not yet realized.

Integration Challenges

Probe storage systems face the same integration challenges as other MEMS such as inkjet print heads, accelerometers, and gyroscopes: the monolithic integration of the micromechanical portion of the system – actuators and probes – and the CMOS electronics. Such integration brings benefits to the overall performance of the system, reducing the parasitics in critical areas such as position sensing from capacitive structures. Monolithic integration is fundamental to enable economically viable mass volume fabrication of probe storage systems.

What can impede monolithic integration is the nature of the different materials and processing techniques applied on the same substrate. For example, processing steps with temperatures greater than 400–450°C cannot be performed on CMOS structures because they would compromise the aluminum interconnect layers.

Wiring solutions are based on a time-multiplexing scheme to address the array row by row [36, 87] much like is usually done in DRAM. With a growing number of probes the maximum current passed through a row or signal line grows. For higher number of probes we must use an electrically stable wiring material capable of carrying high current densities and also having a low resistivity. Despont and coworkers have used a two-level wiring of both gold and nickel in the 32×32 array that is reported in [36]. Schottky diodes formed by doped silicon/nickel interfaces were introduced to avoid cross talk between probes. Bondpads are used for connection to the outside world, as shown in Fig. 3.11. In the 64×64 array, reported in [37], the number of connections has increased to three per probe. In this case a second wafer with the CMOS circuitry is used to which the probes are bonded, as is the case in [87]. An alternative integration with CMOS is a single wafer process described in [167]. Here first the CMOS circuitry is created on top of the last metalization by starting with a chemical mechanical polishing (CMP) step, see Fig. 3.12. Next an insulating layer of 400 nm SiC and a sacrificial oxide layer of 3 μm are

Fig. 3.11 Photograph of a 32×32 array (14×7 mm^2). The probes are interconnected by a 32×32 row/column addressing scheme. A total of 128 bondpads provide connection to the outside world [36]

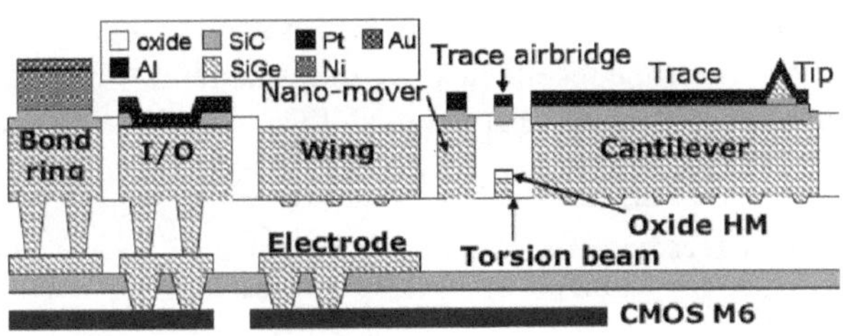

Fig. 3.12 Schematic drawing of a cross section showing the integration of a cantilever array with CMOS circuitry based on a one wafer design [167]

deposited. A structural SiGe of 3 μm is later on deposited to allow definition of the cantilevers. Consequently the cantilevers are etched and the sacrificial oxide is removed to allow cantilever actuation. Others are still working on the integration of the cantilever array with the readout electronics [122, 187].

Positioning Systems

In order to access data at different positions, a positioning system is required. The positioning system should be able to move in the two horizontal directions (x and y) with nanometer precision. Actuation in the vertical z-direction is usually not present in the positioning system, as it is generally part of the probe (array) design. Although arguments can be mounted in favor of having a separate positioning system for each probe, all systems are designed such that there is a single moving platform for the complete device. Most designs feature a stationary probe array and a moving medium sled (scan table)[1], to prevent exciting unwanted vibrations in the probes during servoing and (perhaps more important) to circumvent the problem of how to route many wires to a moving suspended probe array (see also section "Parallel Readout"). The separation of probe array and storage medium with positioning system greatly simplifies the design; the wafers with the probe array and positioning system can be fabricated separately and bonded together afterward.

The movement trajectory usually follows a raster; defining a fast axis x and a slow axis y, the medium sled is scanned back and forth in the x-direction, while making steps in the y-direction. A case can be made to use an outward spiral trajectory similar to a Compact Disc (CD) [91, 109]; however, here we assume a rectangular scan raster that is used in all probe storage systems so far. This means a triangular positioning waveform in the x-direction and a staircase waveform in the y-direction.

Requirements

To fully utilize the storage medium, the displacement range should be at least the probe pitch, currently on the order of 100 μm (see section "Probe Arrays and Parallel Readout"), although this may decrease in the future. If overlapping probe fields are desired, the displacement range should be larger. The required positioning accuracy is on the nanometer scale, depending on the bit size (3–30 nm, see section "Recording Medium") and the allowable bit error rate [145, 164]. For example, Pozidis et al. from IBM showed that at 641 Gbit/in^2, a 2 nm displacement from the track centerline results in a five times larger raw bit error rate. For the 64 × 64 probe array described by Despont et al. [37], a medium sled size of at least 6.4 × 6.4 mm^2

[1] A notable exception is a design by Yang et al. from the Data Storage Institute in Singapore, where the probe array is moved by a one-dimensional linear motor [187].

is required. The areal efficiency is defined as the ratio between medium sled size and total device size.

The performance in terms of access speed is determined by the actuator's maximum acceleration (maximum force divided by mass), sensor bandwidth, and control circuit. The maximum acceleration should be large, not only for faster seek access, but also so that it is possible to quickly reverse motion at the end of each scan line. Because a higher read/write speed will lead to a longer turn-around time, there is an optimal read/write speed and a corresponding optimal data rate [23]. A higher acceleration means a higher continuous data rate can be reached, either by increasing the read speed or by decreasing the turn-around time. Increasing the acceleration capability by reducing the medium sled mass must be done with care, so that the sled is still mechanically rigid enough for reliable probe reading and writing. Increasing the maximum acceleration by increasing the maximum force is limited to what the power budget allows and what is physically possible without damaging the device (e.g., breakdown voltage or electromigration limit).

To perform seek operations or compensate for residual shock forces, the actuation force at any given position must exceed the suspension spring restoring force. The available force equals the actuator force minus the suspension spring force and generally depends on the position of the actuator. Usually, the available force is large for small displacements and smaller for large displacements [47]. The minimum value of the available force is an important measure of how shock-resistant the design is and of how fast the actuator can accelerate. Especially for mobile applications, shock resistance is important [98].

Finally, there may be a certain power limit for the complete positioning system. Although certain types of actuators require very little power for themselves (and seem attractive), their driving or control circuitry may require a lot of power, leading to a high overall energy consumption. For a fair comparison of required power, one should view the positioning system as a whole.

Actuators

Bell et al. [10] have written an extensive review of macro and MEMS actuators and sensors, comparing different kinds with respect to maximum displacement, maximum force, resolution, and resonance frequency. Hubbard et al. [74] published a review of actuators specifically for micro- and nanopositioners. Both articles show that several physical actuation principles may be used in a probe storage device. Not surprisingly then, different types of scanners that have been designed specifically for probe storage are found in the literature. It is yet unclear which actuation type is best suited for probe storage. Research focuses on MEMS actuators because they have a large ratio of in-plane motion range to device volume compared to macro actuators. Agarwal et al. [2] show a comparison in terms of "specific work" (force times travel range divided by footprint); different types of MEMS actuators all score about 10 μN/mm, an order of magnitude lower than conventional milli-actuators.

It shows the challenging problem of creating MEMS actuators with large force, large displacement, and small footprint.

Electromagnetic Actuation

An advantage of electromagnetic scanners is their linearity (linear voltage/current vs. displacement curve), simplifying controller design. Another advantage for mobile probe storage is that an electromagnetic scanner, because it is current driven, can operate at the generally low available voltage of about 3.6 V. A disadvantage is that permanent magnets are needed. Assembling an electromagnetic scanner will therefore be more complicated than assembling, for instance, an electrostatic comb drive scanner. It also means that it is difficult to make the scanner very thin. The energy consumption of electromagnetic scanners is relatively large because of the relatively large required currents.

Figure 3.13 shows the evolution of IBM's electromagnetic scanner designs. In 2000, Rothuizen et al. from IBM reported their proof-of-concept electromagnetic scanner: a five degree of freedom *x*/*y*/*z* scanner (including tilt about the *x*-and *y*-axes) [149] fabricated from silicon and electroplated copper springs and coils. The scanner has a 2×2 cm^2 moving platform held within a 3×3 cm^2 outer frame. It can reach a displacement of 100 μm; however, the required power of about 200 mW is very high. An improved design [150] was reported 2 years later, fabricated from a 200 μm thick SU-8 layer. The design uses a similar coil/magnet configuration. It improves on power dissipation (3 mW at 100 μm displacement), on fabrication cost, and on compactness (the spring system is below the platform), and the spring system provides damping to suppress platform resonance during operation. Two years later, a radical change in design was reported [98, 133]. The new design is fabricated from a 400 μm thick silicon wafer by deep reactive-ion etching through the full thickness of the wafer, the design being an extrusion of a two-dimensional layout. It features a mass-balancing concept to render the system stiff for external shocks while being compliant for actuation, such that the power dissipation is low. The actuator and scan table masses are linked via a rotation point, enforcing their movement in mutually opposite directions: when the actuator moves up, the table moves down and vice versa. External shocks exert inertial forces on the actuator and scan table mass, but, because the directions of the inertial forces are equal, they cancel each other through the rotation point (the scan table and actuator cannot move in the same direction at the same time). Because the springs are 400 μm high (wafer thickness), the stiffness in the *z*-direction is large for passive shock rejection. Coils and magnets are glued manually onto the device. The actuator generates a force of 62 μN/mA. Its power usage at 50 μm displacement is 60 mW (80 mA current); this has been improved to about 7 mA and 2 mW power.[2] The medium sled is 6.8×6.8 mm^2, while the complete device is 16×16 mm^2; the areal efficiencyis about 25% and has decreased dramatically in

[2] Private communication with Mark A. Lantz, IBM Zürich.

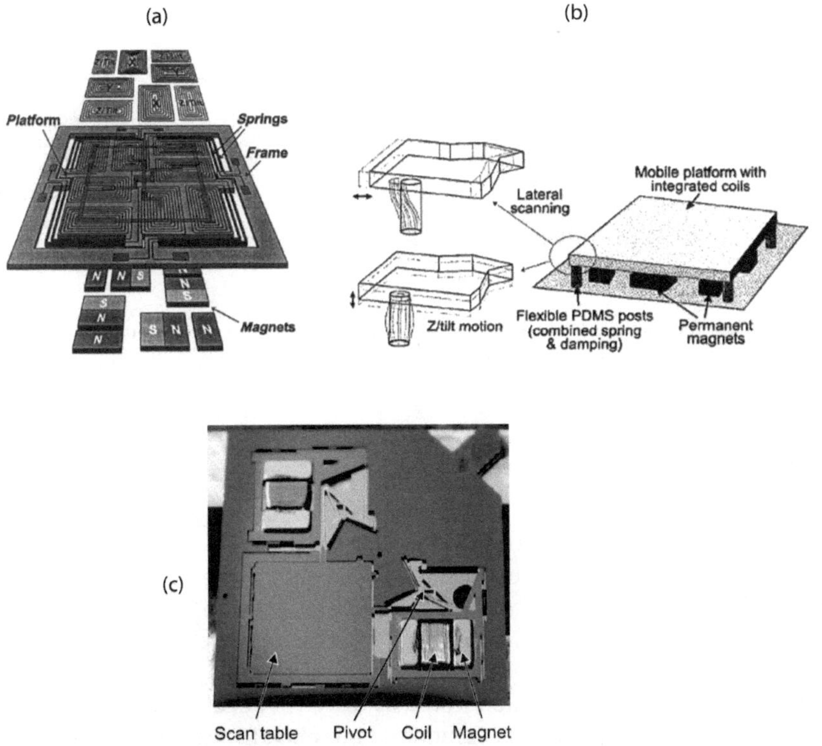

Fig. 3.13 Evolution of IBM's electromagnetic scanner. **(a)** Scan table is 2×2 cm^2 large and has five DOF (2000) [149]; **(b)** Fabricated from SU-8, 1.5×1.5 cm^2 scan table (2002) [150]; **(c)** Mass balanced and *x*/*y* only, 6.8×6.8 mm^2 scan table (2004) [133]

comparison to the earlier designs. The in-plane resonance frequencies lie around 150 Hz; the first out-of-plane resonance frequency lies an order of magnitude higher.

Another electromagnetic scanner was reported in 2001 by Choi et al. from Samsung [29, 30], see Fig. 3.14. The scanner is fabricated from silicon; the coils are made by filling high aspect ratio silicon trenches. The medium sled size equals 5 × 5 mm^2. The displacement of 13 μm at 80 mA is smaller than the scanner by IBM; however, this was measured without top magnets and yokes planned in the design. The measured in-plane resonances are 325 (translational) and 610 Hz (rotational); FEM simulations indicate that the first out-of-plane mode lies at 2160 Hz [29].

A plastic electromagnetic scanner is described by Huang et al. from Seagate; however, not much design and fabrication details are available [73]. The scanner has three DOF: *x*/*y* translation and rotation about the *z*-axis. The ±150 μm displacement range is large, but the resonance frequency is only 70 Hz. The 1% cross talk between the *x*- and *y*-axes is equal to the cross talk in IBM's scanner [98].

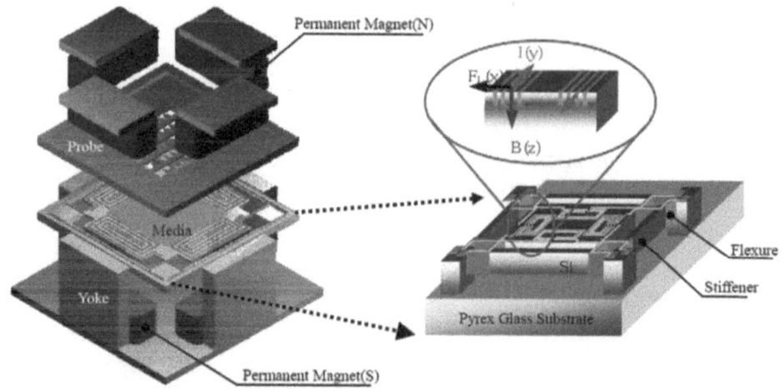

Fig. 3.14 The electromagnetic scanner by Samsung [29]. The size of the total device and scan table is 13×13 and 5×5 mm^2, respectively

Electrostatic Comb Drives

Because of scaling, electrostatic force becomes relatively large at small dimensions. Using the electrostatic force in MEMS devices is therefore attractive, and several different types of electrostatic scanners have been published. The electrostatic comb drive [180] is a popular choice for MEMS actuators, because its design and fabrication are relatively simple and straightforward. Its force, however, is small and many fingers are needed to generate sufficient force, reducing its areal efficiency. Other electrostatic actuators like the dipole surface drive [69] and shuffle drive [181] can generate more force but are more complicated to manufacture, and control is more difficult. Because the electrostatic force is proportional to the square of the applied voltage, the polarity of the applied voltage is unimportant and the force is always attractive. In order to move a stage in both positive and negative directions, two or more electrostatic actuators will be required.[3]

The maximum force of an electrostatic actuator depends on its breakdown voltage (depending on the minimum gap size) and the maximum available voltage. The breakdown voltage is generally quite high (>70 V) and is usually higher than the available voltage.

The main advantage of comb drive actuators is their ease of fabrication. Usually, no assembling is required in contrast to electromagnetic scanners, and the device can be made thinner than an electromagnetic scanner. Although a comb drive is a nonlinear (usually quadratic) actuator, driving one is much simpler than driving other electrostatic actuators. The energy consumption of comb drives is very low; however, the energy consumption of the driving (DC/DC conversion) circuitry must be taken into account for a fair comparison with other scanners. Disadvantages include the (generally) high required voltage and the low areal efficiency due to

[3] A single dipole surface drive or shuffle drive can move in both positive and negative directions. In the sense meant here, they should be thought of as compound actuators consisting of three or more small actuators.

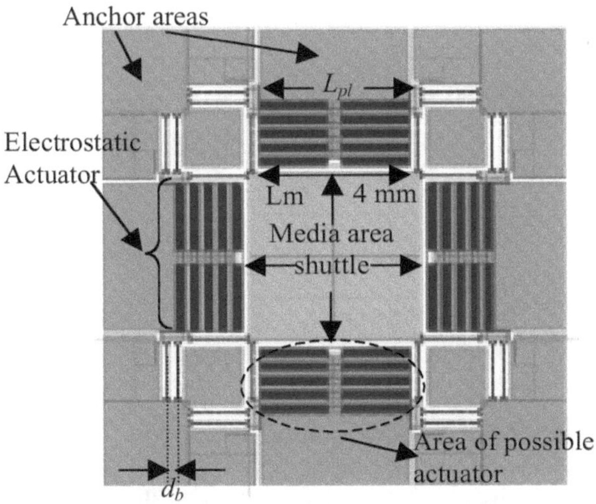

Fig. 3.15 The topology used by Alfaro and Fedder from Carnegie Mellon to find a scanner design with optimal footprint for the required 50 μm stroke [3]

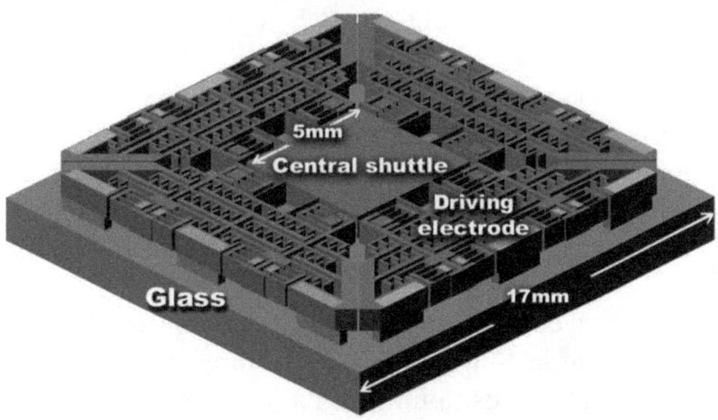

Fig. 3.16 Simplified layout of the micro XY-stage by Kim et al. [83]. The *top* layer is 48 μm thick

the large area needed for comb fingers to generate sufficient force. The force output is highly dependent on the minimum gap size that can be fabricated; reduced gap sizes due to improved fabrication methods will result in stronger comb drive actuators.

Already in 1992, a nanopositioner with integrated tip was published by Yao et al. [189]. It featured electrostatic parallel-plate (gap-closing) actuators, moving the probe instead of the "medium". However, its displacement range of 200 nm at 55 V is very limited.

In 2000, Carley et al. from Carnegie Mellon University described a comb drive actuated scanner for probe storage, which was originally designed for a vibratory-rate gyroscope [22, 23]. The scanner reaches 50 μm displacement at 120 V; it

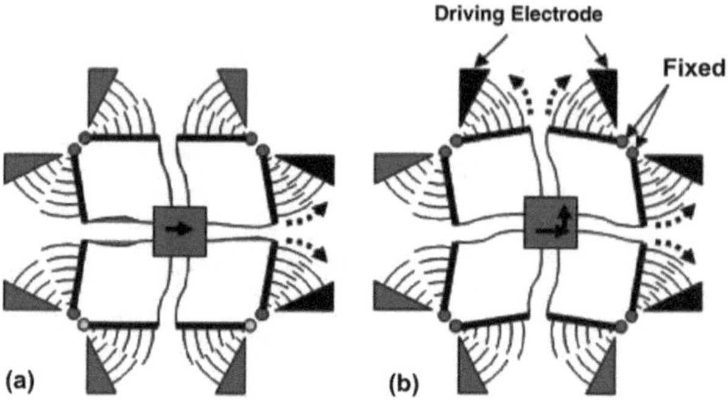

Fig. 3.17 Simplified model of the actuation principle by Kwon et al. [94].
(a) Movement to the right by actuating the rotational comb drives on the *right*;
(b) diagonal movement by actuating the comb drives on the *right* and *top*

contains 800 fingers in total with 500 μm height and 16 μm gap. Alfaro and Fedder from Carnegie Mellon University further improved this design using parametric optimization to optimize the footprint for ±50 μm stroke keeping the topology fixed [3], see Fig. 3.15.

The comb drive scanner reported by Kim et al. uses 27,552 3 μm thick fingers with 48 μm height and 3 μm gap [83], see Fig. 3.16. The scan table is 5×5 mm^2 large and 18 μm static displacement is reached at 13.5 V. Because of the large amount of fingers, the force is large and the required voltage is low; however, it also leads to a low areal efficiency of approximately 11%.

Kwon et al. published an x/y-scanner for optical application, but the design could also be applied to probe storage [94]. The design uses "L"-shaped suspension springs and rotational comb drives (Fig. 3.17). Its displacement range is 55 μm at 40 V. It is unclear whether the suspension is stiff enough for use in probe storage.

The somewhat unconventional positioning system design by Yang et al. from the Data Storage Institute in Singapore [187] features an electrostatic comb drive *x*/*y* scanner [104] for precise positioning and a one-dimensional miniature electromagnetic linear motor for course positioning of the one-dimensional probe array (Fig. 3.18). The comb drive scanner features a scan table of 5×5 mm^2; the areal efficiency is 25%. Simulations indicate static displacement of 20 μm at 55 V, but no measurement results are shown.

All the comb drive scanners mentioned above directly link the actuators to the scan table. This means that in-plane shocks have to be actively compensated for by the actuators. Especially at large displacements, when the available force is low, this leads to a low shock rejection capability. Engelen et al. designed a shock-resistant electrostatic scanner, by replacing the electromagnetic actuators in the mass-balanced scanner by IBM discussed earlier with comb drives [47, 48] (compare Figs. 3.13c and 3.19). In order to increase the comb drive force, modified comb finger shapes are used that increase the force at large displacements, both increasing the maximum stroke and the shock resistance without increasing the maximum voltage. A displacement of 48 μm was reached at 120 V, corresponding to a force of 3.5 mN; the electromagnetic scanner generates 6 mN at 100 mA current.

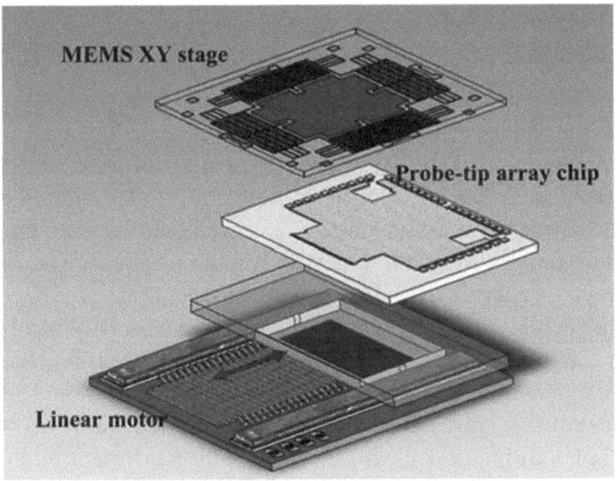

Fig. 3.18 A schematic of the design by Yang et al. from the Data Storage Institute in Singapore, featuring a comb drive XY-stage that carries the storage medium and beneath it a miniature linear motor with a one-dimensional probe array [187]. The MEMS stage's size is 1×1 cm^2

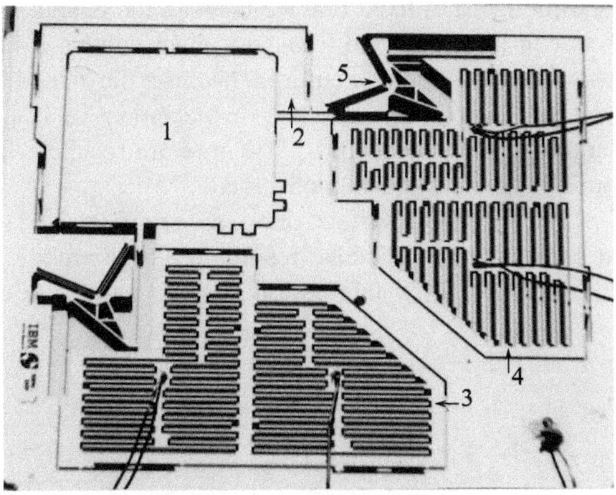

Fig. 3.19 Photograph of a fabricated device (2×2 cm^2). 1: scan table, 2: C-bracket, 3: x-comb drives, 4: y-comb drives, 5: pivoting element [47]. Compare with Fig. 3.13c

The electrostatic scanner designs by Sasaki et al. use mass balancing for *internal* shock force rejection [156]. Inertia forces due to fast acceleration in the y-direction may influence the comb drive gap in the x-direction, which may lead to instability in the x-direction comb drives. To cancel these inertia forces in the y-direction, the scan table is split into two plates of equal mass that are actuated in mutually opposite directions. However, this opposite movement is not mechanically enforced; therefore inertial forces due to external shocks or vibrations are not canceled out. The best of the three investigated designs reached static displacements of 110 μm at 70 V and 90 μm at 125 V for the x- and y-directions, respectively.

Electrostatic Stepper Motors

The dipole surface drive [70] has the advantage of potentially providing much force at low voltages, due to the gap size that is not limited by lithography or deep reactive-ion etching limits. Another important advantage is the high areal efficiency. Disadvantages include the more complex fabrication process and drive circuitry and the significant out-of-plane motion at high forces. To make the design more shock resistant, perhaps a mass-balancing scheme could be used similar to IBM's electromagnetic scanner [98] without decreasing the areal efficiency.

In the "dipole surface drive" design by Hoen et al. [2, 70] from Agilent Laboratories, the actuator is placed under the scan table, greatly increasing the areal efficiency over comb drive designs. Figure 3.20 shows a cross section of this dipole surface drive design, which is similar to a magnetic stepper motor. The bottom of the scan table (translator) features electrode strips, while the stator (to which the scan table wafer is bonded) features electrode strips with a different pitch. By changing the voltages on the electrodes in a special pattern, a force is exerted on the scan table and the scan table will move in small steps of 400 nm determined by the difference in electrode pitch between stator and translator. Smaller displacements are made by adjusting the voltage on one of the electrodes. Operation of the device is very different from a comb drive. Instead of increasing the voltage, here the voltage pattern is shifted with a fixed bias voltage. In the design reported by Agarwal [2], the gap between the stator and translator electrodes is 2.4 μm. A high out-of-plane to in-plane stiffness ratio is required, because the available force is limited by the snap-in voltage (48 V in this case). A displacement of 17 μm was reached with 30 V bias; larger displacements up to 70 μm were reached; however, in that case significant out-of-plane motion was observed.

Like the dipole surface drive, the shuffle drive [181] too can provide much force at low voltages. This also results in a high areal efficiency. Disadvantages include the more complex fabrication process and drive circuitry. Because the device is

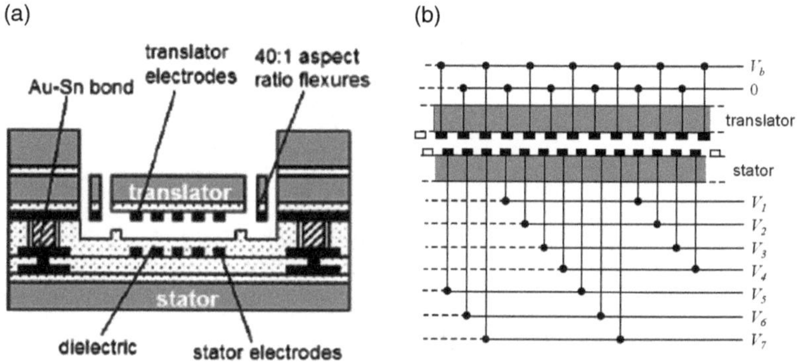

Fig. 3.20 The dipole surface drive actuator by Agilent [2]. **(a)** Schematic cross section; **(b)** wiring diagram of the translator and stator electrodes

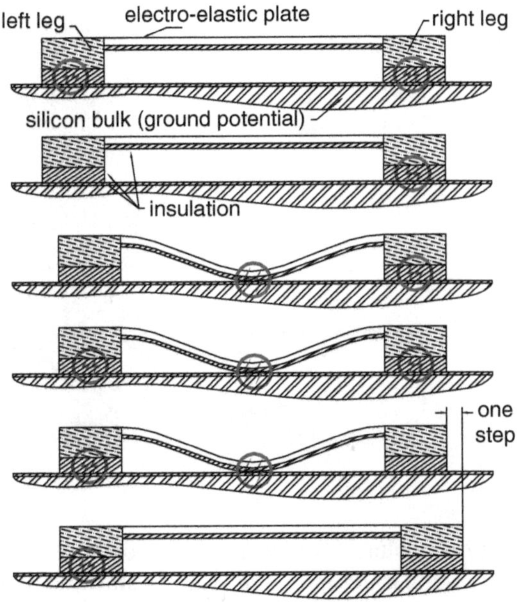

Fig. 3.21 Step sequence of the shuffle motor [140]. The *circles* indicate that a voltage is applied on the corresponding leg or plate

electrostatically clamped to the base plate, it is inherently shock resistant. But when the shock force exceeds the stiction force, there is no way to compensate for it.

The shuffle drive by Tas et al. [181] from the University of Twente is an electrostatic stepper motor. Figure 3.21 shows how the motor makes a step. The step size ranges from 0.6 nm up to 100 nm and is limited to a smaller range for a particular device. The actuator's output force is determined by how much plate deformation is obtained when applying a voltage on the plate. The force is large because the plate is pulled downward like a parallel-plate gap-closing actuator and because the plate acts as a mechanical lever. Sarajlic et al. describe a shuffle motor $200 \times 1500\ \mu m^2$ with a 70 μm displacement range and an output force of 0.45 mN at driving voltages of 65 and 150 V for the plate and clamps, respectively [155]. A $482 \times 482\ \mu m^2$ one-dimensional shuffle motor reached displacements of 60 μm (corresponding to 0.64 mN force) at driving voltages of 45 and 36 V for the plate and clamps, respectively, only being limited by the design layout [154]. Unfortunately, several problems arise due to stiction and friction of the motor's legs [139].

Piezoelectric Actuation

Piezoelectric actuation is commonly used for scanning probe setups. However, common piezo-elements need to be quite large to provide the required displacement range for probe storage. An advantage of using piezo-actuators is that their force is very large and that the areal efficiency can be very high when mechanical amplification is used to increase the displacement range of small piezo-elements. The

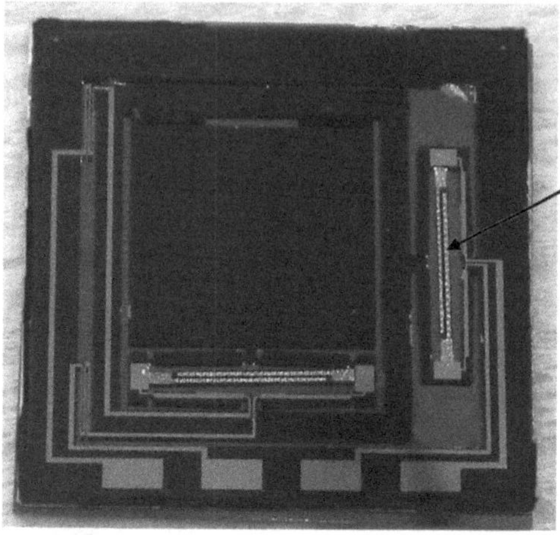

Fig. 3.22 Photograph of the silicon microstage to which PZT- stacked actuators have been attached [51]. The total device is 2×2 cm^2 large

required voltage is reasonable in comparison to electrostatic actuators. However, fabrication is complex and requires attaching the PZT actuators on the silicon stage.

A laser-cut scanning stage fabricated completely from PZT material is described by Zhang et al. [193]. The work continued and the improved design reaches 82 μm at an applied voltage of 70 V [52]. The scanner consists of a silicon MEMS stage, to which PZT actuators have been attached manually (Fig. 3.22). The scan table is 9.5×9.5 mm^2 large. The PZT actuators generate very large forces ($\sim$3 N) and their small displacement is mechanically amplified 20 times to move the scan table. The maximum force on the scan table of 150 mN is very large in comparison to other actuator types.

Sensors and Control

A variety of MEMS displacement sensor types are available for probe storage systems [10]. One of the challenges in a probe storage system is the need for nanometer resolution over 100 μm displacement range. To improve the resolution, a medium-derived error signal can be used [118, 165]. Several probe fields are designated as servo-fields and are written by the device itself with special patterns; these servo-fields are then used to obtain very precise information about the position of the probe array relative to the bit track.

Capacitive sensors using comb drives are an attractive solution when comb drives are already used for actuation. An example of a positioning system using comb drives for actuation and sensing is given by Cheung et al. from the University of California at Berkeley [24]. Pang et al. from DSI in Singapore propose to use the same comb drive for actuation and sensing; however, no measurement results are available yet [132]. Kuijpers [92] from the University of Twente has studied an incremental long-range capacitive displacement sensor based on the capacitance

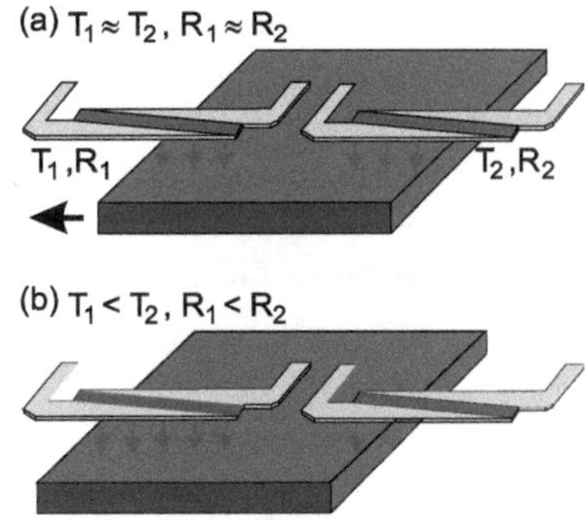

Fig. 3.23 IBM's differential displacement sensor setup using a pair of matched thermal sensors. The sensors are heated with equal power. A difference in overlap with the scan table leads to a difference in cooling between both sensors. This difference is measured in a difference in resistance [97]

change between two periodic structures, and obtained 10 nm resolution with, in principle, unlimited displacement range. However, the bandwidth of the sensor is severely limited by a low-pass filter at the end of the capacitance sensing circuit (a bandwidth of 5 Hz is used in reference [93]). Moreover, the output of the sensor is periodic, and extra circuitry is needed to count how many cycles have been measured.

IBM uses a differential thermal displacement sensor, shown in Fig. 3.23, measuring the change in resistance due to the temperature change when the overlap between heater and scan table changes [97]. It provides 2.1 nm resolution (10 kHz bandwidth) at 10 mW power and a maximum resolution of 0.49 nm at 32 mW. Less than 6 nm of drift over 100 min was measured. The required power makes it less attractive for use in a mobile device. IBM has performed extensive research on the control of their electromagnetic actuator [133, 134, 162, 163]. Different controller architectures have been investigated in order to optimize the time it takes to access a certain part of the medium (the seek time). Using 200 mA as maximum current for the electromagnetic scanner, the fastest seek time for 50 μm movement is on the order of 1.6 ms [134].

For the same reason a comb drive can be used as actuator and sensor, the electrode patterns in a dipole surface drive can also be used for sensing the position. Such a fine line capacitive position sensor integrated with a dipole surface drive is described by Hartwell et al. [63]. The resolution is 0.3 $\text{Å}/\sqrt{\text{Hz}}$. Lee et al. from Seagate use a similar capacitive sensor for their electromagnetic scanner [102]. Like the incremental capacitive sensor by Kuijpers et al., the output signal of both sensors is an ambiguous measure of the position, requiring extra circuitry that keeps count of the number of cycles that have been measured. For the same dipole surface drive Hartwell used, Agarwal et al. describe a capacitive sensor with large plates with a resolution of 0.5 $\text{Å}/\sqrt{\text{Hz}}$ [2]. Because of the large plates, the output signal is no longer periodic and the output signal is a direct measure of the displacement.

Open-loop and closed-loop control for a shuffle motor was investigated by Patrascu et al. [140]; unfortunately only a microscope with camera was available for position measurement, severely limiting the control bandwidth. The actuator position could be measured every 33 ms with 10 nm accuracy. A state-machine controller was designed to generate the correct voltage sequences for making a small or large step in positive or negative direction or standing still. If the measured positioning error is non-zero, the state machine generates correction signals every 33 ms in the form of a number of steps in one direction. Open-loop control resulted in a maximum positioning error of 140 nm. Closed-loop control was shown to work; however, an integrated sensor with much larger bandwidth than the video setup is needed to obtain a fair assessment.

Probe Storage Electronics and System Considerations

Probe storage devices require dedicated, on-board electronics to perform media writing, media reading, media positioning, and generate the data stream to interface the memory to computer systems through standard interfaces such as USB or SATA. Each electronic function depends on the chosen media, read/write mechanism, and MEMS actuator principle.

Due to the intrinsic parallel operation of a probe storage device, each electronic block has an area budget – for instance, the average read/write probe circuit cannot extend beyond the area occupied by a single probe – and a power budget. Specifically, the area budget A_{budget} can be computed as the area subtended by the full swing of the microactuator in both axes of operation $x_{\mathrm{actuation}}$ and $y_{\mathrm{actuation}}$. This is also referred to as the scanning area of a single probe. Probes are then stepped at intervals of the full actuation swing as depicted in Fig. 3.24.

The average power budget for each probe analog front end P_{afe} depends on the number of probes that need to be simultaneously activated. This in first approximation depends on the total memory module data rate $\mathrm{DR}_{\mathrm{tot}}$ divided by the data rate $\mathrm{DR}_{\mathrm{probe}}$ of each individual probe. Other system functions draw power during the memory operation, including the power $P_{\mathrm{actuation}}$ to drive the scanning platform, the power for the channel electronics P_{channel} and the power $P_{\mathrm{controller}}$ required by the controller, DC–DC conversion, and interface electronics. The sum of all these powers cannot exceed the total power P_{total} set by the memory specification that is determined by the market segment targeted by the memory, for example, the Universal Serial Bus standard. Thus the total power can be represented as

$$P_{\mathrm{total}} = P_{\mathrm{actuation}} + P_{\mathrm{channel}} + P_{\mathrm{controller}} + P_{\mathrm{afe}} \cdot \left(\mathrm{DR}_{\mathrm{tot}}/\mathrm{DR}_{\mathrm{probe}}\right)$$

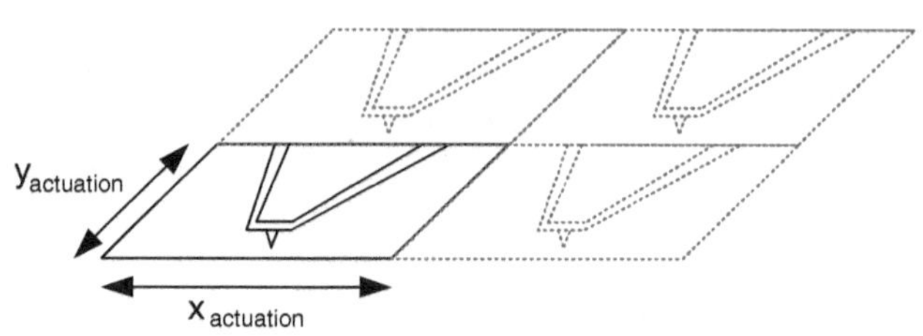

Fig. 3.24 The area budget of the analog front end

Finally, there are cost budgets that determine the technology node (also known as the minimum feature size or technology node) available to the circuit designers. More expensive deep sub-micron processes will indeed provide lower power and smaller area designs. However, using such expensive processes will defeat the cost benefits of probe storage architectures. The technology node should therefore be several generations behind the technology used in competing NAND flash for probe storage to be economically viable, using depreciated manufacturing fabs and overcoming the need of the huge investments for lithographically driven memories as described in the section "Introduction".

Analog Front End

Read/write electronics circuit architectures depend on the read/write mechanism that in turn depends on the media type as seen in the section "Recording Medium". While substantial research work has been carried out to develop different media and read/write mechanisms, only integrated electronic circuit prototypes of the analog front end (AFE) portions have been developed alongside them.

The read/write electronics examples presented in this chapter are specific to thermo-mechanical media. When utilizing this media, typically a polymer, the read/write AFM system needs to reliably form and read back the presence of an indent (representing the bit) by running a current through the selected cantilever. The micromechanical cantilever is modeled in the mechanical, thermal, and electrical domains as a transducer during writing and as a sensor during reading. The block diagram of the read/write channel is described in Fig. 3.25. One of three modes will occur: inactive, write, or read.

In the inactive mode the cantilever is kept at the same potential as the media to avoid accidental pull-in of the cantilever to the surface.

During writing on a PMMA polymer media an indentation is created by applying a current pulse flowing through the resistive portion of the cantilever, heating the

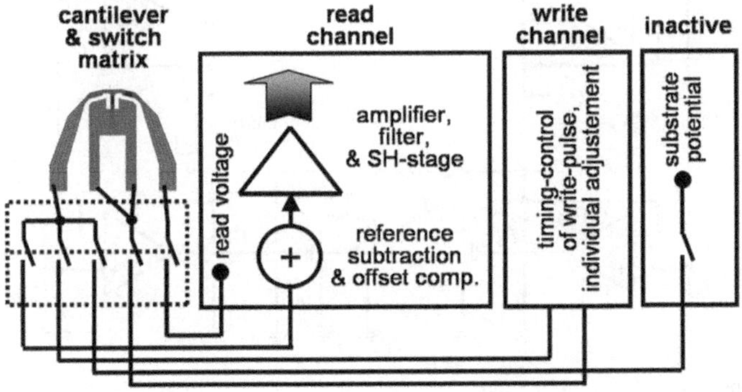

Fig. 3.25 Block diagram of the read/write channel

tip to ≈400°C through the Joule effect while simultaneously applying a capacitive force to move the cantilever toward the medium [90]. The sensitivity and noise of the cantilever depend on the power dissipated in the heating element.

During readback a lower current is run through the cantilever in order to raise its temperature to only 200–250°C. This temperature is not enough to deform the media and create another indentation. The resistivity of the cantilever varies according to the thermal conductance between the cantilever and the medium. There will be more heat dissipated out of the cantilever when the tip is in an indent, causing the temperature of the cantilever to be lower than when the tip is not in an indent. A lower temperature corresponds to a lower cantilever electrical resistance. An analog front end therefore needs to be able to discriminate between a signal varying between two values of resistance, ΔR, compared to the baseline resistivity, R. The signal-to-noise ratio, or $\Delta R/R$, is less than 10^{-3} [61].

The read/write time constant of the heaters is on the order of a few microseconds to 10μs. Due to this relatively low data rate, several hundred channels need to operate in parallel to achieve data rates suitable for mass storage applications, such as the 40 Mbit/s required for blue-ray video streaming or the 480 Mbit/s maximum achievable speed of the USB 2.0 High Speed protocol.

An example of the input stage schematic is represented in Fig. 3.26 where the cantilever is supplied with a differential voltage from two operational amplifiers OP1 and OP2. The current proportional to the signal of the indentation I_{signal} is obtained by subtracting the reference current I_{ref} from the total current in the cantilever. This is done to remove the offset from the readback waveform to avoid the need of high-resolution analog-to-digital conversion. The reference current is generated by biasing a reference element – effectively an unused cantilever – with the same biasing voltage as the cantilever used for readout. Not more than a few hundred microwatts can be budgeted for each probe input stage and the power consumption is dominated by the power consumed in the current path through the cantilever. Another adder and an integrator are added to the path to eliminate the residual DC

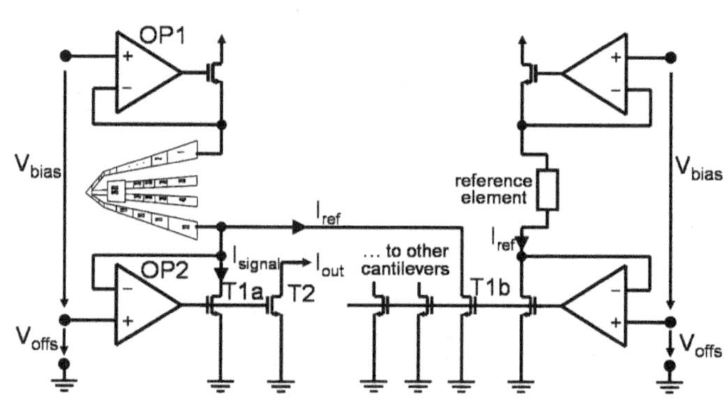

Fig. 3.26 Schematic of the input stage

component. Hagleitner et al. show how, through modeling of the thermoelectric behavior of the micromachined scanning probes and the electrical simulation of the analog front end, an optimum trade-off can be reached between the heating power, the biasing current, and the signal bandwidth.

Probe Channel Electronics

The challenges of read channel designs for probe storage devices are due to the parallel operation of probe read heads as well as novel physical read/write mechanisms that are different from traditional magnetic recording. A typical channel is represented in Fig. 3.27. The area and power budget of the channel section depend on the number of AFE blocks that are shared by the same channel.

Low-power, small-area ADC sections using $\Sigma\Delta$ modulators have been studied in [20] achieving a resolution of 7 bits for AFE signal bandwidths up to 50 kHz with a power consumption of only 14 μW.

The analog-to-digital converter samples the input signal from the AFE at a frequency f_s that is higher than the symbol rate f_{afe}. The ADC output is synchronized to the symbol through a timing recovery loop – a PLL that synchronizes to the zero crossings of the derivative of the readback signal – and an interpolator. These two blocks are represented as SRC (Sampling Rate Converter). The resulting signal is passed to a digital filter that further eliminates residual DC fluctuations and a threshold detector to generate the digital output data stream.

The parallel operation of probe storage presents unique challenges for the channel circuit design: the system can be considered as a multiple-input multiple-output channel (also known as MIMO). Just like in multiuser wireless communication systems, decoding algorithms can potentially exploit the interactions among the input and output signals from multiple tips to improve system performance. However, area and power budgets limit the complexity of each analog front end. Section "Error Correction Coding for Global Jitter" describes where the MIMO parallelism is used to design an optimal detector for global jitter.

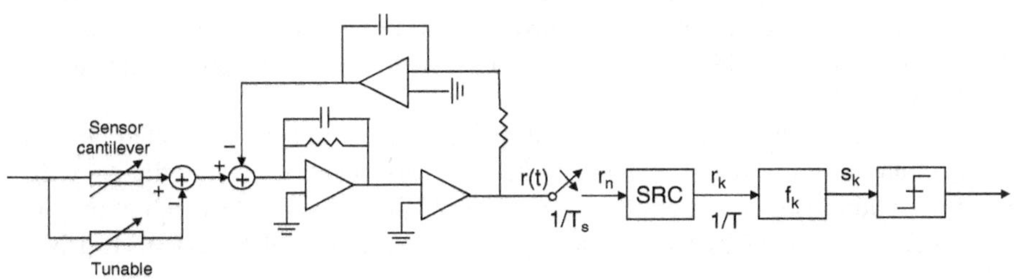

Fig. 3.27 Block diagram of a probe storage channel

Channel algorithms need to be able to recover from electrical non-linearities and cross talk, mechanical tip wear or tip faults while at the same time homogenize the use of the media surface, to avoid that certain areas of the media age more than others. Section "Channel Coding" discusses an RLL code for dealing with partial erasure.

Servo-Control and Actuation Electronics

The positioning systems described in the section "Positioning Systems" require dedicated on-board electronics to control the actuation of the moving media platform. The positioning on the bit needs to be sufficiently precise to avoid excessive raw bit error rate degradation when the probe is off track. Furthermore, since the same probe serves as both read and write transducer, partial erasure of adjacent bit may occur, providing more nonlinear distortion of the readback signal. These factors contribute to increased requirements for error correction algorithms that in turn limit the useful area density.

Electromagnetic actuators have been described in the section "Electromagnetic". They are current driven and generally require low voltages to generate electromagnetic fields through their coils required for actuation. These actuators can be modeled as a two-input (coil currents), two-output (x–y displacements) closed-loop system that uses positioning sensors and data from dedicated servo–bits to generate a position error signal (PES).

Pantazi et al from IBM [135] have demonstrated prototypes of such closed-loop servo-systems having a standard deviation of the total positioning error of 0.7 nm in the x-axis and 0.6 nm in the y-axis. The positioning sensors are two pairs of differential thermal sensors located on the edge of the scan table driven with a constant voltage that heats them at a temperature of about 100 °C as described in the section "Sensors and Control". The electrical resistance varies with the actuator displacement and this resistance change is read back as a variation in the current. The servo-data in this system is recovered by four servo-field bursts, three of which are deliberately placed out of track during formatting. Since the readback signal depends on the centering of the probe vs. the indentation, the PES is obtained by correlating the relative amplitude of these four servo-signals to a pulse shape representing a typical indentation. Such control design schemes that use multiple sensors (thermal sensors and PES) achieve a nanometer-scale positioning precision in a large scan area.

Transconductance amplifier circuit topologies (voltage-controlled current sources) are used to drive electromagnetic actuators but suffer from efficiency limits of the output stages, generally class AB push–pull drivers. Research is ongoing to increase efficiencies by using class D or pulse-width modulation (PWM) driving schemes.

Electrostatic actuators have been described in the section "Electrostatic Comb Drives". The actuation force is proportional to the voltage across the two actuation surfaces. To increase the maximum voltage, DC/DC step-up conversion is used.

Examples of DC/DC converters specifically designed for MEMS applications are a high-voltage CMOS design by Saheb et al. that converts 3 V up to 380 V using an external coil [152] and a standard CMOS design by Hong and El-Gamal that is capable of converting 1.2 to 14.8 V without external components [71].

Coding for Probe Storage

Introduction

The purpose of encoding data recorded on a probe storage device is the same as for any other information storage system: it is to store as much information as possible while being able to retrieve this information with a vanishingly small (typically 10^{-12}) probability of error. In other words, the purpose of encoding information to be stored is to use as much *capacity* offered by the storage device as possible.

As is common for all storage systems, the role of information encoding for probe storage is to eliminate or reduce the effect of various distortions which degrade the quality of the readback waveform. Redundant bits are judiciously added to the information string and this redundancy is exploited at the reading stage to restore the original information [168]. What makes coding research for probe storage so exciting is the unique nature of readback impairments stemming from massive parallelism of the read channel, nonlinearity of recording process at nano-scales, and the nano-scale precision required of the positioning system.

The three main specific causes for coding in probe storage are

- Reliability of probe arrays
- Nonlinear bit erasure effects
- Probe positioning errors

We discuss each of these phenomena together with the corresponding coding schemes designed to counter their detrimental effect on the readback waveform below. The main example we will use in our discussion is the Millipede concept of probe storage developed by IBM, which is based on the principle of thermo-electrical writing and reading, see [46, 112, 183] for details. However, most of our conclusions apply to any nano-scale probe storage device.

Coding for Reliability

The throughput of a thermo-mechanical probe storage is not expected to exceed 1 Mbps per probe [164]. Therefore, in order to meet the throughput requirements typical of a modern storage system (about 3 Gbps for a magnetic hard drive, raising to 6 Gbps in the near future, [115]), a probe storage system must consist of an array of thousands of probes reading and writing in parallel.

Therefore a probe storage device is an example of complex electronic system and its reliability has to be closely examined.[4] The simplest question to ask is what is the lifetime of a probes storage device given the lifetime of a single probe? While the long-term reliability data for nano-scale probes and actuators are not yet available, the lifetime is unlikely to exceed 100 years. Accordingly, let us assume that individual probe fail independently at the constant probability rate

$$\lambda = 10^{-2} \text{ years}^{-1}. \tag{3.1}$$

Given the way probe arrays are fabricated, the independence of probe failures is perhaps an oversimplified yet very useful assumption, as it allows us to make conclusions about the *minimal* amount of coding needed to make the array of probes reliable.

It follows from our assumptions that the probability of a single probe not failing during the interval [0, T] years is

$$P_T(1) = P_0 e^{-\lambda T}, \tag{3.2}$$

where P_0 is the probability that the brand new probe is unbroken. In what follows we will assume that $P_0 = 1$. Then, the probability that the probe does not fail during the first 10 years of operation is close to 1: $P_{10}(1) = e^{-10/100} \approx 0.9$.

In order to achieve a throughput of 1 GB/s, we need an array of 1024 probes reading and writing in parallel is required. If un-coded data are written on the probe storage device, a single probe failure within the array will lead to read/write errors. Therefore, the array fails if at least one probe fails. The probability of the array of N probes not failing for T years is

$$P_T(N) = e^{-\lambda N T}. \tag{3.3}$$

For $T = 10$ years and $N = 1024$ this probability is $e^{-1024/10} \approx 3.4 \times 10^{-45}$, which is essentially zero. In fact, it can be easily seen from the above expression that the mean lifetime of an array of 1024 probes is equal to 100/1024 years which is just over 1 month! Recall that for mobile applications the lifetime of a storage device should be measured in years, whereas an archival storage should have a lifetime of about 50 years!

The conclusion we have just reached is staggering: even if we assume that the probe storage channel is noiseless, a probe storage device with very reliable individual probes is unusable due to a high probability of system failure!

The problem of reliability in large storage systems is not new. For example, redundant arrays of inexpensive disks (RAID's) [141] can achieve very high degree of reliability by adding one or two "parity" disks to the array of up to

[4] All results obtained in the section "Coding for Reliability" have been originally reported in [137].

254 information disks. Parity is generated using Reed–Solomon (RS) code [108] over the Galois field GF(2^8) in such a way that all information can be restored if any two disks of the array fail.

Can we use the same approach and protect the probe storage array against failures by adding redundant probes which will read and write parity information?

The answer to this question is "yes", but the cost of protecting the array of probes measured in terms of storage capacity loss is much higher than for RAID. The reason for that is very simple: failed disks in RAID are replaced immediately; therefore coding only needs to ensure that the probability of more than two disks failing simultaneously is small to avoid the loss of data. Unfortunately, individual nano-scale probes cannot be replaced or repaired. Therefore, sufficient redundancy is needed to prevent information loss during the lifetime of the array.

Assume that the fraction *RN* of all probes in the array read/write information bits and (1–R)*N* probes read/write parity bits. Here $R \leq 1$ is the rate of information storage: we store *R* bits of information per every bit recorded on the medium. The closer the *R* to one, the smaller is the fraction of storage "wasted" on redundant bits.

How small should *R* be to ensure that the probability of information loss is small given that the target lifetime of the device is *T*? Let us assume as we did above that probes in the array fail independently at a constant rate λ. Let us also neglect effects of electronics noise and array positioning errors on the quality of the readback signal. Then we can model the array of probes as a *binary erasure channel*: if the probe is working it outputs crisp zeros and ones, if it fails it outputs nothing. The capacity of this channel is known:

$$C_{\mathrm{BEC}} = p_T(1), \tag{3.4}$$

where $p_T(1)$ is the probability that a single probe does not fail in the interval of $[0, T]$ years.

The answer for the minimal amount of redundancy we need to protect the probe array against failures is given by Shannon's theorem [168]: If $R > C_{\mathrm{BEC}}$ then the probability of error (irrecoverable information loss in the context of probe array) is close to 1 no matter which code is used to generate parity. If on the other hand, $R < C_{\mathrm{BEC}}$, there exists a code of rate *R* such that the probability of information loss is vanishingly small.

Given that $\lambda_R = 10^{-2}$ years^{-1} , we conclude that at least

$$1 - C_{\mathrm{BEC}} = 1 - p_T(1) = 1 - \mathrm{e}^{-10/100} \approx 0.1$$

fraction of storage capacity has to be reserved for parity to ensure small probability of information loss during 10 years of operation. Ten percent of capacity loss to ensure the increase of lifetime from a month to 10 years seems like a very good trade-off. The corresponding probe storage system will be suitable for mobile and backup applications. If, however, we wanted to use probe storage for archival applications with required information lifetime of 50 years, we would need to spend

$$1 - e^{-50/100} \approx 0.4$$

fraction of capacity on parity, which leads to a significant system performance hit!

Having realized how much capacity we need to spend, we now need to construct a practical error correction code which will allow the system to sustain probe failures without any information loss and will operate near capacity of binary erasure channel.

In general Shannon's theorem only tells us that such a code exists, but says nothing about how to construct it.

Fortunately for probe storage, there is a classical code with rate close to capacity which will do the job. Consider Reed–Solomon (RS) code operating on 10 bit symbols. The maximal block size for such a code is $N = 2^{10} - 1 = 1023$ symbols which closely matches the number of probes in the array. (The block size can be reduced via shortening.) If P out of N symbols are parity symbols, RS code can correct up to P erased symbols in known positions. Therefore, if data are encoded with Reed–Solomon code with T parity symbols, information can be retrieved even if up to P probes are broken. The rate of Reed–Solomon code is

$$R = 1 - P/N.$$

Therefore, the minimal number of Reed–Solomon parity bits we need to ensure the required lifetime of the probe storage device is

$$P_{\min} = p_T(1)N.$$

In particular, it follows from the examples considered above that RS(900,1000) code will ensure a lifetime of about 10 years for the probe storage device, whereas RS(600,1000) code is needed for archival probe storage.

Note that in order to realize the full potential of Reed–Solomon codes, the structure of the code block must reflect the fact that a Reed–Solomon decoder operates on m-bit symbols rather than individual bits. To achieve this we must identify m consecutive outputs of a given probe with *one* Reed–Solomon symbol. If $b_{i,t}$ is the bit read by the ith probe at time t, one block of Reed–Solomon code matched to the probe array consisting of 1000 probes looks as follows:

$$\begin{pmatrix} b_{1,t} & b_{2,t} & \cdots & b_{1000,t} \\ b_{1,t-1} & b_{2,t-1} & \cdots & b_{1000,t-1} \\ \cdots & \cdots & \cdots & \cdots \\ b_{1,t-(m-1)} & b_{2,t-(m-1)} & \cdots & b_{1000,t-(m-1)} \end{pmatrix}. \tag{3.5}$$

For the coding scheme we have just described, the probability of irrecoverable information loss is equal to the probability of more than P probes failing in the interval of time $[0, T]$. Therefore, the probability that the probe array survives T years is given by a simple binomial formula

$$P_T^{wr}(N) = \sum_{k=0}^{P} \binom{N}{k} (1 - e^{-\lambda T})^k e^{-\lambda T(N-k)}, \tag{3.6}$$

where $\binom{N}{k}$ is a binomial coefficient. In the limit $N \to \infty$, (3.6) approaches zero if $R > e^{-\lambda T}$ and it approaches one if $R < e^{-\lambda T}$, which is in complete agreement with Shannon's theorem. For finite N we can determine the evolution of survival probability with time by examining (3.6) numerically. For example, to ensure survival of the array of 1024 probes for 50 years with probability 0.992 we need to use RS codes with rate $R = 0.57$. This means that 43% of storage capacity has to be used by parity bits to protect stored information against probe failure.

We calculated the minimal amount of coding redundancy needed to protect a probe storage device against probe failures and constructed a practical code which achieves the capacity of the channel which models probe failures. Our next task is to discuss channel and error correction coding necessary to protect stored information from channel noise stemming from the effect of partial erasure and probe array positioning errors.

Channel Coding

Since the beginning of data storage revolution created by the very first IBM disk drives, modulation codes have been used to maximize storage capacity [7, Chapter 1]. Thermo-mechanical storage is not an exception. Indentations in the medium cannot be placed too close to each other due to the effect of partial erasure: if the probe attempts to write two "ones" in a row, melted plastic displaced from the second indentation will partially fill the first one. The effect is so strong that it can be used to erase data written on the medium thus enabling a truly rewritable thermo-mechanical probe storage device.

The effect of partial erasure can be described by the following nonlinear ISI model:

$$r_k = a_k - \alpha(L) a_k a_{k+1} + \beta(L) a_k a_{k-1} + \text{Noise}, \tag{3.7}$$

where r_k is the sample received at time k, and $a_k \in \{0, 1\}$ is the bit recorded at time k, and $\alpha(L)$ and $\beta(L)$ are positive coefficients which depend on bit spacing L, [146].

As suggested by (3.7), the easiest way to deal with the problem of partial erasure is to store information on the medium using only strings which do not contain two consecutive ones. In other words, one would like to record information using only the strings of bits such that

$$a_k a_{k-1} = 0, \qquad \text{for all } k.$$

The set of all such strings constitutes RLL$(1, \infty)$ code, one of the simplest examples of run-length limited codes [76, 75].

Let us estimate the increase in linear storage density resulting from using such a code. Let W be the width of the indentation left in the medium by a probe. The smallest bit period for which effects of partial erasure do not lead to a significant signal degradation turns out to be close to W. Therefore, the linear information density in the absence of coding is

$$\rho_0 = \frac{1}{W} \text{ bits/m}.$$

Given the same indentation geometry, bit spacing in the presence of RLL code can be halved, see Fig. 3.28. Therefore, the user density is $2/W$, whereas the information density equals

$$\rho_{\text{rll}} = \frac{2R_{\text{rll}}}{W} \text{ bits/m}.$$

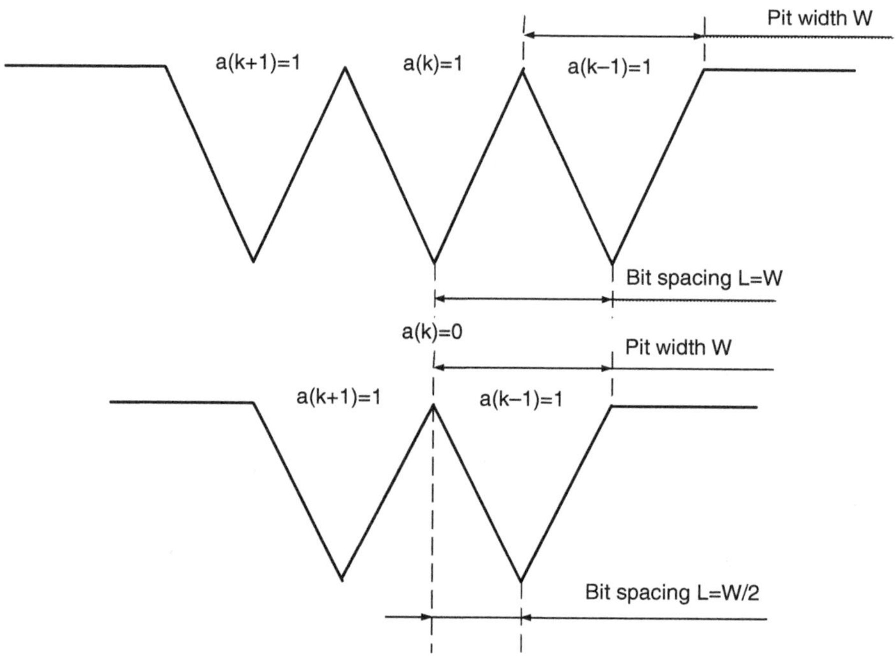

Fig. 3.28 Recording at double user density using RLL$(1, \infty)$ code

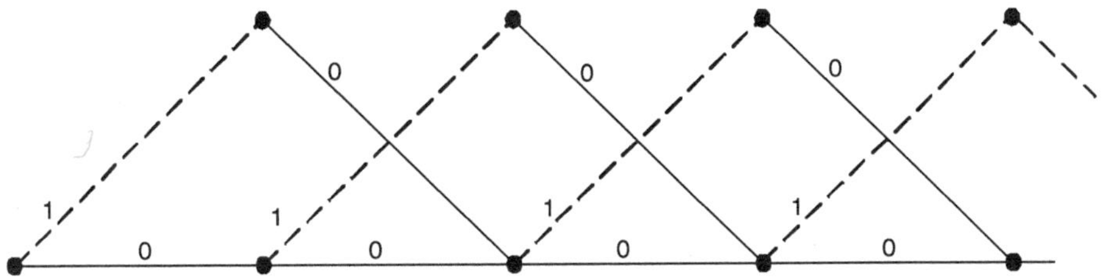

Fig. 3.29 Trellis representation of RLL(1, ∞) sequences

In order to estimate the maximal gain from using the RLL scheme, we have to calculate the information capacity of RLL(1, ∞). We will sketch the capacity calculation here while referring the reader to [75] for details.

All RLL(1, ∞) sequences can be represented as paths on the trellis graph shown in Fig. 3.29.

The trellis can be characterized by the following adjacency matrix:

$$T = \begin{pmatrix} 1 & 1 \\ 1 & 0 \end{pmatrix}, \tag{3.8}$$

which simply says that state 0 of the trellis can be reached from state 0 and 1, whereas state 1 can be reached from state 0 only. Using the adjacency matrix, we can calculate the total number of RLL(1, ∞) sequences of length N as

$$M(N) = \sum_{k=0}^{1} \left(T^N \begin{pmatrix} 1 \\ 0 \end{pmatrix} \right)_k . \tag{3.9}$$

Therefore, for large N,

$$M(N) \approx C \cdot \lambda^N, \tag{3.10}$$

where λ is the largest eigenvalue of matrix T and C is an N-independent constant. Therefore, a long N-bit RLL sequence carries approximately $N \log_2(\lambda)$ bits of information and the capacity of RLL(1, ∞) code is

$$C_{\mathrm{rll}} = \log_2 \lambda. \tag{3.11}$$

The largest eigenvalue of adjacency matrix T is the inverse of the golden ratio:

$$\lambda = \frac{\sqrt{5}+1}{2}. \tag{3.12}$$

Therefore, the maximal linear density increase we can achieve using RLL(1, ∞) coding is

$$\frac{\rho_{\mathrm{rll}}}{\rho_0} = 2\log_2 \lambda \approx 1.4. \tag{3.13}$$

Therefore, we can achieve a 40% increase in information density for thermo-mechanical media using a very simple channel code! Nearly optimal encoding-decoding algorithms for RLL(1, ∞) can be constructed using enumeration encoding, see [75].

The inclusion of the knowledge of constraints into the detection scheme (via the *constraint violation detection* or CVD circuit [138]) can lead to a significant improvement of the detector's performance, albeit that the complexity cost of such an improvement is still under investigation.

Bit spacing in current probe storage prototypes is measured in tens of nanometers, whereas the track pitch is on the order of hundreds of nanometers due to the design of the positioning system. Therefore, nonlinear interference effects are currently restricted to the in-track direction. From a technological point of view, the easiest way to further increase the total information density in probe storage is to increase the track density. It is therefore expected that two-dimensional constraint coding will play an important role for future generations of probe storage. Two-dimensional RLL sequences for probe storage are currently being researched into at the University of Twente [60].

Error Correction Coding for Global Jitter

Yet another distinguishing feature of probe storage is the presence of global jitter–signal degradation due to errors in the array's position which affects *all* probes in the array. The aim of this section is to show how profoundly the performance of traditional error correction code is influenced by global jitter and argue that accounting for the effect of global jitter should be the prime concern for a probe storage ECC designer.[5] Unlike sections "Coding for Reliability" and "Channel Coding", the current section does not suggest a good coding scheme protecting against jitter noise probe storage, as the problem of finding optimal codes for channels affected by global jitter remains largely unsolved.

---

[5]All results presented in the section "Error Correction Codeing for Global Jitter" have been originally reported in [136].

Channel Model

Assume that we use an RLL code similar to the one described in the previous section and information density is low enough for the effects of in-track inter-symbol interference to be neglected. Then the sampled readback signal from the ith probe at a given moment of time can be written as follows:

$$r_i = p(J)a_i + \sigma n_i, i = 1, 2, \ldots N, \tag{3.14}$$

where N is the total number of probes, $a_i \in 0, 1$ is the bit written in position i, J is the sum of global read and write jitter, $p(J)$ is the isolated pulse shape in response to $\ldots 010 \ldots$ written on the medium; $\{\sigma \cdot n_i\}$ is a sequence of random variables which models the combined effect of electronics and media noise.

Extensive experiments performed by IBM for thermo-mechanical storage media [164] demonstrated that electronics/media noise is well modeled by Gaussian random variables. Accordingly, we assume that $\{n_i\}_{1 \leq i \leq N}$ is the set of independent identically distributed Gaussian variables with mean zero and unit variance. Parameter σ is the standard deviation of the resulting additive white Gaussian noise.

IBM experiments using a Millipede-like positioning control loop also demonstrate that jitter J can be treated as a mean-zero Gaussian random variable [164]. Given that similar controls are used, this conclusion applies to other probe storage devices, for instance, phase change media based. Let σ_j be the standard deviation of jitter.

The shape of function p can be extracted from experiment [146]. For the theoretical analysis, the following isolated pulse is easy to work with

$$p(J) = \mathrm{e}^{-\frac{J^2}{W^2}}, \tag{3.15}$$

where W is a parameter related to pulse width.

Asymptotically Optimal Channel Detector in the Presence of Global Jitter

The first problem we would like to solve is as follows: what is the maximally likely set of information bits $a_1, a_2, \ldots, a_N$ for a given output of all probes $r_1, r_2, \ldots, r_N$ at the given moment of time?

In other words, we are trying to find an optimal signal detection scheme for the whole *correlated* array of probes rather than attempting a probe-by-probe signal detection.

Here, we are only considering the inference problem using the information contained in received signals at a given moment of time. This is suboptimal due to the presence of temporal correlations between received signals, but may be necessary due to the system complexity restrictions.

If $N = 1$, the solution is given by the optimal threshold detector described in the section "Probe Storage Electronics and System Considerations", see also [138]. If $N > 1$, the optimal detector is different from the set of independent $N = 1$ detectors, as the outputs of different probe channels are correlated via common jitter.

In general the optimal detector for the channel in (3.14) is quite complicated. If, however, $N \gg 1$ (e.g., $N = 1000$) an asymptotically optimal detector can be built exploiting the law of large numbers, see, e.g., [41 Chapter 1]. Assuming that data bits are independent uniform random variables,[6] the optimal detector can be described as follows:

- Calculate

$$\hat{p} = \frac{2}{N}\sum_{k=1}^{N} r_i. \tag{3.16}$$

- For each i: $1 \le i \le N$, estimate data bit a_i as follows:

$$\hat{a}_i = \begin{cases} 1 & \text{if } r_i > \frac{\hat{p}}{2} \\ 0 & \text{if } r_i \le \frac{\hat{p}}{2} \end{cases}, \tag{3.17}$$

Notice that the described detector requires the addition of N numbers. This may seem like a complex operation if $N \gg 1$, but the complexity *per probe* is still low.

The optimality of detector (3.16), (3.17) is easy to explain heuristically: If $N \gg 1$, due to the law of large numbers,

$$\frac{1}{N}\sum_{i=1}^{N} a_i \approx \mathbb{E}(x_i) = \frac{1}{2},$$

$$\frac{1}{N}\sum_{i=1}^{N} n_i \approx \mathbb{E}(n_i) = 0,$$

where $\mathbb{E}(\cdots)$ stands for the expectation value.

Therefore, summing 3.14 over all i's, we get the following estimate for signal reduction due to jitter:

$$p(J) \approx \frac{2}{N}\sum_{i} r_i.$$

[6] This assumption must be modified if a two-dimensional constraint code is used.

For $N = \infty$, this estimate is exact due to the law of large numbers, but then the optimal detector for channel (3.14) with *known* value of *p(J)* is simply the set of *N*-independent threshold detectors (3.17).

To verify the optimality of detector (3.16), (3.17) for the channel (3.14) carefully, observe that the maximum-likelihood detector must find

$$\hat{\mathbf{a}} = \text{argmax}_{\mathbf{a}} \int_0^1 Pr\left(\mathbf{a} \mid p, \mathbf{r}\right) Pr(p \mid \mathbf{r})\, \mathrm{d}p.$$

As a consequence of the law of large numbers, the conditional probability $\Pr(p \mid \mathbf{r})$ is sharply peaked around the most likely value of reduction factor $p_O(\mathbf{r})$. The Laplace formula [49] then implies that

$$\hat{\mathbf{a}} \approx \text{argmax}_{\mathbf{a}} \Pr\left(\mathbf{a} \mid p_O(\mathbf{r}), \mathbf{r}\right).$$

As it turns out, $p_O(\mathbf{r})$ is given by (3.16), and the maximum-likelihood problem is indeed solved by (3.17) in the limit $N \gg 1$.

So it remains to check that in the limit $N \to \infty$, $P(p \mid \mathbf{r})$ becomes a delta function supported at (3.16). The calculation is based on the application of Central Limit Theorem [41, Chapter 2] and is fairly long. Here we only present the final answer:

$$\Pr(p \mid R) = \frac{1}{Z(R)} \mathrm{e}^{-2N \frac{(p/2-R)^2}{4\sigma^2+p^2}}, \tag{3.18}$$

where $R = \frac{1}{N} \sum_{k=1}^{N} r_i$ and $Z(R)$ is a normalization constant. Therefore, for $N \gg 1$, the distribution is sharply peaked around the value

$$p_O(\mathbf{r}) = 2R$$

and the asymptotic optimality of detector (3.16), (3.17) is proven.

BER curves presented in Figs. 3.30 and 3.31 compare the performance of the detector (3.17) and the performance of *N*-independent optimal threshold detectors for the global jitter channel (3.14). We can draw the following conclusions:

- For $N = 1000$, the asymptotically optimal detector (3.16), (3.17) performs identically to the optimal $N = \infty$ ("Genie") detector.
- Detector (3.16), (3.17) significantly outperforms the classical probe storage detection scheme based on the optimal threshold detectors operating independently on each probe channel.
- The SNR advantage of the detector (3.17) over *N* independent threshold detectors is 0.5–1.0 dB at $\text{BER} = 10^{-4}$ depending on jitter statistics.

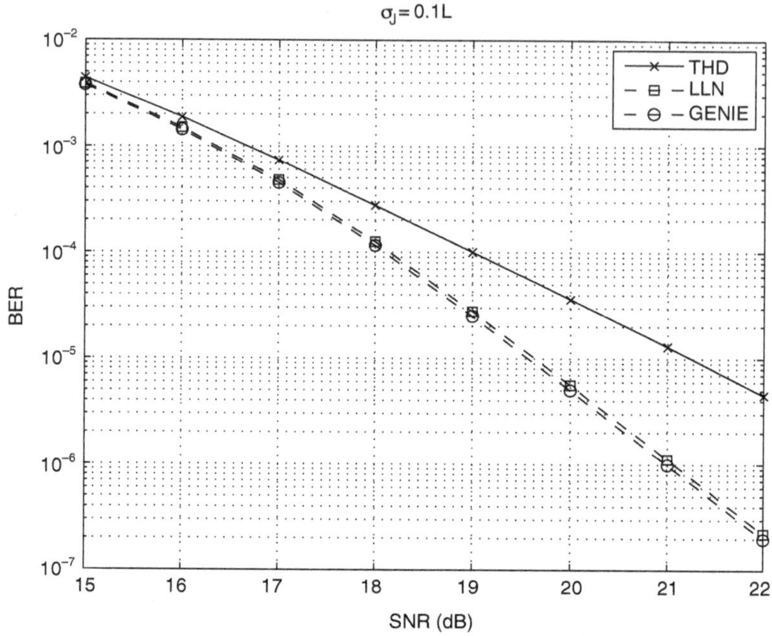

Fig. 3.30 BER for global jitter channel. Jitter strength is 10% of bit spacing. Gaussian mean zero p.d.f. of jitter is assumed. Legend: THD–$N = 1000$ independent optimal threshold detectors; LLN–detector (3.17) for $N = 1000$; Genie–"Genie assisted" detector, which uses the exact value of $p(J)$ rather than its estimate (3.16)

Having solved the problem of signal detection in the presence of global jitter, we can investigate how jitter affects the performance of some classical error correction codes.

Performance of Reed–Solomon Codes on a Global Jitter Channel: Error Floor

As it turns out, the global jitter has very serious consequences for the performance of error correction codes. As we will show below, a straightforward application of Reed–Solomon (RS) codes to global jitter channels *leads to the error floor* in the sector error rate (SER) curve. This means that RS codes *cannot* be used in probe storage without some special modifications (such as interleaving). This modification can be associated with a large hardware cost, making the applicability of RS codes to probe storage very questionable indeed.

Applying large deviations technique [41, Chapter 1] to Reed–Solomon codes we get the following upper bound on sector error (SE) rate conditional on the known value of jitter:

$$\Pr(\mathrm{SE} \mid p) \leq \mathrm{e}^{-SQ(p_{\mathrm{w}},1-p_{\mathrm{w}}\|\tau,1-\tau)}, \tag{3.19}$$

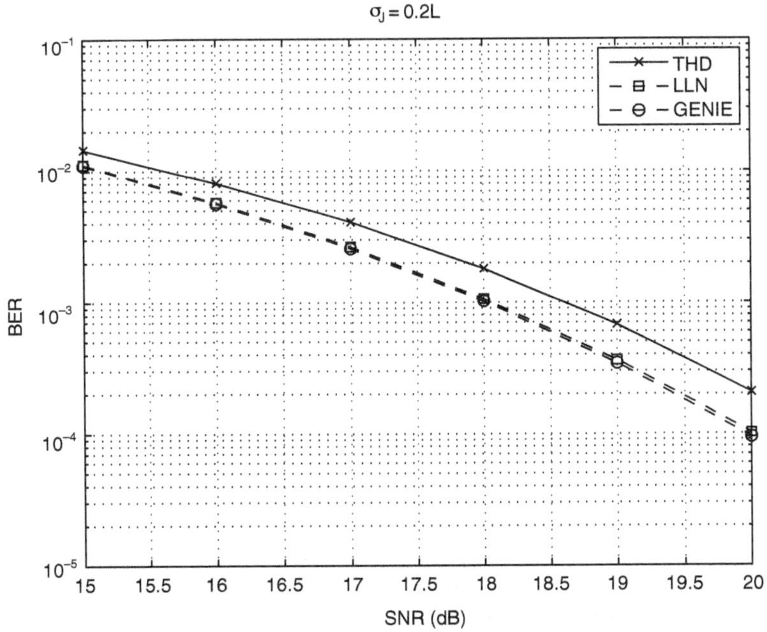

Fig. 3.31 BER for global jitter channel. Jitter strength is 20% of bit spacing. Uniform mean zero p.d.f. of jitter is assumed. Legend: THD–$N = 1000$ independent optimal threshold detectors; LLN–detector (3.17) for $N = 1000$; genie–"Genie-assisted" detector, which uses the exact value of $p(J)$ rather than its estimate (3.16)

where S is the number of RS symbols per block, τ is Reed–Solomon threshold, p_{w} is the probability of symbol error, and $Q(\cdot||\cdot)$ is the Kullback–Leibler divergence or relative entropy, one of the most fundamental objects of information theory [107]

For the optimal threshold detector

$$p_{\mathrm{w}} = 1 - (1 - p_{\mathrm{b}})^n, \tag{3.20}$$

where n is symbol size (typically $n = 8$ bits) and p_{b} is the probability of bit error. For channel (3.14), the probability of bit error, conditional on the pulse depth (a random variable due to global jitter), equals

$$p_{\mathrm{b}}(\mathrm{p}) = \int_{p/2}^{\infty} \frac{\mathrm{d}n}{\sqrt{2\pi\sigma^2}} \mathrm{e}^{-\frac{n^2}{2\sigma^2}}. \tag{3.21}$$

Unconditional sector error rate is therefore bounded as follows:

$$\Pr(\mathrm{SE}) \leq \int_0^1 \rho(p) \mathrm{e}^{-SQ(p_{\mathrm{w}}(p), 1 - p_{\mathrm{w}}(p) || \tau, 1 - \tau)} dp, \tag{3.22}$$

where ρ is the probability density of the reduction factor p.

Expression (3.22) can be analyzed very simply if the distribution of p is such that $p > p_{\min}$ with probability one. This is always true if the distribution of jitter is uniform or truncated Gaussian. Then, provided that

$$p_{\mathrm{w}}(p_{\min}) < \tau,$$

which is always satisfied for high SNR, we get

$$\Pr(\mathrm{SE}) \le \mathrm{e}^{-SQ(p_{\mathrm{w}}(p_{\min}),1-p_{\mathrm{w}}(p_{\min})||\tau,1-\tau)}. \tag{3.23}$$

In the limit $\sigma \to 0$, the right-hand side of the above can be estimated as follows:

$$e^{-SQ(p_{\mathrm{w}}(p_{\min}),1-p_{\mathrm{w}}(p_{\min})||\tau,1-\tau)} \approx e^{-S\left(\frac{p_{\min}^2}{8\sigma^2}-H_2(\tau)\right)}, \tag{3.24}$$

where $H_2(\tau)$ is the binary entropy function, see [107]. We see that the sector error rate suffers a significant degradation in comparison with zero jitter case ($p_{\min} = 1$) as $p_{\min}$ can be close to zero. This suggests that for distributions with $p_{\min} = 0$ the performance of RS codes suffers a catastrophic degradation of performance. An example of such distribution is served by a Gaussian jitter model.

A long calculation gives the following upper bound for the probability error rate in this case:

$$\Pr(\mathrm{SE}) \le C\left(n, S, \tau, \frac{\sigma_j}{W}\right)\sigma^{\frac{W^2}{\sigma_j^2}}, \tag{3.25}$$

where C is a constant independent of σ.

We expect the upper bound (3.25) to be tight in the limit $\sigma \to 0$. If this expectation is correct, it will imply the existence of an *error floor* for Reed–Solomon code for this channel: as SNR goes to infinity, SER decreases only polynomially with the exponent which does not depend on the code rate!

Numerical simulations confirm our conclusions. In Fig. 3.32 sector error rate for rate 0.8 Reed–Solomon code is shown. Uniform jitter with $\sigma_j = 0.2L$ is assumed. Comparing Figs. 3.31 and 3.32 we see that, at the level of noise corresponding to $\mathrm{BER} = 10^{-3}$, the sector error rate is just $SER = 10^{-3}$.

The situation is even worse for Gaussian jitter. In this case, catastrophic jitter events lead (in a complete agreement with the theory developed above) to a pronounced error floor for Reed–Solomon codes, see Fig. 3.33. Moreover, the upper bound (3.25) suggests that the position of the error floor is independent of the code rate!

We conclude that the effect of global jitter on Reed–Solomon code is devastating: SER curves show the error floor rather than waterfall behavior even though an optimal detection and optimal hard-input decoding algorithms are used.

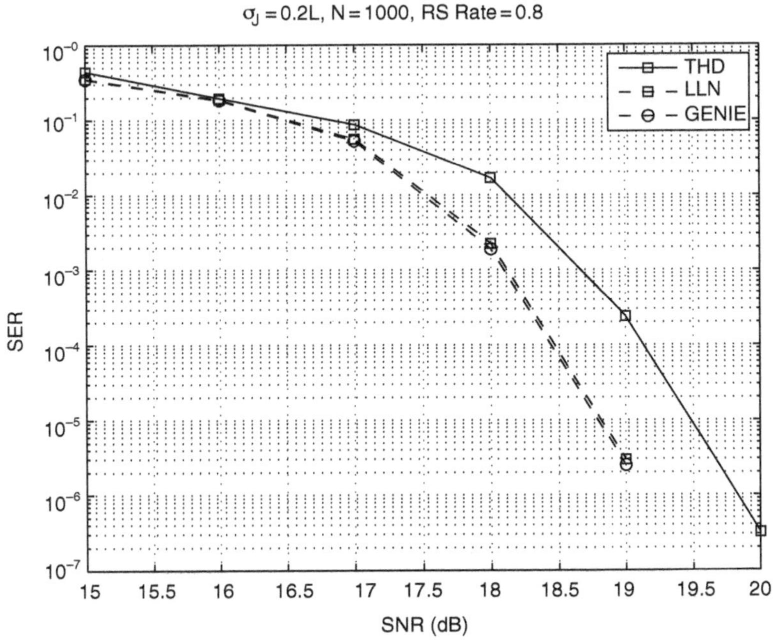

Fig. 3.32 SER for global jitter channel. Jitter strength is 20% of bit spacing. Uniform mean zero p.d.f. of jitter is assumed. Legend: THD–$N = 1000$ independent optimal threshold detectors connected to RS decoder; LLN–detector (3.17) for $N = 1000$ connected to RS decoder; genie–"genie-assisted" detector, which uses the exact value of $p(J)$ rather than its estimate (3.16), connected to RS decoder

Are there any ways of combating effects of global jitter using coding? At the moment, we can only present a tentative answer. First, one can try to encode also the output of *each* probe with a powerful Reed–Solomon code to try and spread information between many time slices thus mitigating effects of strong jitter on a single time slice. One potential drawback of this approach is that global jitter also has long time correlations (tens of sample periods); therefore the decoder for a sufficiently deeply interleaved Reed–Solomon code will have a very high complexity. The advantage of this approach is that it can also solve the reliability problem posted in the section "Coding for Reliability".

One can also try using soft information about jitter produced by the optimal detector to build an effective LDPC encoder for probe storage [144]. One immediate question is to calculate the capacity of channel (3.14) to see what are the achievable rates for a good code for global jitter.

Despite a somewhat pessimistic nature of result we presented here, our feeling is that there is a lot of new and beautiful information theory hidden behind the channel model (3.14). The discovery of this theory will in our opinion help the advancement of not only probe storage, but any type of nano-scale storage device based on the reader moving over the storage medium.

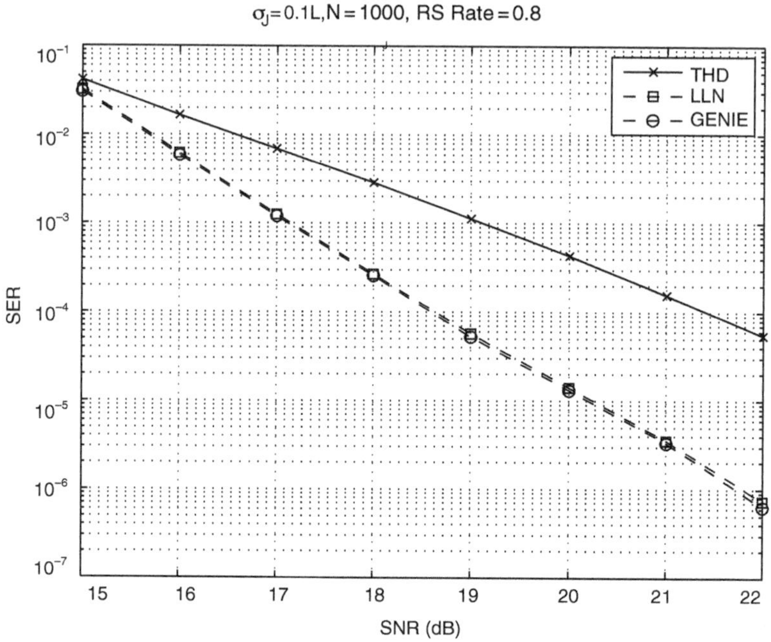

Fig. 3.33 SER for global jitter channel. Jitter strength is 10% of bit spacing. Gaussian mean zero p.d.f. of jitter is assumed. Legend: THD–$N = 1000$ independent optimal threshold detectors connected to RS decoder; LLN–detector (3.17) for $N = 1000$ connected to RS decoder; genie–"genie-assisted" detector, which uses the exact value of $p(J)$ rather than its estimate (3.16), connected to RS decoder

File system

The previous sections elaborated on the main components of a probe storage device. In the remainder of this chapter we address the integration of a probe storage device in computer systems. Specifically, we detail the way a file system needs to organize user data on the storage medium to enhance the quality of service. Quality of service encompasses timing performance, energy consumption, and capacity.

Data Layout Basics

The data layout of a storage device is concerned with the way user data are organized on the storage medium of the device. This organization can be split into two levels: low-level data layout and logical data layout The low-level layout clusters a certain number of bits into larger storage granularity called the sector. The logical data layout assigns logical numbers to sectors on the medium such that each sector is uniquely addressable.

Figure 3.34 shows a two-dimensional look of the storage medium of a simple probe storage device of 4 × 4 probes. Each probe reads and writes data in its exclusive storage field. As shown in Fig. 3.34a, contiguous bits along the y-direction are clustered to form larger storage units. Each storage unit contains overhead bits in addition to user data in order to enable data accessibility. Figure 3.34b shows the

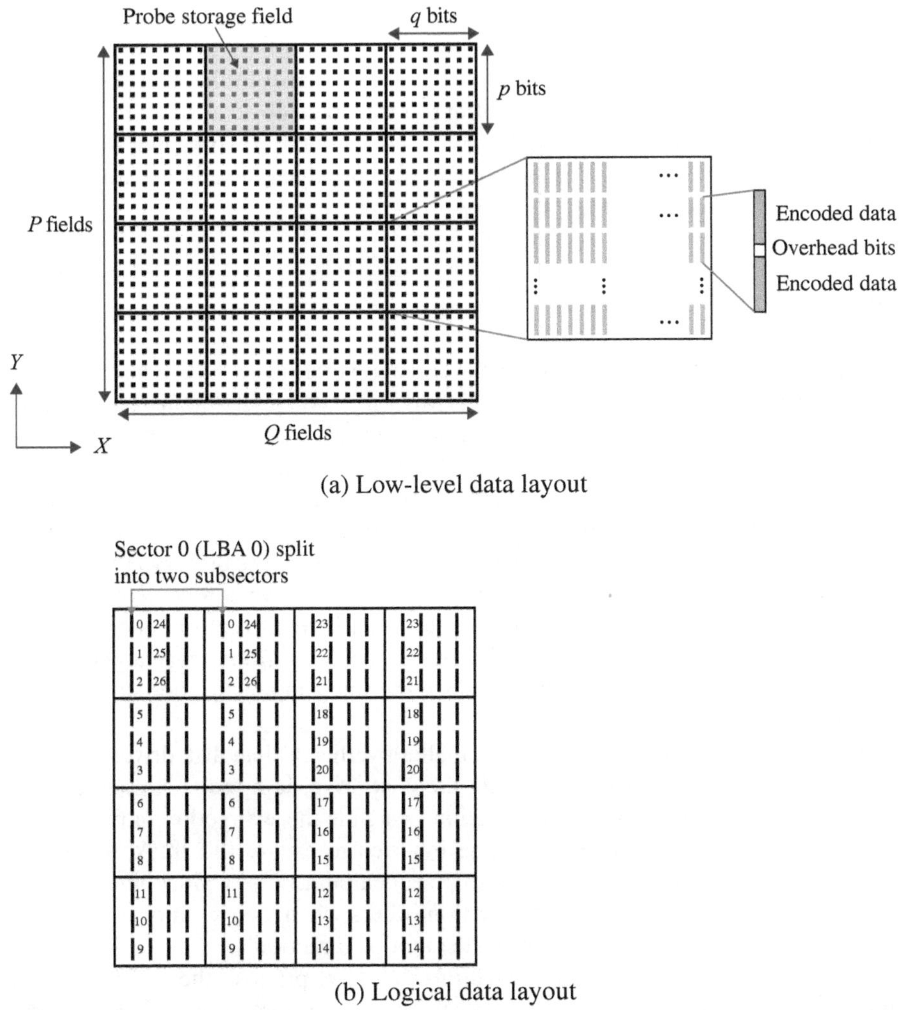

Fig. 3.34 A two-dimensional look at the storage medium of a probe storage device split logically into $P \times Q$ storage fields, each is exclusively accessible by a single probe [79, Chapter 3, PP. 39–43]. (**a**) Bits are clustered together into a sector. Each sector contains user data and additional overhead bits for error correction, for example. (**b**) The corresponding logical data layout, which assigns consecutive LBAs to contiguous sectors. The depicted logical data layout follows the cylinder-mode address-mapping scheme from the disk drive

corresponding logical data layout, where each sector is striped across two probes and is assigned a unique number.

A storage device receives requests from the computer system demanding sectors by their identifier. These identifiers are called Logical Block Addresses. Internally, the device translates an LBA to its corresponding Physical Block Address (PBA) to locate the sector physically. The address mapping from an LBA to a PBA varies between devices and is tailored to enhance certain aspects of the device, such as the throughput.

Two common types of address mapping are known from disk drives: cylinder mode and serpentine mode [77, chapter 18, pp. 655–656]. In the next section, we discuss works that study the data layout in probe storage devices, drawing from the analogy with disk drives. We follow up with a discussion of works that investigate the two dimensionality of probe storage. We discuss dynamic address mapping in probe storage devices in the last section.

Conventional Data Layout

To the best of our knowledge, the first work that studies the data layout in probe storage devices was carried out at Carnegie Mellon University [59, 160, 161]. Griffin et al. [59] draw on the analogy between disk drives and probe storage devices. They import the data-layout terminology from disk drives: cylinder, track, and sector.

Later, Schlosser et al. [161] discuss the "unwritten contract" of address mapping shared between disk drives from different disk vendors. In this contract, they argue that all address schemes adopted in different disk drives preserve the sequentiality. That is, sectors laying contiguously on a track receive consecutive numbers, whereas numbering may differ at track and cylinder boundaries from one vendor or production batch to another. As a result, data are accessed from physically contiguous places on the storage medium to reduce the interruptions to the one-direction mechanical motion.

Although a probe storage device has smaller overheads than disk drives, preserving sequentiality is still needed. Like in the disk drive, in a probe storage device, the seek time is dominated by one direction, which is the x-direction in probe storage devices. This is due to settling time incurred to ensure full track alignment. Laying data along one direction speeds up sequential accesses, since interruptions are reduced. An interruption to the y-motion occurs when the sled moves along the x-direction to change track and reverses the motion direction.

Borrowing from the disk drive, Griffin et al. [59] propose the cylinder mode address-mapping scheme for probe storage devices. Figure 3.34b shows address mapping following the cylinder mode: cylinder by cylinder. For sequential accesses, the cylinder mode incurs fewer seeks along the x-direction compared to the serpentine mode, resulting in better performance. The cylinder mode scheme enables high throughput for sequential accesses found in streaming applications.

Another work from the University of California at Santa Cruz addresses the data layout from the device dimensionality viewpoint. Sivan-Zimet [172] studies the influence of a set of physical parameters on the timing performance of probe storage devices. Sivan-Zimet investigates the influence of the number of probes, dimensions of a probe storage field, and the arrangement of probes. The study is performed under real-world traces. She recommends designing with the physical parameters set with certain ranges to make probe storage competitive.

The previous works preserve the sector size of 512 bytes in probe storage devices as in disk drives. Schlosser [160] stripes a sector across 64 probes. In his MEMS G2 model, where 640 probes are simultaneously active, 10 sectors are accessible at a time. On the contrary, Sivan-Zimet [172] *stripes* a sector across all 320 simultaneously active probes in her model. Khatib et al. [82] study the striping problem and show that the number of accessible sectors can be tuned based on the workload to enhance the performance of a probe storage device. That is, neither 10 sectors nor 1 sector results always in the best performance.

Two-Dimensional Data Layout

A probe storage device bears resemblance to disk drives in many aspects, so that similar treatment as done with the address-mapping scheme is justifiable. Nonetheless, a probe storage device exhibits a unique characteristic that is worthy of exploitation. A probe storage device has an array of probes, out of which a large set can operate in parallel. Although a probe storage device does the best when it moves in one direction, the existence of an array of probes extends the access dimensionality by a second orthogonal dimension.

Exploiting this unique characteristic, Yu et al. [192] propose probe storage for database applications. Here, two-dimensional data sets (e.g., tables of columns and rows in a database repository) are striped elegantly across the two-dimensional data layout of a probe storage device. Probes can be switched on and off selectively depending on the requested data. Switching off probes avoids retrieving irrelevant data, which reduces the energy consumption as well as cache pollution.

On the same basis, Lin et al. [103] study the effectiveness of switching probes selectively to update file system metadata. Because metadata updates are relatively very small (i.e., a few bytes) compared to a whole sector, enhancement in performance and reduction in energy consumption are significant, if probes are switched on selectively.

Work carried out at the University of Twente addresses the two dimensionality of the data layout. Khatib et al. [80, 81] study the influence of the data layout on the response time, energy consumption, and the capacity of a probe storage device. Mobile and streaming applications are investigated. Khatib et al. formulate the data layout of a probe storage device with three parameters and make the case to format the layout based on the expected workload. The parameters are

(i) *The total number of active probes (N)*: How many probes should operate in parallel?

(ii) *Sector parallelism (M)*: How many sectors should be simultaneously accessible from the device?
(iii) *Sector size (S):* Should the conventional sector size of 512 bytes stay the same in probe storage devices?

Figure 3.35 depicts a simple probe storage device with two possible configurations of the data-layout parameters. The first configuration stripes a hypothetically 16 bit sector across one probe only and can access two sectors simultaneously. The second configuration stripes the sector across two probes, but it still can access two sectors simultaneously, because it doubles the number of active probes.

The straightforward configuration of the three parameters would be to (1) operate all probes simultaneously to gain peak throughput, (2) access one sector at a time to maximize bandwidth utilization, and (3) keep the sector size intact to access useful data only. That way, a probe storage device bears complete resemblance to the disk drive.

Simulation with these configurations, however, shows that none of the three design targets (i.e., response time, energy consumption, and capacity) of a probe storage device reaches optimality. In fact, the targets compete and trade-offs must be made. Increasing the sector parallelism and the sector size gives the designer the opportunity to explore areas of the design space that exhibit small trade-offs. Research shows that the effective capacity increases and energy consumption decreases, while response time decreases.

Figure 3.36 shows a two-dimensional look at a design space of a probe storage device. Pareto-optimal configurations of the previous three parameters exist. These are `M-20-BE` and `M-20-BP`, where "M" denotes MEMS, 20 corresponds to the

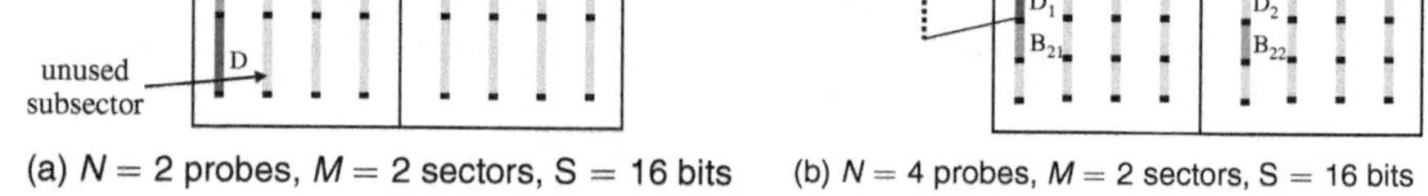

Fig. 3.35 Two possible configurations of the three data-layout parameters: the total number of active probes, sector parallelism, and sector size. The figure shows a simplified probe storage device, where in (**a**) file A fits in one sector, whereas file B is split over two sectors, B_1 and B_2. The single sector of A in (**a**) is split into sectors A_1 and A_2 in (**b**) when doubling the number of probes per sector (the same goes for B_1 which is split into sectors B_{11} and B_{12}). The figure shows two configurations. The first configuration shown in (**a**) uses two out of four probes simultaneously, each accessing a 16-bit sector at a time. By using twice as many active probes as in (**b**), a probe accesses only 8 bits per sector, so that four probes access two sectors in total simultaneously

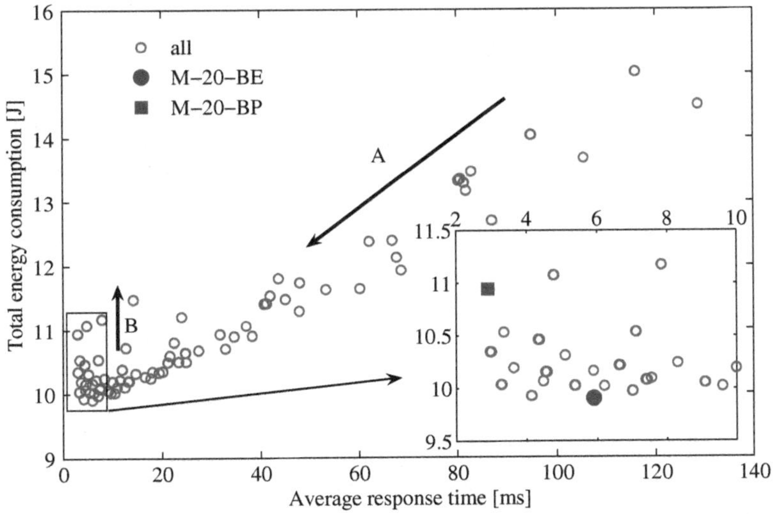

Fig. 3.36 Trade-offs between the energy consumption and the response time for the configurations of the data-layout parameters studied for a mobile workload

nominal throughput of 20 MB/s, "BP" denotes best performance, and "BE" denotes best energy. Unlike disk drives, these configurations exhibit large sector parallelism and sector size (1 sector, 4 KB) and (16 sectors, 4 KB), respectively.

Summarizing, is that significant enhancement in the quality of service of a probe storage device is attainable, provided it is treated differently from the disk drive.

Dynamic Address Mapping

Recent advancement in probe storage has revealed wear of probes [12, 16, 65]. A probe wears by the repetitive operation of writing (and possibly reading) data. A probe can lose parts or collect particles (section "Tip Wear" gives more details). In either case, its functionality degrades to a point that it is considered worn out.

Integrating a probe storage device into the computer system results in exercising its probes with different loads. The difference in load across probes results in uneven wear across the probes. Uneven wear leads to premature expiry of probes, resulting in serious consequences for the reliability and performance of a probe storage device.

Khatib et al. [81] address the challenge of uneven probe wear in probe storage devices. Wear leveling policies are devised that maintain an even load across probes and thus even wear. Khatib et al. show that wear in probe storage is tackled differently and in a simpler way than in flash memory. Due to wear-leveling dynamic PBA to LBA mapping is needed as in flash memory.

A probe storage device must maintain a mapping pad to store the correspondence between LBA and PBA. Every time a request arrives at the device, its

LBA is assigned to the PBA the wear-leveling policy decides for. The mapping pad is consulted by successive requests to look-up the corresponding PBA of their LBA.

Acknowledgments The authors wish to acknowledge Andrew Pauza and Marilyn Armand for useful input to the medium section; Angeliki Pantazi and Christoph Hagleitner (IBM Zürich) for suggestions to the electronics section; Tom Parnell (Siglead Europe), Haralampos (Haris) Pozidis (IBM Zürich), David Wright and Purav Shah (University of Exeter), and Joost van Honschoten (University of Twente); Mark A. Lantz (IBM Zürich) for providing the latest details about their electromagnetic scanner and tip wear; colleagues of the University of California at Santa Cruz for the fruitful discussions.

References

1. Abelmann L, Bolhuis T, Hoexum AM, Krijnen GJM, Lodder JC (2003) Large capacity probe recording using storage robots. IEEE Proc Sci Measure Technol 150:218–221. doi: 10.1049/ip-smt:20030693
2. Agarwal M, Dutton DT, Theil JA, Bai Q, Par E, Hoen S (2006) Characterization of a dipole surface drive actuator with large travel and force. J Microelectromech Syst 15(6):1726–1734. doi: 10.1109/JMEMS.2006.883886
3. Alfaro JF, Fedder GK (2002) Actuation for probe-based mass data storage. In: 2002 International conference on modeling and simulation of microsystems – MSM 2002, San Juan, Puerto Rico, pp 202–205
4. Algre E, Gaudin G, Bsiesy A, Nozieres JP (2005) Improved patterned media for probe-based heat assisted magnetic recording. IEEE Trans Magn 41(10):2857–2859. doi: 10.1109/TMAG.2005.854680
5. Alvarez M, Tamayo J (2005) Optical sequential readout of microcantilever arrays for biological detection. Sens Actuators B 106(2):687–690. doi: 10.1016/j.snb.2004.09.016
6. Aziz MM, Wright CD (2005) A slope-theory approach to electrical probe recording on phase-change media. J Appl Phys 97(10):103537. doi: 10.1063/1.1904156
7. Bane Vasic EK (2004) Coding and signal processing for magnetic recording systems. CRC press, Boca Raton, FL
8. Bao H, Li X (2008) A heater-integrated scanning probe microscopy probe array with different tip radii for study of micro-nanosize effects on silicon-tip/polymer-film friction. Rev Sci Instrum 79(3):033701. doi: 10.1063/1.2885682
9. Bauschlicher Jr CW, So CR (2001) Using hydrogen and chlorine on Si(111) to store data. Chem Phys Lett 333(1–2):1–5. doi: 10.1016/S0009-2614(00)01340-3
10. Bell DJ, Lu TJ, Fleck NA, Spearing SM (2005) MEMS actuators and sensors: observations on their performance and selection for purpose. J Micromech Microeng 15(7):153–164. doi: 10.1088/0960-1317/15/7/022
11. Bennewitz R, Crain JN, Kirakosian A, Lin JL, McChesney JL, Petrovykh DY, Himpsel FJ (2002) Atomic scale memory at a silicon surface. Nanotechnology 13:499–502. doi: 10.1088/0957-4484/13/4/312
12. Berger R, Cheng Y, Forch R, Gotsmann B, Gutmann JS, Pakula T, Rietzler U, Schartl W, Schmidt M, Strack A, Windeln J, Butt HJ (2007) Nanowear on polymer films of different architecture. Langmuir 23(6):3150–3156. doi: 10.1021/la0620399
13. Bhaskaran H, Sebastian A, Despont M (2009) Nanoscale PtSi tips for conducting probe technologies. IEEE Trans Nanotechnol 8(1):128–131. doi: 10.1109/TNANO.2008.2005199
14. Bhaskaran H, Sebastian A, Drechsler U, Despont M (2009) Encapsulated tips for reliable nanoscale conduction in scanning probe technologies. Nanotechnology 20(10):105701. doi: 10.1088/0957-4484/20/10/105701

15. Bhaskaran H, Sebastian A, Pauza A, Pozidis H, Despont M (2009) Nanoscale phase transformation in $Ge_2Sb_2Te_5$ using encapsulated scanning probes and retraction force microscopy. Rev Sci Instrum 80(8):083701. doi: 10.1063/1.3204449
16. Bhushan B, Kwak KJ, Palacio M (2008) Nanotribology and nanomechanics of AFM probe-based data recording technology. J Phys Condens Matter 20(36):365207 (34pp). doi: 10.1088/0953-8984/20/36/365207
17. Bichet O, Wright CD, Samson Y, Gidon S (2004) Local characterization and transformation of phase-change media by scanning thermal probes. J Appl Phys 95(5):2360–2364. doi: 10.1063/1.1644899
18. Binnig G, Rohrer H (2000) Scanning tunneling microscopy. IBM J Res Dev 44(1–2): 279–293
19. Binnig G, Despont M, Drechsler U, Häberle W, Lutwyche M, Vettiger P, Mamin HJ, Chui BW, Kenny TW (1999) Ultrahigh-density atomic force microscopy data storage with erase capability. Appl Phys Lett 74(9):1329–1331. doi: 10.1063/1.123540
20. Bonan J, Hagleitner C, Aboushady H (2006) Low-power cell-level ADC for a MEMS-based parallel scanning-probe storage device. In: Proceedings of the 32nd European solid-state circuits conference, Montreux, Switzerland, pp 239–242
21. Bustillo JM, Howe RT, Muller RS (1998) Surface micromachining for microelectromechanical systems. Proc IEEE 86(8):1552–1573. doi: 10.1109/5.704260
22. Carley LR, Bain JA, Fedder GK, Greve DW, Guillou DF, Lu MSC, Mukherjee T, Santhanam S, Abelmann L, Min S (2000) Single-chip computers with microelectromechanical systems-based magnetic memory. J Appl Phys 87(9 pt 3):6680–6685. doi: 10.1063/1.372807
23. Carley LR, Ganger G, Guillou DF, Nagle D (2001) System design considerations for MEMS-actuated magnetic-probe-based mass storage. IEEE Trans Magn 37:657–662. doi: 10.1109/20.917597
24. Cheung P, Horowitz R, Rowe RT (1996) Design, fabrication, position sensing, and control of an electrostatically-driven polysilicon microactuator. IEEE Trans Magn 32(1):122–128. doi: 10.1109/20.477561
25. Cho Y, Fujimoto K, Hiranaga Y, Wagatsuma Y, Onoe A, Terabe K, Kitamura K (2002) Tbit/inch^2 ferroelectric data storage based on scanning nonlinear dielectric microscopy. Appl Phys Lett 81(23):4401–4403. doi: 10.1063/1.1526916
26. Cho Y, Fujimoto K, Hiranaga Y, Wagatsuma Y, Onoe A, Terabe K, Kitamura K (2003) Tbit/inch^2 data storage using scanning nonlinear dielectric microscopy. Ferroelectrics 292:51–58. doi: 10.1080/00150190390222817
27. Cho Y, Fujimoto K, Hiranaga Y, Wagatsuma Y, Onoe A, Terabe K, Kitamura K (2003) Terabit inch^{-2} ferroelectric data storage using scanning nonlinear dielectric microscopy nanodomain engineering system. Nanotechnology 14(6):637–642. doi: 10.1088/0957-4484/14/6/314
28. Cho Y, Hiranaga Y, Fujimoto K, Wagatsuma Y, Onoe A (2004) Fundamental study on ferroelectric data storage with the density above 1 Tbit/inch^2 using congruent lithium tantalate. Integr Ferroelectr 61:77–81. doi: 10.1080/10584580490458810
29. Choi JJ, Park H, Kim KY, Jeon JU (2001) Electromagnetic micro x-y stage for probe-based data storage. J Semiconductor Technol Sci 1:84–93
30. Choi JJ, Park H, Kim KY, Jeon JU (2001) Electromagnetic micro x-y stage with very thick Cu coil for probe-based mass data storage device. In: Proceedings of SPIE, vol 4334, Newport Beach, CA, pp 363–371. doi: 10.1117/12.436622
31. Chong NB, Yang J, Mou J, Guo G (2005) Thermo-mechanical data bit formation of small-sized microcantilever probe tip array. In: Proceedings of SPIE – The international society for optical engineering, Singapore, Singapore, vol 5852 part I, pp 270–275. doi: 10.1117/12.621530
32. Cooper EB, Manalis SR, Fang H, Dai H, Matsumoto K, Minne SC, Hunt T, Quate CF (1999) Terabit-per-square-inch data storage with the atomic force microscope. Appl Phys Lett 75(22):3566–3568. doi: 10.1063/1.125390

33. Craus CB, Onoue T, Ramstock K, Geerts WJMA, Siekman MH, Abelmann L, Lodder JC (2005) A read and write element for magnetic probe recording. J Phys D: Appl Phys 38: 363–370. doi: 10.1088/0022-3727/38/3/002
34. Cuberes MT, Schlittler RR, Gimzewski JK (1996) Room-temperature repositioning of individual C_{60} molecules at Cu steps: operation of a molecular counting device. Appl Phys Lett 69(20):3016–3018. doi: 10.1063/1.116824
35. David Wright C, Armand M, Aziz MM (2006) Terabit-per-square-inch data storage using phase-change media and scanning electrical nanoprobes. IEEE Trans Nanotechnol 5(1): 50–61. doi: 10.1109/TNANO.2005.861400
36. Despont M, Brugger J, Drechsler U, Durig U, Haberle W, Lutwyche M, Rothuizen H, Stutz R, Widmer R, Binnig G, Rohrer H, Vettiger P (2000) VLSI-NEMS chip for parallel AFM data storage. Sens Actuators A 80(2):100–107. doi: 10.1016/S0924-4247(99)00254-X
37. Despont M, Drechsler U, Yu R, Pogge H, Vettiger P (2004) Wafer-scale microdevice transfer/interconnect: its application in an AFM-based data-storage system. J Microelectromech Syst 13(6):895–901. doi: 10.1109/JMEMS.2004.835769
38. Drechsler U, Burer N, Despont M, Dürig U, Gotsmann B, Robin F, Vettiger P (2003) Cantilevers with nano-heaters for thermomechanical storage application. Microelectron Eng 67–68:397–404. doi: 10.1016/S0167-9317(03)00095-9
39. Dürig U (2005) Fundamentals of micromechanical thermoelectric sensors. J Appl Phys 98(4):044,906. doi: 10.1063/1.2006968
40. Dürig U, Cross G, Despont M, Drechsler U, Häberle W, Lutwyche M, Rothuizen H, Stutz R, Widmer R, Vettiger P, Binnig GK, King WP, Goodson KE (2000) “Millipede” – an AFM data storage system at the frontier of nanotribology. Tribol Lett 9(1–2):25–32. doi: 10.1023/A:1018844124754
41. Durret R (2005) Probability: theory and examples, 3rd edn. Brooks/Cole, Pacific Grove, CA
42. Eagle SC, Fedder GK (1999) Writing nanometer-scale pits in sputtered carbon films using the scanning tunneling microscope. Appl Phys Lett 74(25):3902–3903. doi: 10.1063/1.124218
43. Eigler DM, Schweizer EK (1990) Positioning single atoms with a scanning tunnelling microscope. Nature 344(6266):524–526. doi: 10.1038/344524a0
44. El-Sayed RT, Carley LR (2003) Performance analysis of a 0.3-Tb/in^2 low-power MFM-based scanning-probe device. IEEE Trans Magn 39(6):3566–3574. doi: 10.1109/TMAG.2003.819457
45. El-Sayed RT, Carley LR (2004) Analytical and micromagnetic-based modeling of quantization noise in MFM-based pulse-width-modulation perpendicular recording. IEEE Trans Magn 40(4 II):2326–2328. doi: 10.1109/TMAG.2004.829172
46. Eleftheriou E, Antonakopoulos T, Binnig GK, Cherubini G, Despont M, Dholakia A, Durig U, Lantz MA, Pozidis H, Rothuizen HE, Vettiger P (2003) Millipede – a MEMS-based scanning-probe data-storage system. IEEE Trans Magn 39(2 I):938–945. doi: 10.1109/TMAG.2003.808953
47. Engelen JBC, Lantz MA, Rothuizen HE, Abelmann L, Elwenspoek MC (2009) Improved performance of large stroke comb-drive actuators by using a stepped finger shape. In: Proceedings of 15th international conference on solid-state sensors and actuators (Transducers ’09), pp 1762–1765. doi: 10.1109/SENSOR.2009.5285744
48. Engelen JBC, Rothuizen HE, Drechsler U, Stutz R, Despont M, Abelmann L, Lantz MA (2009) A mass-balanced through-wafer electrostatic *x*/*y*-scanner for probe data storage. Microelectron Eng 86:1230–1233. doi: 10.1016/j.mee.2008.11.032
49. Erdelyi A (1956) Asymptotic Expansions. Dover, New York, NY
50. Esashi M, Ono T, Yoshida S (2005) Multiprobe systems for data storage and other applications. In: Proceedings of IEEE Sensors, vol 2005, Irvine, CA, United States, pp 258–259. doi: 10.1109/ICSENS.2005.1597685

51. Faizul MS, Ono T, Esashi M (2009) A large displacement piezodriven silicon XY-microstage. In: Proceedings of the 15th international conference on solid-state sensors and actuators (Transducers '09), pp 2405–2408. Denver, CO, USA. doi:10.1109/SENSOR.2009.5285429
52. Faizul MS, Ono T, Esashi M (2009) Modeling and experimental validation of the performance of a silicon XY-microstage driven by PZT actuators. J Micromech Microeng 19(9):095004. doi: 10.1088/0960-1317/19/9/095004
53. Fontana R *et al* (2009) Millions of square inches (MSI) comparisons & implications for magnetic and solid state storage class memories. In: INSIC annual meeting symposium, Santa Clara, CA, USA
54. Franke K, Besold J, Haessler W, Seegebarth C (1994) Modification and detection of domains on ferroelectric PZT films by scanning force microscopy. Surf Sci 302(1–2):L283–L288. doi: 10.1016/0039-6028(94)91089-8
55. Gidon S, Lemonnier O, Rolland B, Bichet O, Dressler C, Samson Y (2004) Electrical probe storage using joule heating in phase change media. Appl Phys Lett 85(26):6392–6394. doi: 10.1063/1.1834718
56. Gotoh T, Sugawara K, Tanaka K (2004) Minimal phase-change marks produced in amorphous $Ge_2Sb_2Te_5$ films. Jpn J Appl Phys 43(6 B):L818–L821. doi: 10.1143/JJAP.43.L818
57. Gotsmann B, Duerig U, Frommer J, Hawker CJ (2006) Exploiting chemical switching in a diels-alder polymer for nanoscale probe lithography and data storage. Adv Funct Mater 16(11):1499–1505. doi: 10.1002/adfm.200500724
58. Gotsmann B, Duerig UT, Sills S, Frommer J, Hawker CJ (2006) Controlling nanowear in a polymer by confining segmental relaxation. Nano Lett 6(2):296–300. doi: 10.1021/nl0520563
59. Griffin JL, Schlosser SW, Ganger GR, Nagle DF (2000) Operating system management of MEMS-based storage devices. In: Proceedings of the 4th symposium on operating systems design & implementation, San Diego, CA, 23–25 October., pp 227–242. URL http://citeseer.ist.psu.edu/griffin00operating.html
60. Groenland JPJ, Abelmann L (2007) Two-dimensional coding for probe recording on magnetic patterned media. IEEE Trans Magn 43(6):2307–2309. doi: 10.1109/TMAG.2007.893137
61. Hagleitner C, Bonaccio T, Rothuizen H, Lienemann J, Wiesmann D, Cherubini G, Korvink JG, Eleftheriou E (2007) Modeling, design, and verification for the analog front-end of a MEMS-based parallel scanning-probe storage device. IEEE J Solid-State Circ 42(8):1779–1789. doi: 10.1109/JSSC.2007.900287
62. Hamann HF, O'Boyle M, Martin YC, Rooks M, Wickramasinghe HK (2006) Ultra-high-density phase-change storage and memory. Nat Mater 5(5):383–387. doi: 10.1038/nmat1627
63. Hartwell PG, Walmsley RG, Fasen DJ, Hoen S (2004) Integrated position sensing for control of XY actuator. In: Rocha D, Sarro P, Vellekoop MJ (eds) Proceedings of IEEE Sensors 2004, vol 3, Vienna, pp 1407–1410. doi: 10.1109/ICSENS.2004.1426448
64. Hidaka T, Maruyama T, Saitoh M, Mikoshiba N, Shimizu M, Shiosaki T, Wills LA, Hiskes R, Dicarolis SA, Amano J (1996) Formation and observation of 50 nm polarized domains in $pbzr_{1-x}ti_xo_3$ thin film using scanning probe microscope. Appl Phys Lett 68(17):2358–2359. doi: 10.1063/1.115857
65. Hinz M, Marti O, Gotsmann B, Lantz MA, Durig U (2008) High resolution vacuum scanning thermal microscopy of HfO_2 and SiO_2. Appl Phys Lett 92(4):043122. doi: 10.1063/1.2840186
66. Hiranaga Y, Fujimoto K, Cho Y, Wagatsuma Y, Onoe A, Terabe K, Kitamura K (2002) Nano-sized inverted domain formation in stoichiometric $litao_3$ single crystal using scanning nonlinear dielectric microscopy. Integr Ferroelectr 49:203–209. doi: 10.1080/10584580215489
67. Hiranaga Y, Cho Y, Fujimoto K, Wagatsuma Y, Onoe A (2003) Ultrahigh-density ferroelectric data storage using scanning nonlinear dielectric microscopy. Jpn J Appl Phys Part 1: Regular Papers Short Notes Rev Papers 42(9 B):6050–6054. doi: 10.1143/JJAP.42.6050

68. Hiranaga Y, Fujimoto K, Tanaka K, Wagatsuma Y, Cho Y (2007) Study on SNDM ferroelectric probe memory. Rec Electrical Communi Engi Conversazione Tohoku Univ 75(1):107–110
69. Hoen S, Merchant P, Koke G, Williams J (1997) Electrostatic surface drives: Theoretical considerations and fabrication. In: Proceedings of the 9th international conference on solid-state sensors and actuators (Transducers '97), vol 1, IEEE, Chicago, IL, USA, pp 41–44. doi: 10.1109/SENSOR.1997.613576
70. Hoen S, Bai Q, Harley JA, Horsley DA, Matta F, Verhoeven T, Williams J, Williams KR (2003) A high-performance dipole surface drive for large travel and force. In: TRANSDUCERS 2003, 12th international conference on solid-state sensors, actuators and microsystems, vol 1, pp 344–347. doi: 10.1109/SENSOR.2003.1395506
71. Hong DS, El-Gamal MN (2003) Low operating voltage and short settling time CMOS charge pump for MEMS applications. In: ISCAS '03 proceedings of the 2003 international symposium on circuits and systems, vol 5, pp 281–284. doi: 10.1109/ISCAS.2003.1206254
72. Hosaka S, Koyanagi H, Kikukawa A, Miyamoto M, Imura R, Ushiyama J (1995) Fabrication of nanometer-scale structures on insulators and in magnetic materials using a scanning probe microscope. J Vac Sci Technol B: Microelectron Proc Phenomena 13(3):1307–1311. doi: 10.1116/1.587843
73. Huang X, Lee JI, Ramakrishnan N, Bedillion M, Chu P (2010) Nano-positioning of an electromagnetic scanner with a MEMS capacitive sensor. Mechatronics 20:27–34. doi: 10.1016/j.mechatronics.2009.06.005
74. Hubbard NB, Culpepper ML, Howell LL (2006) Actuators for micropositioners and nanopositioners. Appl Mech Rev 59(1–6):324–334. doi: 10.1115/1.2345371
75. Immink KAS (2004) Codes for mass data storage systems. Shannon Publishing Foundation, Rotterdam, The Netherlands
76. Immink KAS, Siegel PH, Wolf JK (1998) Codes for digital recorders. IEEE Trans Inf Theory 44(6):2260–2299
77. Jacob B, Ng SW, Wang DT (2008) Memory systems (Cache, DRAM, Disk). Morgan Kaufmann, Burlington, MA
78. Kado H, Tohda T (1997) Nanometer-scale erasable recording using atomic force microscope on phase change media. Jpn J Appl Phys Part 1: Regular Papers Short Notes Rev Papers 36(1B):523–525
79. Khatib MG (2009) MEMS-based storage devices – Integration in energy-constrained mobile systems. PhD thesis, University of Twente. doi: 10.3990/1.9789036528474
80. Khatib MG, van Dijk HW (2009) Fast configuration of MEMS-based storage devices for streaming applications. In: Proceedings of the 2009 IEE/ACM/IFIP workshop on embedded systems for real-time multimedia (ESTIMedia 2009), Grenoble, France
81. Khatib MG, Hartel PH (2009) Policies for probe-wear leveling in MEMS-based storage devices. In: Proceedings of the 17th IEEE/ACM international symposium on modeling, analysis and simulation of computer and telecommunications systems (MASCOTS 2009), Lausanne, Switzerland, pp 152–161
82. Hhatib MG, Hartel PH (2010) Optimizing MEMS-based storage devices for mobile battery-powered systems. ACM Trans. Storage, 6, 1, Article 1 (March 2010):37. doi: 10.1145/1714454.1714455. http://doi.acm.org/10.1145/1714454.1714455
83. Kim CH, Jeong HM, Jeon JU, Kim YK (2003) Silicon micro XY-stage with a large area shuttle and no-etching holes for SPM-based data storage. J Microelectromech Syst 12(4):470–478. doi: 10.1109/JMEMS.2003.809960
84. Kim Y, Cho Y, Hong S, Buhlmann S, Park H, Min DK, Kim SH, No K (2006) Correlation between grain size and domain size distributions in ferroelectric media for probe storage applications. Appl Phys Lett 89(16):162907. doi: 10.1063/1.2363942
85. Kim YS, Sunyong Lee C, Jin WH, Jang S, Nam HJ, Bu JU (2005) 100×100 thermo-piezoelectric cantilever array for SPM nano-data-storage application. Sens and Mater 17(2):57–63

86. Kim YS, Nam HJ, Jang S, Lee CS, Jin WH, Cho IJ, Bu JU, Chang SI, Yoon E (2006) Wafer-level transfer of thermo-piezoelectric Si_3N_4 cantilever array on a CMOS circuit for high density probe-based data storage. In: Proceedings of the IEEE international conference on micro electro mechanical systems (MEMS), Istanbul, Turkey, pp 922–925. doi: 10.1109/MEMSYS.2006.1627951
87. Kim YS, Jang S, Lee C, Jin WH, Cho IJ, Ha MH, Nam HJ, Bu JU, Chang SI, Yoon E (2007) Thermo-piezoelectric Si_3N_4 cantilever array on CMOS circuit for high density probe-based data storage. Sens Actuators A 135(1):67–72. doi: 10.1016/j.sna.2006.10.021
88. King WP, Kenny TW, Goodson KE, Cross G, Despont M, Durig U, Rothuizen H, Binnig GK, Vettiger P (2001) Atomic force microscope cantilevers for combined thermomechanical data writing and reading. Appl Phys Lett 78(9):1300–1302. doi: 10.1063/1.1351846
89. King WP, Kenny TW, Goodson KE, Cross GLW, Despont M, Durig UT, Rothuizen H, Binnig G, Vettiger P (2002) Design of atomic force microscope cantilevers for combined thermomechanical writing and thermal reading in array operation. J Microelectromech Syst 11(6):765–774. doi: 10.1109/JMEMS.2002.803283
90. Knoll A, Bachtold P, Bonan J, Cherubini G, Despont M, Drechsler U, Durig U, Gotsmann B, Haberle W, Hagleitner C, Jubin D, Lantz MA, Pantazi A, Pozidis H, Rothuizen H, Sebastian A, Stutz R, Vettiger P, Wiesmann D, Eleftheriou ES (2006) Integrating nanotechnology into a working storage device. Microelectron Eng 83(4–9 SPEC ISS):1692–1697. doi: 10.1016/j.mee.2006.01.206
91. Kotsopoulos AG, Antonakopoulos T (2010) Nanopositioning using the spiral of archimedes: the probe-based storage case. Mechatronics 20:273–280. doi: 10.1016/j.mechatronics.2009.12.004
92. Kuijpers AA (2004) Micromachined capacitive long-range displacement sensor for nanopositioning of microactuator systems. PhD thesis, University of Twente, Enschede, The Netherlands
93. Kuijpers AA, Krijnen GJM, Wiegerink RJ, Lammerink TSJ, Elwenspoek M (2006) A micromachined capacitive incremental position sensor: Part 2. Experimental assessment. J Micromech Microeng 16:S125–S134. doi: 10.1088/0960-1317/16/6/S19
94. Kwon HN, Lee JH, Takahashi K, Toshiyoshi H (2006) MicroXY stages with spider-leg actuators for two-dimensional optical scanning. Sens Actuators A 130–131:468–477. doi:10.1016/j.sna.2005.10.037
95. Lang HP, Berger R, Andreoli C, Brugger J, Despont M, Vettiger P, Gerber Ch, Gimzewski JK, Ramseyer JP, Meyer E, Güntherodt HJ (1998) Sequential position readout from arrays of micromechanical cantilever sensors. Appl Phys Lett 72(3):383–385. doi: 10.1063/1.120749
96. Lantz MA, Gotsmann B, Dürig UT, Vettiger P, Nakayama Y, Shimizu T, Tokumoto H (2003) Carbon nanotube tips for thermomechanical data storage. Appl Phys Lett 83(6):1266–1268. doi: 10.1063/1.1600835
97. Lantz MA, Binnig GK, Despont M, Drechsler U (2005) A micromechanical thermal displacement sensor with nanometre resolution. Nanotechnology 16(8):1089–1094, doi: 10.1088/0957-4484/16/8/016
98. Lantz MA, Rothuizen HE, Drechsler U, Häberle W, Despont M (2007) A vibration resistant nanopositioner for mobile parallel-probe storage applications. J Microelectromech Syst 16(1):130–139. doi: 10.1109/JMEMS.2006.886032
99. Lantz MA, Wiesmann D, Gotsmann B (2009) Dynamic superlubricity and the elimination of wear on the nanoscale. Nat Nanotechnol 4(9):586–591. doi: 10.1038/nnano.2009.199
100. Lee CS, Nam HJ, Kim YS, Jin WH, Cho SM, uk Bu J (2003) Microcantilevers integrated with heaters and piezoelectric detectors for nano data-storage application. Appl Phys Lett 83(23):4839–4841. doi: 10.1063/1.1633009
101. Lee DW, Ono T, Esashi M (2002) Electrical and thermal recording techniques using a heater integrated microprobe. J Micromech Microeng 12(6):841–848. doi: 10.1088/0960-1317/12/6/315
102. Lee JI, Huang X, Chu PB (2009) Nanoprecision MEMS capacitive sensor for linear and rotational positioning. J Microelectromech Syst 18(3):660–670. doi: 10.1109/JMEMS.2009.2016275

103. Lin Y, Brandt SA, Long DDE, Miller EL (2002) Power conservation strategies for MEMS-based storage devices. In: Proceedings of the 10th IEEE international symposium on modeling, analysis and simulation of computer and telecommunications systems, 2002, MASCOTS 2002, pp 53–62. Fort Worth, TX, USA. doi: 10.1109/MASCOT.2002.1167060
104. Lu Y, Pang CK, Chen J, Zhu H, Yang JP, Mou JQ, Guo GX, Chen BM, Lee TH (2005) Design, fabrication and control of a micro X-Y stage with large ultra-thin film recoding media platform. In: Proceedings of the 2005 IEEE/ASME international conference on advanced intelligent mechatronics, pp 19–24. Monterey, CA, USA. doi: 10.1109/AIM.2005.1500959
105. Lutwyche M, Andreoli C, Binnig G, Brugger J, Drechsler U, Haberle W, Rohrer H, Rothuizen H, Vettiger P, Yaralioglu G, Quate C (1999) 5×5 2D AFM cantilever arrays a first step towards a terabit storage device. Sens Actuators A 73(1–2):89–94. doi: 10.1016/S0924-4247(98)00259-3
106. Lutwyche MI, Despont M, Drechsler U, Dürig U, Häberle W, Rothuizen H, Stutz R, Widmer R, Binnig GK, Vettiger P (2000) Highly parallel data storage system based on scanning probe arrays. Appl Phys Lett 77(20):3299–3301. doi: 10.1063/1.1326486
107. Mackay D (2003) Information theory, inference and learning algorithms. Cambridge University Press, Cambridge
108. MacWIlliams FJ, Sloane NJA (1977) The theory of error-correcting codes. North-Holland, Amsterdam
109. Mahmood IA, Moheimani SOR (2009) Fast spiral-scan atomic force microscopy. Nanotechnology 20(36):365,503. doi: 10.1088/0957-4484/20/36/365503
110. Mamin HJ, Rugar D (1992) Thermomechanical writing with an atomic force microscope tip. Appl Phys Lett 61(8):1003–1005. doi: 10.1063/1.108460
111. Mamin HJ, Terris BD, Fan LS, Hoen S, Barrett RC, Rugar D (1995) High-density data storage using proximal probe techniques. IBM J Res Dev 39(6):681–699
112. Mamin HJ, Ried RP, Terris BD, Rugar D (1999) High-density data storage based on the atomic force microscope. Proc IEEE 87(6):1014–1027. doi: 10.1109/5.763314
113. Manalis S, Babcock K, Massie J, Elings V, Dugas M (1995) Submicron studies of recording media using thin-film magnetic scanning probes. Appl Phys Lett 66(19):2585–2587. doi: 10.1063/1.113509
114. Maruyama T, Saitoh M, Sakai I, Hidaka T, Yano Y, Noguchi T (1998) Growth and characterization of 10-nm-thick c-axis oriented epitaxial $PbZr_{0.25}Ti_{0.75}O3$ thin films on (100)Si substrate. Appl Phys Lett 73(24):3524–3526. doi: 10.1063/1.122824
115. Marvell Inc press release (January 2010) Marvell 6Gb/s SATA controllers power next generation motherboards. URL http://www.marvell.com/company/news/press_detail.html?release%ID=1371. Accessed 31 Jan 2010
116. Meyer G, Amer NM (1988) Erratum: novel optical approach to atomic force microscopy (Appl Phys Lett (1988) 53; 1045). Appl Phys Lett 53(24):2400–2402. doi: 10.1063/1.100425
117. Meyer G, Amer NM (1988) Novel optical approach to atomic force microscopy. Appl Phys Lett 53(12):1045–1047. doi: 10.1063/1.100061
118. Min DK, Hong S (2005) Servo and tracking algorithm for a probe storage system. IEEE Trans Magn 41(2):855–859. doi: 10.1109/TMAG.2004.840347
119. Minne SC, Adams JD, Yaralioglu G, Manalis SR, Atalar A, Quate CF (1998) Centimeter scale atomic force microscope imaging and lithography. Appl Phys Lett 73(12):1742–1744. doi: 10.1063/1.122263
120. Naberhuis S (2002) Probe-based recording technology. J Magn Magn Mater 249(3): 447–451. doi: 10.1016/S0304-8853(02)00459-6
121. Nakamura J, Miyamoto M, Hosaka S, Koyanagi H (1995) High-density thermomagnetic recording method using a scanning tunneling microscope. J Appl Phys 77(2):779–781. doi: 10.1063/1.359000
122. Nam HJ, Kim YS, Lee CS, Jin WH, Jang SS, Cho IJ, Bu JU, Choi WB, Choi SW (2007) Silicon nitride cantilever array integrated with silicon heaters and piezoelectric detectors for probe-based data storage. Sens Actuators A 134(2):329–333. doi: 10.1016/j.sna.2006.05.030

123. Nicolau DV (2004) Simulation of the chemical storage of data via metal-ligand chelation. Current Appl Phys 4(2–4):312–315
124. Nishimura T, Iyoki M, Sadayama S (2002) Observation of recording pits on phase-change film using a scanning probe microscope. Ultramicroscopy 91(1–4):119–126, doi: 10.1016/S0304-3991(02)00090-6
125. Ohkubo T, Kishigami J, Yanagisawa K, Kaneko R (1991) Submicron magnetizing and its detection based on the point magnetic recording concept. IEEE Trans Magn 27(6 pt 2): 5286–5288. doi: 10.1109/20.278814
126. Ohkubo T, Kishigami J, Yanagisawa K, Kaneko R (1993) Sub-micron magnetizing and its detection based on the point magnetic recording concept. NTT R&D 42(4):545–556
127. Ohkubo T, Yanagisawa K, Kaneko R, Kishigami J (1993) Magnetic force microscopy for high-density point magnetic recording. Electron Commun Jpn Part II: Electronics (English translation of Denshi Tsushin Gakkai Ronbunshi) 76(5):94–103
128. Ohkubo T, Maeda Y, Koshimoto Y (1995) Point magnetic recording using a force microscope tip on Co–Cr perpendicular media with compositionally separated microstructures. IEICE Trans Electron E78-C(11): 1523–1529. URL http://ci.nii.ac.jp/naid/110003210782/
129. Onoue T, Siekman MH, Abelmann L, Lodder JC (2004) Probe recording on CoNi/Pt multilayered thin films by using an MFM tip. J Magn Magn Mater 272–276:2317–2318. doi: 10.1016/j.jmmm.2003.12.940
130. Onoue T, Siekman MH, Abelmann L, Lodder JC (2005) Heat-assisted magnetic probe recording on a CoNi/Pt multilayered film. J Magn Magn Mater 287(SPEC ISS):501–506. doi: 10.1016/j.jmmm.2004.10.083
131. Onoue T, Siekman H, Abelmann L, Lodder JC (2008) Heat-assisted magnetic probe recording onto a thin film with perpendicular magnetic anisotropy. J Appl Phys D 41(15):155008. doi: 10.1088/0022-3727/41/15/155008
132. Pang CK, Lu Y, Li C, Chen J, Zhu H, Yang J, Mou J, Guo G, Chen BM, Lee TH (2009) Design, fabrication, sensor fusion, and control of a micro X-Y stage media platform for probe-based storage systems. Mechatronics 19:1158–1168. doi: 10.1016/j.mechatronics. 2009.03.005
133. Pantazi A, Lantz MA, Cherubini G, Pozidis H, Eleftheriou E (2004) A servomechanism for a micro-electromechanical-system-based scanning-probe data storage device. Nanotechnology 15(10):612–621. doi: 10.1088/0957-4484/15/10/019
134. Pantazi A, Sebastian A, Cherubini G, Lantz MA, Pozidis H, Rothuizen HE, Eleftheriou E (2007) Control of MEMS-based scanning-probe data-storage devices. IEEE Trans Control Syst T 15(5):824–841. doi: 10.1109/TCST.2006.890286
135. Pantazi A, Sebastian A, Antonakopoulos TA, Bächtold P, Bonaccio AR, Bonan J, Cherubini G, Despont M, DiPietro RA, Drechsler U, Dürig U, Gotsmann B, Häberle W, Hagleitner C, Hedrick JL, Jubin D, Knoll A, Lantz MA, Pentarakis J, Pozidis H, Pratt RC, Rothuizen HE, Stutz R, Varsamou M, Weismann D, Eleftheriou E (2008) Probe-based ultrahigh-density storage technology. IBM J Res Dev 52(4–5):493–511. doi: 10.1147/rd.524.0493
136. Parnell T, Zaboronski O (2009) Month 36 periodic activity report. Work package 4, task 4.4. Technical report, Publications of EU FP6 project ProTeM
137. Parnell T, Wright D, Zaboronski O (2007) Month 12 periodic activity report. Work package 4, task 4.4. Technical report, Publications of EU FP6 project ProTeM
138. Parnell T, Pozidis H, Zaboronski O (2010) Performance evaluation of the probe storage channel. In: Proceedings of IEEE Globecom. Miami, FL, USA
139. Patrascu M (2006) Characterization, modeling and control of the μWalker – a micro actuator for data storage. PhD thesis, University of Twente, Enschede, The Netherlands
140. Patrascu M, Stramigioli S, de Boer MJ, Krijnen GJM (2007) Nanometer range closed-loop control of a stepper micro-motor for data storage. In: Proceedings of 2007 ASME international mechanical engineering congress and exposition, seattle, WAS, USA, ASME, Seattle, pp 1–9

141. Patterson DA, Gibson GA, Katz R (1988) A case for redundant arrays of inexpensive disks. In: SIGMOD international conference on data management, Chicago, IL, pp 109–116
142. Pertsev NA, Rodriguez Contreras J, Kukhar VG, Hermanns B, Kohlstedt H, Waser R (2003) Coercive field of ultrathin $Pb(Zr_{0.52}Ti_{0.48})O_3$ epitaxial films. Appl Phys Lett 83(16): 3356–3358. doi: 10.1063/1.1621731
143. Pires D, Gotsmann B, Porro F, Wiesmann D, Duerig U, Knoll A (2009) Ultraflat templated polymer surfaces. Langmuir 25(9):5141–5145. doi: 10.1021/la804191m
144. Pozidis H, Cherubini G (2009) Probabilistic data detection for probe-based storage channels in the presence of jitter. In: Proceedings of IEEE ICC. Dresden, Germany
145. Pozidis H, Häberle W, Wiesmann D, Drechsler U, Despont M, Albrecht TR, Eleftheriou E (2004) Demonstration of thermomechanical recording at 641 Gbit/in^2. IEEE Trans Magn 40(4):2531–2536. doi: 10.1109/TMAG.2004.830470
146. Pozidis H, Bächtold P, Cherubini G, Eleftheriou E, Hagleitner C, Pantazi A, Sebastian A (2005) Signal processing for probe storage. In: ICASSP, IEEE international conference on acoustics, speech and signal processing – Proceedings, pp 745–748. Philadelphia, PA, United States. doi: 10.1109/ICASSP.2005.1416411
147. Repp J, Meyer G, Olsson FE, Persson M (2004) Controlling the charge state of individual gold adatoms. Science 305(5683):493–495. doi: 10.1126/science.1099557
148. Rosi M, Bauschlicher Jr CW (2001) Using hydrogen and chlorine on Si(1 1 1) to store data, an improved model. Chem Phys Lett 347(4–6):291–296. doi: 10.1016/S0009-2614(01)01060-0
149. Rothuizen H, Drechsler U, Genolet G, Haberle W, Lutwyche M, Stutz R, Widmer R, Vettiger P (2000) Fabrication of a micromachined magnetic *X/Y/Z* scanner for parallel scanning probe applications. Microelectron Eng 53 s(1):509–512. doi: 10.1016/S0167-9317(00)00366-X
150. Rothuizen H, Despont M, Drechsler U, Genolet G, Haberle W, Lutwyche M, Stutz R, Vettiger P (2002) Compact copper/epoxy-based electromagnetic scanner for scanning probe applications. In: Proceedings of the IEEE micro electro mechanical systems (MEMS), Las Vegas, NV, pp 582–585. doi: 10.1109/MEMSYS.2002.984338
151. Ruigrok JJM, Coehoorn R, Cumpson SR, Kesteren HW (2000) Disk recording beyond 100 Gb/in^2: Hybrid recording? J Appl Phys 87(9):5398–5403. doi: 10.1063/1.373356
152. Saheb JF, Richard JF, Sawan M, Meingan R, Savaria Y (2007) System integration of high voltage electrostatic MEMS actuators. Analog Integr Circuits Signal Proc 53(1):27–34. doi: 10.1007/s10470-006-9020-x
153. Sahoo DR, Haberle W, Bachtold P, Sebastian A, Pozidis H, Eleftheriou E (2008) On intermittent-contact mode sensing using electrostatically-actuated micro-cantilevers with integrated thermal sensors. In: 2008 American control conference, ACC, Seattle, WA, pp 2034–2039. doi: 10.1109/ACC.2008.4586792
154. Sarajlic E, Berenschot JW, Fujita H, Krijnen GJM, Elwenspoek MC (2005) Bidirectional electrostatic linear shuffle motor with two degrees of freedom. In: 18th IEEE international conference on micro electro mechanical systems, 2005, Miami, IEEE Computer Society Press, Los Alamitos, CA, pp 391–394. doi: 10.1109/MEMSYS.2005.1453949
155. Sarajlic E, Berenschot JW, Tas NR, Fujita H, Krijnen GJM, Elwenspoek MC (2005) High performance bidirectional electrostatic inchworm motor fabricated by trench isolation technology. In: Proceedings of the 13th international conference on solid-state sensors, actuators and microsystems, vol 1, Seoul, IEEE Computer Society Press, Los Alamitos, CA, pp 53–56. doi: 10.1109/SENSOR.2005.1496357
156. Sasaki M, Bono F, Hane K (2008) Large-displacement micro-*XY*-stage with paired moving plates. Jpn J Appl Phys 47(4 PART 2):3226–3231. doi: 10.1143/JJAP.47.3226
157. Saurenbach F, Terris BD (1990) Imaging of ferroelectric domain walls by force microscopy. Appl Phys Lett 56(17):1703–1705. doi: 10.1063/1.103122
158. Saurenbach F, Terris BD (1992) Electrostatic writing and imaging using a force microscope. IEEE Trans Ind Appl 28(1):256–260. doi: 10.1109/28.120239
159. Schaffhauser D (2008) A storage technology that breaks Moore's Law. Computer World www.computerworld.com. URL http://www.computerworld.com/action/article.do?command=printArticleBasic&articleId=9068318. Accessed 25 Aug 2008

160. Schlosser SW (2004) Using MEMS-based storage devices in computer systems. PhD thesis, Carnegie Mellon University, Pittsburgh, PA, report Nr. CMU-PDL-04-104
161. Schlosser SW, Ganger GR (2004) MEMS-based storage devices and standard disk interfaces: A square peg in a round hole? In: FAST'04: Proceedings of the 3rd USENIX conference on file and storage technologies, USENIX Association, Berkeley, CA, USA, pp 87–100
162. Sebastian A, Pantazi A, Cherubini G, Eleftheriou E, Lantz MA, Pozidis H (2005) Nanopositioning for probe storage. In: Proceedings of the American control conference, vol 6, Portland, OR, United States, pp 4181–4186. doi: 10.1109/ACC.2005.1470634
163. Sebastian A, Pantazi A, Cherubini G, Lantz M, Rothuizen H, Pozidis H, Eleftheriou E (2006) Towards faster data access: seek operations in MEMS-based storage devices. In: 2006 IEEE international conference on control applications, Munich, Germany, pp 283–288. doi: 10.1109/CACSD-CCA-ISIC.2006.4776660
164. Sebastian A, Pantazi A, Pozidis H (2007) Jitter investigation and performance evaluation of a small-scale probe storage device prototype. In: 50th annual IEEE global telecommunications conference, GLOBECOM 2007, Washington, DC, pp 288–293. doi: 10.1109/GLOCOM.2007.61
165. Sebastian A, Pantazi A, Reza Moheimani SO, Pozidis H, Eleftheriou E (2008) Achieving subnanometer precision in a MEMS-based storage device during self-servo write process. IEEE Trans Nanotechnol 7(5):586–595. doi: 10.1109/TNANO.2008.926441
166. Seigler MA, Challener WA, Gage E, Gokemeijer N, Ju G, Lu B, Pelhos K, Peng C, Rottmayer RE, Yang X, Zhou H, Rausch T (2008) Integrated heat assisted magnetic recording head: Design and recording demonstration. IEEE Trans Magn 44(1):119–124. doi: 10.1109/TMAG.2007.911029
167. Severi S, Heck J, Chou TKA, Belov N, Park JS, Harrar D, Jain A, Hoof RV, Bois BD, Coster JD, Pedreira OV, Willegems M, Vaes J, Jamieson G, Haspeslagh L, Adams D, Rao V, Decoutere S, Witvrouw A (2009) CMOS-integrated poly-SiGe cantilevers with read/write system for probe storage device. In: Proceedings of 15th International conference. on solid-state sensors and actuators (Transducers '09), Denver, CO, USA. doi: 10.1109/SENSOR.2009.5285430
168. Shannon CE (July 1948, p.379; Oct 1948, p.623) A mathematical theory of communication. Bell Syst Technical J, vol 27
169. Shaw GA, Trethewey JS, Johnson AD, Drugan WJ, Crone WC (2005) Thermomechanical high-density data storage in a metallic material via the shape-memory effect. Adv Mater 17(9):1123–1127. doi: 10.1002/adma.200400942
170. Shi DX, Ma LP, Xie SS, Pang SJ (2000) Nanometer-scale data storage on 3-phenyl-1-ureidonitrile thin film using scanning tunneling microscopy. J Vac Sci Technol B 18(3):1187–1189. doi: 10.1116/1.591357
171. Shin H, Lee KM, Moon WK, Jeon JU, Lim G, Pak YE, Park JH, Yoon KH (2000) Application of polarized domains in ferroelectric thin films using scanning probe microscope. IEEE Trans Ultrason Ferr Freq Control 47(4):801–807. doi: 10.1109/58.852061
172. Sivan-Zimet M (2001) Workload based optimization of probe-based storage. Master's thesis, University of California, Santa Cruz, CA, USA
173. Sulchek T, Grow RJ, Yaralioglu GG, Minne SC, Quate CF, Manalis SR, Kiraz A, Aydine A, Atalar A (2001) Parallel atomic force microscopy with optical interferometric detection. Appl Phys Lett 78(12):1787–1789. doi: 10.1063/1.1352697
174. Takahashi H, Ono T, Cho Y, Esashi M (2004) Diamond probe for ultra-high-density ferroelectric data storage based on scanning nonlinear dielectric microscopy. In: Proceedings of the IEEE international conference on micro electro mechanical systems (MEMS), Maastricht, the Netherlands, pp 536–539. doi: 10.1109/MEMS.2004.1290640
175. Takahashi H, Ono T, Onoe A, Cho Y, Esashi M (2006) A diamond-tip probe with silicon-based piezoresistive strain gauge for high-density data storage using scanning nonlinear dielectric microscopy. J Micromech Microeng 16(8):1620–1624. doi: 10.1088/0960-1317/16/8/025

176. Takahashi H, Onoe A, Ono T, Cho Y, Esashi M (2006) High-density ferroelectric recording using diamond probe by scanning nonlinear dielectric microscopy. Jpn J Appl Phys, Part 1: Regular Papers Short Notes Rev Papers 45(3 A):1530–1533. doi: 10.1143/JJAP.45.1530
177. Takahashi H, Mimura Y, Mori S, Ishimori M, Onoe A, Ono T, Esashi M (2007) Multi-probe with metallic tips for ferroelectric recording probe storage. In: TRANSDUCERS '07 & Eurosensors XXI. 2007 14th international conference on solid-state sensors, actuators and microsystems, Lyon, France, pp 2509–2512. doi: 10.1109/SENSOR.2007.4300681
178. Takimoto K, Kuroda R, Shido S, Yasuda S, Matsuda H, Eguchi K, Nakagiri T (1997) Writing and reading bit arrays for information storage using conductance change of a Langmuir-Blodgett film induced by scanning tunneling microscopy. J Vac Sci Technol B 15(4):1429–1431. doi: 10.1116/1.589466
179. Tanaka K, Hiranaga Y, Cho Y (2008) Study of servo tracking technique for ferroelectric data storage system. Rec Electrical Commun Eng Conversazione Tohoku Univ 76(1): 384–385
180. Tang WC, Nguyen TCH, Howe RT (1989) Laterally driven polysilicon resonant microstructures. Sens Actuators 20(1–2):25–32. doi: 10.1016/0250-6874(89)87098-2
181. Tas N, Wissink J, Sander L, Lammerink T, Elwenspoek MC (1998) Modeling, design and testing of the electrostatic shuffle motor. Sens Actuators A 70(1–2):171–178. doi: 10.1016/S0924-4247(98)00129-0
182. Tohda T, Kado H (1995) Ultra-high density recording on phase change material using an atomic force microscope. Nat Tech Rep 41(6):31–36
183. Vettiger P, Cross G, Despont M, Drechsler U, Dürig U, Gotsmann B, Häberle W, Lantz MA, Rothuizen HE, Stutz R, Binnig GK (2002) The "millipede"-nanotechnology entering data storage. IEEE Trans Nanotechnol 1(1):39–54. doi: 10.1109/TNANO.2002.1005425
184. Watanuki O, Sonobe Y, Tsuji S, Sai F (1991) Small magnetic patterns written with a scanning tunneling microscope. IEEE Trans Magn 27(6 pt 2):5289 – 5291. doi: 10.1109/20.278815
185. Wiesmann D, Rawling C, Vecchion R, Porro F, Gotsmann B, Knoll A, Pires D, Duering U (2009) Multi tbit/in¡sup¿2¡/sup¿storage densities with thermomechanical probes. Nano Lett 9(9):3171–3176. doi: 10.1021/nl9013666
186. Wright CD, Armand M, Aziz MM, Senkader S, Yu W (2003) Understanding the electro-thermal and phase-transformation processes in phase-change materials for data storage applications. In: Materials research society symposium – Proceedings, vol 803, Boston, MA, United States, pp 61–72
187. Yang JP, Mou JQ, Chong NB, Lu Y, Zhu H, Jiang Q, Kim WG, Chen J, Guo GX, Ong EH (2007) Probe recording technology using novel MEMS devices. Microsyst Technol 13: 733–740, doi: 10.1007/s00542-006-0267-z
188. Yang Z, Yu Y, Li X, Bao H (2006) Nano-mechanical electro-thermal probe array used for high-density storage based on NEMS technology. Microelectron Reliabil 46(5–6):805–810. doi: 10.1016/j.microrel.2005.07.117
189. Yao JJ, Arney SC, MacDonald NC (1992) Fabrication of high frequency two-dimensional nanoactuators for scanned probe devices. J Microelectromech Syst 1(1):14–22. doi: 10.1109/84.128051
190. Yoshida S, Ono T, Oi S, Esashi M (2005) Reversible electrical modification on conductive polymer for proximity probe data storage. Nanotechnology 16(11):2516–2520. doi: 10.1088/0957-4484/16/11/009
191. Yoshida S, Ono T, Esashi M (2007) Conductive polymer patterned media for scanning multiprobe data storage. Nanotechnol 18(50):505302. doi: 10.1088/0957-4484/18/50/505302
192. Yu H, Agrawal D, Abbadi AE (2003) Tabular placement of relational data on MEMS-based storage devices. In: VLDB'2003: Proceedings of the 29th international conference on very large data bases, pp 680–693. Berlin, Germany
193. Zhang D, Chang C, Ono T, Esashi M (2003) A piezodriven XY-microstage for multiprobe nanorecording. Sens Actuators A 108(1–3):230–233. doi: 10.1016/S0924-4247(03)00373-X

194. Zhang L, Bain JA, Zhu JG, Abelmann L, Onoue T (2004) A model for mark size dependence on field emission voltage in heat-assisted magnetic probe recording on CoNi/Pt multilayers. IEEE Trans Magn 40(4):2549–2551. doi: 10.1109/TMAG.2004.830220
195. Zhang L, Bain JA, Zhu JG, Abelmann L, Onoue T (2006) Characterization of heat-assisted magnetic probe recording on CoNi/Pt multilayers. J Magn Magn Mater 305(1):16–23. doi: 10.1016/j.jmmm.2005.11.022
196. Zhang L, Bain JA, Zhu JG, Abelmann L, Onoue T (2006) Dynamic domain motion of thermal-magnetically formed marks on CoNi/Pt multilayers. J Appl Phys 100(053901):1–5. doi: 10.1063/1.2336505
197. Zhang L, Bain JA, Zhu JG, Abelmann L, Onoue T (2006) The effect of external magnetic field on mark size in heat-assisted probe recording on CoNi/Pt multilayers. J Appl Phys 99(023902):1–5. doi: 10.1063/1.2162272
198. Zhang L, Bain JA, Zhu JG, Abelmann L, Onoue T (2006) Heat-assisted magnetic probe recording on a granular CoNi/Pt multilayered film. J Phys D: Appl Phys 39:2485–2487. doi: 10.1088/0022-3727/39/12/002
199. Zhang L, Bain JA, Zhu JG, Abelmann L, Onoue T (2006) A model of heat transfer in STM-based magnetic recording on CoNi/Pt multilayers. Physica B 381:204–208. doi: 10.1016/j.physb.2006.01.007
200. Zhang L, Bain JA, Zhu JG, Abelmann L, Onoue T (2007) The role of MFM signal in mark size measurement in probe-based magnetic recording on CoNi/Pt multilayers. Phys B Condens Matter 387(1–2):328–332. doi: 10.1016/j.physb.2006.04.028
201. Zhao Y, Johns E, Forrester M (2008) A MEMS read-write head for ferroelectric probe storage. In: Proceedings of MEMS 2008, Tucson, AZ, USA, pp 152–155. doi: 10.1109/MEMSYS.2008.4443615
202. Zhong ZY, Zhang L, Zhang HW (2007) Demonstration of close-bit writing in probe storage on magnetic perpendicular media. EPJ Appl Phys 38(3):259–262. doi: 10.1051/epjap:2007071
203. Zybill CE, Li B, Koch F, Graf T (2000) Substrate influence on the domain structure of (111) PZT $PbTi_{0.75}Zr_{0.25}O_3$ films. Phys Status Solidi (A) Appl Res 177(1):303–309, doi: 10.1002/(SICI)1521-396X(200001)177:1<303::AID-PSSA303>3.0.CO;2-G

Chapter 4
Modern Hard Disk Drive Systems: Fundamentals and Future Trends

Tong Zhang, George Mathew, Hao Zhong, and Rino Micheloni

Abstract The objective of this chapter is to provide the readers with a basic technical overview of modern hard disk drives. As a mainstream data storage media, magnetic recording hard disk drives have been enjoying a steady increase in areal recording density for more than four decades, with a rate at least as dramatic as what has been predicted by Moore's Law for the increase of transistor density in integrated circuits. Thanks to continuous advances in recording materials, read and write heads, and read channel signal processing and error correction coding, the areal density increase in hard disk drives appears to be well on track and is steadily moving toward achieving the milestone of 1 Tb/in^2. This chapter will provide an overview of modern hard disk drive systems with the focus on recording channel modeling and advanced signal processing and coding techniques being used in current design practice.

Keywords Hard disk drive · Magnetic recording · Read channel

Hard Disk Drive Systems

Arguably, the two most important technologies that enable today's information age are integrated circuits and magnetic recording. With more than 100 years of non-stopping development [11], magnetic recording has enabled a multibillion dollar industry. Today, magnetic recording systems, particularly hard disk drives, are being used virtually everywhere and play a crucial role in numerous computing and communication applications. During the evolution of hard disk drives, the most noticeable metric is probably *areal storage density*. Since IBM introduced the first

T. Zhang (✉)
Electrical, Computer and Systems Engineering Department, Rensselaer Polytechnic Institute, NY, USA
e-mail: tzhang@ecse.rpi.edu

G. Campardo et al. (eds.), *Memory Mass Storage*,
DOI 10.1007/978-3-642-14752-4_4,

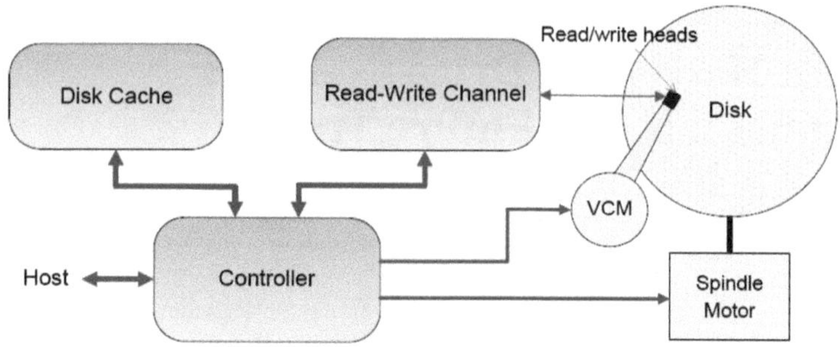

Fig. 4.1 Illustration of a hard disk drive structure

commercial hard disk drive with an area storage density of 2 Kb/in^2 in 1957, the areal storage density compound annual growth rate (CAGR) has been 30% for the first 35 years, jumped to 60% with the introduction of magnetoresistive (MR) heads in 1992, and further increased to around 100% with the introduction of giant magnetoresistive (GMR) heads in the late 1990s. In 2008, hard disk drives with an areal storage density of 610 Gb/in^2 were successfully demonstrated. Such areal storage density growth is enabled by continuous improvements across a variety of technological aspects, mainly including (i) thinner magnetic media with better properties, (ii) better and smaller write and read heads with smaller spacing between heads and media, (iii) more powerful magnetic recording read channel with sophisticated signal processing and coding, and (iv) more accurate head positioning servo.

Figure 4.1 illustrates a simplified hard disk drive structure. The read/write heads and magnetic recording disk are certainly the two most important components and largely determine the overall hard disk drive system performance. The motor controller drives both the spindle motor that rotates the disk and the voice coil motor (VCM) that rotates the actuator on which the read/write heads are mounted. The read–write channel carries out sophisticated digital and analog data processing, geared to the specific characteristics of recording media and read/write heads, to write data to and read data from the disk. The hard disk controller handles a variety of control and management functions such as interface between the hard disk drive and the host computer, disk cache management, and error recovery and fault management, and the disk cache is used as a buffer between host and physical magnetic recording media to improve the overall hard disk drive system performance. This section aims to provide a brief overview of major components in modern hard disk drives.

Hard Disk Drive Recording Media

A recording media is a rotating disk with ferromagnetic surface plane that, in current practice, is a uniform magnetic film in which each bit is stored across a few hundred magnetic grains that are magnetically isolated. As illustrated in Fig. 4.2, the media magnetic anisotropy can be either oriented in the recording medium plane, referred

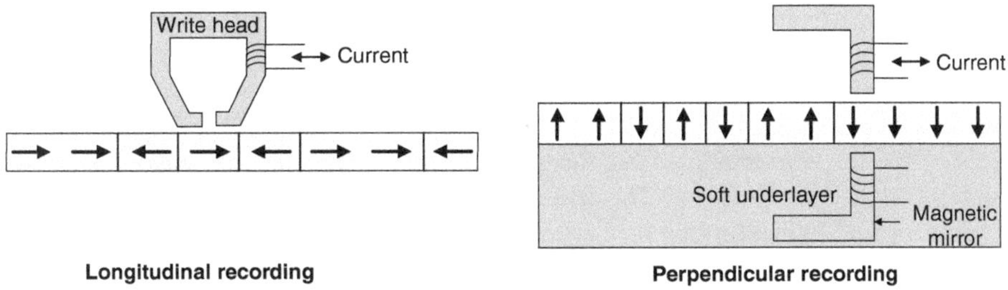

Fig. 4.2 Illustration of longitudinal recording and perpendicular recording

to as longitudinal recording, or aligned perpendicular to the recording medium plane, referred to as perpendicular recording [27].

The essential objective of magnetic recording media design is to maximize the areal density (i.e., minimize the footprint of each bit on the surface of storage media) and meanwhile ensure sufficient magnetization reliability. Magnetization reliability depends on the product of anisotropy constant K_u and volume V of the magnetic grain, called anisotropy energy $K_u V$. A larger anisotropy energy $K_u V$ means a higher energy barrier for switching the magnetization direction. As we reduce the grain volume V to increase the areal storage density, the anisotropy energy will accordingly reduce. If it becomes even comparable to the thermal energy $k_B T$, where k_B is the Boltzmann constant and T is the temperature, the magnetization of grains may not be able to guarantee its stability over a long period such as 10 years. In general, hard disk drives should maintain a large ratio (e.g., 60 and larger) between anisotropy energy $K_u V$ and thermal energy $k_B T$. In theory, perpendicular recording could achieve a higher storage density than its longitudinal counterpart for two main reasons: (i) By aligning the magnetic anisotropy perpendicular to the media surface, we could reduce the footprint of each bit while maintaining a sufficiently large magnetic region with the use of a thicker film; (ii) Aided with the soft magnetic underlayer (SUL) in perpendicular recording as illustrated in Fig. 4.2, the write head field gradient can be stronger, which makes it possible to employ storage material with a higher anisotropy constant K_u. Moreover, as the areal density increases, the demagnetization field of the magnetic grains, which is in opposition to the magnetization of the grains, will decrease. Another advantage of perpendicular recording is that under the same areal density, it is able to produce a larger playback signal compared with longitudinal recording, leading to better signal-to-noise ratio (SNR).

In spite of the above theoretical advantages of perpendicular recording, practical realization of suitable perpendicular recording media is much more complicated than its longitudinal counterpart [9]. As a result, longitudinal recording has been dominating the commercial hard disk drives until recently when the industry began to ship perpendicular recording hard disk drives in 2005. Since then, perpendicular recording has gained an ever increasing momentum, e.g., Hitachi demonstrated 610 Gb/in^2 perpendicular recording in 2008. Today, perpendicular recording is widely considered as a viable technology to approach 1 Tb/in^2 storage density, and the entire industry is quickly switching from longitudinal recording to perpendicular recording.

Currently, several techniques are being actively researched to further enhance the potential of perpendicular recording, among which two promising candidates are heat-assisted magnetic recording [57] and bit-patterned media [32]. In heat-assisted magnetic recording, magnetic properties of recording media monotonically depend on the temperature, and the magnetization can be more easily switched under a higher temperature. The write process exploits this effect to facilitate the data recording by heating up the recording media to a sufficiently high temperature while the write head field is applied and then rapidly cooling down the heated region to an ambient temperature.

In bit-patterned media, the conventional uniform magnetic thin film is replaced by a uniform array of isolated single-domain magnetic islands, each island stores one bit. Since the volume of each island can be much larger than the magnetic grains used in conventional recording media, bit-patterned media can achieve a much better thermal stability. Moreover, by pre-defining the position of each bit, bit-patterned media is not subject to bit position/boundary randomness as in conventional technology. To achieve a sufficiently high-areal storage density, these isolated single-domain magnetic islands should be very closely packed and each island should have a small footprint, e.g., see Fig. 4.3. One of the most critical challenge of bit-patterned media is how to economically realize such a fine-grained patterning, for which several advanced lithography technologies such as nano-imprint or X-ray lithography and self-assembly technologies are currently being actively explored. Accurate synchronization of the write data to the array of magnetic islands is another challenge.

Regardless of the specific recording technology, data on each disk is always organized in a hierarchical structure. First, the disk is partitioned into concentric circles, each circle is called a *track*, and then each track is further partitioned into a certain number of *sectors*. Sector is the basic indivisible storage unit in hard disk drives. Today, each sector contains 512-byte user data, while the industry now is in a transition to a 4096-byte user data sector format. Since the disk rotates at a constant speed, if all the tracks use the same maximum possible recording density, read and write heads will have to read from and write to different tracks at different speeds. Given the hundreds of thousands of tracks on one disk, such an ideal scheme will

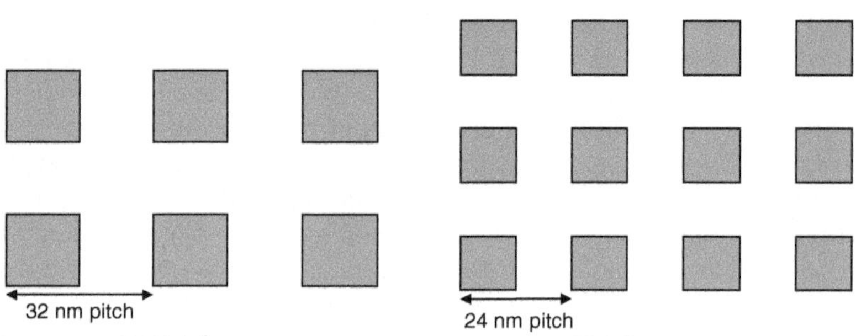

Fig. 4.3 Illustration of the required island pitch to achieve 600 Gb/in^2 and 1 Tb/in^2 storage density

make read and write circuit prohibitively complex. One simple solution used in the early days is that enforce all the tracks have the exactly same number of sectors; hence read and write circuits always operate under the same speed regardless the position of heads on the disk. This nevertheless resulted in a significant loss of storage capacity. As a compromise, a technique called zone-bit recording is typically used today. The basic idea is to partition the disk surface into a few concentric zones, and all the tracks within the same zone have the same number of sectors.

Magnetic Recording Heads

Magnetic recording write heads are responsible to write channel signals on the recording medium, and read heads are responsible to sense and convert the information on the medium to an electrical signal. Although the same inductive transducer was used to both write and read during the early days, modern hard disk drives use separate inductive transducer write heads and flux-sensing read heads. An inductive write head contains a core of magnetically soft material around which a coil of wire is wrapped. As illustrated in Fig. 4.2, in longitudinal recording, the ring-shaped core has a very short gap and, when a current passes through the coil, the fringing flux of the gap is used to write the media, leading to the magnetization oriented in the recording medium plane; while in perpendicular recording, the medium is located in the gap formed by the write head and the underlying SUL, leading to the magnetization perpendicular to the recording medium plane.

Both longitudinal and perpendicular recording use the same basic read head design that utilizes the MR or GMR effect (i.e., the MR or GMR sensor electrical resistance changes as a function of externally applied magnetic field) to sense magnetization transition recorded on the media. By steering a constant current through the MR or GMR sensor, the resistance change is converted to a voltage signal that is fed to the read channel. In both longitudinal and perpendicular recording, read head aims to sense the change of magnetization between adjacent magnetization regions. Let $g(t)$ represent the isolated transition response, i.e., the signal pulse due to a single change in magnetization direction; the noiseless readback waveform can be expressed as

$$z(t) = \sum_k (b_k - b_{k-1})\, g\,(t - kT),$$

where b_k represents the bits recorded and T is the spacing of a single bit. In longitudinal recording, the isolated transition response $g(t)$ is typically modeled as a simple single-parameter Lorentzian pulse:

$$g(t) = \frac{1}{1 + \left(\dfrac{2t}{\mathrm{PW50}}\right)^2},$$

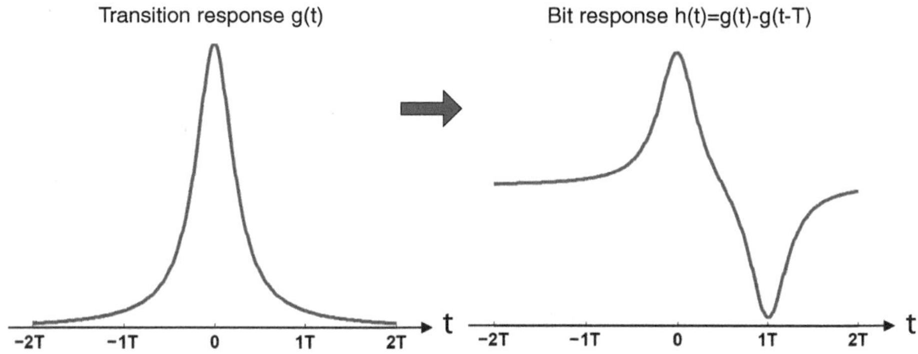

Fig. 4.4 Transition response and the corresponding bit response shapes in longitudinal recording

where the parameter PW50 represents the pulse width at the 50% of the maximum amplitude. In perpendicular recording, the isolated transition response $g(t)$ can be approximated as

$$g(t) = \mathrm{erf}\left(\frac{\sqrt{\ln 16} \cdot t}{\mathrm{PW50}}\right),$$

where erf(·) is the error function defined as $\mathrm{erf}(x) = \left(2/\sqrt{\pi}\right)\int_0^x \mathrm{e}^{-z^2}\mathrm{d}z$ and PW50 is the pulse width of the derivative of $g(t)$ at the 50% of the maximum amplitude. Equivalently, the noiseless readback waveform can also be expressed in terms of the bit response $h(t) = g(t)–g(t–T)$ as

$$z(t) = \sum_k b_k h\,(t - kT).$$

Figures 4.4 and 4.5 illustrate the corresponding waveforms of *g(t)* and *h(t)* in the longitudinal recording and perpendicular recording, respectively.

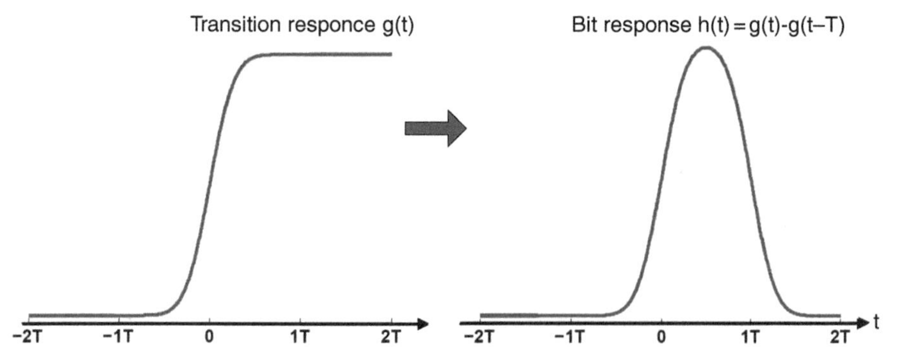

Fig. 4.5 Transition response and the corresponding bit response shapes in perpendicular recording

It is clear from the above figures that the bit response is (much) longer than the spacing between two adjacent bits, T. This means the signal sensed by the read head at each bit position depends on both the current bit and the bits before/after the current bit, which is referred to as inter-symbol interference (ISI). Typically, the significance of ISI is represented by the normalized density D, which is defined as the ratio between PW50 and T, i.e., $D = \text{PW50}/T$.

Finally, we note that one critical parameter affecting the write/read performance, and hence areal storage density is the spacing between the heads and media, referred to as *head fly height*. To sustain the continuous growth of storage density, the head fly height has been consistently shrinking and is only a few nanometers in modern hard disk drives [12]. To maintain such a small head fly height without direct friction between heads and disk, air bearing is used to make the heads ride hydrodynamically on a cushion of air over the disk surface. Moreover, a very thin layer of polymeric lubricant layer is deposited on the disk surface to protect the magnetic material from potential scratches.

Read–Write Channel

As the interface between the read/write heads and hard disk drive controller, read–write channel performs necessary data processing, geared to the characteristics of recording media and heads, to ensure data storage integrity in the presence of inevitable mechanical, magnetic, and electronic inaccuracy and noises. Figure 4.6 shows a simplified structure of a write channel, which mainly carries out modulation encoding, write pre-compensation, and write current driving. Modulation encoding imposes a certain constraints into the bit stream to be recorded in order to facilitate the timing recovery and signal detection during the read process. The most well-known constraints are the run length constraints, which is realized by run length limited (RLL) coding. With two parameters d and k, an RLL(d, k) code enforces that any two 1 s are separated by at least d and not more than k 0 s. Since a 1 is recorded on the hard disk as a magnetization transition, the d-constraint could limit the significance of ISI, and the k-constraint aims to provide sufficiently frequent transitions to improve timing recovery and automatic gain control during the read process. Besides run length constraints, modulation encoding may also enforce some other constraints in order to improve signal detection performance. For example, read channel employs PRML (partial response maximum likelihood) signal detection that uses a Viterbi detector. The detection errors of a Viterbi detector can be dominated by several different error patterns. Hence, certain constraints

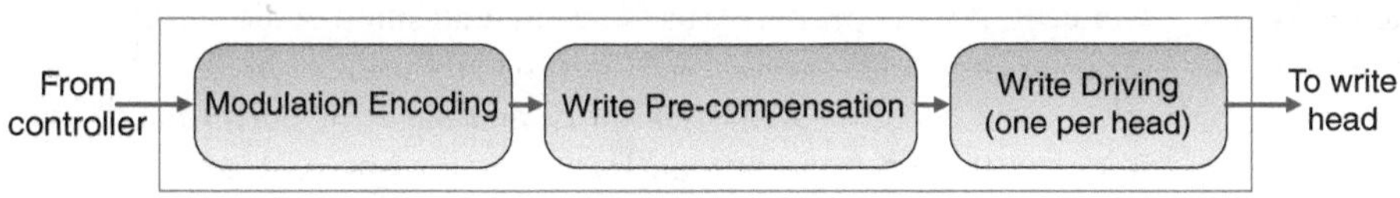

Fig. 4.6 Illustration of write channel structure

can be imposed into the bit stream, which can help to capture the occurrence of such dominant error events during the read process. In this case, the simplest constraint is the parity check constraint. Over the years, a significant amount of research efforts have been devoted to developing various modulation coding techniques for hard disk drives, and interested readers are referred to [24] for detailed discussions.

The purpose of write pre-compensation is to help mitigate nonlinear ISI. If two magnetization transitions are very close, the demagnetizing field from the previous transition will cause the current transition to be shifted during writing. The amount of shift depends on the pattern of transitions immediately preceding the current transition. The basic idea of write pre-compensation is to intentionally introduce a time shift of the write transition during the write process to compensate the effect of the nonlinear transition shift due to previous transitions. Finally, the write current is delivered to the write head through a write current driver in write channel.

Compared with write channel, read channel has a much more complicated structure. The sensed readback signal is amplified by the preamplifier, and then passes the automatic gain control (AGC) circuit to adjust the peak value and a continuous time low-pass anti-aliasing filter. The subsequent ADC (analog-to-digital converter) and timing recovery circuits sample the readback signal and send the digitized signal to the following signal processing and coding functions. Read channel will be discussed in more detail in sections "Modeling of Magnetic Recording Channels" and "Signal Detection and Decoding for Magnetic Recording Channel."

Controller

As the brain of hard disk drives, the controller ensures the entire hard disk drive functions and responds to the host computer correctly. The major functions handled by the hard disk controller include (1) interfacing with the host computer, (2) motor control, (3) managing cache memory, and (4) error correction coding and defect management.

Host Interface

The controller interfaces with the host computer, which can be a server, personal computer, or microcontroller in a consumer product, through a standard data communication protocol. Over the years, a variety of industry standard protocols have been developed, such as serial and parallel ATA and SCSI [26], which defines physical signal transfers, required registers, and command sets. ATA and SCSI have different costs vs. performance trade-off and hence are being used in different systems. With a lower cost, ATA is primarily used in low-end small-scale systems with only few devices to be connected, while SCSI is typically used in high-end larger-scale computing systems because of its flexibility and superior performance in a multitasking and/or multi-user environment. Early versions of ATA and SCSI are parallel with an 8-bit or 16-bit data bus, which has been recently replaced by their serial counterparts that can reduce the cable-bulk and cost and meanwhile increase data transfer rate (e.g., up to 6 Gbit/s data transfer rate can be realized by serial ATA and SCSI).

To reduce design cost and improve flexibility, a hard disk drive controller may support multiple interface protocols. Moreover, it is not uncommon for a controller to embed certain special purpose hardware to handle a small set of critical interface protocol commands, in order to ensure system performance and reduce the load of main processor in the controller.

Motor Control

As illustrated in Fig. 4.1, the controller controls two motors, spindle motor and VCM. The spindle motor spins the disk at a constant rotational speed, while the VCM aims to move the read/write heads to the desired location on the disk very precisely. The controller controls the operation of these two motors through a feedback loop by constantly monitoring the current status/position of the disk and heads based on which it generates control signals to the motors. To facilitate accurate and fast head positioning, periodic servo fields are inserted on the disk surface, as illustrated in Fig. 4.7.

Those servo fields are written on the disk surface once when the disk is manufactured; hence during the run-time, the controller must ensure the servo fields will never be accidentally overwritten by the write head. Since the servo fields occupy the disk surface that would be otherwise available to record user data, there is a design trade-off on how many servo fields should be embedded. In most hard disk drives, the servo fields typically occupy 5–10% of the disk surface.

Cache Memory Management

Modern hard disk drives contain a disk cache DRAM (dynamic random access memory), typically ranging from 8 to 64 MB, as a buffer to streamline the data transfer between the host and the hard disk. The appropriate use of cache can greatly improve the hard disk drive response time and help reduce hard disk drive energy

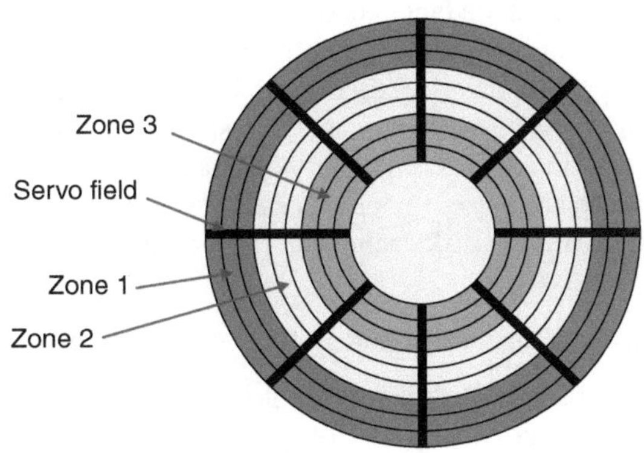

Fig. 4.7 Illustration of a disk with eight servo fields along each track and three zones

consumption. The disk cache can serve for both write and read I/O requests. In the context of write caching, the write data from the host are not immediately recorded on the disk; instead the data are transferred to the disk cache. Since DRAM has a much less write latency than hard disk, the controller can respond to the host to acknowledge the write much earlier, thus allowing the host to quickly continue its own operations without a long wait. However, before being actually recorded on the disk, the cached write data are *dirty* and will result in data storage integrity failure if power is somehow lost. When the write cache becomes (almost) full, a part or all of the data will be flushed to the hard disk, preferably during the idle period of hard disk to minimize the impact on the hard disk drive performance. Besides the obvious advantage of accelerating write acknowledge, the use of write cache can bring the following two advantages: (i) Given a relatively large amount of data to be written, the hard disk drive controller has a high flexibility to schedule the order of the write operations to minimize the hard disk drive write energy consumption and/or latency. By exploiting the proximity among all the data to be written, an appropriate write order scheduling can effectively reduce the overall disk/head rotation time and energy consumption. Over the years, many scheduling algorithms have been developed and their effectiveness has been well demonstrated (e.g., see [8, 17, 66]). (ii) For applications that tend to frequently update the same data, these multiple writes can be filtered out by the write cache, leading to a reduced total number of actual disk recording operations.

Read cache holds a very small portion of the data stored on the disk, which may be soon read by the host with a relatively high probability. Hence, for each host read command, the controller always checks whether the requested data already reside in the disk cache; if yes (i.e., a cache hit), then the controller simply fetches the data from the cache instead of executing the actual disk read operation. Clearly, if the cache hit rate is high enough, read cache can largely improve the overall hard disk drive system performance. To improve the read cache hit rate, good prefetching algorithms should be used for effectively exploiting the temporal and spatial locality presented in the host data access (e.g., see [53, 60]).

In today's hard disk drives, a unified DRAM is used to serve both write caching and read caching, in which the percentages of memory dedicated to read caching and write caching are dynamically configurable. This could make the disk cache dynamically adaptable to the run-time work-loads with different read and write characteristics. Finally, we note that although disk cache predominantly uses DRAM in current practice, there have been much recent discussions on using non-volatile memory, particularly NAND flash memory, to replace DRAM as disk cache in hard disk drives. The potential advantages include write caching without data integrity issue, faster host boot process by pre-loading critical operating system files in the non-volatile disk cache, and reduced overall hard disk drive power consumption.

Error Correction and Defect Management

All the hard disk drives employ error-correcting codes (ECCs) to ensure their storage integrity. The key of ECC is to appropriately introduce a certain amount of redundant information, which can be exploited to detect and correct errors in the retrieved data. Over the past several decades, a very rich family of various ECCs and their efficient encoding/decoding have been developed [6, 40], and ECC has been playing a critical role in numerous data storage and communication systems. Reed–Solomon (RS) codes [65] are used as ECCs in most of today's hard disk drives. RS codes are a class of non-binary linear block codes defined over Galois Fields (GFs). With an underlying GF(2^m), an (n, k) RS code (i.e., the codeword contains n m-bit symbols out of which k symbols carry the user data and the other n–k symbols are redundant data) satisfies $n \leq 2^m - 1$ and $n - k \geq 2t$, where t is the number of erroneous symbols that can be corrected. The objective of RS decoding is to identify the locations of the erroneous symbols and recover their correct values. Because of the use of Viterbi detection in the read channel and bursty nature of recording media defects, the errors in the output of read channel tend to be bursty. The popularity of RS codes in hard disk drives mainly attributes to their non-binary nature and hence good burst error correction capability (i.e., since each symbol contains m bits, an RS code can correct up to $m{\cdot}t$-bit burst errors). Moreover, there are efficient decoding algorithms for RS codes, e.g., the well-known Berlekamp–Massey and Euclidean algorithms [6, 40], which makes it possible to satisfy the decoding speed requirement at reasonable silicon and energy cost. Furthermore, it is possible that the locations of some erroneous symbols can be identified beforehand by the read channel. Such errors with known locations are called *erasures*. Let t and s denote the number of errors and erasures, decoding of an (n, k) RS code will succeed as long as $n{-}k{\geq}2t{-}s$. Although the error correction capability of RS codes can be directly improved by increasing the amount of coding redundancy (i.e., decreasing the code rate k/n), this will nevertheless reduce the disk surface used to store real user data. Hence, the code rate has to be carefully selected so that the overall effective storage density can be maximized while satisfying the desired storage reliability. As a result, hard disk drives typically use RS codes with relatively high code rates, e.g., 9/10 and higher.

As the areal storage density continues to grow, the raw bit error rate (BER) after the read channel tends to increase, which in turn demands the use of more powerful ECCs. One simple solution is to use a longer RS code, i.e., its codeword length is longer. In general, to achieve the same target decoding failure rate, the longer the codeword length is, the higher code rate can be used. This has motivated the industry to switch from the traditional 512-byte user data per sector to 4096-byte user data per sector. However, it should be pointed out that the decoder implementation complexity and decoding latency will accordingly increase as we increase the codeword length. Encouraged by the success of iterative codes such as Turbo codes [4] and LDPC (low-density parity-check) codes [16] in wireless communications, researchers have been heavily investigating on applying such iterative codes to hard disk drives, which may complement with or even completely replace RS codes. However, due to the lack of accurate analytical methods, it remains a challenge to predict the error-correcting performance of such iterative codes under the very low decoding failure rates as demanded by hard disk drives (e.g., 10^{-10} and below). Moreover, decoders of these iterative codes require the read channel signal detector to deliver soft output, which may largely increase the read channel implementation complexity and power consumption.

To further assure the data storage integrity, a cyclic redundancy checksum (CRC) is also generated and recorded together with each sector of user data. This can be used to identify those undetected ECC decoding failures. In case of ECC decoding failures detected by either ECC itself or CRC check, the controller will issue a retry command to reread the same sector. Such reread may be tried several times, and various read head and read channel parameters may be adjusted. Due to the obviously big impact on the system performance, hard disk drives demand a very low retry rate of about 10^{-14}.

No matter how powerful the ECC is, there are always some sectors that contain too many defects to be successfully corrected by the ECC. Such sectors are called defective sectors. Those too many defects may be due to several possible causes, such as disk surface scratches, insufficient magnetic coating material at some spots, and deterioration of magnetic materials. Moreover, regardless to their causes, defects can be either primary defects, which are detected during the hard disk manufacturing, or grown defects, which develop during the lifetime of the hard disk drives.

Once a defective sector is identified, the controller should manipulate the logical to physical address mapping in such a way that this defective sector will never be used to store data. In case of defective sectors detected during manufacturing, the controller simply removes all those defective sectors from the physical address space and makes the non-defective sectors to cover a continuous physical address space. However, in case of defective sectors detected in the field, the controller has to relocate the physical address of each defective sector to another spare sector.

Modeling of Magnetic Recording Channels

Read channel is a critical block in hard disk drives. It is responsible for performing a variety of operations including writing the data bits onto the magnetic medium and recovering the stored data from the medium during readback. The main modules in a read channel chip are [see Fig. 4.8] encoders for protecting the user data bits against distortions in the recording channel, a write precompensation module for minimizing media-induced non-linear distortions during the write process, a loop module for compensating the readback signal for timing/gain/offset errors, a filtering and equalization module for minimizing noise and equalizing the playback signal to a desired shape, a detection module for recovering the stored data bits from the equalized signal, and decoders for decoding the detected data into the original user data bits. To get the best performance from a hard disk drive, these modules in the read channel should be optimally configured according to the specific characteristics of the recording channel consisting of the recording medium and read–write heads. This necessitates the need for accurate modeling of the recording channel so that the resulting channel model encompasses all the key characteristics of heads and media from a signal processing perspective.

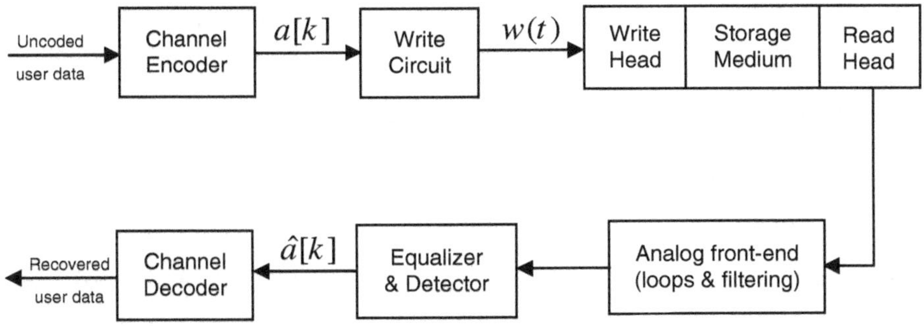

Fig. 4.8 Block schematic of a hard disk drive system

While technological innovations in heads and media are mainly responsible for the growth in storage capacity in hard disk drives, the role of advanced coding and signal processing techniques realized through read channels is critical not only for supporting the advanced heads–media but also as a cost-efficient means for enhancing storage capacity. As conventional magnetic recording technology used in hard disk drives is approaching physical limits and as alternative magnetic recording technologies are being explored, the role of advanced read channel techniques becomes even more critical in maintaining the growth in storage capacity. This underscores the utmost importance of accurate modeling of the recoding channel, and this section deals with this topic.

Key Characteristics of Recording Channel

The various characteristics of recording channel can be classified into categories such as linear vs. non-linear, deterministic vs. stochastic, data-dependent vs. data-independent (here, "data" refer to the data bits written on the medium), stationary vs. non-stationary, with memory vs. without memory, and write-path vs. read-path. We shall briefly list the key channel characteristics here and we will elaborate on these in subsequent sections.

- Bit response: This is the response of the recording channel to a single bit at its input. Duration of this response indicates the memory of the channel. Impulse response and step response are equivalent representation of the recording channel in place of bit response. Ideal readback signal (i.e., noiseless and distortionless) is obtained by convolving the encoded data bits with bit response.
- Electronics noise: This noise is generated by electronic devices, such as preamplifier that conditions the readback signal before passing it to the read channel. Shot noise, thermal noise, and flicker noise constitute the main components of electronics noise [63]. It is data independent and is often modeled as a white noise, recognizing the dominance of shot/thermal noises since flicker noise is

important only at very low frequencies. The most simplified model of recording channel consists of a signal term due to bit response and an additive white noise term due to electronics noise.

- Head noise: There are two types of head noise: Barkhausen noise and Johnson noise [63]. The former arises from fluctuations of magnetic domain walls of the head core material and the latter arises from resistive dissipation in the head. These are also data-independent noises.
- Medium noise: This noise is generated by the recording medium and has several components. The main components are transition noise, DC noise, and modulation noise [5, 68]. Transition noise occurs due to fluctuation in the grain magnetization at the written transition boundaries, and it manifests in the form of position-jitter and width-variation in the written transition, resulting in jitter noise and width-variation noise, respectively. Clearly, transition noise is dependent on the written data and is also non-stationary. DC noise occurs due to randomness in grain sizes and grain center locations, and it is a data-independent stationary noise. Modulation noise occurs due to magnetic fluctuations taking place between magnetic transitions and is hence a non-stationary data-dependent noise. Texture noise is an important form of modulation noise and is associated with the mechanical texture of the disk substrate.
- Write-related non-linear distortions: There are several non-linear distortions that occur during the writing process. The important distortions are non-linear transition shift (NLTS), partial erasure (PE), transition broadening (TB), and overwrite [48, 51, 62]. NLTS and TB refer to shifts and broadening, respectively, of written transitions due to the influence of the demagnetizing field from previous transitions on the head field writing the current transition. Bandwidth limitations of the write-path also contribute to NLTS [3]. PE refers to partial erasing of a transition and the resulting loss in signal amplitude when closely spaced transitions are written. Overwrite or hard transition shift (HTS) refers to shift in written transitions due to the influence of old information residing on the medium. These distortions are clearly data dependent and approximately deterministic in nature.
- Read-related non-linear distortions: Two major non-linear distortions during readback are that due to the non-linear transfer function of the magneto-resistive (MR) read-head and that due to the presence of asperities on the medium [59, 72]. When MR head comes in contact with the medium because of the presence of an asperity, a resistance transient is induced in the head resulting in a sudden change in the readback signal and is referred to as MR thermal asperity (TA). The non-linearity of the transfer function of a given MR head is clearly a data-independent deterministic distortion. The occurrence and intensity of TA distortions are data independent and random in nature, due to randomness associated with the location and origin of asperities.
- Media defects: The presence of defects on the medium causes the readback signal to be attenuated (known as drop-outs), and the degree/profile of attenuation and duration of this depend on the severity of the defect [61]. The occurrence and intensity of drop-outs are data independent and random in nature, due to randomness associated with the location and origin of defects.

In the next few sections, we will elaborate on each of the above characteristic. Our emphasis will be on providing simple and sufficiently accurate models for representing each of these in numerical simulation of the recording channel. Before we get into the details, we shall first describe some general principles that would help to clarify the similarity between recording channels and communication channels and lay the foundation for the discussion in subsequent sections.

Channel Modeling Preliminaries

Figure 4.8 shows the essential blocks of a hard disk drive from a signal processing perspective. Here, $a[k]$ denotes the channel encoded data bits in $\{-1,+1\}$ format, $w(t)$ denotes the current waveform supplied to the write head, and $\hat{a}[k]$ denotes the data bits recovered/detected from the readback signal. The signal path from the input of "write head" to the output of "read-head" is what is referred to as "recording channel" and this is what we need to model for simulation and design of read channels.

Hard disk drives make use of the so-called saturation recording for storing binary data bits on magnetic medium. That is, bit "+1" or "-1" is stored by magnetizing a selected small region on the medium in one direction or in the other. This is done by supplying current pulses of appropriate polarity to the inductive write-head so that the resulting head field will be in the correct direction to magnetize the medium. Thus, there is a one-to-one correspondence between the magnetization pattern on the medium and the sequence of data bits. Upon readback, a MR read-head senses the magnetization stored on the medium, converts that into a voltage signal, and gives it to the preamplifier. The ideal readback signal can be mathematically represented as [48]

$$r_{\mathrm{s}}(t) = \frac{\mathrm{d}w(t)}{\mathrm{d}t} \otimes h_{\mathrm{s}}(t) \qquad \text{with} \qquad w(t) = \sum_k a[k]s\,(t-kT), \tag{4.1}$$

where $h_s(t)$ denotes a readback system response which is related to the head-medium parameters, $w(t)$ denotes the current supplied to the write-head, T denotes the duration of one bit, and $s(t)$ denotes an ideal unit-amplitude rectangular current pulse of duration T located at $t = 0$. Using the expression for the write current waveform, we can obtain

$$r_{\mathrm{s}}(t) = \sum_k a[k]h_{\mathrm{b}}\,(t-kT), \tag{4.2}$$

where

$$h_{\mathrm{b}}(t) = h_{\mathrm{s}}(t) - h_{\mathrm{s}}\,(t-kT)\,. \tag{4.3}$$

Here, $h_{\mathrm{b}}(t)$ is known as "bit response" of the recording channel since it is the read-head output for a single bit "+1" at the input of write circuit. It is also called "pulse response" since a single bit at write circuit input corresponds to pulse $s(t)$ of one bit duration at the input of write-head. Correspondingly, $h_s(t)$ is known as "step response" of the channel since the pulse $s(t)$ is obtained by subtracting a unit step input shifted by T from the unshifted one. Finally, "impulse response" $h_i(t)$ of the channel, which is the readback response when a Dirac delta function is applied at write-head input, is given by the time-derivative of step response as

$$h_i(t) = \frac{dh_s(t)}{dt}. \tag{4.4}$$

Observe from (4.2) that the magnetic recording channel resembles a binary pulse amplitude modulated (PAM) base-band communication channel.

Linear Channel Models

The description of a noiseless linear recording channel is complete if its bit response, step response, or impulse response are specified. All of these three responses are equivalent. Commonly used models for step response and corresponding impulse response in perpendicular recording are [18, 61, 69]

$$h_{s1}(t) = A_0 \mathrm{erf}\left(\frac{2\sqrt{\ln 2}}{\mathrm{PW}_{50}}t\right), \quad h_{i1}(t) = A_0 \frac{4\sqrt{\ln 2}}{\sqrt{\pi}\mathrm{PW}_{50}} \exp\left(-\frac{4\ln 2}{\mathrm{PW}_{50}^2}t\right), \tag{4.5}$$

$$h_{s2}(t) = A_0 \tanh\left(\frac{\ln 3}{T_{50}}t\right), \quad h_{i2}(t) = A_0 \frac{\ln 3}{T_{50}} \mathrm{sech}^2\left(\frac{\ln 3}{T_{50}}t\right), \tag{4.6}$$

$$h_{s3}(t) = A_0 \tan^{-1}\left(\frac{2}{T_{50}}t\right), \quad h_{i3}(t) = A_0 \frac{2 \cdot T_{50}}{T_{50}^2 + 4t^2}, \tag{4.7}$$

where T_{50} denotes the time required for the step response to rise from –50% to + 50% of the saturation amplitude and PW_{50} denotes the pulse-width of the impulse response at 50% of its peak amplitude. The ratios T_{50}/T and PW_{50}/T are used as measures of the normalized linear bit density from a signal processing perspective. The higher these numbers are, the larger will be the memory of the channel and ISI caused by the channel, and vice versa. Figure 4.9 shows the step response and the resulting bit response for the error function model given in (2.5) for PW_{50}/T equal to 1.5 and 3.0. Observe that as PW_{50}/T increases, the length of ISI increases and amplitude of the bit response decreases.

Head Noise and Electronics Noise

The Barkhausen-type head noise arises in both inductive and MR heads. As mentioned already, it is caused by large changes in domain wall structure in the thin films in response to magnetic fields or mechanical stresses. This noise can be minimized

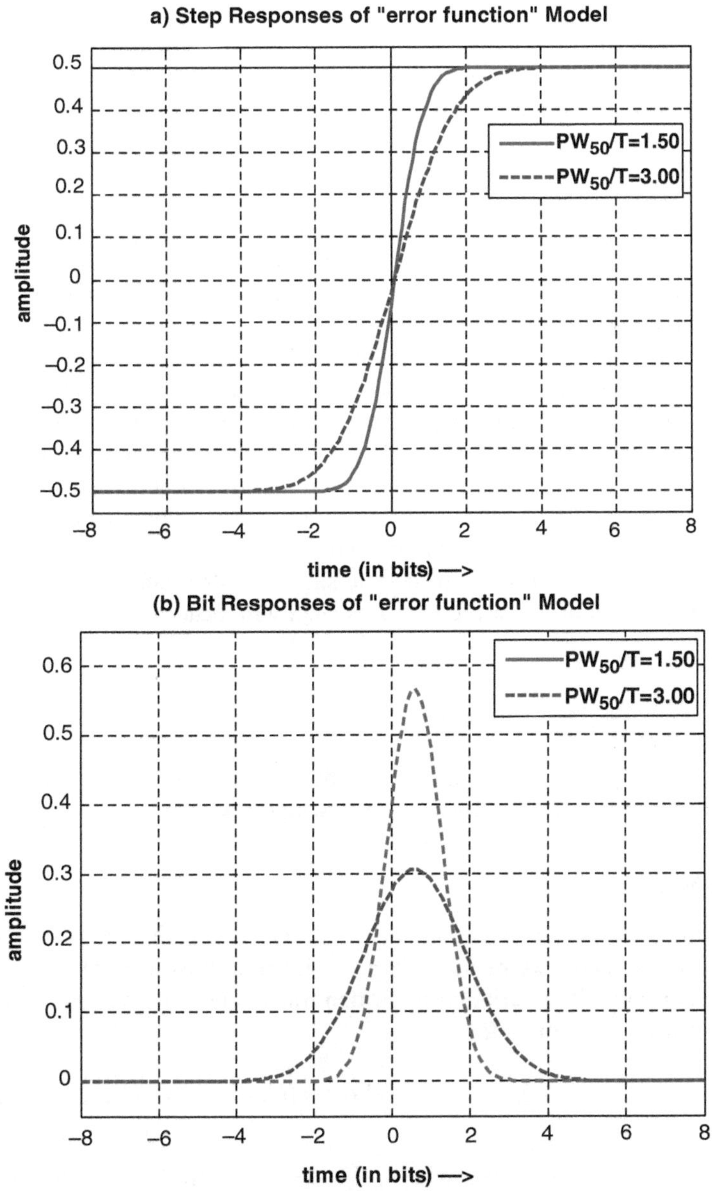

Fig. 4.9 Error function model for perpendicular recording channel with normalized linear bit density 1.5 and 3.0. (**a**) Step response and (**b**) bit response

by careful micro-magnetic design of the head [47]. The source for Johnson (or thermal)-type head noise is the resistive component of the head impedance. The resistive part dissipates energy and generates noise [47]. The root mean square (rms) value of the resulting Johnson noise is given by

$$V_{\mathrm{rms}} = \sqrt{4k_{\mathrm{b}}T_0\mathrm{R}e(Z)\Delta f}, \tag{4.8}$$

where k_b denotes Boltzmann's constant, T_0 denotes the absolute temperature, $\mathrm{R}e(Z)$ denotes the real part (i.e., resistive component) of the head impedance, and Δf denotes bandwidth. In MR heads, Johnson noise is determined almost solely by the electrical resistance of the MR sensor.

Electronics noise, which is a random signal characteristic of all electronic devices, is caused by random fluctuations in time of the electric charge carriers. As mentioned already, the main components of electronics noise in magnetic recording are thermal (or, Johnson) noise and shot noise. In a given device, thermal noise can be described by an equation similar to (4.8), while shot noise is related to the average current through the device. Over the frequency range of interest in practical systems, the combined thermal and shot noise can be modeled by an additive white Gaussian process $\varepsilon_e(t)$ with double-sided power spectral density by $N_e/2$. The resulting model of linear recording channel with head/electronics noise can be given as

$$r_1(t) = \sum_k a[k]h_b\,(t - kT) + \varepsilon_e(t). \tag{4.9}$$

The flicker noise component, which arises due to device imperfections, has power spectral density that is proportional to $1/f^{\alpha}$ with $0.5 < \alpha < 2$.

Medium Noise

As highlighted earlier, the main components of media-generated noise are transition noise, DC noise, and texture noise. Since transition noise can be modeled using position-jitter and width-variation in written transitions [71], readback signal with transition noise can be modeled as

$$r(t) = \sum_k b[k]h_s\,(t - kT + \tau[k], W + \Delta W[k]), \tag{4.10}$$

where $b[k] = a[k] - a[k-1]$, with $b[k] \in \{-2, 0, +2\}$, denotes the sequence of written transitions; W stands for PW_{50} or T_{50}; and $\tau[k]$ and $\Delta W[k]$ denote the position-jitter and width-variation, respectively, in the transition at kth bit instant. When the width-variation and position-jitter are sufficiently small, an approximate expression for transition noise can be obtained from (4.10) using first-order Taylor series approximation of $h_{\mathrm{s}}\,(t - kT + \tau[k], W + \Delta W[k])$ resulting in [48]

$$n_{\mathrm{m}}(t) = \sum_k b[k]\,\{\tau[k]h_{\mathrm{i}}\,(t - kT) + \Delta W[k]h_{\mathrm{w}}\,(t - kT)\} \tag{4.11}$$

with

$$h_{\rm i}\,(t) = \frac{\partial h_{\rm s}(t)}{\partial t} \qquad \text{and} \qquad h_{\rm w}\,(t) = \frac{\partial h_{\rm s}(t)}{\partial W}, \tag{4.12}$$

where $h_{\rm i}\,(t)$ and $h_{\rm w}\,(t)$ are the impulse response and width response, respectively, of the recording channel. Extension of this model using second-order Taylor approximation is straightforward [48]. It is easy to observe from Eqs. (4.10), (4.11), and (4.12) that (i) transition noise depends on the written data sequence, (ii) it is strong in areas where there are more transitions and vice-versa, and (iii) it can be characterized by specifying the distributions of position-jitter $\tau[k]$ and width-variation $\Delta W[k]$ and step response $h_{\rm s}\,(t)$. It is common to model $\tau[k]$ and $\Delta W[k]$ as mutually independent Gaussian random variables that are independent of the data sequence $b[k]$. At very high linear recording densities, the distribution of $\tau[k]$ may need to be modified to support high jitter percentages with restricted distribution tails.

DC noise is modeled as a data-independent stationary colored noise with power spectral density given by that of channel impulse response [29]. The resulting expression for DC noise can be given by

$$n_{\rm d}(t) = \int_{-\infty}^{\infty} \varepsilon_{\rm d}(u) h_{\rm i}\,(t-u)\,{\rm d}u, \tag{4.13}$$

where $\varepsilon_{\rm d}(t)$ is a Gaussian white noise process with double-sided power spectral density $N_\varepsilon/2$. Using (4.9), (4.10), (4.11), (4.12), and (4.13), the model of linear recording channel with head/electronics noise and media noise can be given as

$$r(t) = \sum_k b[k] h_{\rm s}\,(t - kT + \tau[k], W + \Delta W[k]) + \varepsilon_e(t) + n_{\rm d}(t), \tag{4.14}$$

$$\approx \sum_k b[k] h_{\rm s}\,(t - kT, W) + n_{\rm m}(t) + \varepsilon_{\rm e}(t) + n_{\rm d}(t). \tag{4.15}$$

Write-Related Non-linear Distortions

As recording density increases, nearby magnetic transitions begin to interact and this leads to breaking down the linear nature of the channel. As introduced briefly earlier, the important non-linear distortions are NLTS, PE, TB, and HTS. A model of the noiseless readback signal incorporating these distortions can be given as [37, 52]

$$\begin{aligned} r_{\rm n}(t) = {} & \sum_k \tilde{b}[k] h_{\rm s}\left(t - kT + \frac{\Delta_0}{4} b[k]\,(1 - a[k]) + \frac{\Delta_1}{4} b[k] b[k-1]\right) \\ & + \frac{\Delta_b}{4} \sum_k \tilde{b}[k] \cdot b[k] b[k-1] h_{\rm s}''\,(t - kT), \end{aligned} \tag{4.16}$$

where $\tilde{b}[k]$ is a modified form of $b[k]$ due to PE given by

$$\left.\begin{aligned}\tilde{b}[k] &= \gamma^2 b[k] \text{ if } \; b[k+1] \neq 0 \text{ and } \; b[k-1] \neq 0\\ &= \gamma b[k] \text{ if } \; b[k+1] \neq 0 \text{ or } \; b[k-1] \neq 0\\ &= b[k] \text{ if } \; b[k+1] = 0 \text{ and } \; b[k-1] = 0.\end{aligned}\right\}. \tag{4.17}$$

Here, $0 < \gamma < 1$ denotes amplitude loss in transition due to PE, Δ_0 denotes the shift in current transition due to pre-existing magnetization in the direction of $a[k] = +1$ (i.e., HTS), Δ_1 denotes the shift in current transition due to a transition one-bit earlier (i.e., NLTS), and Δ_b models the broadening of current transition due to NLTS arising from immediately preceding transition. For the sake of simplicity, (4.16) considers the NLTS from the transition one-bit earlier only. Generalizing this to the case of patterns of transitions preceding the current transition is straightforward [39]. When transition shifts are small, (4.16) can be simplified using first-order Taylor series as

$$\begin{aligned} r_n(t) \approx & \sum_k \tilde{b}[k] h_s(t-kT) + \frac{\Delta_0}{4} \sum_k \tilde{b}[k] b[k] (1-a[k]) h_i(t-kT) \\ & + \frac{1}{4} \sum_k \tilde{b}[k] b[k] b[k-1] \left\{ \Delta_1 h_i(t-kT) + \Delta_b h_s''(t-kT) \right\}. \end{aligned} \tag{4.18}$$

Readback Non-linear Distortions

MR non-linearity and MR thermal asperity (TA) are the two major readback non-linear distortions. Non-linear transfer function of MR head manifests asymmetry and saturation [41, 72]. Asymmetry arises due to not choosing the operating point of the MR head in the center of the linear region, and saturation arises when input signal amplitude exceeds the linear region in both directions. If $r(t)$ denotes the output of ideal linear MR head, then the output of non-linear MR head can be modeled as

$$\tilde{r}(t) = r(t) + \alpha r^2(t) + \beta r^3(t), \tag{4.19}$$

where α and β denote the amount of signal asymmetry and stripe saturation, respectively.

MR TA caused by contact between MR head and a raised defect on the medium manifests in the readback signal as a change in baseline that is characterized by fast rise time, large peak amplitude, and approximately exponential decay [13, 59]. The resulting model for the baseline shift caused by TA can be given by

$$\left.\begin{aligned} c(t) &= A_0 t / T_r & &\text{for } 0 \leq t \leq T_r\\ &= A_0 \exp\left(-(t-T_r)/T_d\right) & &\text{for } T_r < t \leq T_f\end{aligned}\right\}, \tag{4.20}$$

where A_0 denotes the peak TA amplitude, T_r and T_d denote the rise time and decay time-constant, respectively, and T_f denotes the duration of TA. This model for baseline shift is added to the normal readback signal to create readback signal with TA.

The peak amplitude in TA signal is proportional to the maximum average temperature rise in the stripe, and the decay time constant is related to the read-gap thickness. While this type of contact-based MR TAs constitutes the majority of occurrences of MR TAs, there are also non-contact-type MR TAs [59]. Manifestation or signature of non-contact MR TAs in readback signal is quite different from that of contact MR TAs.

Media Defects

Most commonly encountered media defects are characterized by loss of amplitude and are known as drop-outs. These are caused mainly by various types of defects on the medium and/or loose particles in the head-medium interface. The most common model used for modeling drop-outs modulates the amplitude of the readback signal based on certain profile [58, 61]. Parameters of the model include the depth of amplitude loss and duration of drop-out. The loss of amplitude may be modeled as sudden drop in amplitude or a continuous variation in amplitude over the defect length.

Generalized Channel Models

The models that we have considered in the above sections are closely tied to the physical mechanism underlying each of the distortion. There have also been other models proposed in the literature which are developed with the objective of capturing multiple distortions with a single generalized model. The data-dependent auto-regressive (DDAR) model [31] and Volterra series model [22] are typical examples for generalized modeling. While these models are very appealing from a signal processing perspective, it is difficult to relate the model parameters directly to physical mechanisms.

The DDAR model is very extensively used in magnetic recording. It is a concise parametric representation of the recording channel and provides comprehensive modeling coverage. The reason for its wide acceptance can be attributed to the several advantages it offers. It is able to model linear/non-linear deterministic distortions and stochastic noises with only a few parameters. Since there are only a few parameters, estimation of model parameters can be done rather fast with good accuracy. Most importantly, the form of the DDAR model points to the form of data detector that is optimum for such models [30]. We recall here the causal version of this model, while the non-causal version is available in [30]. Based on this model, the bit-rate sampled output of the channel is given by

$$z[k] = y\left(a_{k-I}^{k}\right) + v[k], \tag{4.21}$$

$$v[k] = \sum_{i=1}^{L} g_i\left(a_{k-I}^{k}\right) \cdot v[k-i] + w\left(a_{k-I}^{k}\right), \tag{4.22}$$

where $y\left(a_{k-I}^{k}\right)$ models the noiseless output of the channel which depends on the $I+1$ data bits $a_{k-I}^{k}=\{a[k-I],\ a[k-I+1],\ \cdots,\ a[k]\}$, I is the data memory length, and $v[k]$ models the additive noise component. The signal component $y\left(a_{k-I}^{k}\right)$ is constructed as a look-up table to incorporate both linear and non-linear channel distortions. The noise component $v[k]$ is the output of a data-dependent auto-regressive (AR) filter whose parameters, i.e., coefficients $g_i\left(a_{k-I}^{k}\right)$, $i=1,2,\ldots,L$, and standard deviation $\sigma_w\left(a_{k-I}^{k}\right)$ of uncorrelated Gaussian excitation $w[k]$, are dependent on the data bits a_{k-I}^{k}. Here, L denotes the Markov memory length of the noise model. Clearly, the noise $v[k]$ is both data-dependent and correlated, and hence non-stationary. The parameters of this DDAR model, i.e., $\left\{y\left(a_{k-I}^{k}\right),\ g_1\left(a_{k-I}^{k}\right),\ g_2\left(a_{k-I}^{k}\right),\ \cdots,\ g_L\left(a_{k-I}^{k}\right),\ \sigma_w\left(a_{k-I}^{k}\right)\right\}$ for all possible a_{k-I}^{k}, can be estimated using data-dependent mean and covariances and solving the Yule–Walker equations [31].

The Volterra series model [22] models the non-linear portion of the channel output as the sum of the outputs of non-linear kernels, where each kernel is driven by a product of selected data bits. The shape of the kernels and associated bit-products at their inputs depend on the types of non-linear distortions being modeled. As an example, considering non-linear distortions that result in non-liner combination of $\{a[k],\ a[k-1],\ a[k-2],\ a[k-3]\}$ at k th instant, the resulting Volterra model for the non-linear distortion components in channel output can be given by

$$\tilde{r}_n(t)=\sum_k\left\{\begin{array}{l} a[k]a[k-1]h_{2,1}(t-kT)+\ a[k]a[k-2]h_{2,2}(t-kT)\\ +\ a[k]a[k-3]h_{2,3}(t-kT)\\ +\ a[k]a[k-1]a[k-2]h_{3,1}(t-kT)\\ +\ a[k]a[k-1]a[k-3]h_{3,2}(t-kT)\\ +\ a[k]a[k-2]a[k-3]h_{3,3}(t-kT)\\ +\ a[k]a[k-1]a[k-2]a[k-3]h_{4,1}(t-kT) \end{array}\right\}, \tag{4.23}$$

where $h_{\mathrm{i,j}}(t)$ denotes the kernels associated with the various bit-products. These kernels can be estimated from measured data, e.g., by an adaptive training approach or through systematic cross-correlation approaches. It is straightforward to see that by adding the linear component and stochastic noise components to $\tilde{r}_{\mathrm{n}}(t)$ in (4.23), one can obtain a complete model for the channel output.

Signal-to-Noise Ratio

Signal-to-noise ratio (SNR) is the fundamental metric used for assessing the effectiveness or advantage of coding and/or signal processing approaches. The purpose of SNR is to determine the amount of noise required to achieve a specified level of error rate performance for a given signal strength. Consequently, to ensure fair comparison, it is important that the SNR definition is independent of the code rate of the channel code and recording density. In magnetic recording systems, this requirement is complicated by the fact that media noise is dependent on the written data bits.

A SNR definition that is widely used in magnetic recording [49, 69] is as follows:

$$\mathrm{SNR} = \frac{E_\mathrm{i}}{N_0 + M_0}, \qquad N_\mathrm{mix} = \frac{M_0}{N_0 + M_0}. \tag{4.24}$$

where E_i is the energy of the channel impulse response, $N_0/2$ is the power spectral density of electronics noise (white Gaussian), $M_0/2$ is the average energy of the media noise associated with each transition, and N_mix denotes the ratio of medium noise power to the total in-band noise power. Clearly, the quantities $\{E_\mathrm{i},\ N_0,\ M_0\}$ are independent of the code rate and linear density. For the first-order model for transition noise given in (4.11), the expression for M_0 can be obtained as

$$M_0 = 2\sigma_\tau^2 \int_{-\infty}^{\infty} h_\mathrm{i}^2(t)\mathrm{d}t + 2\sigma_\mathrm{w}^2 \int_{-\infty}^{\infty} h_\mathrm{w}^2(t)\mathrm{d}t, \tag{4.25}$$

where σ_τ^2 and σ_w^2 are the variances of write-jitter $\tau[k]$ and width-variation $\Delta W[k]$, respectively. In the presence of DC noise, the denominators in (4.24) should be replaced by $N_0 + M_0 + D_0$ where D_0 denotes the DC noise power normalized by a reference bandwidth that is independent of code rate and linear density. To make the media noise specification complete, it is also necessary to define the ratio of D_0 to $N_0 + M_0 + D_0$.

Equalization for Magnetic Recording Channels

In this section, we briefly discuss the role of equalization in magnetic recording channels and the various strategies used for equalization. Figure 4.10 shows the schematic of a recording channel followed by equalizer and detector. Data bits $a[k] \in \{-1, +1\}$ written to the medium result in a readback signal $r(t)$. The read-head output $r(t)$ is conditioned using preamplifier and the analog front-end in read channel. The conditioned signal $x(t)$ is sampled at bit-rate $1/T$ to result in $x[k] = x(kT + t_0)$ where T denotes the duration of one bit and t_0 denotes the sampling phase. The sampled output $x[k]$ is filtered using an equalizer and the equalized sequence $y[k]$ is acted upon by a detector to recover the written data bits. One could also sample $x(t)$ at a rate higher than the bit-rate and accordingly use an equalizer with fractionally spaced taps to ensure that aliasing caused by sampling

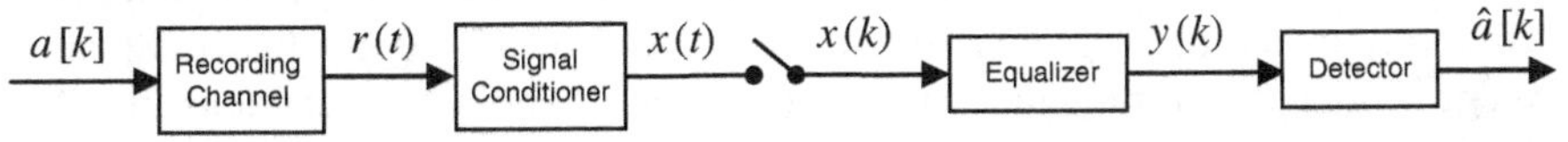

Fig. 4.10 Recording channel with equalizer and detector

does not affect equalizer performance. In this section, we assume that $x(t)$ is sufficiently band-limited so that bit-rate sampling results in no or negligible aliasing in $x[k]$, and hence the equalizer performance can be considered to be independent of the sampling phase t_0. Therefore, without loss of generality, we set $t_0 = 0$ in this section.

It is clear from the previous section that the head output $r(t)$ contains linear ISI, non-linear distortions, media noise, and electronics noise. We assume that mitigation approaches such as write precompensation and asymmetry correction have been used so that non-linear distortion can be assumed negligible in $x(t)$. The main purpose of equalizer is to shape the signal component in $x[k]$ into a form suitable for detection of the data bits by detector, while minimizing the noise as much as possible. In other words, the function of equalizer is closely tied to the technique used for detecting the data bits. In modern read channels, equalizer is implemented as a T-spaced finite impulse response (FIR) filter with $N_w + 1$ coefficients given by w_i for $i = 0, 1, 2, \ldots, N_w$. In the rest of this section, we will briefly discuss the various equalization strategies and the different techniques used for designing the equalizer.

Equalization Strategies

There are mainly three types of equalization strategies: full response (FR) equalization, partial response (PR) equalization, and decision feedback (DF) equalization. Assuming that $x[k]$ is free from non-linear distortions, we can express it as

$$x[k] = h_0 a[k] + \sum_{i \neq 0} h_i a[k-i] + \tilde{\eta}[k], \tag{4.26}$$

where h_0 and h_i for $i \neq 0$ denote the linear ISI in unequalized channel and $\tilde{\eta}[k]$ denotes the noise part. The equalizer output $y[k]$ can be expressed as

$$y[k] = g_0 a[k] + \sum_{i \neq 0} g_i a[k-i] + \eta[k],$$

where g_0 and g_i for $i \neq 0$ denote equalized linear ISI and $\eta[k]$ denotes the corresponding noise part.

The objective of FR equalization is to "fully" cancel the ISI caused by the finite bandwidth of the recording channel, i.e., to make $g_i = 0$ for $i \neq 0$ so that the equalizer output consists of only a scalar multiple of the current data bit $a[k]$ and noise. Thus, the equalized channel (i.e., the cascade of recording channel, signal conditioner and equalizer) behaves like an ideal infinite bandwidth channel with flat magnitude response. Consequently, the detector can be a simple threshold detector to discern the sign of $y[k]$ since the mean of $\eta[k]$ is 0. However, this equalization strategy is not useful for magnetic recording except at very low linear densities. This is because, in frequency domain, the FR equalizer behaves as inverse of the channel transfer function. This results in excessive noise enhancement at medium and

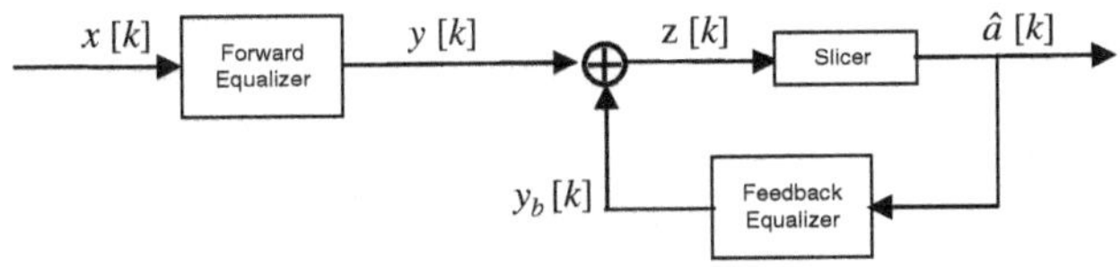

Fig. 4.11 Decision-feedback equalizer

high densities since the magnetic recording channel becomes more and more band limited as density increases [54].

The objective of DF equalization is also to fully cancel the ISI, but this is implemented using "decision-feedback" so as to achieve FR equalization without suffering from undesirable noise enhancement. Figure 4.11 shows the schematic of DF equalizer. The equalization task is divided between forward and feedback equalizers. Forward equalizer is designed to cancel non-causal ISI, i.e., $g_i = 0$ for all $i < 0$, in its output $y[k]$. Thus, forward equalizer output can be expressed as

$$y[k] = g_0 a[k] + \sum_{i>0} g_i a[k-i] + \eta[k]. \tag{4.27}$$

Since the second term on the RHS corresponds to ISI from past bits (i.e., causal ISI), the feedback equalizer is configured to compute and cancel this ISI using detector decisions, assuming correct decisions. The detector input and output are given by

$$z[k] = y[k] - y_b[k] \qquad \text{and} \qquad \hat{a}[k] = \text{sgn}\{z[k]\}, \tag{4.28}$$

where $y_b[k] = \sum_{i>0} g_i \hat{a}[k-i]$. Design of finite-length DF equalizers is addressed in [46]. The advantage of DF equalization over FR equalization arises from the fact that cancellation of non-causal ISI can be accomplished with no or minimum noise enhancement [1]. The main disadvantage of DF equalization is the possibility of error propagation, i.e., the presence of decision errors at detector output results in incorrect cancellation of causal ISI at detector input which leads to the possibility of more decision errors. Error propagation is one of the main reasons why DF equalization is not used in magnetic recording applications. Another reason for its unpopularity in magnetic recording is the limitation in slicer-detector to support sophisticated coding and detection methods. Nevertheless, there have been a few extensions to enhance its detection capability [2, 25].

PR equalization addresses the noise enhancement problem without using decision feedback. This is done by requiring the equalizer to shape the channel response into a specified response called "PR target" whose duration spans multiple bits, which is unlike FR equalization where the underlying PR target spans only one bit. Since PR equalization allows for certain amount of ISI (in the form of PR target) to remain at equalizer output, noise enhancement can be minimized by choosing the PR target to be spectrally similar to the unequalized channel response [54].

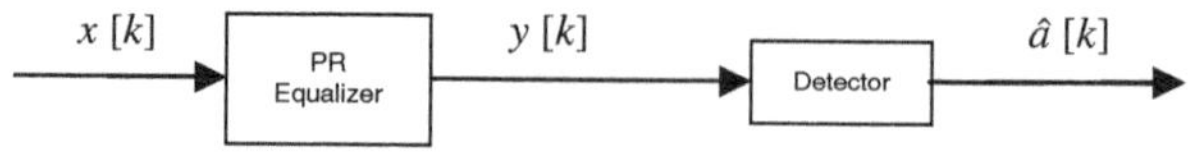

Fig. 4.12 Partial response (PR) equalizer with detector

Knowledge of the PR target is used in the detector to provide near-optimum detection performance. Viterbi detector or its advanced versions are used for data detection [30]. Let $\{g_0, g_1, \ldots, g_{N_g}\}$ be the PR target. The output of PR equalizer can be given by (see Fig. 4.12)

$$y[k] = \sum_{i=0}^{N_g} g_i a[k-i] + \eta[k], \tag{4.29}$$

where $\eta[k]$ denotes noise and residual ISI, where residual ISI is the difference between unequalized channel and PR target. Since the hardware complexity of the Viterbi-like detectors is exponential in the number of coefficients in PR target, the choice of PR target is governed by the trade-off between spectral match of PR target to unequalized channel and hardware complexity of detector.

Approaches for Equalizer Design

Since PR equalization is the approach used in all magnetic recording systems currently, we restrict our discussion on equalizer design to that for designing PR equalizer. There are mainly two approaches for designing equalizers: zero-forcing (ZF) approach and mean square (MS) approach. Whereas the ZF approach aims to minimize residual ISI, the objective of MS approach is to minimize the mean square value of the sum of residual ISI and noise. The MS approach is the most commonly used in disk drives.

The error at PR equalizer output for a given PR target can be given by

$$e[k] = y[k] - d[k] = e_{\text{isi}}[k] + e_{\text{nse}}[k], \tag{4.30}$$

where

$$y[k] = \sum_{i=0}^{N_w} w_i x[k-i] = \mathbf{w}^T \mathbf{x}[k], \qquad d[k] = \sum_{i=0}^{N_g} g_i a[k-i] = \mathbf{g}^T \mathbf{a}[k], \tag{4.31}$$

$d[k]$ denotes the desired output of PR target; $e_{\text{isi}}[k]$ and $e_{\text{nse}}[k]$ denote residual ISI and noise, respectively, in the error $e[k]$; $\mathbf{w} = [w_0, w_1, \cdots, w_{N_w}]^T$ denotes the equalizer coefficients, $x[k] = [\mathbf{x}[k], x[k-1], \cdots, x[k-N_w]]^T$ and $\mathbf{a}[k] = [a[k], a[k-1], \cdots, a[k-N_g]]^T$ denote the inputs of equalizer and PR target filters, respectively, at kth time instant, and superscript T denotes matrix transpose. ZF equalization amounts to, e.g., minimizing $E[e_{\text{isi}}^2[k]]$ and MS approach amounts to minimizing $E[e^2[k]]$ where $E[\cdot]$ denotes statistical expectation operator. Clearly, second-order statistics of equalizer input and data bits are sufficient to determine the optimum equalizers under ZF and MS criteria. For example, the optimum MS equalizer is given by

$$\boldsymbol{w}_{\text{opt}} = \boldsymbol{R}_{xx}^{-1}\boldsymbol{r}_{xd}, \tag{4.32}$$

where $\boldsymbol{R}_{xx} = E\left[\boldsymbol{x}[k]\boldsymbol{x}^T[k]\right]$ and $\boldsymbol{r}_{xd} = E\left[d[k]\boldsymbol{x}[k]\right]$ are the correlation matrix of equalizer input and cross-correlation vector between equalizer input and target output, respectively.

While PR equalization is usually exercised with pre-specified PR targets [18, 29, 54], the design of good PR targets is very important to guarantee near-optimum detection performance since the choice of PR target determines the noise characteristics at detector input and the type of dominant error events in detection. As a result, an important problem in equalization is joint design of equalizer and PR target. To prevent trivial solution (i.e., zero values for equalizer and target), joint optimization has to be done under some constraints imposed on equalizer and/or target [36, 50]. We shall illustrate this using the so-called monic constraint on target (i.e., $g_0 = 1$). For example, the optimum equalizer and monic-constrained target under MS criterion are given by

$$\boldsymbol{w}_{\text{opt}} = \boldsymbol{R}_{xx}^{-1}\boldsymbol{R}_{xa}\,\boldsymbol{g}_{\text{opt}} \quad \text{and} \quad \boldsymbol{g}_{\text{opt}} = \frac{1}{\boldsymbol{u}^T\boldsymbol{R}_o^{-1}\boldsymbol{u}}\boldsymbol{R}_o^{-1}\boldsymbol{u}, \tag{4.33}$$

where $\boldsymbol{R}_o = \boldsymbol{R}_a - \boldsymbol{R}_{xa}^T\boldsymbol{R}_x^{-1}\boldsymbol{R}_{xa}$ and $\boldsymbol{u} = [1, 0, \cdots, 0]^T$.

In disk drives, design of equalizer and target using closed-form expressions of the type, (4.33) are not a recommended option in view of computational complexity as well as the necessity to keep track of the changing characteristics of the recording channel across heads and media. Therefore, it is very common to use adaptive approaches [55] for designing equalizer. The adaptive version of MS approach is obtained by implementing the well-known gradient descent approach for minimizing the squared error $e^2[k]$. The resulting algorithm is traditionally known as LMS (least mean square) algorithm. As an example, the LMS adaptive algorithm for designing equalizer and monic-constrained target can be given by [14]

$$\left.\begin{aligned} e[k] &= \boldsymbol{w}^T[k-1]\boldsymbol{x}[k] - \boldsymbol{g}^T[k-1]\boldsymbol{a}[k] \\ \boldsymbol{w}[k] &= \boldsymbol{w}[k-1] - \mu_w e[k]\boldsymbol{x}[k] \\ \boldsymbol{g}[k] &= \boldsymbol{g}[k-1] + \mu_g e[k]\boldsymbol{a}[k] \\ g_0[k] &= 1 \end{aligned}\right\}, \tag{4.34}$$

where $\boldsymbol{w}[k]$ and $\boldsymbol{g}[k]$ denote the coefficient vectors of equalizer and target, respectively, at kth time instant, and μ_w and μ_g are positive scalars that control the adaptation rate.

It should be kept in mind that the ZF and MS approaches for equalizer design do not guarantee that detection performance will be optimized under all conditions.

For example, equalizer and target that result in minimum mean square error do not necessarily result in minimum bit error rate unless the noise at Viterbi detector input is white Gaussian and residual ISI is 0. Nevertheless, practical systems resort to MS and ZF approaches since the approaches that minimize error rate are usually quite expensive to implement while the solutions of MS and ZF approaches are usually near the optimum. We would like to close this section by pointing out certain references that present error rate minimizing approaches for equalizer design. Reference [38] presents an analytical approach for designing optimum PR target and equalizer by maximizing the detection SNR of Viterbi detector – which is equivalent to minimizing bit error rate. Reference [56] presents an adaptive algorithm that minimizes bit error rate for designing equalizer.

Signal Detection and Decoding for Magnetic Recording Channel

As pointed out earlier, ECCs have played an important role in achieving reliable data transmission. A good overview of various error correction codes can be found in [40]. In hard disk drive, the Reed–Solomon (RS) coding has been used for decades. Recently, the industry is replacing RS codes with low-density parity-check (LDPC) codes and iterative detection/decoding which can achieve more coding gain. ECC provides error correction capability at the cost of adding redundancy into information bits. This leads to additional reduction of magnetic recording density. Given a constant rotation speed, the sampling rate 1/T is accordingly increased due to more bits are compacted in the unit length of track. As a result, the energy decreases in the pulse response because the isolated pulses p(t) and −p(t) get closer and cancel each other more. Moreover, the increase in sampling rate expands the signal bandwidth which introduces more noise energy. So, the overall SNR is decreased. By adopting the LDPC codes into read channel, the error correction capability can be dramatically improved in terms of sector failure rate, as shown in Fig. 4.13.

In this section, we mainly focussed on soft in soft out (SISO) channel detection, soft LDPC decoding, and iterative detection and decoding.

Channel Detector

The read channel is typically equalized as target which can be described in terms of the partial response polynomial:

$$H(D) = h_0 + h_1 \cdot D + h_2 \cdot D^2 + \cdot + h_\nu \cdot D^\nu,$$

where D indicates one sample delay. The target coefficients h_i are often integers to simplify the implementation of the read channel.

Given the channel response polynomial, the channel can be treated as rate a trellis code with 2^ν states. So, a trellis decoder can be employed to detect the

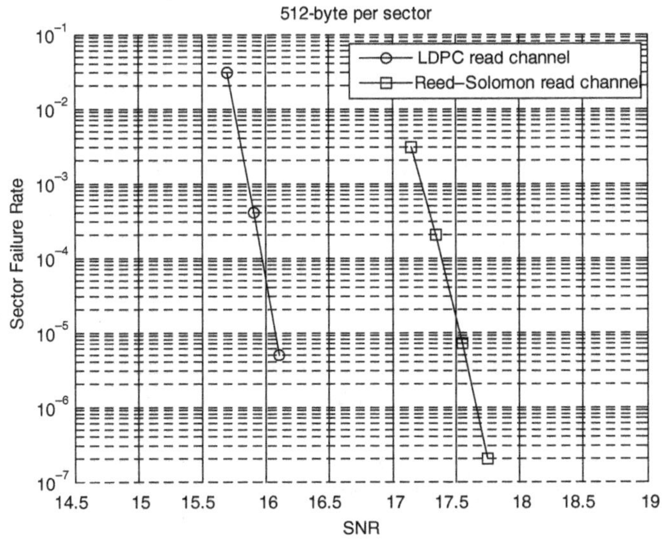

Fig. 4.13 Error correction comparison between RS and LDPC

recorded sequence. Hard decision Viterbi detector has been used since early 1990s. The detected bit sequence is passed to a Reed–Solomon decoder for further error correction. However, to exploit the powerful error correction capability of LDPC decoder, soft output information as a measure of a posteriori probability (APP) is required from the channel detector. As a result, detection algorithms which provide soft output are required.

There are two types of soft output detectors. The first one is maximum a posteriori (MAP) detector that implements the Bahl–Cocke–Jelinek–Raviv (BCJR) algorithm [7], which is often referred as Max-Log-MAP algorithm when it is approximated and implemented in the log domain. The other one is maximum likelihood sequence detector that employ soft output Viterbi algorithm (SOVA) [28, 70]. The former one provides better soft output quality at the cost of higher computation complexity. Moreover, we note that the SOVA can be modified [15] so that it can perform equivalently well as Max-Log-MAP algorithm.

SOVA Detector

Conventional Viterbi algorithm takes soft quantized information as input and outputs hard decision sequence. It has been recognized that performance can be improved if the reliability of each decision bit can be obtained for further use by outer code. As a matter of fact, soft output Viterbi algorithm has been known for long time, though it was not used in real applications due to the high computational complexity. With the progress on reduction of algorithm complexity and advance of VLSI technology, soft output detectors/decoders are now playing an important role in channel coding, equalization, and signal detection. In this section, the VLSI architecture of high-speed SOVA is presented.

The process of SOVA as shown in Fig. 4.14a can be divided into two stages: survivor processing and update processing. The survivor processing is similar to conventional hard decision VA. First, the branch metric of each possible state transition is computed by branch metric calculation unit (BMCU). Then, the branch metrics are added to state metrics, which denote of weighting of the state transition sequence. This is performed by Add-Compare-Select unit (ACSU), which also compares the accumulated state metrics of the paths entering into same state and selects the one with minimum metric value. At last, the most likely output sequence is found out through the survivor memory unit (SMU). The update processing aims to evaluate the difference between the survivor path and competing path and update the reliability of the decision. The diagram of SOVA architecture is shown in Fig. 4.14b. The BMCU, ACSU, and SMU form a VA detector, which is employed to determine the survivor states and the output at depth L. The hard decision and metric difference of each state are computed by ACSU and stored in FIFO as a delay line. The path comparison unit (PCU) tracks the hard decision through the final survivor (decision)

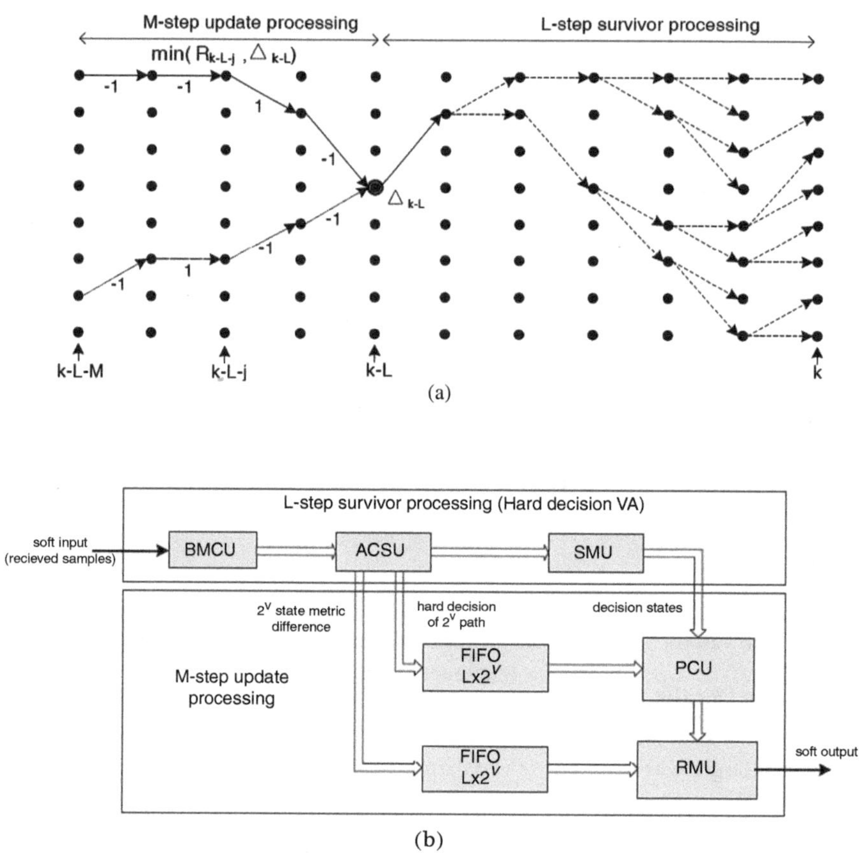

Fig. 4.14 SOVA detector

path and the competing path. If the hard decision is different, the reliability measure unit (RMU) updates the reliability value R_{K-L-j} by $\min(R_{K-L-j}, \Delta_{K-L})$, where Δ_{K-L} is the metric difference between the decision path and competing path.

Log-Max-MAP Detector

The Max-Log-Map detector is usually implemented via sliding window approach [7]. To further improve the throughput, we can use multiple sliding windows sub-detectors that operate in parallel, leading to a parallel sliding window detector.

The number of sub-detectors depends on the desired trade-off between silicon area and throughput. Each sub-detector has a structure as shown in Fig. 4.15. Each sub-detector consists of one branch metric unit (BMU), one forward recursion unit (FRU), two backward recursion units (BRU), one log-likelihood-ratio (LLR) calculation unit, one memory bank Gamma Memory consisting of 4 single-port memory blocks to store the branch metrics computed by BMU, and one memory bank Alpha Memory consisting of 2 single-port memory blocks to store the forward state metrics computed by FRU. The operation can be briefly described as follows:

1. Branch Metrics Calculation: Upon receiving the symbol y_k with additive noise, BMU calculates the branch metrics of the transition from state s' to state s as:

$$\gamma_k(s', s) = L(y_k|u_k) + L(u_k),$$

where u_k is the transmitted symbol, y_k is received sample, and L is log-likelihood ratio.

2. Forward Recursion: FRU computes the forward state metrics α_k as

$$\alpha_k(s) = \max_{s'}(\alpha_{k-1}(s') + \gamma_k(s', s)).$$

3. Backward Recursion and Soft output Calculation: With the available branch metrics and forward state metrics, the BRUs compute backward state metrics β_{k-1} as

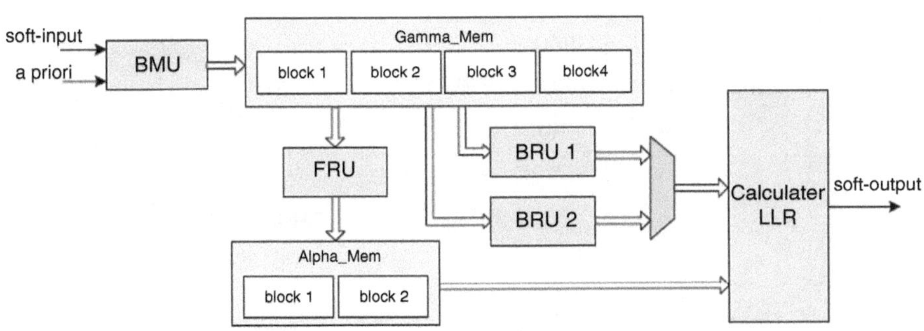

Fig. 4.15 Sliding window max-log-map detector

$$\beta_{k-1}(s') = \max_{s}(\beta_k(s) + \gamma_k(s', s)).$$

When all the needed α, β, and γ are ready, the LLR calculation unit works out the a posteriori LLR of the information bit u_k as

$$L(u_k|y) = \max_{\substack{(s',\,s)\\ u_k=+1}} (\alpha_{k-1}(s') + \gamma_k(s',\,s) + \beta_k(s)) - \max_{\substack{(s',s)\\ u_k=-1}} (\alpha_{k-1}(s') + \gamma_k(s', s) + \beta_k(s)).$$

The operation sequences of each component are illustrated in Fig. 4.16. For the purpose of simplicity, the operation of LLR calculation unit is merged into BRUs and the operation of BRUs is divided into two stages: BRU_{1a} and BRU_{2a} do not generate LLR output, and BRU_{1b} and BRU_{2b} generate LLR output. In Fig. 4.16, the vertical axis represents time, and the horizontal axis represents the memory access address. From $t = 0$, in every L clock cycles, BMU calculates L of γ values to store them in the corresponding block of Gamma Memory (each block has L words). From $t = 2L$, FRU performs forward recursions to calculate L of α values in every L clock cycles and store them in Alpha Memory. At the same time BRU_1 begins backward recursions to accumulate backward state metrics for path mergence. From $t = 3L$, BRU_1 continues backward recursion and soft output is calculated every clock cycle, while BRU_2 starts backward recursion to accumulate βs in parallel. Then BRU_1 and BRU_2 alternate every L clock cycles to generate for soft-out calculation and accumulate for path mergence. The detection latency is $3L$ clock cycles.

LDPC Codes and Decoding

Invented by Gallager [16] in 1962, LDPC codes were largely neglected for several decades due to their high computation complexity. Inspired by remarkable

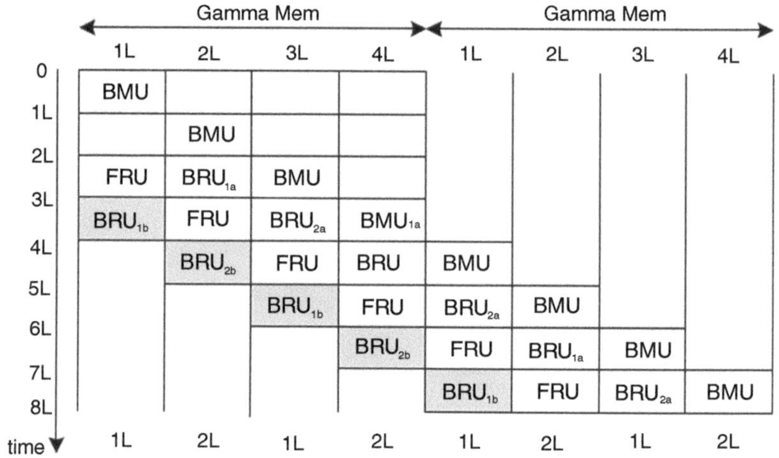

Fig. 4.16 Scheduling for sliding window max-log-map detector

success of Turbo codes in early 1990s, MacKay and Neal [44] and Wiberg [64] re-discovered LDPC codes in 1996. Since then, LDPC codes have been attracting tremendous research interest because of their excellent error-correcting performance and highly parallel decoding scheme. Today, many industry standards begin to adopt LDPC codes, such as digital video broadcasting (DVB-S2) for satellite video, IEEE 802.3 for 10G Ethernet, IEEE 802.16 for Wireless Metropolitan Area Network (WiMAX), IEEE 802.11 for Wireless Local Area Network (WiFi), IEEE 802.12 for Wireless Personal Area Network (WPAN), and IEEE 802.20 for mobile Broadband Wireless Access Networks (MBWA).

An LDPC code is defined as the null space of a parity check matrix H. The name comes from the characteristic of H in which the number of 1 s is much less than the number of 0s. There are two common representations of LDPC codes. Like all linear block codes they can be described via matrices. The other one is a graphical representation in which each row of H is represented by a check node; each column is represented by a variable node; and each "1" element is represented by an edge connecting between the corresponding check and variable nodes. For example, an $M \times N$ sparse parity check matrix H can be represented by a bipartite graph consisting of M check nodes in one set, N variable nodes in the other set and L edges connecting between the two sets of nodes, where L is the total number of 1s in H. The bipartite graph representation greatly facilitates the development and illustration of the decoding algorithm. If the number of 1s in both the rows and the columns of the parity check matrix is same, i.e., all the variable nodes have the same degree and all the check node also have the same degree, the LDPC codes are called regular LDPC codes. On the contrary, LDPC codes which do not have a constant number of non-zero entries in the rows or in the columns of parity check matrix are called irregular LDPC codes. Following the notation in [42], they are specified by the weight distribution of rows $\rho(x)$ and the weight distribution of column $\lambda(x)$:

$$\rho(x) = \sum_{i=2}^{d_c} \rho_i x^{i-1},$$

$$\lambda(x) = \sum_{i=2}^{d_v} \lambda_i x^{i-1},$$

where ρ_i and λ_i denote the proportion of rows and columns with weight i; while d_c and d_v are the maximum row weight and column weight, respectively. Given the parity check matrix $H_{\mathrm{M}\times\mathrm{N}}$, the rate R of LDPC codes is defined by $R > R_{\mathrm{d}} = (N - M)/N$. $R_{\mathrm{d}} = R$ if the parity check matrix has full rank.

Decoding Algorithms

Generally, the LDPC decoding algorithms can be categorized into two classes: hard decision decoding and soft decision decoding. The latter schemes provide superior performance at the cost of higher complexity. The commonly known hard decision algorithms include majority-logic (MLG) decoding, bit-flipping decoding, and weighted BF or MLG decoding, while the sum-product algorithm (SPA) and min-sum algorithm are the two most popular soft decision decoding schemes. Since the hard decision decoding is relatively simple and the decoding performance is inferior to soft decision, we focus only on the soft decision decoding. The readers are referred to [34, 43] for various hard decision decoding algorithms.

The soft decision decoding of LDPC codes is also called iterative message passing or belief-passing (BP) decoding [35, 43] that directly matches the code bipartite graph as illustrated in Fig. 4.17: After each variable node is initialized with the input channel message, the decoding messages are iteratively computed by all the variable nodes and check nodes and exchanged through the edges between the neighboring nodes. When the decoding is performed in log domain, it is called iterative Log-BP decoding algorithm.

It is well known that various soft decision LDPC decoding algorithms may fall into two categories, including Sum-Product algorithm (SPA) and Min-Sum algorithm. Depending on the decoder scheduling, they can be further categorized as non-layered and layered decoding algorithm [20, 23, 45]. Min-Sum algorithm is commonly used because its check node processing approximation may potentially lead to significant silicon area savings from two perspectives: (i) the logic complexity may be reduced due to the elimination of the function log[tanh(/2)] that is typically implemented as lookup-table (LUT) in hardware and (ii) more importantly, the size of memory for decoding message storage may be reduced due to the possible compact representation of check-to-variable messages. However, this memory saving potential has not been fully exploited by existing high-speed partially parallel Min-Sum decoders, although such potential has been pointed out in some serial Min-Sum decoding schemes [19, 67]. Moreover, for partially parallel decoder design, the conventional Min-Sum algorithm formulation results in explicit implementation of a sorter in each check node processing unit, which will make the potential of logic silicon area saving quickly diminish and result in an essential speed bottleneck as the code rate (or, more specifically, the row weight of parity check matrix) increases.

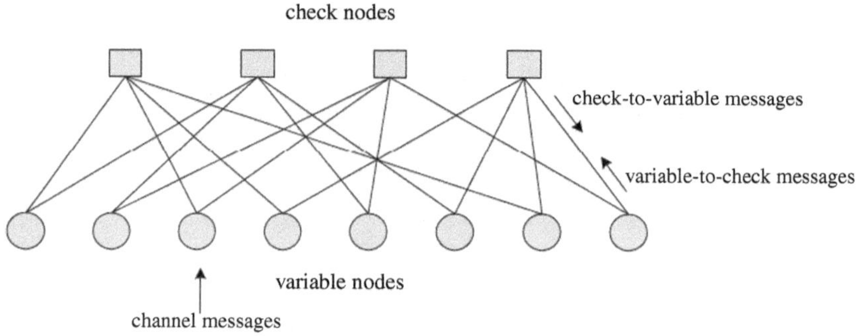

Fig. 4.17 Illustration of LDPC code bipartite graph

Before presenting the decoding algorithm, some definitions are introduced as follows: Let H denotes the $M \times N$ parity check matrix and $H_{\mathrm{m,n}}$ denote the entry of H at the position (m, n). The set of bits n that participate in parity check m is defined as $N(m) = \{n : H_{\mathrm{m,n}} = 1\}$, and the set of parity check m in which bit n participates is defined as $M(n) = \{m : H_{m,\mathrm{n}} = 1\}$. The set N($m$) with bit n excluded is denoted as $N(m)\backslash n$, and the set $M(n)$ with check m excluded as $M(n)\backslash m$. The channel message, variable-to-check message, check-to-variable message, and posterior log-likelihood ration (LLR) are denoted as γ_n, $\alpha^i_{m,n}$, $\beta^i_{m,n}$, and λ^i_n, respectively, where the superscript i is iteration index.

The main difference between Sum-Product algorithm and Min-Sum algorithm lies in the check node processing, i.e., the check node processing in SPA is realized as

$$\beta_{m,n} = \Phi\left(\sum_{n' \in \mathrm{N}(m)\backslash n} \Phi(|\alpha_{m,n'}|)\right) \prod_{n' \in \mathrm{N}(m)\backslash n} \mathrm{sign}(\alpha_{m,n'}),$$

where$\Phi(x) \equiv -\log[\tanh(x/2)]$. The check node process in Min-Sum algorithm is approximated as

$$\beta_{m,n} = \min_{n' \in \mathrm{N}(m)\backslash n} (\alpha_{m,n'}) \prod_{n' \in \mathrm{N}(m)\backslash n} \mathrm{sign}(\alpha_{m,n'}).$$

Therefore, the function $\Phi(x)$, which is typically implemented as look-up table (LUT) in hardware, is eliminated in Min-Sum algorithm. The conventional Min-Sum algorithm formulation is described as follows:

Intialization: $\alpha^0_{m,n} = \gamma_n$;
for *i= 0 to i_{max} or convergence to codeword* **do**
  forall *check nodes c_m, $m \in \{1,\ldots,M\}$* **do**
    $mag\left(\beta^i_{m,n}\right) = \min_{n' \in N(m)\backslash n}\left\{\left|\alpha^{i-1}_{m,n'}\right|\right\}$;
    $sign\left(\beta^i_{m,n}\right) = \prod_{n' \in N(m)\backslash n} sign\left(\alpha^{i-1}_{m,n'}\right)$;
  end
  forall *variable nodes v_n, $n \in \{1,\ldots,N\}$* **do**
    $\lambda^i_n = \gamma_n + \sum_{m \in M(n)} \beta^i_{m,n}$;
    $\alpha^i_{m,n} = \lambda^i_n - \beta^i_{m,n}$;
  end
end
Output the decoded bits as sign $\left(\lambda^i_n\right)$

Due to the check node processing approximation, the check-to-variable messages from each check node only have two different magnitudes (i.e., the minimum and the second minimum ones among the magnitudes of all the variable-to-check messages entering into this check node), no matter how large the check node degree is. Meanwhile, the check node processing approximation eliminates the function $\Phi(x)$. Intuitively, these two features may be leveraged to reduce the storage and logic silicon area. However, a direct realization of partially parallel decoders based on the above conventional Min-Sum algorithm formulation may not be able to effectively materialize such silicon area saving potential for following two main reasons:

1. In spite of much less check-to-variable messages storage requirement, the total number of distinct variable-to-check messages always equals to the total number of 1s in the parity check matrix. A direct realization of partially parallel decoders may have to provide explicit storage for these variable-to-check messages, leading to the same (or similar) storage requirement as in its SPA decoder counterpart.
2. The direct realization of partially parallel decoders tends to implement parallel-input parallel-output check node processing units that use a sorter to search the two minimum ones among all the incoming variable-to-check messages. As code rate increases, the silicon area overhead incurred by sorters will quickly increase.

To solve the above two issues, a transformed Min-Sum algorithm is described in the following. Although it is mathematically equivalent to the original Min-Sum algorithm, its formulation and execution order make it straightforward to realize silicon area savings at VLSI architecture level. In particular, this algorithm transformation has two key features, including (1) the check node processing and variable node processing are interleaved in such a way that each newly updated variable-to-check message may be directly absorbed by check node processing units without being intermediately stored and (2) the check node processing is sequentialized so that the explicit implementation of a sorter is eliminated. To generate the outgoing check-to-variable messages from each check node (i.e., $\{|\beta_{m,n}|, n \in \mathrm{N}(m)\}$), the sequentialized check node processing n|, n 2 N(m)}), only needs to keep track of the two minimum magnitudes, i.e., $\min1_m$ and $\min2_m$ where $\min1_m \leq \min2_m$, among the input variable-to-check messages, the sign $s_{m,n}$ of each input variable-to-check message and $S_m = \prod s_{m,n}$, and the variable node index I_m representing which variable node provides the message with the minimum magnitude.

---

for *i= 0 to* i_{max} *or convergence to codeword* **do**

  forall *variable nodes* v_n, $n \in \{1, \ldots, N\}$ **do**

    if *i= 0* **then**

      $\alpha^i_{m,n} = \gamma_n;$

    else

      $$\beta^i_{m,n} = \begin{cases} S^i_m \cdot s^i_{m,n} \cdot min1^i_m & n \neq I^i_m \\ S^i_m \cdot s^i_{m,n} \cdot min2^i_m & \text{otherwise} \end{cases}$$

      $\lambda^i_n = \gamma_n + \sum_{m \in M(n)} \beta^i_{m,n};$

      $\alpha^i_{m,n} = \lambda^i_n - \beta^i_{m,n};$

    end

Initialize $min1_m^{i+1} = min2_m^{i+1} = +\infty, S_m^{i+1} = 1$;
forall *check nodes* c_m, $m \in M(n)$ do
  if $\left|\alpha_{m,n}^i\right| < min1_m^{i+1}$ **then**
    $min1_m^{i+1} = \left|\alpha_{m,n}^i\right|$;
    $I_m^{i+1} = n$;
  else if $\left|\alpha_{m,n}^i\right| < min2_m^{i+1}$ **then**
    $min2_m^{i+1} = \left|\alpha_{m,n}^i\right|$;
  $s_{m,n}^{i+1} = sign\left(\alpha_{m,n}^i\right)$;
  $S_m^{i+1} = S_m^{i+1} \cdot s_{m,n}^{i+1}$;
end
end
end
Output the decoded bits as sign $\left(\lambda_n^i\right)$

Additionally, we note that the Min-Sum algorithm is generally less sensitive to quantization errors compared to SPA and hence may enable the use of smaller finite word-length. For example, it has been shown in [10] that a 4-bit quantization of decoding messages may be sufficient to avoid error floor, although it may result in about 0.2 dB performance loss compared to 6-bit quantization.

QC-LDPC Code and Decoder Architecture

For efficient hardware realization, the LDPC code is typically constructed in quasi-cyclic (QC) form. The parity check matrix of a QC-LDPC code consists of an $m_b \times n_b$ array of $p \times p$ square circulant matrices. So the $m_b{\cdot}p\times\, n_b\, p$ parity check matrix can be represented by a bipartite graph as shown in Fig. 4.18, where each group of consecutive p rows (or columns) is represented by a set of p check (or variable) nodes. Notice that the cyclic permutation blocks realize message passing between adjacent variable and check node groups through simple cyclic permutations based on the quasi-cyclic parity check matrix structure.

In each decoding iteration, the decoding scheduling directly follows the above transformed formulation. This can be illustrated in Fig. 4.19 and explained as follows. One out of the total n_b variable node sets is processed at one time and all the check nodes are processed in a serial manner interleaved with the variable node processing. In Fig. 4.19a the first set of variable nodes is processed and feeds variable-to-check messages to all the check nodes for partial check node processing. Then the decoding moves to the second step, as shown in Fig. 4.19b, where the second set of variable nodes is processed and feeds variable-to-check messages to all the check nodes for further partial check node processing. Once all the variable nodes are processed within present iteration, as shown in Fig. 4.19c, all the check nodes also receive all the input variable-to-check messages and finish the check node processing for present iteration. Then all the check-to-variable messages will be fed to the variable nodes for the next iteration. Figure 4.19 clearly illustrates the two desirable features of this proposed Min-Sum algorithm transformation, i.e., the

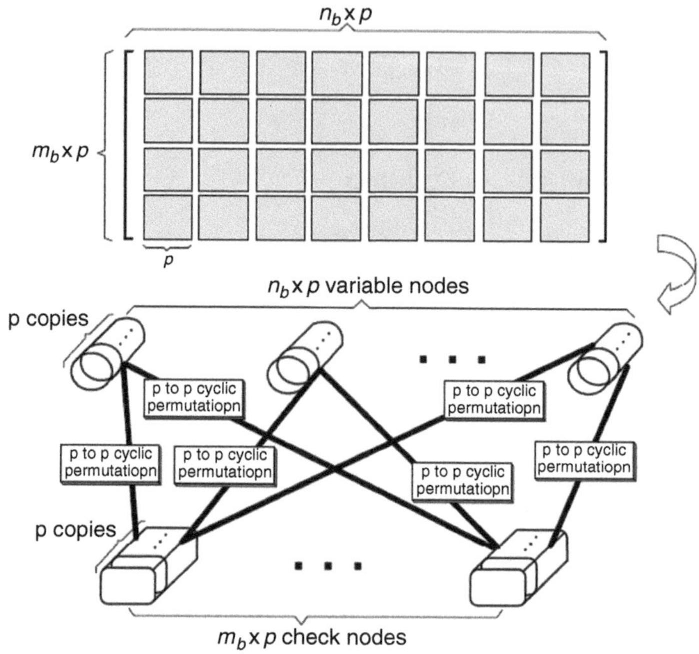

Fig. 4.18 Quasi-cyclic LDPC and graph mapping

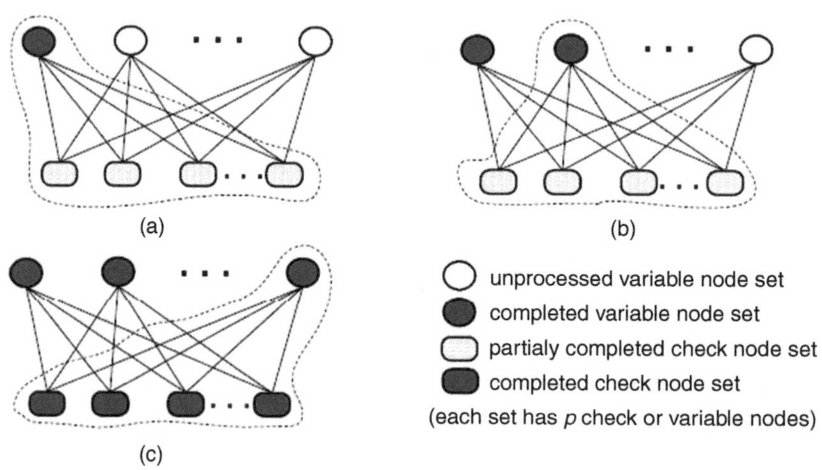

Fig. 4.19 Illustration of decoding scheduling

obviation of explicit storage of variable-to-check decoding messages and concurrent operation of check node processing and variable node processing.

Figure 4.20 shows the corresponding partially parallel QC-LDPC decoder architecture. It contains p variable node processing units (VNUs) that process one set of

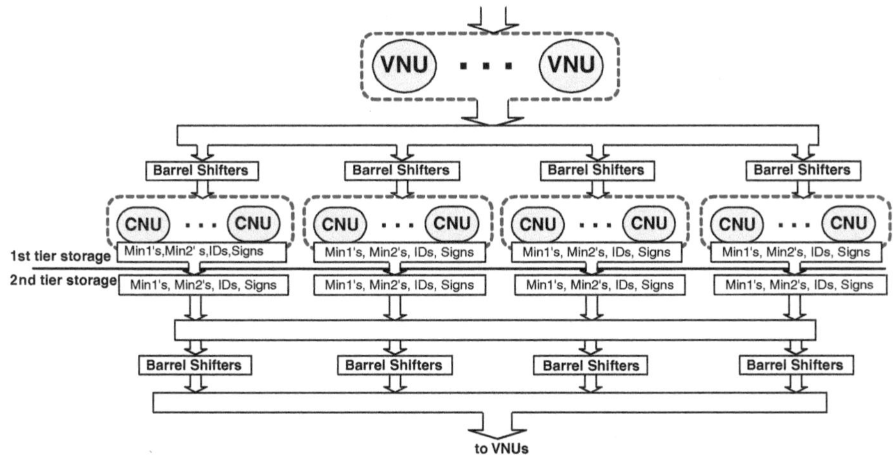

Fig. 4.20 Corresponding partially parallel QC-LDPC decoder architecture

p variable nodes in each clock cycle, and $m \cdot p$ check node processing units (CNUs) that process all the $m \cdot p$ check nodes in n clock cycles. The variable-to-check messages computed in each clock cycle are directly routed to the $m \cdot p$ CNUs via the barrel shifters which are configured according to the structures of circulant matrices. Each CNU serially updates $\min 1_m$, $\min 2_m$, $s_{m,n}$, S_m, and I_m that are kept in the first set storage. After n clock cycles, all the check-to-variable messages are computed and then passed to the second set storage that provide input to all the VNUs. Note that the storage for the signs in the two tiers can be shared using first in first out buffers (FIFOs). The reverse routing from CNUs to VNUs is also realized by barrel shifters. The data path is naturally pipelined so that all the VNUs and CNUs can work concurrently. All the variable-to-check messages are immediately absorbed by CNUs and hence do not require storage, while only two minimum magnitudes, one location index, and signs need to be stored to generate the check-to-variable messages from each check node.

For layered decoding algorithm, only p check node processing units need to be implemented; however it needs m times of cycles to complete all layers for a full iteration. Typically, the layered decoding converges faster, so the number of iterations can be reduced.

Iterative Detection/Decoding

Following the Turbo principle [21, 33], an iterative soft detection and LDPC decoding system has the block diagram as shown in Fig. 4.21. The soft-in-soft-out (SISO) channel detector takes received sample and generates reliability or log-likelihood ratio (LLR) of binary decision. Taking max-log-map detector as an example, it maximizes the APP $p(u_k \mid y)$ given the observed sequence y, i.e.,

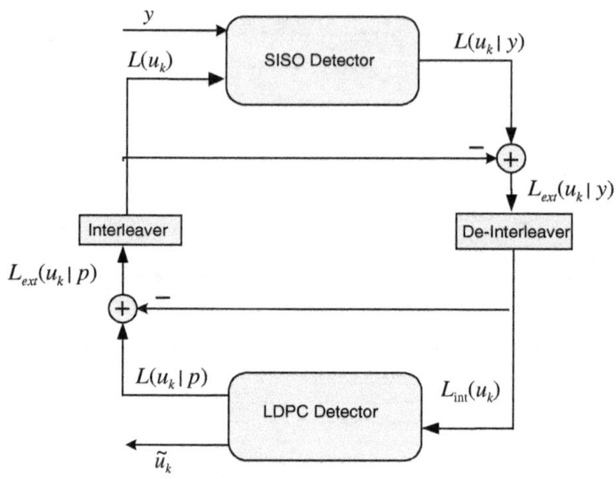

Fig. 4.21 Iterative detection and decoding

$$\hat{u}_k = \arg\max_{a \in \{0,1\}} P(u_k = a \mid y),$$

and calculate the log-likelihood ratio, i.e.,

$$L(u_k \mid y) = \ln \frac{p(u_k = 0 \mid y)}{p(u_k = 1 \mid y)}.$$

The probability $P(u_k = a \mid y)$, where $a \in \{0, 1\}$, can be written as

$$P(u_k = a \mid y) = \sum_{\forall u : u_k = a} p(u \mid y) = \sum_{\forall u : u_k = a} \frac{p(y \mid u) p(u)}{y}.$$

The probability $P(\boldsymbol{u})$ is the a priori probability of sequence $\mathbf{u}$, which can be used to derive knowledge about the source producing the bit u_k. Assuming the bits in $\boldsymbol{u}$ are independent, i.e., the probability $P(\boldsymbol{u})$ factors as $\prod_{k=1}^{K} p(u_k)$, the LLR can be reformulated as follows:

$$\begin{aligned} L(u_k \mid y) &= \ln \frac{\sum_{\forall u : u_k = 0} p(y \mid u) \prod_{i=1}^{K} p(u_k)}{\sum_{\forall u : u_k = 1} p(y \mid u) \prod_{i=1}^{K} p(u_k)} \\ &= \ln \frac{\sum_{\forall u : u_k = 0} p(y \mid u) \prod_{i=1:i \neq k}^{K} p(u_k)}{\sum_{\forall u : u_k = 1} p(y \mid u) \prod_{i=1:i \neq k}^{K} p(u_k)} + L(u_k) \\ &= L_{\text{ext}}(u_k \mid y) + L(u_k) \end{aligned}.$$

The term $L_{ext}(u_k \mid y)$is the extrinsic information about u_k contained in $\mathbf{y}$, while the $L(u_k)$ is the a priori information about u_k. The extrinsic information is key point in iterative detection and decoding, which is also called Turbo equalization.

The extrinsic information $L_{ext}(u_k \mid y)$ is passed through a de-interleaver and fed to LDPC decoder as intrinsic message $L_{int}(u_k)$. The LDPC decoder performs message passing decoding and generates soft output $L(u_k \mid p)$ and hard decision $\tilde{u}_k$. By subtracting intrinsic message from the soft output, the extrinsic message $L_{ext}(u_k \mid p)$ is obtained. The extrinsic information means that the soft output is not the current best estimate of the reliability of the decision, but is instead the new information of current best estimate excluding the intrinsic soft-input. The extrinsic messages $L_{ext}(u_k \mid p)$ go through interleaver and pass to the detector as prior information $L(u_k)$ for next iteration.

Typically the iteration between the detector and decoder is called global iteration, while the iteration in LDPC decoding is called local iteration. The number of local and global iterations is determined by the trade-off between error correction performance, hardware implementation complexity and throughput. Usually, more iteration can help error correction performance at cost of hardware resource.

Conclusions

This chapter briefly discusses the basics of modern hard disk drives and overall system organization. As the hard disk drive areal storage density is being pushing toward the 1 Tb/in^2 with the introduction of new technologies such as perpendicular recording and patterned media, powerful read channel designs are becoming increasingly crucial. Hence, this chapter further elaborates on the magnetic recording read channels including modeling and signal processing and coding. In spite of recent significant progress of solid-state drives based on NAND flash memory, areal density scaling and cost reduction of hard disk drives will surely continue with a rate at least same as solid-state drives and will remain the mainstream mass data storage technology for the foreseeable future.

References

1. Belfiore C, Park J (1979) Decision feedback equalization. Proc IEEE 67:1143–1156
2. Bergmans J, Voorman J, Wong-Lam H (1997) Dual decision feedback equalizer. IEEE Trans Commun 45:514–518
3. Bergmans J, Voorman J, Wong-Lam H (1999) Structure and adjustment of a novel write precompensation scheme. IEEE Trans Magn 35:2053–2059
4. Berrou C, Glavieux A, Thitimajshima P (1993) Near shannon limit error-correcting coding and decoding: Turbo-codes. In: Proceedings of ICC'93, Geneva, Switzerland, pp 1064–1070
5. Bertram H, Takeo A, Jin Z, Fu C (2006) Transition shape analysis from medium noise measurements. IEEE Trans Magn 42:1620–1625
6. Blahut RE (1984) Theory and practice of error control codes. Addison Wesley, New York, NY
7. Boutillon E, Gross WJ, Gulak PG (2003) VLSI architectures for the MAP algorithm. IEEE Trans Commun 51:175–185

8. Bruno J, Brustoloni J, Gabber E, Ozden B, Silberschatz A (1999) Disk scheduling with quality of service guarantees. In: Proceedings of IEEE International Conference on Multimedia Computing and Systems, Florence, Italy, pp 400–405
9. Cain W, Payne A, Baldwinson M, Hempstead R (1996) Challenges in the practical implementation of perpendicular magnetic recording. IEEE Trans Magn 32:97–102
10. Chen J, Dholakia A, Eleftheriou E, Fossorier M, Hu XY (2005) Reduced-complexity decoding of LDPC codes. IEEE Trans Commun 53:1288–1299
11. Daniel E, Mee C, Clark M (1999) Magnetic recording: the first 100 years. IEEE Press
12. Elerath J (2007) Hard disk drives: the good, the bad and the ugly! ACM Queue 5(6):28–37
13. Erden M, Kurtas E (2004) Thermal asperity detection and cancellation in perpendicular magnetic recording systems. IEEE Trans Magn 40:1732–1737
14. Farhang-Boroujeny B (1999) Adaptive filters: theory and applications, chap. 6. Wiley, New York, NY
15. Fossorier M, Burkert F, Lin S, Hagenauer J (1998) On the equivalence between SOVA and Max-Log-MAP decodings. IEEE Commun Lett 2(5):137–139
16. Gallager RG (1962) Low-density parity-check codes. IRE Trans Inf Theory IT-8:21–28
17. Geist R, Daniel S (1987) A continuum of disk scheduling algorithms. ACM Trans Comput Syst 5:77–92
18. Gopalaswamy S, McEwen P (2001) Read channel issues in perpendicular magnetic recording. IEEE Trans Magn 37:1929–1931
19. Guilloud F, Boutillon E, Danger J (2003) l-min decoding algorithm of regular and irregular codes. In: Proceedings of the 3nd international symposium on turbo codes and related topics, Brest, France, pp 451–454
20. Gunnam K, Choi G, Yeary MB: A parallel VLSI architecture for layered decoding for array LDPC codes. In: Proceedings of International Conference on VLSI Design, Andhra Pradesh, India, pp 738–743
21. Hagenauer J (1997) The Turbo principle: tutorial introduction and state of the art. In: Proceedings of the international symposium on turbo codes, Brest, France, pp 1–11
22. Herrman R (1990) Volterra modeling of digital magnetic saturation recording. IEEE Trans Magn 26:2125–2127
23. Hocevar D (2004) A reduced complexity decoder architecture via layered decoding of LDPC codes IEEE workshop on signal processing systems (SIPS), Texas, USA, pp 107–112
24. Immink KS, Siegel P, Wolf J (1998) Codes for digital recorders. IEEE Trans Inf Theory 44:2260–2299
25. Indukumar K, Lee Y, Mathew G (1999) Performance comparison of modified multilevel decision feedback equalization detectors. IEEE Trans Magn 35:594–604
26. Information Technology Standards. I.C.: http://www.incits.org/ (2010)
27. Iwasaki S, Nakamura Y (1977) An analysis for the magnetization mode for high density magnetic recording. IEEE Trans Magn 13:1272–1277
28. Joeressen O, Meyr H (1995) A 40 Mb/s soft-output Viterbi decoder. IEEE J Solid State Circuits 30:812–818
29. Kaitsu I, Inamura R, Toda J, Morita T (2006) Ultra high density perpendicular magnetic recording technologies. Fujitsu Sci Tech J 42:122–130
30. Kavcic A, Moura J (2000) The Viterbi algorithm and Markov noise memory. IEEE Trans Inf Theory 46:291–301
31. Kavcic A, Patapoutian A (1999) A signal-dependent autoregressive channel model. IEEE Trans Magn 35:2316–2318
32. Kikitsu A, Kamata Y, Sakurai M, Naito K (2007) Recent progress of patterned media. IEEE Trans Magn 43:3685–3688
33. Koetter R, Singer AC, Tuchler M (2004) Turbo equalization. IEEE Signal Process Mag 21: 67–80
34. Kou Y, Lin S, Fossorier MPC (2001) Low-density parity-check codes based on finite geometries: a rediscovery and new results. IEEE Trans Inf Theory 47:2711–2736

35. Kschischang FR, Frey BJ, Loeliger HA (2001) Factor graphs and the sum-product algorithm. IEEE Trans Inf Theory 47:498–519
36. Lee I, Cioffi J (1997) Equalized maximum likelihood receiver with a unit energy constraint. IEEE Trans Magnetics 33:855–862
37. Lee I, Yamauchi T, Cioffi J (1994) Modified maximum likelihood sequence estimation in a simple partial erasure model. In: Proceedings Of IEEE Internal Conference on Communications (ICC), LA, USA, pp 245–249
38. Li C, Mathew G (2006) Analytical solution for optimum partial-response target in viterbi-based receivers. IEEE Trans Commun 54:1715–1719
39. Lim F, Kavcic A (2005) Optimal precompensation for partial erasure and nonlinear transition shift in magnetic recording using dynamic programming. In: Proceedings of IEEE International Conference on Global Telecommunications (GLOBECOM), Missouri, USA, pp 58–62
40. Lin S, Costello DJ (2004) Error control coding: fundamentals and applications. 2nd edn. Prentice Hall
41. Liu B, Lee Y, Mutoh H, Garg H (1997) The effect of MR head nonlinerity on MDFE and PRML performance. IEEE Trans Magn 33:2764–2766
42. Luby M, Mitzenmacher M, Shokrollahi A, Spielman D (1998) Analysis of low density codes and improved designs using irregular graphs. In: Proceedings of 30th Annual ACM Symposium Theory of Computing, Texas, USA, pp 249–258
43. MacKay DJC (1999): Good error-correcting codes based on very sparse matrices. IEEE Trans Inf Theory 45:399–431
44. MacKay DJC, Neal RM (1996) Near Shannon limit performance of low density parity check codes. Electron Lett 32:1645–1646
45. Mansour M, Shanbhag N (2006) A 640-Mb/s 2048-bit programmable LDPC decoder chip. IEEE J Solid State Circuits 41:684–698
46. Mathew G, Farhang-Boroujeny B, Wood R (1997) Design of multilevel decision feedback equalizers. IEEE Trans Magn 33:4528–4542
47. Mee C, Daniel E (1996) Magnetic recording technology, vol 3, chap. 8. McGraw Hill, New York
48. Moon J (1998) The role of SP in data-storage. IEEE Signal Process Mag 15:54–72
49. Moon J (2000) Signal-to-noise ratio definition for magnetic recording channels with transition noise. IEEE Trans Magn 36:3881–3883
50. Moon J, Zeng W (1995) Equalization for maximum likelihood detectors. IEEE Trans Magn 31:1083–1088
51. Moon J, Zhu J (1991) Nonlinear effects of transition broadening. IEEE Trans Magn 27: 4831–4833
52. Palmer D, Ziperovich P, Wood R, Howell T (1987) Identification of nonlinear write effects using pseudo-random sequences. IEEE Trans Magn 23:2377–2379
53. Patterson R, Gibson G, Ginting E, Stodolsky D, Zelenka J (1995) Informed prefetching and caching. In: Proceedings of ACM symposium on operating systems principles, Colorado, USA, pp 79–95
54. Proakis J (2005) Partial response equalization with application to high density magnetic recording channels. In: Vasic B, Kurtas EM (eds) Coding and signal processing fro magnetic recording systems. CRC Press, 8:1–23
55. Qureshi S (1985) Adaptive equalization. Proc IEEE 73:1349–1387
56. Riani J, van Beneden S, Bergmans J, Immink A (2007) Near minimum-BER equalizer adaptation for PRML systems. IEEE Trans Commun 55:2316–2327
57. Rottmayer R, Batra S, Buechel D, Challener W, Hohlfeld J, Kubota Y, Li L, Lu B, Mihalcea C, Mountfield K, Pelhos K, Peng C, Rausch T, Seigler M, Weller D, Yang X (2006) Heat-assisted magnetic recording. IEEE Trans Magn 42:2417–2421
58. Sarigoz F, Song H, Kumar B, Bain J (2001) Dropout-tolerant read channels. IEEE J Selected Areas Commun 19:744–755

59. Sawatzky E (1998) Thermal asperities: MR heads face new dangers. Data Storage Mag 5:49–54
60. Smith A (1985) Disk cachemiss ratio analysis and design considerations. ACM Trans Comput Syst 3:161–203
61. Tan W, Cruz J (2005) Detection of media defects in perpendicular magnetic recording channels. IEEE Trans Magn 41:2956–2958
62. Taratorin A, Klaassen K (2006) Media saturation and overwrite in perpendicular recording. IEEE Trans Magn 42:157–162
63. Wang S, Taratorin A (1999) Magnetic information storage technology, chap. 9. Academic, New York, NY
64. Wiberg N (1996) Codes and decoding on general graphs. PhD dissertation, Linkoping University, Sweden
65. Wicker SB, Bhargava VK (1994) Reed-Solomon codes and their applications. IEEE Press, New York, NY
66. Worthington B, Ganger G, Patt Y (1994) Scheduling algorithms for modern disk drives. In: Proceedings of ACM SIGMETRICS conference on Measurement and modeling of computer systems, Tennessee, USA, pp 241–251
67. Wu Z, Burd G (2005) Equation based LDPC decoder for inter symbol interference channels. In: Proceedings of IEEE International Conference on Acoustics, Speech, and Signal Processing (ICASSP), pp 757–760
68. Xing X, Bertram H (1999) Error rate analysis of partial response channels in the presence of texture noise. IEEE Trans Magn 35:2070–2079
69. Yang X, Kurtas E (2005) Signal and noise generation for magnetic recording channel simulations. In: Vasic B, Kurtas EM (eds) Coding and signal processing fro magnetic recording systems, CRC Press, Boca Raton, FL, 6:1–20
70. Yeo E, Augsburger S, Davis W, Nikolic B (2003) A 500 Mb/s soft-output Viterbi decoder. IEEE J Solid-State Circuits 38:1234–1241
71. Yuan S, Bertram H (1992) Statistical data analysis of magnetic recording noise mechanisms. IEEE Trans Magn 28:84–92
72. Ziperovich P (1991) Performance degradation of PRML channels due to nonlinear distortions. IEEE Trans Magn 27:4825–4827

Chapter 5
Introduction to SSD

Massimo Iaculo, Francesco Falanga, and Ornella Vitale

Abstract Solid-state drives (SSD) represents the state of the art *of mass storage solutions*. They can be looked at as the evolution of Hard Disks which eventually "lost" the mechanical arm moving onto a rotating magnetic disk in favor of using *solid-state* nonvolatile *memory* chips to store *data*. This simple difference makes an SSD the preferred choice if read/write access time, power consumption and dimensions are the key requirements of the application. Cost and endurance are instead the reasons why traditional Hard Disks still survive in specific areas. The chapter starts with an historical evolution of the Hard Disk and moves then on its main features and parameters. Interface communication protocol is presented in more detail, being it inherited as it is by the SSDs. A glance on the Internal SSD HW and FW architecture is given with a particular focus on FTL (Flash Translation Layer) which assumes the use of NAND flash as NVM memory. Finally a brief analysis of market-segment coverage possibilities of SSD is presented.

Keywords Solid state memory · NAND flash · Performance · Endurance · Cost · Market requirements

Evolution

A quarter of century has passed since IBM introduced its 350 *Disk Storage Unit*, the very first hard disk.

Conceived by the famous American inventor Reynold Johnson, IBM employee at that time, the 350 DSU was announced on September 4, 1956, and marketed 10 days later together with the 305 RAMAC, the Big Blue computer (Fig. 5.1).

M. Iaculo (✉)
Micron, WSG Department, Via Remo de Feo, 1, 80022 Arzano (NA), Italy
e-mail: miaculo@micron.com

G. Campardo et al. (eds.), *Memory Mass Storage*,
DOI 10.1007/978-3-642-14752-4_5, © Springer-Verlag Berlin Heidelberg 2011

Fig. 5.1 1956 – The first hard disk is born; it could store 5 MB and was 1.7 m high

It had 50 magnetic disks, 24 in. each, which provided totally a capacity of 5 MB. This was astonishing for that period considering that 30 years later, in 1983, the pc IBM XT came out with a 10 MB hard disk.

As far as technologic and economic profile is concerned, between the two there is an abyss: pc XT drive had the same dimension as current devices, the 350 DSU instead was a "container" 1.5 m long, 1.7 m high, and 0.7 m deep.

It ran at 1,200 rpm with a throughput of 8.5 KBps. It was leased for $3,200 per month.

Five years later, with IBM 1301, each surface had its own head and access time reduced from 800 ms down to 180 ms (Fig. 5.2). This eliminated the time necessary for the arm to pull the head out of one disk and move up or down to a new disk. Track seeking was faster as well, since the head would hold somewhere in the middle instead of going back and starting from the disk edge every time.

Each module was made of 20 disks and 40 recording surfaces, with 250 tracks per surface. It could store 25 MB of data and spun at 1,800 rpm with a performance of 87.9 KBps.

The market price of 1301 was $115,500 but it could be leased at $2,100 per month.

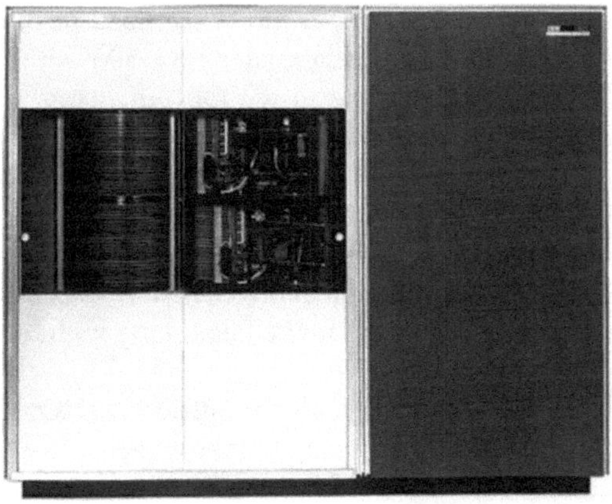

Fig. 5.2 1961 – IBM 1301

In 1963, IBM invented the head uplifting mechanism through the air. In 1973, the technique was first introduced with the 3340 Winchester, the same name as the famous rifle "30–30 Winchester" since it was provided with two disks capable of saving 30 MB each (Fig. 5.3). The heads would run on a 0.5-μm-thick air film. The industry adopted Winchester technology for the next 20 years.

The first model designed for personal computer was the Seagate ST-506 made in 1980 (Fig. 5.4). It could store up to 5 MB of data, had a diameter of 5.25 in., and was

Fig. 5.3 1973 – Modern disks ancestor: IBM 3340 Winchester

Fig. 5.4 1980 – Seagate Technology produces the first 5 in. hard disk. In this model, power consumption was also reduced

provided with a stepper motor for head positioning (voice coil control would only be introduced few years later). ST-506 equipped AT&T personal computers with 286 processors made in the Olivetti plants of Scarmagno, after the collaboration of the Italian company with the American multinational. At the same time, OPE (Olivetti Peripheral Equipment), an associate company, supplied hard disks for M24 computer; historically, this company was the only one in Europe to commit for the design and the production of this kind of peripheral.

A similar, although delayed, historical path can be detected for solid-state storage systems. There were many false starts with nonvolatile semiconductor technologies which did not survive.

In the late 1970s, silicon nitride EAROMs (electrically alterable ROMs) were marketed by a company called General Instruments. Unfortunately, after about 3 years, it became clear that the extrapolated data life of 10 years would not be achieved in practice. As a result, this product was dropped by users and did not survive in the market.

In 1976, Dataram sold an SSD called BULK CORE which attached to minicomputers from Modular Computer Systems and emulated hard disks made by DEC and Data General. Each chassis held 8× 256 k × 18 RAM modules and had a capacity of 2 MB.

In 1978, a gigabyte of RAM SSD would have cost $1 million. Texas Memory Systems introduced a 16 KB RAM-based solid-state disk system designed to accelerate field seismic data acquisition for oil companies.

In the early 1980s, Intel's 1 M bit bubble memory created a lot of excitement as a new nonvolatile solid-state memory technology (Fig. 5.5). Intel shipped design kits and boards to developers using this technology, which was positioned as a solid-state floppy disk. But it failed to be scalable or cost effective. Intel spun off the magnetic division in 1987 to Memtech (who later made flash SSDs), but bubble memory dropped into oblivion.

In 1985, Curtis introduced the ROMDISK, the first SSD for the original IBM PC.

In 1990, NEC marketed 5.25" SCSI SSDs using internal-battery-backed RAM.

In 1995, two SSD products aimed at the Sun server market:

Fig. 5.5 Bubble memory

- T8000 was an 80 MB, 10 MBps SSD on a single-slot SBus card, made by Colorado-based CERAM. Units in multiple slots could be chained to appear as a single SSD up to 960 MB. Performance was 2,000 IOPs.
- SAM-2000 was a rackmount SSD up to 8 GB, with 500 MBps internal bandwidth made by Texas Memory Systems. The transfer rate through the SBus adapter was 22 MBps. Other bus interfaces included VMEbus and HIPPI.

In June 2001, Adtron shipped the world's highest capacity 3.5" flash SSD. The S35PC had 14 GB capacity and cost $42,000.

In 2005, M-Systems announced availability of the industry's highest-capacity 2.5" SATA SSD with 128 GB of storage.

Physical DATA Organization into a Hard Disk

Digital information is saved on hard disk platter respecting certain physical allocation principles. Thanks to these storing schemes it is possible to reach the reading/writing zone with a very high speed. The most popular is CHS system, acronym of Cylinder/Head/Sector. This scheme assigns a physical address to each file, whereas this address is the set of numbers assigned to each of the following physical parts (Fig. 5.6).

Platter

Hard disks are supplied with one or more disks whose surface, covered with ferromagnetic material, can store data. Each disk side is called "platter" and is associated with a unique number.

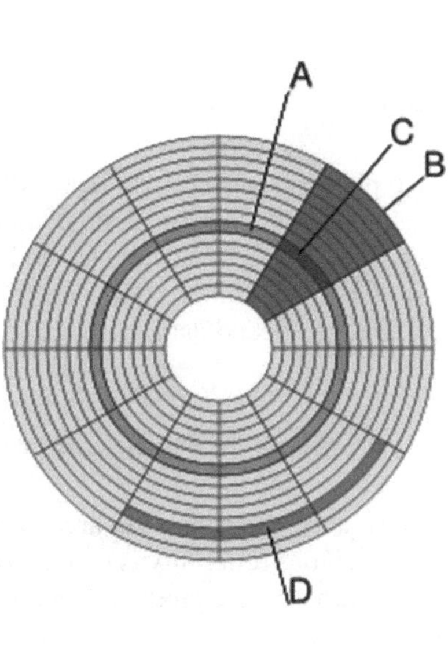

Fig. 5.6 Platter surface structure. A: Trace; B: Sector; C: Trace sector; D: Cluster

Trace (A)

Plates can be thought as a set of concentric rings named traces, each being identified by a unique number. The same ring of different plates will have the same number.

Cylinder

A cylinder is the set of the same trace out of all the hard disk plates.

Sector (B)

A sector is the area included in between two radii. Sectors in a plate have all the same dimension and each is associated with a unique number.

Trace Sector (C)

It is the portion of a trace in between two radii.

Cluster (D)

It is a set of consecutive trace sectors.

Block

All the sectors of both plates of a disk form a block.

Head

Each plate is scanned by a head which reads and writes data on the surface. There is one head for each magnetic platter surface on the spindle, mounted on a common arm. Thus, all the heads are located onto the same cylinder at a given time.

A CHS coordinate indicates which cylinder, head, and sector is necessary to retrieve the needed data.

Hard Disk Key Parameters

Performance and reliability are two key factors in determining the quality of a hard disk. As far as reliability is concerned, there is not a reference comparison table; the only criteria are the producer's reputation and price. The main, performance-related, features are

- Capacity
- Access time
- Transfer speed
- Power consumption

Capacity, usually expressed in gigabyte, is the most important aspect for consumers in making a choice on the model to buy. Modern hard disks overcome 1 TB capability.

Access time (or seek time), often neglected, is instead the most important criterion for the quality of a hard disk. This is determined by the time the head takes to position on a track and the disk rotational speed (rpm) that affects the time to reach the sector where the file is located.

The faster is the disk rotation the greater is the recording DATA density on the platter surface and, proportionally, the higher is the transfer speed.

Typical seek time ranges from 2 ms for high-end server drives to 15 ms for miniature drives.

Data transfer speed is determined by two processes: data transfer from the disk to the volatile internal buffer and data transfer from the data buffer to the external world, through the interface. A typical 7,200 rpm desktop hard drive has a sustained "disk-to-buffer" data transfer rate of about 70 MBps. This rate depends on the track location, so it will be highest for data on the outer tracks (where there are more data sectors) and lower toward the inner tracks (where there are fewer data sectors). A widely used interface is SATA, which can send about 300 MBps from the internal buffer to the external world and thus is still meaningfully greater than today's disk-to-buffer transfer rates. Data transfer rate (read/write) can be measured by writing a large file to disk using special file generator tools, then reading back the file. Transfer rate can be influenced by file system fragmentation and the layout of the files.

Moving to a Solid-State Drive

A Solid-State Drive (SSD) is sometimes more improperly referred to as Solid-State Disk although no rotating disk is present. It is a mass memory device which uses solid-state memory for data storage instead of classical rotating disk which may possibly become obsolete early in the future (Fig. 5.7).

Solid-state devices are currently based upon NAND-type flash memory which uses the "tunnel" effect to modify the electronic state of transistor cells; for that reason no mechanical and magnetic parts (disks, heads, drives) are needed, introducing remarkable advantages:

No noise
Lower likelihood of breakage
Lower operating power consumption

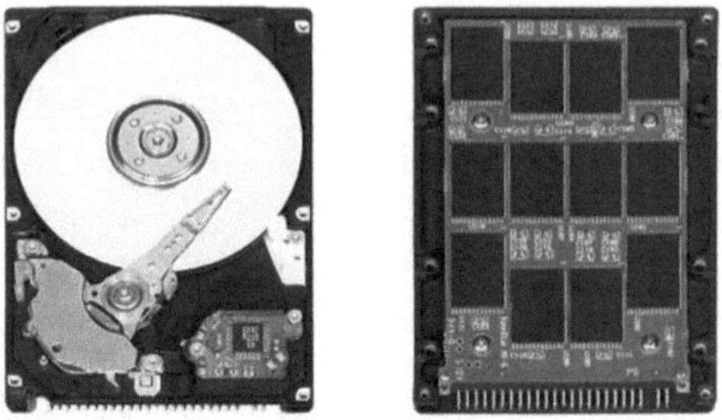

Fig. 5.7 An insight of an solid state disk

Lower read/write access time: in the order of 100 μs instead of 5 ms
Higher shock resistance: some SSD makers declare shock resistance up to 1,500 g
Lower heating

There are two main disadvantages:

Higher price per bit, as much as 12 times the traditional cost of an HD.
Lower endurance due to limited number of write/erase cycles of flash memories currently in the order of 10,000 per block.

Both issues seem to be however next to be solved. New technologies are taking flash memory to higher density, saving the cost per bit. Furthermore, the endurance is being improved taking advantage of controller wear leveling routines able to spread the memory usage uniformly over all the available space.

A point immediately emerges when analyzing SSD performances: the writing speed lower than the reading speed, along with its strong dependence on file dimension you are going to write and the defragmentation status. The reason is to be found in the intrinsic NAND architecture which has a minimum erasable block (approximately megabytes) much greater than minimum readable/writable sector (approximately kilobytes) and a sector overwriting prohibition unless the block is erased first.

Operating System's File Systems typically operate with sector of 4 K for their accesses. This means a robust logical to physical translation layer to manage proper data allocation.

Inside SSD

In RAM SSD, nonvolatility is achieved embedding internal battery systems and disk backup systems. For flash SSD, nonvolatility is achieved due to the inherent nonvolatility property of flash memory.

In the following, we will have a glance at flash-based SSD internal architecture.

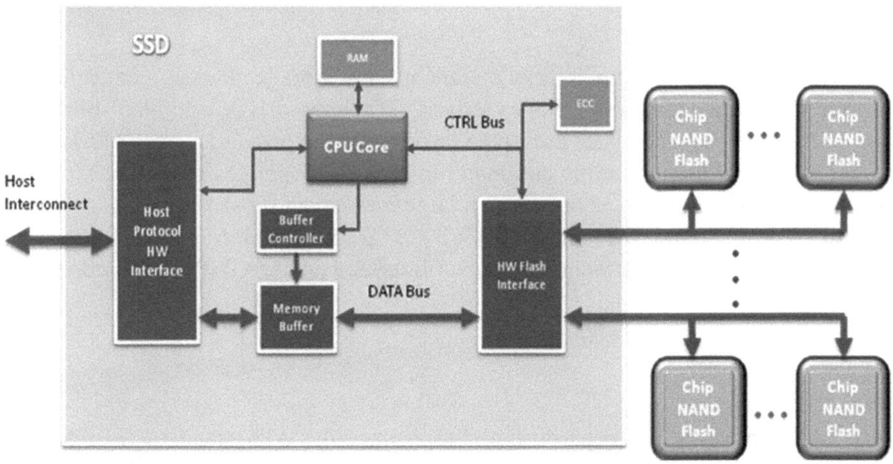

As described in the previous picture, an SSD is mainly composed of RAM (both Static and Dynamic Random Access memory) volatile memory and many NAND flash nonvolatile memory chips (SLC, MLC, or both) [2]. High-end flash storage systems, addressing enterprise solutions, consist of a mixture of DDR[1] RAM (Double Data Rate Random Access Memory) memory (as cache) and flash memory as storage.

In more detail we can locate in the scheme the following main components.

Host Protocol HW Interface

An SSD inherits hard disk drive interface (ATA, SATA, or SCISI) so that it is easy to replace HD with SSD in most systems. This way the hosting system does not know it is communicating to an SSD or an HDD.

More specifically the interface is the part dedicated to the "translation" from standard ATA, SATA, or SCISI read/write/erase command into NAND specific operations.

The most common SSD interface will be briefly described in the following sections.

[1]DRAM – Dynamic Random Access Memory is used to store volatile information on computers. DRAM is made up of many cells and each cell is referred to as a bit. A cell contains a capacitor and a transistor. Since computer machine language is made up of 1 s and 0 s, it has the value of 1 when active and 0 when inactive.

SDRAM or Synchronous Random Access Memory is the result of DRAM evolution. This type of memory synchronizes the input and output signals with the system board. Its speed ratings are in MHz. SDRAM transmits every clock count at a specific time.

DDR RAM (or Double Data Rate Random Access Memory) transmits twice every clock count. This makes DDR RAM twice as fast as SDRAM.

Parallel ATA

Parallel ATA (PATA) is an interface standard for storage devices connection of peripheral such as hard disks, floppy drives, CD-ROM drives, and also solid-state drives (Fig. 5.8). The interface presented the drive to the host as an array of 512-byte blocks with a relatively simple command interface. It operates at up to 133 MBps and requires very low power but needs a large number of pins.

The PATA standard is managed by X3/INCITS committee. It uses the underlying AT Attachment and AT Attachment Packet Interface (ATA/ATAPI) standards [1].

SATA

Serial ATA interface has been designed to replace the ATA standard (Fig. 5.9). It uses the same low-level command set, but at physical level the SATA host and devices communicate via a high-speed serial cable over two pairs of conductors + grounds.

From the performances point of view there are two revisions. SATA 1.5 Gbps communicates at a rate of 1.5 Gbps, but taking into account the 8b/10b encoding overhead the effective uncoded transfer rate is of 1.2 Gbps (about 143 MBps). SATA

Fig. 5.8 Parallel ATA cable interface

Fig. 5.9 SATA cable interface

specification revision 2.0 doubles instead the transmission rate up to 3.0 Gbps. The most updated SATA specifications define data transfer rates as high as 6.0 Gbps per device.

PCI Express

PCI Express (Peripheral Component Interconnect Express), or more simply PCIe, is a standard for computer expansion card. PCIe 2.1 is the latest standard for expansion cards that are currently available. It supports 500 MBps per lane and has very low power consumption. The PCIe bus can be thought of as a high-speed serial bus as a replacement of the old (parallel) PCI/PCI-X bus. In terms of protocol structure, the PCIe communication is encapsulated in packets. Indeed, PCI Express is a layered protocol, consisting of a *transaction layer*, a *data link layer*, and a *physical layer*. The data link layer is subdivided to include a media access control (MAC) sublayer. The physical layer is subdivided into logical and electrical sublayers. The physical logical sublayer contains a physical coding sublayer (PCS). (The terms used refer to IEEE 802 conventions of networking protocol.)

SCISI

SCISI (Small Computer System Interface) is another common standard for physical connection and data transferring between computers and peripheral devices. The SCSI standards specify electrical interfaces, protocol structure, and command set. The SCSI interface is most commonly used for high end hard disk drivers in enterprise applications, but it can connect a wide range of other devices, including SSD.

RAM Buffer

The RAM buffer caches data coming from/to the system host during write or read operations, respectively, and is normally managed by a buffer controller that interacts with the core microcontroller. The CPU core will use the hardware flash interface that acts like a DMA (Direct Memory Access) to speed up data transfer time toward/from the flash memory channels and the buffer memory. The RAM buffer implementation can range from a simple circular buffer structure to a complex cache mechanism. The filling state of this cache is orchestrated by caching management algorithms normally running on the CPU core.

CPU Core

SSD normally embeds at least a 32 bit RISC core to execute relevant firmware processes with the aim of managing all the SSD activities.

Hardware Flash Interface

The hardware flash interface is designed to rightly interface flash NAND devices in terms of bus dimension and timing. It is mainly a multichannel DMA (Direct Memory Access) toward the flash raid to achieve the target-sustained throughput.

The SSD controller is connected to *N* groups of flash, each group on a different independent channel.

As shown in Fig. 5.10, signals between controller and flash are collected in the bus.

- At least two I/O bus 8 o 16 bits (dual-channel) multiplexed are reserved for command, address, and data. As shown in the picture, more flash can be linked to at the same channel.
- Control Bus includes flash control signals (CLE, command latch enable; ALE, address latch enable; RE, read enable; WE, write enable) for each channel.
- *N* Chip Enable (CEx, $x = 1$ to *N*) one for each chip e *N* ready/busy (R/Bx, $x = 1$ to *N*), allow independent flash chip management.

Particularly, the possibility to manage *N* R/B signals ($N > 1$) allows to access more NAND chips in parallel for both read and write operations, increasing the final throughput.

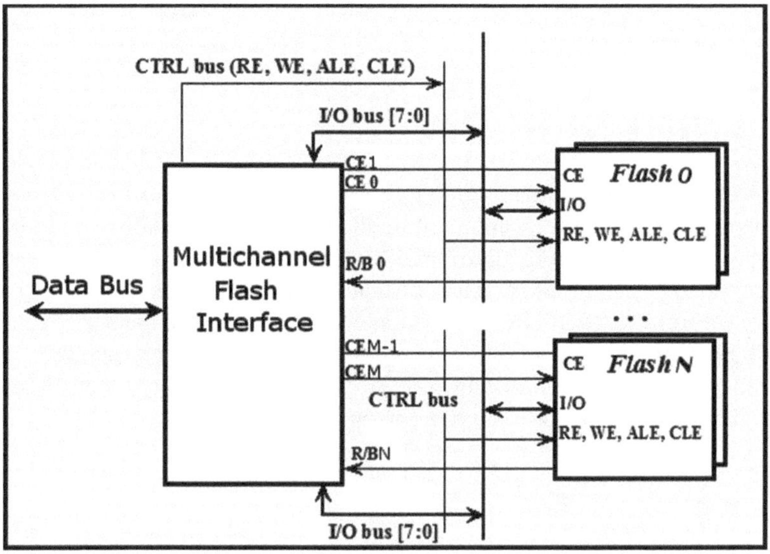

Fig. 5.10 Flash SSD controller connections

Error Correction Code

Flash memories are not error-free. The probability of error depends on both device technology and usage.

Stress induced by read/write/erase operations tends to reduce the cell capability of keeping charge into the floating gate and preserving information.

This stress is amplified by the matrix structure of flash memories which causes interference phenomena between cells. This defect causes wrong bits on original user data.

The SSD controller can make the probability of error of the system lower than that of the physical mean. In fact, besides the above-described Wear Levelling and Read Disturb Management techniques, the Error Correction Code (ECC) is used in order to maximize the system lifespan.

The ECC is a redundant code which is added to data during the program operation and checked during the read phase, in order to detect errors and, if possible, correct them. As the correction capability is related to redundant code size and algorithm complexity, the system architect has to cope with the trade-off between system reliability and controller overhead in order to choose the proper ECC.

Usually, ECC codes require complex elaboration of data blocks making this calculation implemented by firmware (in charge of the core CPU) unfeasible. This brings to implement this calculation with an ECC hardware module.

Understanding I/O

Access Time

We have seen how access time is determined in an HD and how it affects performances. In SSD context, the *access time* or *latency* is the average time lag between a request for information stored and its delivery [7].

Taking apart the cases of cache hits (when the data to be retrieved are presented into the volatile RAM buffer), typical SSD access times are in the order of hundreds of microseconds, with respect to few milliseconds of the HD.

SSD IOPS

In SSD, a common definition to provide a measure of IO capability is IOPS: I/O operations per second [3]. This parameter is strongly dependent on testing conditions; for this reason it has to be provided along with additional information.

Definitions on the following will help to clarify the point.

Chunk size: It is the amount of data to be written/read in a single access.

Sustained accesses: Continuous sustained accesses. The total amount of transferred data is in the order of the SSD capacity.

Burst accesses: Limited number of accesses with a total amount of data transferred at least one order of magnitude smaller than the system capacity. This kind of access normally benefits of cache hits and background operations [5].

Random accesses: The addresses of read/write operations are chosen randomly out of the whole addressable space.

Sequential accesses: The addresses of read/write operations are chosen sequentially [6].

IOPS measured values strongly varies changing the chunk and access mode as defined above.

Normally, HDD manufacturers rarely declare their IOPS performance with respect to the different access mode. The customer should be aware that typical declared values are related to sequential access modes. Similarly, flash SSD makers mainly refer to read performance and often based on cache hits assumption (best case); considering that write performances are worse than reading, declaring a "single" IOPS metric based upon reading can be misleading.

Sequential IOPS are not enough to provide system capability in the "real world" because real application usage models do not result in sequential disk access. For example, for database transactions we can identify two main features: small data chunk (averaging around 8 K) and access addresses are random.

Most people think of these solid-state devices as just another hard disk, but their performance characteristics can be very different.

Firmware Architecture of an SSD

Flash Translation Layer

The Flash Translation Layer (FTL) translates a logical address written or read by the host in a NAND physical position (Fig. 5.11).

NAND memory can be programmed with a page granularity and erased with a block one. Furthermore, a block erase operation is required before reprogramming a written page.

The FTL makes these NAND limitations transparent to the host. Data can be written and read with a fixed protocol regardless of the features (page size, block size, SLC or MLC technology, number of memory chips, etc.) of the physical memory which stores them.

The FTL enables the SSD to mimic a hard disk drive. In fact, it is able to emulate sector rewrites and erase operations with a granularity smaller than a NAND block. These operations are managed by dynamically changing the physical address associated with each logical address (logical to physical address remapping). So, if the host deletes or rewrites data, the controller does not have to erase the physical block which contains it. The FTL just writes data in a different address, updates the information related to the physical position of that data, and considers the old position

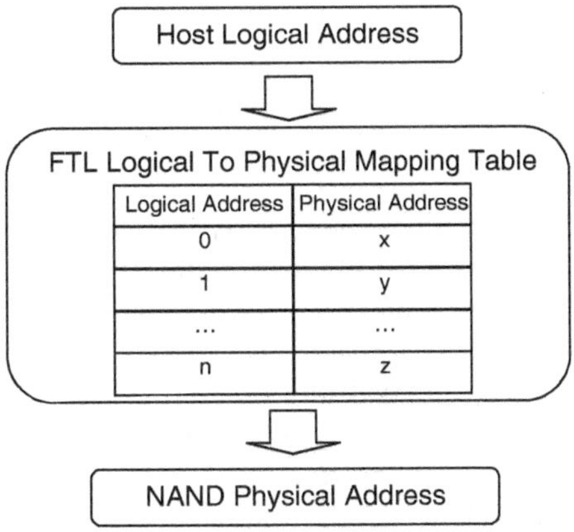

Fig. 5.11 Logical to physical remap

as invalid or dirty. When the controller needs a free NAND block, it performs a garbage collection (or defrag) operation in order to recover dirty space: valid pages, if present, of a selected block are copied into a swap (free) block and then the dirty block is erased.

Different techniques can be used to select the block to be defragmented. The choice is usually based on the number of dirty pages in the block and the number of erase operations already performed on it, in order to optimize device performance and endurance.

Caching techniques can also be used by the controller in order to reduce the number of garbage collection operations and then increase both the device endurance and performance.

The FTL architectures can manage the logical to physical remap with different granularities (sector, NAND page or multiple of it, NAND block or multiple of it). The choice depends on the trade-off between performance and resource consumption: a fine granularity leads to better performance but requires more resources than a coarse granularity.

A hybrid solution can be used in order to combine the competitive performance proper to block remap with low cost (in terms of RAM and CPU power) proper to page remap. In these solutions, some cache blocks are used to temporarily store incoming data. Cache blocks are managed with a page remap, so each sector can be written in any physical position. Garbage collection operations are performed, when needed, in order to merge data belonging to the same logical block (a logical block is a rage of logical addresses with a size equal to or multiple of a NAND physical block size).

For any FTL architecture, the frequency of garbage collection activities depends on host usage model: a defrag operation can be never required if the host writes sequentially all logical addresses while it can be required at each host access for random writes.

In order to improve system performance, some garbage collection operations can be performed in background.

Performance

Unlike hard disk, SSDs have write performance very different from read performance. Furthermore, both read and write performance are strongly dependent on the access type.

For sequential read and write operations, SSDs have performance similar to HDDs. As NAND flash memory does not have a seek delay, SSDs have better performance than HHDs for random reads. However, the characteristics of NAND flash memory, which requires a block erase before reprogramming a page, make the SSDs worse than HDDs for random writes.

SSDs usually contain several NAND chips which can be programmed and read in parallel in order to make the system performance better than raw NAND performance. This parallelism can be performed at different levels.

If the NAND chips share both control logic and data bus (single-channel architecture), the parallelism can be obtained through the interleave (Fig. 5.12) [2]. It exploits the NAND read/write operation latency: while a NAND is busy the bus is used to transfer data to/from another NAND. This technique can be used to increase the system throughput up to the NAND interconnect bandwidth.

If the NAND chips share control logic but have different data buses (multichannel architecture), the controller can send a read/write request to all chips at the same time (Fig. 5.13). So, if compared with the single-channel performance, the global system throughput for sequential operations is increased by a factor equal to the number of channels without a significant complexity increase.

If the NAND chips are fully independent from each other (fully connected architecture), they can receive separate flows of requests and concurrently process them (Fig. 5.14). The drawback of this architecture is the cost of making the controller able to manage these independent flows of requests.

Read/write performance can be further increased through the use of a cache RAM memory. In order to guarantee data reliability, the system has to be provided with an internal battery which allows flushing data from RAM to flash, in case of an external power loss.

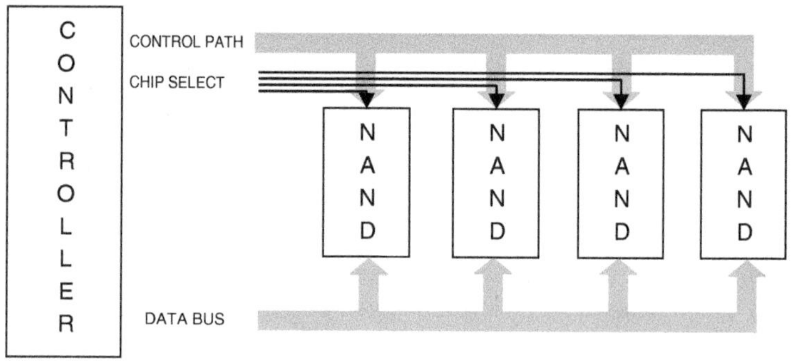

Fig. 5.12 Single-channel architecture

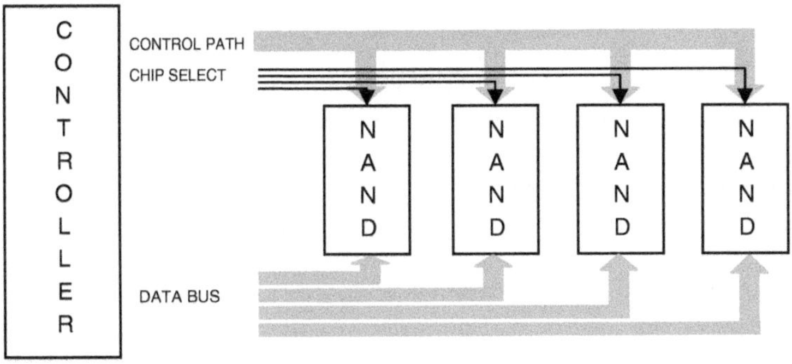

Fig. 5.13 Multichannel architecture

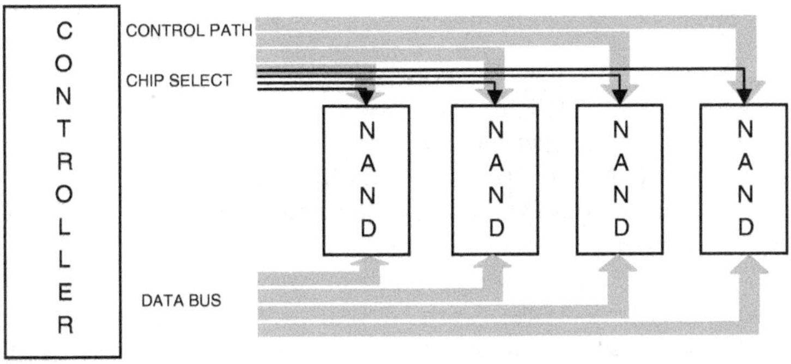

Fig. 5.14 Fully connected architecture

As previously said, the main SSD limitation is the random write performance, which is mainly related to the frequency of defrag operation.

The cache RAM can improve burst random write, but a proper data management is required to have a sustained improvement.

A fine granularity remap, for example, allows writing data sequentially into the NAND, regardless of their logical address.

Furthermore, a pool of spare blocks can be used to delay and, when possible, optimize the defrag operation. These blocks can be preerased in background to be ready to receive incoming data.

Bad Block Management

Flash memories can contain bad blocks, that is, blocks which have some bits whose value is not changeable. This problem can happen both during the production process and during device usage.

Bad blocks are marked as invalid by programming a fixed pattern in the block spare area.

In order to guarantee data reliability, firmware has to avoid using bad blocks, so it has to be able to detect these blocks and properly manage them.

The bad block management can be based on different techniques. However, the different approaches can be classified into two main groups: skip method and spare block method.

The former is integrated in the FTL and affects the block allocation phase: when a free block is required the FTL selects a valid block discarding the invalid ones (Fig. 5.15).

The latter is transparent to the FTL which is able to use only a part of the NAND addressable space but always sees an array of valid blocks: when a visible block becomes bad it is replaced with another block belonging to the reserved (that is not visible to the FTL) part of the NAND array (Fig. 5.16).

Wear Levelling

Flash memories have a limited program/erase cycle capability. The maximum number of program/erase cycles per block for current technologies usually varies from 3k cycles for MLC NANDs to 100k cycles for SLC NANDs [4].

When a block reaches this cycling limit it becomes unreliable because some cells begin losing their program/erase capability.

The SSD controller usually reserves a part of NAND addressable space (not visible to the host) in order to replace worn-out blocks. When all the spare blocks wear out, the device becomes no more writable/erasable even if the other NAND blocks are still good (Fig. 5.17).

In order to extend the SSD lifetime, the controller has to guarantee a uniform usage of all available flash blocks, regardless of the host usage model.

This uniform erase distribution is performed by the wear levelling algorithm.

The wear levelling can involve either dynamic data or both static and dynamic data. Dynamic data are defined as any data updated by the host, while static data are defined as data written once by the host and never changed, such as operating system files, look-up tables, and executable files.

Dynamic wear levelling assures a uniform usage of all available free blocks: when a new block is required in order to write host data it selects the less used block belonging to the pool of free blocks (Fig. 5.18).

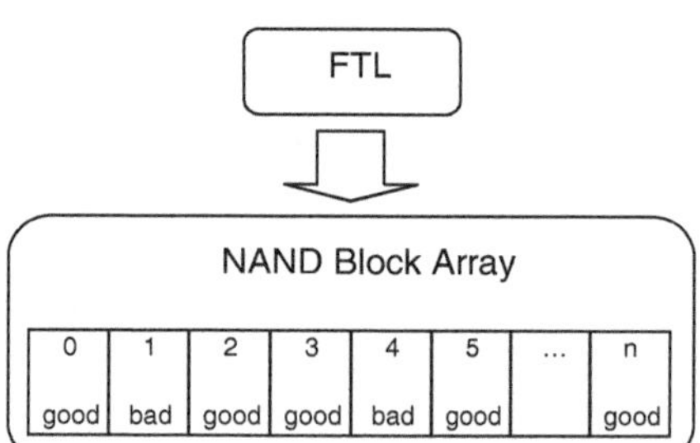

Fig. 5.15 BBM – Skip method

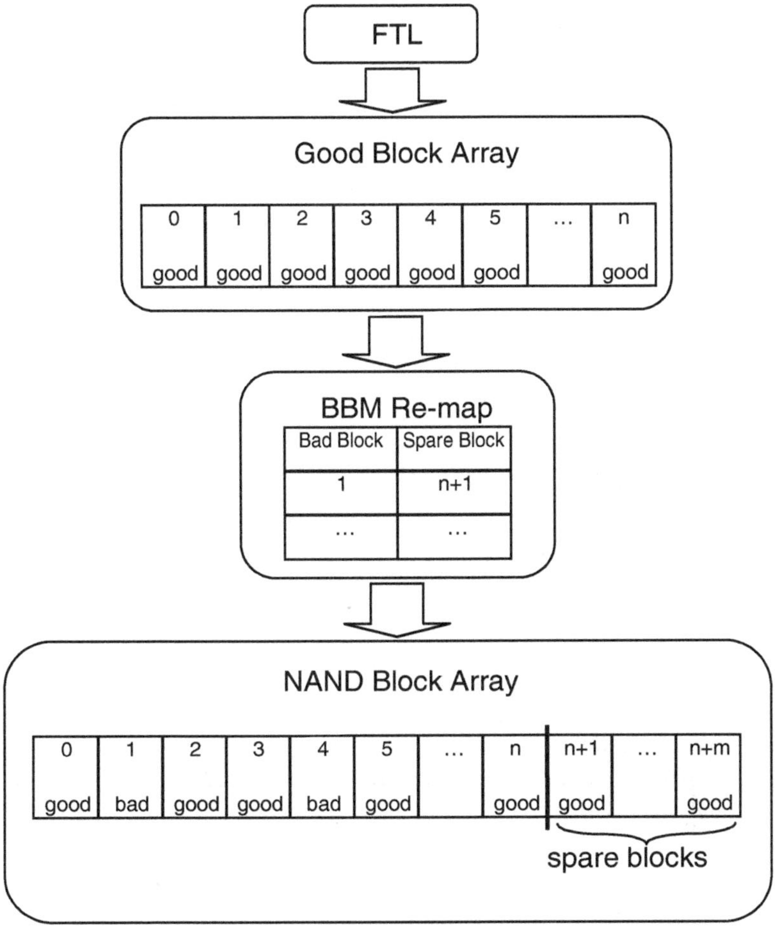

Fig. 5.16 BBM – Spare block method

Fig. 5.17 No wear levelling

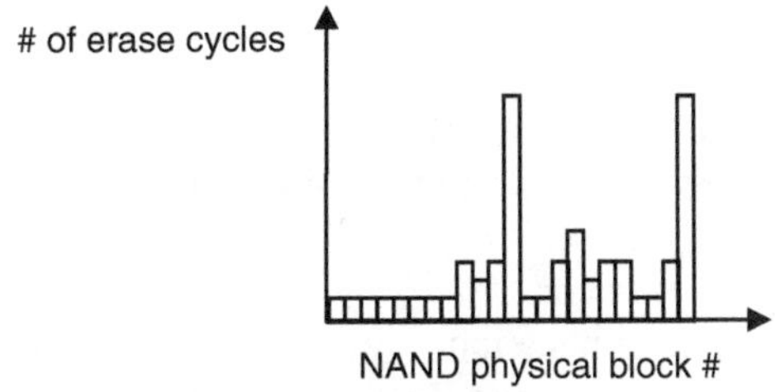

This operation can be performed through either a table which contains an erase counter for each physical block or an ordered queue in which the most recently erased block is in the last position.

Static wear levelling involves blocks which contain valid data.

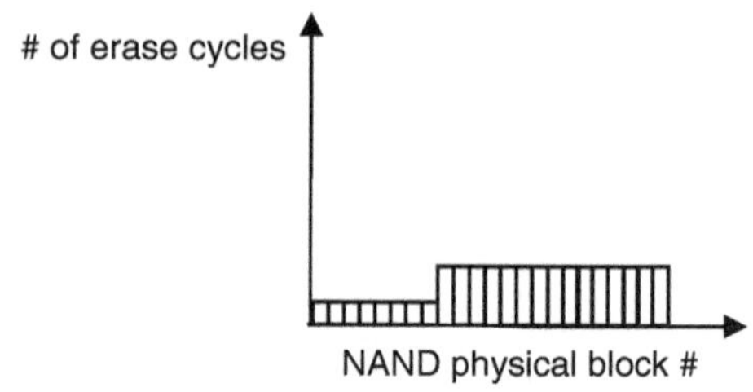

Fig. 5.18 Dynamic wear levelling

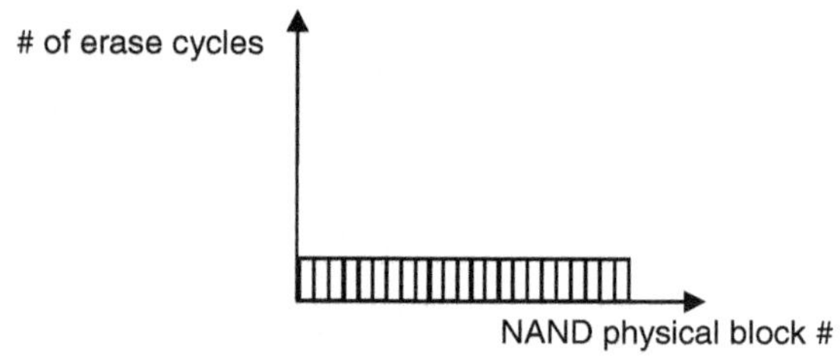

Fig. 5.19 Static wear levelling

It selects the written block with the least number of erases and copies its data into the most frequency erased free block.

Static wear levelling algorithm is usually based on a threshold method: a remap operation happens when the difference between the number of erase performed on a free block and that performed on a programmed block goes over the threshold value (Fig. 5.19).

Read Disturbance Management

Read operations on NAND memories do not cause the device to wear out but can corrupt stored data.

In fact, a NAND page read operation can affect data stored in other NAND pages belonging to the same physical block: the value of one or more bits can be changed from "1" to "0," that is, some not programmed bits can be unintentionally programmed.

For application where read operations are very frequent and data are rarely updated, a proper Read Disturbance Management is required in order to avoid data loss.

The controller, if required, has to refresh data, that is, it has to reprogram overread data into a new physical position and erase the old block.

As the read disturbance problem does not damage physical cells, the erased block can be used again to store data.

Endurance

The SSD endurance is defined as *the maximum quantity of data writable by a host on the device before a failure takes place.*

It is important to note that endurance is not just a function of the physical storage media but it is strictly dependent on controller architecture and host access type.

The controller can have an influence on the global endurance of the system through Flash Translation Layer strategy, Bad Block Management, Error Correction Code, and Wear Levelling algorithms.

SSD endurance can be calculated by using the following formula:

$$\text{Endurance} = \frac{\text{NAND Reliability} \times \text{Device Capacity} \times \text{Wear Levelling Efficiency}}{\text{Write Amplification Factor}}.$$

NAND Reliability is the maximum number of program/erase cycles which can be performed on each NAND block without overcoming the retention and error rate limits tolerated by the system.

Device Capacity is the NAND space used for host data remap. This value is usually smaller than NAND size but bigger than the SSD capacity.

Wear Levelling Efficiency is a parameter which quantifies the Wear Levelling capability of distributing the program/erase operations over all the NAND available space in order to maximize the usage of each block before the system wears out.

Write Amplification is the ratio between data programmed by the controller in NAND and data written by the host.

This value is usually greater than one because of both NAND technology limitations (page granularity for program operation and block granularity for erase operation) and data management algorithms implemented by the controller in order to increase system performance and reliability (FTL, Wear Levelling, Read Disturb Management, etc.). The basic elements to be taken into account in order to evaluate this parameter are:

- NAND Page Size: the program operations on NAND Flash are done per page so if the size of data programmed by the host is not a multiple of NAND page, some space is wasted;
- Block Size: the erase operations are done by blocks so a garbage collection operation is required in order to erase only a part of data contained in NAND block;
- Data Management Overhead: FTL, Wear Levelling and Read Disturb Management algorithms perform some additional write/erase operations.

Write Amplification is strongly dependent on the nature of access to disk; usually, it is maximum for small chunk random writes and minimum for huge chunk sequentially writes. So, some representative usage models have to be used in order to have the SSD endurance parameters for a selected set of applications.

Market Segments

The main aim of SSD is to offer a valid alternative to traditional HD.

Storage capacity is requested to increase by 50–60% every year. In addition, more aggressive power consumption and throughput requirements are arising into the top priority list because of mobile market growth. These new requirements cannot be completely fulfilled by the traditional mechanical-based system.

As seen before, HDD technology is a 50-year-old proven technology and is used as the primary storage device in a number of applications today. But computing has become increasingly "mobile" in the form of notebooks, portable media players, and cell phones so that alternatives, mostly based on NAND, to mechanical HDDs have emerged as potential challengers.

Storage systems can generally be characterized by capacity, read/write performance, form factor, endurance, reliability, shock resistance, and cost. Each market segment requires a different set of the above-mentioned parameter values.

SSDs are currently found in a number of different applications where other storage technologies are not able to fulfill market requirements, for example, shock resistance for military application. SSDs are, though more slowly for cost reason, being introduced in mass market.

In the following, an overview of the main market segments is presented along with some use details [8].

Personal Computer

Today, a number of notebook PCs are based on SSD solutions. SSDs are a very suitable solution in segments like NetBook and ultraportable notebook where form factor, mobility, and battery life are basic factors.

Gaming PCs are another area that can benefit from the use of SSDs. These systems can take advantage of an SSD's fast access time to accelerate load times and increase performance to create a better end-user experience. Capacity for this kind of segments ranges from 25 to 128 GB and above. In this segment, the cost per gigabyte is more important. Consumer-based SSDs are generally constructed using multilevel cell NAND because it is cheaper, but also less durable and reliable.

Industrial

Industrial systems include a wide range of applications such as pill-dispensing machines in hospitals, robotics on an automotive assembly line, and highly resistant sensors in remote oil and gas pipelines.

A first important requirement, all the above-mentioned share, is to cope with high shock and vibration, extreme temperature fluctuation, and hostile environments, as well as long product life and ease of maintenance in the field.

Typical market applications are

- Data recorders and remote data collection systems
- Test and measurement equipment
- Industrial workstations
- Robotic systems
- Automated inspection equipment

All share the following requirements:

- High performance and reliability in harsh conditions
- Fast data transfer
- Rugged, small footprint form factors
- Power anomaly management
- Ability to forecast drive life

Military

Military applications require the highest performance and reliability. Whether deployed on the battlefield in a wearable computer or an Abrams tank or in a digital surveillance system in a Blackhawk helicopter, performance must be flawless in the harshest climates and conditions.

Typical market applications are

- Mission data recorders
- Troop and field computers
- Unmanned vehicle control
- Launch control systems
- GPS communication systems

All share the following requirements:

- Failure-proof performance in extreme environments
- Multiyear product life cycle
- Meet military compliance standards
- Rugged, small footprint form factors
- Security options to prevent data and software IP theft

Enterprise

Although SSD is becoming a standard option for laptop and desktop storage, its impact on enterprise server storage is not completely assessed.

Provisioning server storage capability is challenging. It requires optimizing for the performance, capacity, power, and reliability needs of the expected workload, all while minimizing financial costs.

Replacing disks by SSDs is not a cost-effective option due to the low capacity per dollar of SSDs. Depending on the workload, the capacity per dollar of SSDs needs to increase by a factor of 3–3,000 for an SSD-based solution to break even with a disk-based solution. Thus, without a large increase in SSD capacity per dollar, only the smallest volumes, such as system boot volumes, can be cost-effectively migrated to SSDs.

To cope with cost issue, alternatives to complete replacement of disks by SSDs can be found such as use of SSDs as an intermediate tier between disks and DRAM.

Typical market applications are

- Web Server

All share the following requirements:

- Capacity per dollar
- Performance
- Power consumption
- Reliability

References

1. Kzierok CM (2001-04-17) The PC Guide: Overview and history of the IDE/ATA interface. http://www.pcguide.com/ref/hdd/if/ide/over.htm
2. Agrawal N, Prabhakaran V, Wobber T, Davis JD, Manasse M, Panigrahy R. (2008) Design tradeoffs for SSD performance. Microsoft Research, Silicon Valley, University of Wisconsin-Madison. To appear in the Proceedings of the USENIX Technical Conference, June 2008
3. Bowen J (2007) Flash in the enterprise. Texas Memory Systems White Paper
4. SiliconSystems (2007) Increasing flash SSD reliability. StorageSearch.com
5. Casey M.Solid state file-caching for performance and scalability. StorageSearch.com
6. Kerekes Z (2009) The Problem with write IOPS – in flash SSDs. StorageSearch.com
7. Hutsell W (2008) An In-depth Look at the Ramsan-500 Cached Flash Solid State Disk. Texas Memory Systems
8. Goldstein H (January, 2006), Loser too little, too soon. IEEE Spectrum (spectrum.ieee.org)

Chapter 6
Packaging Trends and Technology in Wireless and SSD Applications

Aldo Losavio, Davide Codegoni, Maria Luisa Polignano, Federica Zanderigo, and Luca Zanotti

Abstract Hand-handled, wireless world represents the most challenging environment for package technology where all the system performances must be densely stacked. Chip scale package (CSP) is the generic term for packages approaching chip size, and the fine-pitch versions of BGA have become the most widely used kind of CSP for system miniaturization. With the introduction of the new feature, where heterogeneous semiconductor devices are stacked together within a single package working at extremely high frequencies, the package design in terms of system simulation – mechanical, electrical, and thermal – is becoming one of most important development activities for delivering robust system solutions. In this chapter, the package historical trends against the system evolution will be discussed, analyzing the principal integration challenging. Among them, this chapter will focus on thin die thickness trend, taking into account the new process technology and the related impact on the device characteristics. This is considered as one of the most important back-end technologies, enabling the new era of the package integration and miniaturization. New interconnection technologies among dies will also be reviewed and discussed by deeply analyzing the features of through silicon via process. This process will allow the new interconnection scheme for microelectronics for the next decade.

Keywords Package · Ball grid array · Through silicon via (TSV) · Die thinning · Backgrinding · Polishing · Silicon damage · Data retention · Ion contamination · Multichip (MCP) · Wireless application · Non volatile memory · Solid state drive · Silicon defects

Ball Grid Array Evolution into Wireless Applications

Lead-type packaging technology was at one time the basis of semiconductor packaging. and the main purpose of the package was to protect the semiconductor

A. Losavio (✉)
Micron, Back End Operation, Via Olivetti 2, 20041 Agrate Brianza (MB), Italy
e-mail: alosavio@micron.com

G. Campardo et al. (eds.), *Memory Mass Storage*,
DOI 10.1007/978-3-642-14752-4_6,

device from the environment. Prevention of thermal stress and moisture-induced corrosion after attachment to the circuit board were the most important issues that had to be addressed. Since then, advances in packaging materials, chip-surface protection materials, and package structure design technology have greatly improved resistance to heat and moisture—and packages have become smaller and thinner. The new needs being engendered by higher-pin (lead) counts, severe stacked 3-D structures, and higher speeds (as demanded by high-frequency products) are determining the functions required of the package. In the past, package terminals were arranged around the periphery of the lead-type package, causing the package area to grow with the square of the number of leads as the pin count increased. Thus for high-pin-count packages, those with more than 250 pins, the package size became much larger than the chip size, and retaining small sizes became difficult from the standpoint of assembly and production technology. However, with an area-array in which the terminals are arranged on the rear surface of the package in a rectangular grid, the package area is proportional to the number of pins.

Therefore, an area array type package is essential to achieve small multipin packages. On the other hand, with low-pin-count packages, an area-array type package is not very effective in reducing the area compared with a peripheral-lead type package, and there is not sufficient benefit are to justify use of a new-configuration package. However, the today's higher-speed requirements are leading to a change in conditions. Inductance and capacitance of leads inside the package are a cause of signal noise, and this will create a need to shorten lead (interconnect) length inside the package.

Area-arrays are better than peripheral-head type packages for shortening the interconnect length within the package because lead length can be reduced from the bonding pad on the chip to the connection with the circuit board. Thus area-arrays are now being usedto meet high-speed requirements even low-pin count packages.

One of the most important functions of packaging is to serve as a scale converter between the chip and the board. Actually, Ball Grid Arrays (BGA) having solder balls with coarse pitches of 0.5 mm or more are now in general use as connection leads that make for ease of assembly. If we make an even more detailed study of the purpose of packaging, we find that the terminal pitch is the most important element to consider when designing area-array packages. In general, when the pitch is large, the package size becomes large, and when the pitch is small connection to the board becomes difficult. We believe that today's interconnect guidelines specify less than 0.1 μm for chips, but generally several hundred micrometers on assembly circuit boards. Pads for connection to the chip require a spacing of about 100 μm; lands for connection to the board require a spacing of 0.5 mm or more.

Thus the purpose of a package is not only to protect a chip; we must also consider its function as a scale converter. Formerly packaging technology was designed to make a container to encapsulate a single semiconductor device. The first packaging technology evolution was to become a means to provide the optimum solution in combination with the assembly board material and design. That is, the function demanded of packaging technology has changed from providing a package as an individual component to packaging that includes board assembly; in other words the technology has expanded to provide a system solution. This will become clearer when further BGA evolution, MultiChip BGA and Package-on-Package technology, is reviewed and discussed. With the introduction of the new feature, where heterogeneous semiconductor devices are stacked together within a single package working at extremely high frequencies, the package design in terms of system simulation—mechanical, electrical, and thermal—is becoming one of most important development activities for delivering robust system solutions.

BGA vs. QFP Features

Today's Quad Flat Pack (QFP) and the Ball Grid Array (BGA) packages both offer a large number of I/Os, as required by modern IC technology. The BGA concept has been received very favorably owing to its inherent, potential benefits for surface mount production. In order to accommodate the increasing number of I/Os needed, the peripheral QFP technology is forced into an ever-finer lead pitch with thinner and more fragile leads. The BGA, taking advantage of the area under the package for the solder sphere interconnections, satisfies the I/O demand using a far coarser pitch.

The relationship between BGA and QFP package size and I/O count can be better understood from the following example. A typical 0.65 mm (25.6 mil) fine-pitch QFP with 160 leads measures 28×28-mm. Modern portable electronics requiring the same number of leads in a package 14×14-mm ends up with 0.3-mm (11.8-mil) pitch with a space between the leads of only 0.15 mm (6 mils). Alternatively, increasing the number of I/Os while retaining the 0.65-mm pitch means, e.g., 232 leads in a 40×40-mm body. A 27×27-mm plastic BGA (PBGA) houses 225 I/Os with a coarse 1.5-mm pitch. The distance between adjacent spheres of solder is approximately 0.8 mm.

The more I/Os needed, the better suited the BGA in terms of package size, since the dimensions only grow as the square root of the I/O count for a given pitch rather than linearly as is the case for QFPs. For reasons of cost, however, plastic BGAs are the only real alternative for most consumer applications, especially for wireless applications. There were some doubts about the reliability of the leadless feature of the BGA package in harsh environments, e.g., in hand-held applications, where strong mechanical resistance is required. Industry has made an enormous effort in the last decade to provide a robust solder joint reliability solution, even with smaller pitches (down to a 0.4-mm pitch). Today BGAs can to overcome the severe mechanical/thermal shock restriction connected to different types of electronic applications (from automotive to wireless). This development is due manly to the effort to improve board level reliability, increasing the solder joint resistance by changing and/or optimizing the solder ball metallurgy and the substrate finishing properties.

Driving Forces for Using BGAs

Replacing QFPs with BGAs not only means that higher pin counts or smaller packages can be achieved, but also that a considerably higher manufacturing process yield can be reached. Even though a choice between the BGA and QFP technologies seems easy from a production point of view, the alternatives still have to be considered and all pertinent issues should be carefully addressed. To summarize, the driving forces can be listed (not in order of importance) as follows:

- Flexibility of design of multiple stacking products.
- Savings in the printed circuit board area required per function.
- Increased electrical performance.
- Lower overall production costs.
- Reliability constraints in specific applications.

Consideration of all these features and careful analyses of the pros and cons of adopting BGA technology has resulted in the Ball Grid Array structure being almost universally recognized as the best way to support the never-ending race to further miniaturize electronic equipment and applications. This will be clearer after the discussion on Multi Chip Stacked BGA products and its evolution into the Multi Package Stacked solutions, such as the Package-on-Package (POP) technology.

Chip Scale Package Technology

Chip Scale Package (CSP) is the generic term for packages approaching chip size, and many different types have been proposed. Among area-array-type packages, fine-pitch versions of BGA have become the most widely used kind of CSP for system miniaturization, mostly for wireless applications. Thus we believe that this is the most effective package today for making smaller electronic devices. In general, there are variations in the external configuration of BGA-type CSPs in combinations of package size and solder ball pitch, but there are a multitude of variations in package structure, including those related to reliability, manufacturability, and design of interconnects within the package. At present, these CSPs are used in mobile phones and similar applications using high interconnect density boards, where small size is essential. From the standpoint of test socket technology and assembly board interconnect design, solder ball pitch is now at the stage where stable functional technology now exists for 0.4–0.8 mm.

The most important determinants with regard to reliability are solder-process heat resistance and thermal-cycle life after assembly, moisture resistance, and mechanical stress resistance. The thickness of CSP packages makes it difficult to maintain a low level of moisture absorption by the package before assembly, and the danger of cracks occurring during reflow soldering is higher than with earlier types of packages. Also, it is more difficult to maintain reliability after assembly than for earlier types because CSP packages are so small and thin.

The semiconductor industry's demand for higher levels of integration, lower costs, and a growing awareness of complete system configuration through SiP solutions is enabling wide acceptance of multichip packaging. With the increasing focus on the importance of form factor reduction, especially in mobile and wireless consumer products, stacked die packages are becoming more common in cell phones, digital cameras, and hand-held devices.

Electronic packaging techniques that use only planar dimensions are no longer effective enough to answer the requirements of continually shrinking personal electronic products. Remarkable size and weight reductions have been reported using 3-D packaging instead of packaging in planar dimensions. In addition, 3-D interconnection delivers other significant advantages. The interconnection lengths can be kept shorter and the connectivity between active parts in a 3-D package is better than in planar packages.

The reductions in noise and delay owing to the shorter interconnection lengths are advantageous for high-speed operations. However, packaging large amounts of silicon in a small footprint provides various stress-related challenges to product reliability.

The Next IC Packaging Challenges

In this era of communications and entertainment, growth of consumer electronics is exploding. The silicon solutions driving these products are more highly integrated than ever before, as advancements in process technology are delivering System-on-a-Chip (SoC) solutions that are smaller, faster, and lower in cost.

These trends, along with the broad range of back end equipments emerging, require many different kinds of new IC package types to meet specific applications or special markets. Increased device complexity will generate an explosion of new creative technology packaging solutions, and in some markets and applications, packaging technology will become a key differentiator when making purchasing decisions.

To that end, there are a number of general packaging technology challenges the industry must address in the coming years. Addressing the issues below will take packaging technology to a new level, including migration toward ultraminiaturization; growth of package on package; modularization; package scaling to match silicon scaling; nanostructures; 3-D IC interconnects; optical input/interconnect and heterogeneous integration.

Packaging Advanced CMOS Technology with Copper and Ultralow K Dielectrics

The biggest challenge facing packaging in the next 5 years is the industry's ability to develop solutions for packaging both advanced CMOS with copper and ultralow K dielectrics. Both the rapid adoption and high-volume ramp of advanced CMOS silicon with copper interconnects and low-K dielectrics have had a major impact on packaging. The industry must move toward compliant chip-to-chip substrate interconnects (such as through silicon via process, see the next paragraph) and low-stress packaging solutions, along with mechanical reinforcements under pads. The ability to perform better characterization is essential, as is development of interconnect ILD materials to address interfacial adhesion and improve fracture toughness. Thermal connectivity with strained silicon is potentially four to five times worse than with bulk silicon. The industry has to better understand this issue and address the associated potential problems. Developing wafer thinning and dicing solutions for wafers with copper and ultralow K dielectrics will also be required.

Scaling for More Die-Per-Wafer and New Interconnect Technology

Innovative packaging solutions will be required in order to continue scaling down the size and area for product input/output and the pad ring in CMOS nodes beyond the 30-nm process. As metal systems continue to thin, IR drop and EM issues will become more prominent and new interconnect alternatives will have to be considered. Techniques such as stud bumping, dual bonding on same pad, thinner aluminum wire, copper wire, bond on PO, copper over anything, and bond wire jumpers are all potential solutions.

In addition, ultrafine pitch and compliant interconnect solutions such as micro bumps, coated-wire bonding processes, a fine pitch with less than 50 μm wire bonding, and Au stud bumping to thinned dies will also be needed within this time frame. Possibly industry will see the end of the solder bump for ultrafine flip chip interconnect applications. Other technologies that will be explored are room temperature Cu–Cu interconnection and fine-pitch pad alignment; 3-D IC interconnect; wafer-to-wafer; die-to-wafer, through silicon vias and Cu nails.

SoC and SiP Integration and Functionality: Automated System Level Co-Design Tools

Product designers are continuously facing the question of whether they should design new products using SoC or System-in-Package (SiP) or both. This is especially true in the mobile handset market, where consumers are increasing their reliance on the handset to listen to music, take and send pictures, record and watch video, play games, and more. Operating multiple applications places an increased strain on system memory requirements and is driving the need for SiP solutions. Functionality increases with time for SiP and SoCs, and cost is typically the deciding factor for SoC vs. SiP at any given time. Nowadays, the ultimate trend is to reinforce the role of SiP vs. the SiC solution. In this case the main factors are mostly related to the product supply chain and flexibility (both in terms of technical and economic points of view).

The typical example is represented by the Package-on-Package solution, where instead of using a single package that includes the entire set of dies (microprocessor and memories), technically feasible now, smaller footprint and lower package cost, the industry preferred to use two separate packages, increasing the integration complexity, the cost, and the footprint but keeping the two market segments separated (memories and microprocessor). This will discussed at the end of this section.

Another critical gap the industry must close is faster development and availability of 3-D Chip-Package-System Co-Design Automation for electrical, thermal, and mechanical compatibility. Development of cost-effective package on package modules is also critical, as is the impact on manufacturing business models and testability.

The industry is also experiencing challenges in rolling out complex SoC products that include RF and analog integration around digital cores. Digital and RF integration into smaller modules requires closer placement of the digital and RF dies, which has raised interference issues between the two signals that must be resolved; integrated shielding and flip chip solutions could help address the problems. Modeling tools with more predictable resultsand the ability to allow virtual packaging of advanced silicon and systems are essential. System and component level reliability failure modes and acceleration factors must be well understood, and there will be an increased need for modeling and design tools to predict manufacturability and reliability performance.

Higher Performance and Higher Thermal Density

A change from the use of wide/slow busses to high-speed serial I/O will drive the need for low-cost, matched transmission line characteristic packages, and new thermal management techniques will be implemented to address cooling. We are currently exploring methods such as RF shields and expect those to become an integral part of packaging for future consumer electronics and portable products. Techniques such as the use of heat pipes and vents will ultimately be used in these systems to move heat to the system enclosure. Additionally, multicore micro processors will drive die size up and increase demands for high-performance multichip modules. Nontraditional thermal management solutions, such as liquid cooling and compact and solid state refrigeration will be needed but at significantly lower cost.

The industry is experiencing increased demand for low-cost, high-performance packages, and development of high-performance wire bond and low-cost flip chip packaging is essential. A better understanding of fundamental issues related to lead-free packaging, e.g., metallurgical life prediction models, will help on this front. In addition, hot-spot minimization will drive the development of a high thermal conductivity die that will attach directly to a Cu spreading plate in organic substrate BGA packages.

Package-on-Package Approach

One of the most attractive even if not a definitive way to integrate multiple functionalities within a limited footprint is the Package-on Package (POP) solution. The preferred approach is to stack logic (multimedia microprocessor) in a package and place another package for multimemory on top of it. This new emerging technology is becoming even more popular especially where a very high integration level is required, i.e., in handset and multimedia mobile phones.

Package-on-Package has gained a lot of market acceptance through standardization and is now is produced in significant quantities.

Why is POP becoming so popular? Certainly, standard MCP has lowest assembly cost: one package substrate, smallest form factor, and leveraging the existing assembly infrastructure. However, standard MCP may not have lowest total cost: cost of test and test access issues, Known Good Die (KGD) issues, business flow, and device supply chain matter.

If all these parameters are correctly taken into account, then, in most cases, the lowest cost for integrating devices is Package-on-Package. The Package-on-Package solution is made up of two packages: (1) the top package normally devoted to stack memories (nonvolatile and volatile memory parts) based on conventional array molded stacked die FBGA, but larger ball size and pitch and thinner mold body; (2) the bottom package to house the multimedia processor with a specific mold technology, which must leave the top perimeter of the package free for top package attach. The POP technology development is driven primarily by the following:

Reducing Package Size

The POP footprint reduction impacts strongly on the space required for the application board. Furthermore, the planar dimensions also impact the POP package costs, which are normally dominated by the substrate costs (in this case two substrates). However, the need to exchange more electrical signals between the two packages (higher parallelism) and between the multimedia microprocessor and the application board (up to 700 signals) requires finer ball pitches (today in the range of 0.5–0.4 mm). This requirement has a serious impact on the final overall POP integration, especially on the board mount yield.

Reducing Overall Mounted Height

From a wireless perspective, the overall mounted height is one of the most important POP challenges of the immediate future for two main reasons: (1) there will still be a significant demand to lower the overall height (most of the requests are to stay within 1.4 mm); and (2) there is an increased complexity for the TOP package in terms of heterogeneous memory content, NAND flash memory, and managed NAND devices, DRAM DDR1 and DDR2, NOR flash memory, and PSRAM. These driving forces are pushing the assembly capabilities to extreme integration solutions: very thin die attach materials, very thin dies (>50 μm) with the associated wafer thinning technologies (polishing, ultrafine Poligrind, texturized dry polishing), new molding processes, and heterogeneous devices with different form factors. All of these developments will definitely be called for in the relatively near future.

Board Mount Yield

The board mount yield parameters in the BGA space are normally considered a minor matter. However, using the POP approach, the board yield where the top package has to be mounted on top of the bottom one can turn this into a serious issue because the package warpage of the top and bottom packages can be very different from one another. This seriousness of this problem increases as the package footprint becomes larger (current POP solutions range from 10×10 to 15×15 mm^2), which requires a good match in terms of coplanarity between the two packages.

Higher Performance and Device Integration

The new improved characteristics of the memory devices placed in the TOP package, in particular the higher I/O parallelism and the working frequencies, which are driven mainly by the DDR features, require a higher degree of integration in the design of the two set of products (multimedia processor and memories). Specifically the signal integrity and power delivery analyses of the electrical signals traveling through the POP structure must be done carefully taking all the features of the entire system TOP plus BOTTOM into account.

Package-on-Package is one of the most important drivers for hand-held, wireless applications. This new approach is becoming the real reason behind the assembly technology developments: die thinning, assembly materials and processes, substrate material and design, heterogeneous device integration, and co-design of different devices.

It is certain that this will not be the final step in terms of device integration, owing to the emerging interconnection technologies that will be presented into the next section (i.e., Through Silicon Via). However, owing to its enormous flexibility this solution might dominate the wireless application field.

Back Grinding and Wafer Finishing Techniques for Nonvolatile Memories

The semiconductor roadmap indicates more and "Moore" stringent development not only in terms of design shrinkage and lithographical requirements, but also in packaging, seeking for ultrathin dies (~40 μm thick); for instance, Smart-Card chip thickness is now of the order of 100 μm and nonvolatile memories are stacked with 50-μm die thicknesses each, whereas stand alone flash EEPROMs are actually thinned down to about 130 μm. Moreover the Smart Label (RF ID/transponders) market has been growing fast in the last few years responding to various developments and applications requiring tags, from clothes to banknotes, from libraries to superstores, and from waste disposal organizations to airport baggage tracking.

The potential of ultrathin wafers has three main areas of application: thin and flat packages, flexible electronic modules, and better cooling of power circuits.

Beyond that, the stacking of ultrathin chips offers the opening of different development scenarios for highly integrated logic circuits, broad-band, linear, and high-frequency circuits, and cost benefits over an elaborate mix of technologies on one chip.

Ultrathin chips allow the introduction of innovative assembly techniques, e.g., for paper thin and flexible packages, Smart Cards, and Smart Labels, as well as chip-in-paper and chip-in-board embedded electronics. Chips for those applications

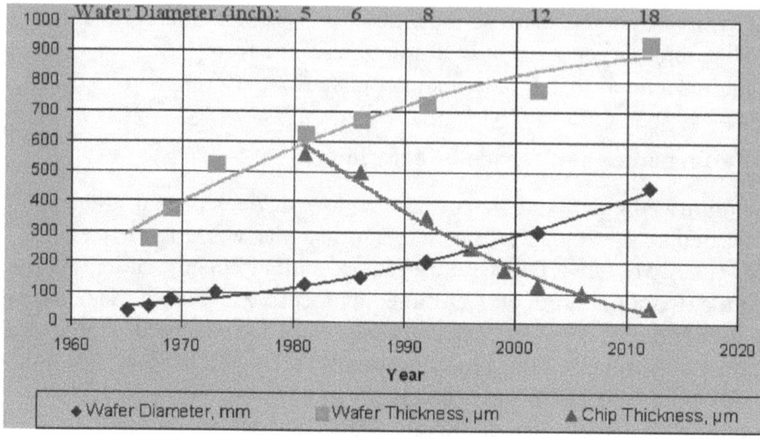

Fig. 6.1 Wafer and IC die thickness vs. wafer diameter roadmap

are 75–100 μm thick, values destined to drop to 50 μm in the near future. The trend is critical from the back-end point of view, considering that the wafer diameter is always increasing (from 150 to 200 to 300 mm, with plans for 450 mm), so that back-grinding to thinner wafers with larger diameters requires strong handling capability and process statistical repeatability. Flexibility, fragility, and warp are key parameters in processing ultrathin wafers and chips. Although thin wafers bend like paper foil, they can suffer mechanical damage during handling. Suitable equipment for ultrathin wafers must have industry standard techniques for smooth and gentle movement, processing, transportation, and final chip placement.

The shift toward these requirements needs not only the development of equipment able to process large wafer diameters, but also a complete review of the order of the back-end operation: wafer and die edge smoothing, burn-in steps, EWS inkless probing with automatic wafer mapping, and datum recording, packing, shipping, assembling, and final testing.

We now proceed to an overview of the different possible back-grinding and wafer-finishing technologies, with look at possible process flow options.

Figure 6.1 represents the trend over time of wafer size and thickness and die thickness, respectively, predicting development up to 450-mm-diameter wafers, with a thickness of 925 μm, and a final chip thickness around 50 μm and less.

Back-Grinding Process Flow

The silicon wafer thinning process is done via mechanical back-grinding (BG).

A classical BG process flow used for Non Volatile Memory (NVM) devices can be summarized as follows:

(a) Front surface wafer protection with stick foil.
(b) Grinding in two steps: with a rough and, subsequently, a fine grit mesh grinding wheel.

(c) Back-side stress relief (by acid mixture, by dry or chemical-mechanical polishing, by low-temperature plasma treatment).
(d) Protective tape removal.
(e) Front surface wet or dry cleaning.
(f) Metrology and quality checks [thickness, Total Thickness Variation (TTV), warp, quality optical inspection, etc.].
(g) Electrical Wafer Sort1 (EWS, "writing").
(h) Burn-in to force eventual charge loss phenomenon.
(i) Electrical Wafer Sort2 (EWS, "reading").
(j) Mounting on frame.
(k) Dicing.
(l) Assembly (pick and place, die-attach, wiring, packaging into the molding compound).
(m) Final testing.

The process flow just described is used when two EWS tests with a bake in between are necessary, as for semiconductor devices containing NVM on 8" wafers down to 170–150 μm. For less thick 8" or 12" wafers EWS usually is done before starting BG operations. In that case automatic die recording is performed through mapping recognition and only the good dice undergo assembly and are process through to final testing.

For other IC devices that require only one EWS testing level, special probing machines are available with suitable handling and electrical tests of thin wafers, already sawed or not, lodged on the mounting tape frame, can be carried out.

In cases in which a back-side stress relief is performed through a relatively high-temperature plasma treatment, with the wafer not sufficiently cooled on the wafer holder, a suitable heat-resistant protective tape has to be laminated to the front-side at the beginning of the BG operations, otherwise it has to be removed before starting such plasma step and the process flow just described changes therefore.

Back-Grinding Technology

Wafer back-grinding begins with a step (known as "coarse," "gross," or "rough" grinding), which removes the most of the silicon. The second step, called "fine" grinding, removes only 15–20 μm of silicon to give smooth out the roughness and provide more wafer mechanical strength. At the end of the overall BG process, a wafer finishing treatment is applied to the wafer back-side (plasma-dry or by polishing, dry or chemical-mechanical one, wet with acid mixture [1, 2]).

Before starting the BG operation, a protective sticky foil, generally "pressure sensitive" or "UV" type, is applied to the wafer front side.

Pressure-sensitive tapes can by used for back-grinding down to about 150 μm, whereas UV tapes are generally indicated for thinner wafers or when polyimide is used as a buffer coating on the final passivation inorganic layer. In fact the adhesion strength of the UV-tape is much reduced after UV exposure and the tape can be easily removed once the adhesion strength is decreased (e.g., from 300 N/inch before UV treatment to 10 N/inch after irradiation). It is important to have good tape-to-wafer-edge adhesion in case the wafer finishing is done with an acid solution to prevent acid infiltration between wafer and tape, which could corrode the external die metal wire bonding pads. A tape-less grinding process has also been offered to the market.

There has to be particular attention given to the back-grinding process setup in order to avoid slippery friction and heat generation between the fine finishing wheel and the wafer owing to a weak grip between the two. This weakness allows wheel resin to be released on the wafer's rear surface, coloring it brown and causing serious scratching (such a phenomenon is misnamed "wafer-burning"). If that should occur a constant electrical current has to be applied to the spindles to avoid burning effects, which is achieved by the right combination of wheels (mesh) and grinding parameters (e.g., grinding wheel air cut, spark out, and escape out steps), as well as good mechanical setting of the wafer chuck inclination [3].

Figure 6.2 shows the two possibilities for back-grinding, called "in-feed" and "cross-feed" modes. In the in-feed mode a fine grinding spindle rotates in the same direction as the rough grinding one, whereas in the cross-feed mode the two spindles rotate in opposite directions. In any case both modes are used in production environment, so the right choice between the two depends on the particular application and the type of substrate, even if the cross-feed mode can facilitate the initial grip between the wafer and the Z2 shaft.

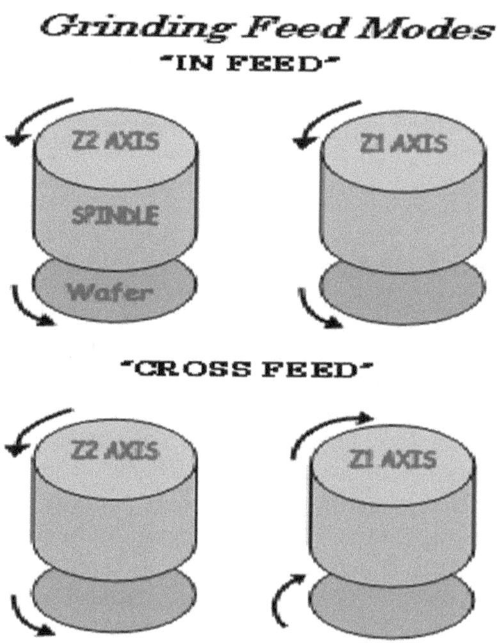

Fig. 6.2 Schematic representation of in-feed and cross-feed BG modes

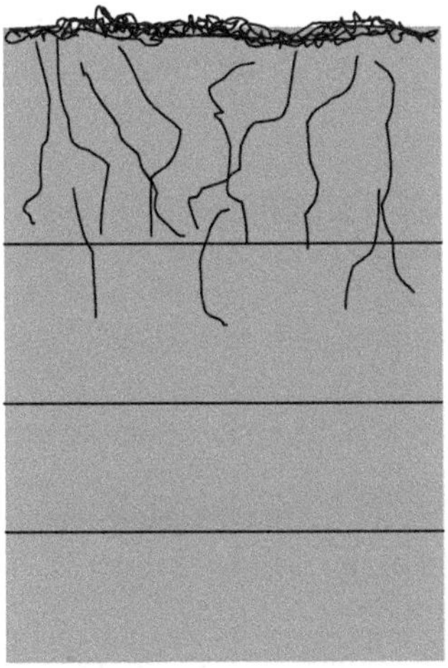

Fig. 6.3 Sketch of the back-side damage mechanism caused by back-grinding

Grinding operations leave the surface with some damage [4]. Figure 6.3 shows a qualitative classification of five different wafer zones after back-grinding:

(a) The first zone is generally considered as "poly-silicon" and, depending to the grinding wheel mesh, can reach a thickness of about 2 μm; some authors have further subdivided this region into amorphous silicon and poly-silicon itself layers [5].
(b) A textured region shows the presence of micro-cracks.
(c) The third zone is in an intermediate condition, where the microcrack presence ends.
(d) At the fourth level the silicon crystal is still stressed by elastic deformation.
(e) In the last belt the monocrystalline bulk zone is completely recovered.

All these factors have to be taken into account for quality assurance and application.

Another technique worth mentioning is the so-called Dicing before Grinding (DBG) or "Dicing by Thinning": wafers are partially sawed by a rotating blade starting from the front-side into scribe lines to a thickness equal to the target thickness and then laminated with sticky foil and generally processed in back-grinding equipment. When the fine-grinding wheel reaches the previously obtained trenches, the wafers are automatically separated into chips lying on the BG tape [6]. The chips are then transferred to the mounting tape; UV tape is preferred.

Back-side polishing or plasma etch can be then performed to remove silicon crystal damage induced by grinding in order to increase wafer and die mechanical strength.

Dicing by a thinning process is also possible in the case of plasma thinning after grinding, with the sawing step taking place during the plasma finishing, and this has some potential advantages, such as:

- Risk of wafer breakage due to warp, handling, and edge chipping is minimized.
- By simply carrying out a precut step, it can be added as an option, to a standard BG followed by stick foil peeling after the wafer has been mounted on the frame.
- Higher dicing throughput (no pass through cut, without risk of the blade clogging into the tape).
- Less dicing damage and chipping (soft sawing approach) with consequent lower risk of die breakage.
- A minor relief step can be added after grinding by dry polishing (without slurry) or plasma etching, with respect to that performed in the classic process.

Table 6.1 lists the principal thinning methods applied in silicon IC manufacturing lines.

Table 6.1 Principal back-grinding technologies

Grinding only
Grinding + wet etching
Grinding + wet polishing
Grinding + dry polishing
Grinding + plasma etching in vacuum
Grinding + atmospheric plasma etching
Dicing before grinding (DBG)
DBG + dry polishing
DBG + plasma etching
Grinding + dicing by thinning with plasma

Back-Side Finishing Processes

In addition to finishing the wafer rear surface with chemical spinning etch, it is also possible to use polishing, with cluster-integrated tools, just after BG operations, or stand alone stations. The polishing operation removes a few microns of silicon (1–3 μm) for a gain in wafer and die mechanical strength, quite similar to that obtained with the etch processes by acid mixture.

Depending on the availability of a grinding-plus-polishing process wafer level electrical testing can be carried out either before back-grinding or on thinned wafers already affixed to the dicing mount tape. In the latter case the testing process cannot predict burn-in between the electrical wafer sort steps, but eventual final yield discrimination (i.e., detection of eventual electrical yield loss due to back-grinding operations themselves) is done at the final testing level.

Back-side wafer finishing with dry etching can be performed with either an atmospheric pressure jet or vacuum plasmas; the characterizing equipment parameters are the throughput and the temperature reached during the process. That can eliminate the possibility to use a plastic back-grinding tape inside the plasma chamber, removing it just after back-grinding: in fact in case of back-side plasma etching, stress relief and tape removal steps can be performed in reversed order in the process sequence, because usual BG tapes cannot stand up to the high temperature generated in the plasma chamber. Low-temperature plasma back-side etch processes at around 60°C, maintaining the grinding tape on the wafer top, are in any case available besides BG tapes compatible over 100°C of usage temperature.

Plasma solution has been also developed for integration in dicing prior to the grinding process, which has the advantage of reducing the sawing damage in chip sidewalls (healing effect).

In any case attention should be paid to uniformity and surface cosmetics, so a possible compromise in consideration of quality can be choice to set a very small removal amount.

The wafer rear surface finishing condition (the surface must be free of mechanical damage, hazes, scratches, stain, "comets", and, possibly, mirror-like) is important for quality in outgoing and die attach compatibility.

For some power devices that require rapid heat transfer, back-side surface chemical and physical properties have strong influence on the adhesion with metallization layers, such as Cr/Ni/Au stack. For example, with dry polishing a mirror-like back-side aspect is attained with 1-μm removal. Slurry chemical mechanical polishing also leaves a mirror surface if a proper rinse and dry operation is performed to remove slurry residuals. An opaque back-side appearance is typical of plasma-finished back-side wafers, but the degree of opacity can be varied according to the parameters of the plasma.

To get an idea about roughness, experimental data by AFM are reported in Table 6.2. The numbers typically vary from a minimum of 0.4 nm for polishing techniques to a maximum of 60 nm for plasma techniques, depending on the process. Visually, opaque zones have a roughness several times higher with respect to mirror-like zones. As the plasma process is mostly anisotropic, the tracks left by the grinding wheels are still visible to the eye.

Note that, unlike other finishing methods, the experimental plasma process increases surface roughness with the amount of silicon removal.

Table 6.2 Rrms Order of Magnitude (AFM Technique)

Process	Rrms (nm)
Dry polish 1 μm	2
Dry polish 2 μm	2
Dry polish 3 μm	0.7
Wet polish 2 μm	2.2
Wet polish 3 μm	0.4
Plasma etch 10 μm	43
Plasma etch 20 μm	58
Plasma etch 50 μm	60
Wet etching 3 μm	11.5
Wet etching 20 μm	1

Wet Etch

Wafer back-side finishing is performed after back-grinding mainly to relieve wafer stress, to increase its mechanical resistance, to remove crystal defects due to grinding, and to make the surface clean and with a repeatable aspect (reflectivity) useful for process control.

Stress relief equipment uses acid mixtures to remove silicon from the wafer back-side after back-grinding. That kind of tool can be based, e.g., on spin etching, with the wafer lying on a rotating chuck or exploiting the acid solution meniscus effect directly onto the wafer surface that undergoes the stress relief treatment. The typical mix is composed of an FNPS solution that is a blend of hydrofluoric, nitric, phosphoric, and sulfuric acids: according to the percentage of each component, etch rate and final roughness can be varied to attain the desired surface aspect. If the surface has to be covered by metallization, a high roughness is preferred to improve adhesion. Back-side metallization is required in case of power ICs for electrical grounding or heat dissipation. It is noteworthy that roughness is not the only parameter to be checked for adhesion: oxidation levels or the presence of spurious carbon and fluorine compound left on the surface can also compromise the metal-to-silicon adhesion strength.

After the acid solution is delivered onto the wafer by the medium dispenser, a DIW (deionized water) rinse is performed followed by a nitrogen drying step.

According to the chosen overall back-grinding process flow the stress relief step can be executed with or without back-grinding tape; usually the standard process is done with tape on in order to avoid acid leaking onto the edge of the wafer front-side surface.

Depending on the system capability, the same acid mixture bath can be utilized several times, but it obviously loses its etch rate effectiveness with the repeated use, which has to be accounted for by automatically increasing the acid mixture dispensing time so that the amount of silicon removal remains constant until the mixture is washed off and a fresh new solution is loaded into the equipment chemical delivery system.

In Fig. 6.4 three SEM pictures show how the wafer surface becomes increasing smoother as the silicon is removed by spin etch up to 20 μm.

For the sake of completeness, Fig. 6.5 shows the SEM images of a silicon surface treated with a different ratio of FNPS solution, in order to obtain a higher roughness for back-side metallization.

Chemical Mechanical Polishing

In the case of CMP back-side finishing, the slurry used can be composed of silica particles and ammonia in aqueous solution, whereas, with respect to the standard CMP used, e.g., in intermediate dielectric planarization, the polishing pad can be smaller than usual, i.e., 150 mm, for both 200- and 300-mm wafers.

After grinding, the wafer back-side finishing removal with acceptable equipment throughput is typically less than 3 μm. In a cluster tool, the wafer is then optionally

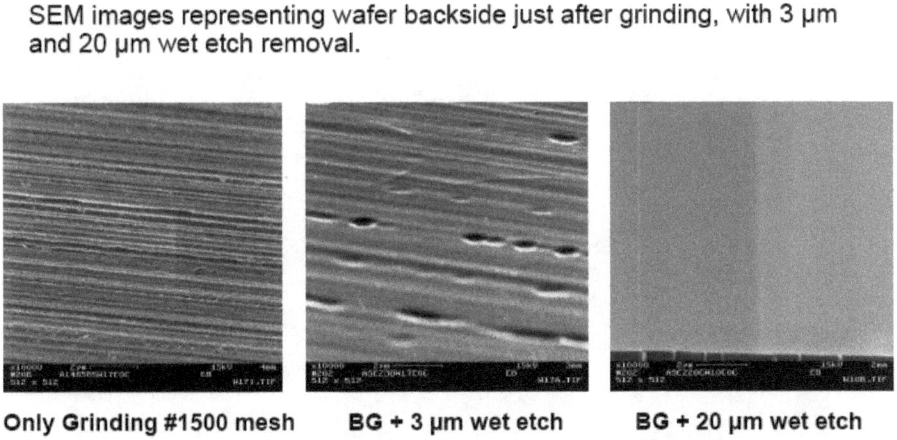

Fig. 6.4 Surface roughness after back-grinding, 3- and 20-μm removal with FNPS solution

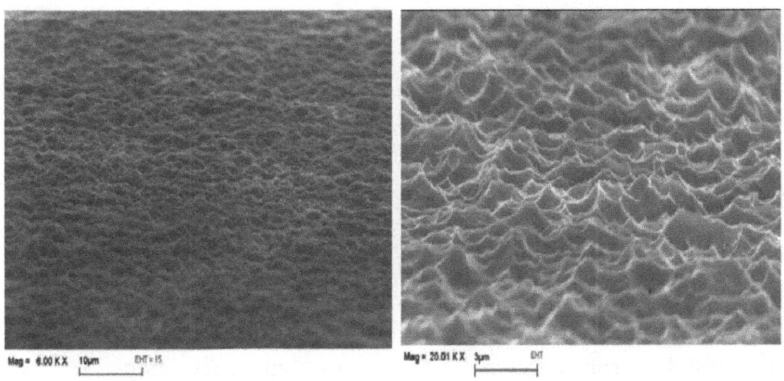

Fig. 6.5 Surface roughness obtained through a special wet etch solution for metallization

UV irradiated (depending on tape type), and brought to the mounter/peeler station: the wafer is first placed on a dicing frame and then the back-grinding tape is peeled off; in this way no thin wafers are left without protection tape in any time, so there is little risk of breakage. On the other hand, stand alone systems (taping, BG, back-side finishing and detaping) offer flexibility in production as thanks to their modularity overall workability is not interrupted if one of them is down.

Bar code or OCR reading and label application are further tool options and the addition of an infrared or interferometer probe, with the advantage of being contactless, can provide wafer thickness and TTV measurements.

A polishing process implies the management of light chemistry slurry, in terms of both in situ mixture preparation and chemical disposal that can be drained together with the grinding disposal line (water + silicon particles).

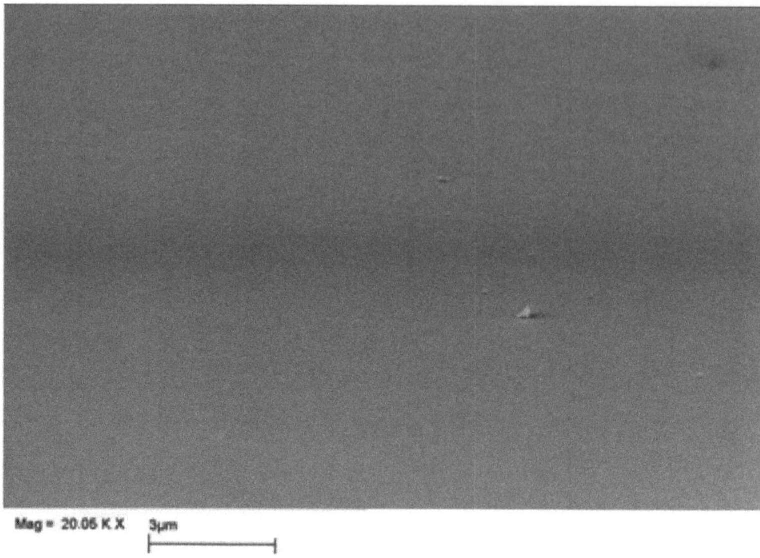

Fig. 6.6 3 μm CMP removal

A polishing step can be adjusted mainly through the following factors:

- Polishing time and pressure.
- Alternate movement between pad and wafer, to optimize uniformity.
- Slurry mixture pH.

The polishing removal check is performed together with the check on the final wafer thickness by probes and periodically on control test chips. The back-side surface after CMP is very smooth and generally does not offer optimal characteristics for back-side metallization deposition (in fact a certain grade of roughness is required to avoid delamination).

Figure 6.6 showing an SEM image of an example of silicon surface after chemical-mechanical polishing is reported to highlight the smoothness level.

Dry Polishing

Dry polishing is obtained through a particular pad operating on the wafer backside in an alternate plus spinning movement. During the process no water is used, but both subsequently the chuck and the wafers are cleaned with de-ionized water.

A dry polishing technique is commonly used for standard background wafers or in combination with Dicing before Grinding (DBG) and it can provide a sufficient level of mechanical strength both for the die and the wafer. It is used in the production of stand alone nonvolatiles memories (NAND or NOR) or stacked memories in combination with RAMs.

Fig. 6.7 Dry polishing (3-μm removal)

In Fig. 6.7 an SEM image of a silicon surface after dry polishing is shown.

Plasma Finishing

Plasma for back-side finishing can be generated in a vacuum chamber with RF or microwave sources or through jets in atmospheric pressure.

In case of vacuum chambers, the RF or microwave energy gets into the reactor and generates radicals, ions, and UV radiation from the reaction gases (e.g., SF_6 and N_2O). Radicals and ions get to the substrate surface on the electrode and react according to the following equation:

$$2SF_6 + 3Si + 4N_2O \Leftrightarrow 3SiF_4 + 2SO_2 + 4N_2$$

The volatile reaction by-products are pumped away by root-pumps. Additionally RF power is applied to the wafer electrode (bias). Due to this the ions are accelerated to the wafer surface and the etch rate is increased.

During the dry etch process the wafer is kept in a homogeneous temperature distribution by a helium- or water-cooled vacuum chuck. This is necessary in order to control wafer temperature because the reaction described above is exothermal. The actual etch rate is influenced by wafer temperature, process pressure, and gas ratio.

In the DBG technique, plasma etching is used to obtain the "ideal chip" in terms of mechanical stress resistance: plasma can enter in between the scribing lines and

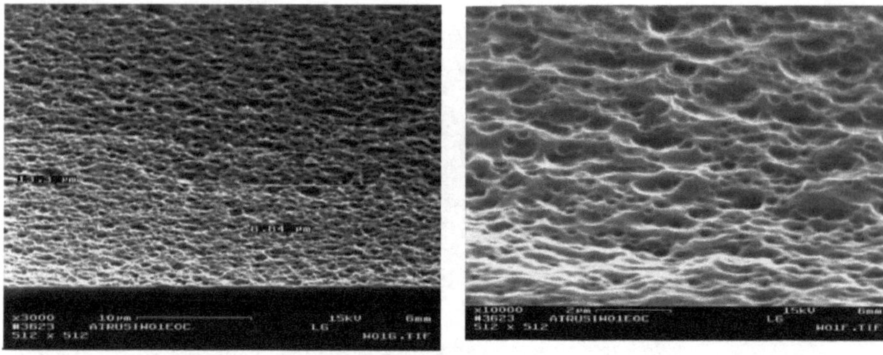

Fig. 6.8 Back-side conditions (by SEM) after a 20-μm plasma removal at two different magnifications

minimize damage from both chip back-side and side walls, removing dicing sawing marks and defects ("healing effect").

Figure 6.8 shows the SEM images of a "porous-like" surface left by a special plasma baskside finishing process, where microcraters are visible.

Die Strength

The various back-side finishing processes are compared considering their possible applicable settings; for this reason, for instance, plasma and wet etching have been chosen up to 50 μm and up to 20 μm, respectively, whereas polishing processes have been set up to 3 μm removal owing to acceptable throughput and cost of ownership criteria.

Die strength tests are performed using a one-point load method (Fig. 6.9) that discriminates thinning process quality, whereas the four-point bending test (same picture, right-hand drawing) normally includes dicing and chipping effects at side walls [2].

The equipment used for one-point load die strength is usually made up of an automated column coupled with a dynamometer, recording peak load values at the silicon breakage point, on the top of which is mounted a conical tip. With this method it is possible to measure mechanical strength of dies down to 150–170 μm of thickness: Weibull breakage distribution (Fig. 6.10), obtained on chips 3.57 by 4.70 mm^2, shows that within the described test method applied, there is a big difference between grinding alone and stress relieved samples, regardless of the finishing process (plasma or wet etching, CMP, dry polish, etc.).

None of the stress relief techniques show important strength differences up to 5 N of applied force. Hence it can be argued that it is sufficient to apply few micron polishing, that is a removal thickness with acceptable equipment productivity, to strongly improve die strength at levels comparable to those obtained by plasma or acid process by which higher amounts of silicon are usually taken off.

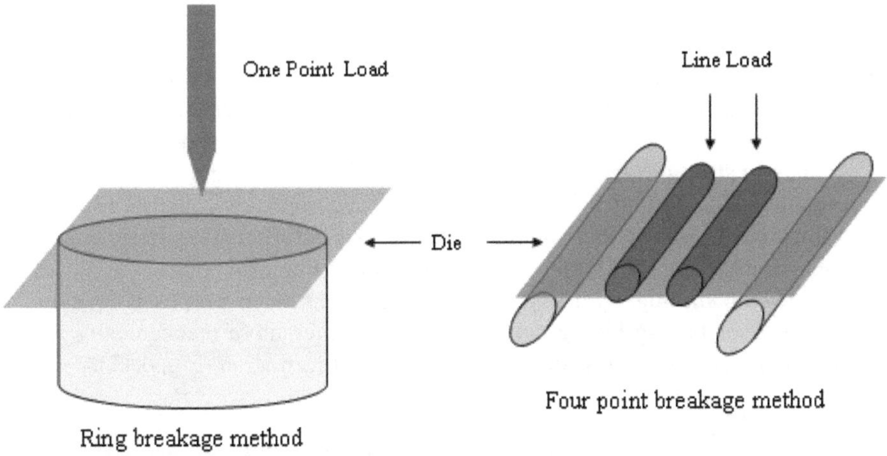

Fig. 6.9 Supports used in die mechanical strength measurements

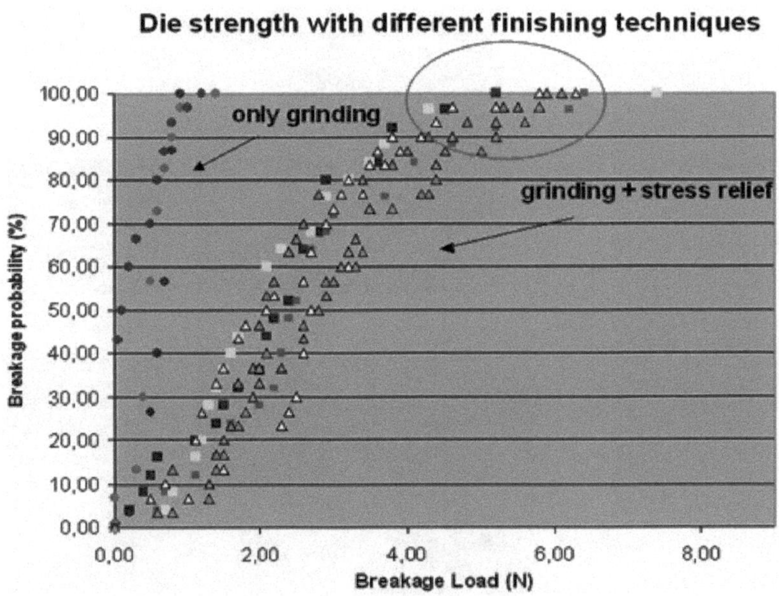

Fig. 6.10 Weibull distribution of die strength with different finishing techniques (vacuum or atmospheric plasma etch, wet etch, CMP, dry polishing, etc.) showing an increase in mechanical resistance after back-side finishing treatment with respect to simple grinding ("one-point load" measurements)

Wafer Edge

Under particular process conditions, wafer edge can be controlled and shaped close to a round cross-section, with improved wafer strength. For very thin wafers, wafer handling inside and outside the equipment and during packing and shipment phases is critical, so a strong wafer edge is essential in order to withstand small knocks and edge rubbing. The optimal solution is to achieve a completely rounded edge, similar to thick virgin wafers (but smaller in scale), but unfortunately after mechanical grinding the edge is quite sharp (knife-blade shaped; Fig. 6.11).

The idea of starting with asymmetrical edge has been proposed to avoid a blade edge after back-grinding and to make the wafer more mechanically robust by minimizing edge chipping and consequent defect formation (Fig. 6.12).

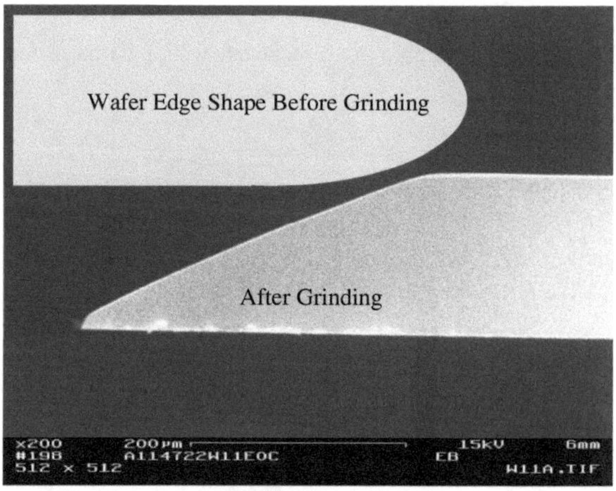

Fig. 6.11 After grinding, wafer edge has a knife-blade shape

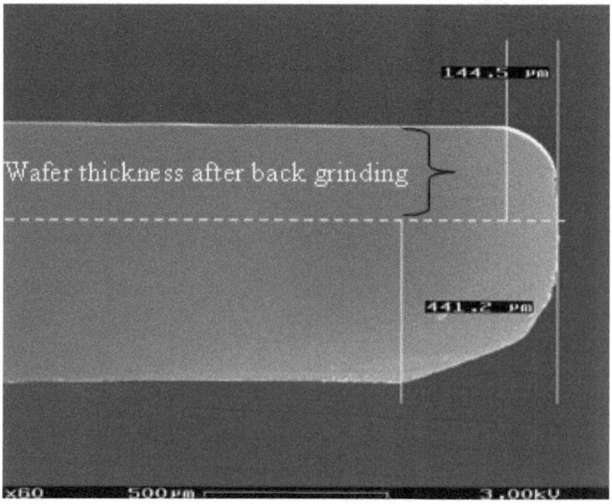

Fig. 6.12 Asymmetrical border to avoid sharp edge after back grinding

It is noteworthy that if the equipment output has the wafer mounted on the dicing frame, on the one hand, the wafer edge shape becomes less important because of the improved hand-ability that results from the protection mount tape and from the external metal ring that protects the wafer edge from most possible bumps. On the other hand, edge integrity remains crucial because any microcrack could propagate over time.

Equipment for final edge rounding is available on the market using abrasive tape, plasma technology, or laser [7, 8].

Process Control

The primary process parameters that must be characterized in the R&D phase and controlled during manufacturing by Statistical Process Control (SPC) are:

- Average final thickness.
- Total Thickness Variation (TTV), within wafers and wafer to wafer.
- Warp.
- Front and back-side superficial organic and inorganic contamination level.
- Wafer rear surface crystal damage (μ-cracks through silicon), detectable by SEM/TEM.
- Wafer rear surface reflectivity.
- Surface roughness.
- Edge integrity and shape.

Wafer thickness can be measured and averaged by capacitive probes in continuous mode or by taking a fixed number of points on the wafer. Another thickness measurement method is offered by infrared interferometry, which can be performed even if the wafer is lying on the mount tape [9].

Total thickness variation TTV comes directly from the thickness measurements by subtracting the minimum from the maximum value.

Wafer warp can be measured through a conventional laser flatness tester [5], but if the wafer lies on a flat wafer holder during the measurement one must distinguish between its intrinsic warp and gravitational effects, which tend to flatten very thin slices. Bow and warp can be obtained from the instruments used to measure the thickness and the effects of gravitation can be eliminated through special SW.

In of the search for process repeatability, taking into account the front and back end of the line compatibility requirements, periodic checks of contamination, both organic and inorganic, are suggested. Most common elemental analysis techniques are represented by AES, ToF SIMS, XPS.

Crystal damage to the back-side can first be detected by through optical microscope inspection and then in a deeper mode by SEM and TEM after a special sample preparation procedure. Eventual cross and stain defects are detected in this way.

Optical interferometers used to measure thickness and refractive index of dielectric films can be also utilized to measure silicon reflectivity, which is correlated with the roughness or surface cleanliness level. Roughness, detected by a mechanical stylus profile-meter or by atomic force microscope, is a key parameter for monitoring grinding and wafer finishing process repeatability [10]. Finally, automatically or using a specially trained operator, wafer edge is checked optically to detect and classify edge defects, such as passing through ones, and eventual chipping sizes.

Front-Side Wafer Finishing, Testing, and the Back End of Line

The wafer back grinding process requires protecting the silicon wafer surface with a sticking grinding tape ("sticky foil"). The adhesive material between the tape and the wafer's front-side can leave glue residuals such as particles and an ultrathin organic layer for a thickness of few nanometers [11]. Such organic contamination would be detrimental for a good adhesion between the final passivation film and the package plastic molding compound and leading to local package delamination, especially at the silicon die corners, where stress effects are known to be maximized. Different empirical and analytical techniques are available to study the surface initial conditions of the final passivation and an oxygen plasma cleaning process can be helpful in overcoming such constraints in a manufacturing environment. In fact oxygen burns out the organic particles and the "mono-layer" left on the wafer front-side after sticky foil release. Indeed, the adhesion between wafer final passivation and package molding compound is linked to the kind of materials used [mainly final passivation silicon nitride and phosphosilicate glass (PSG), protection tapes, adhesive, resin molding compound] and to wafer surface cleanliness status.

Among the different analysis techniques are water droplet contact angle, ToF-SIMS for elemental analysis, plastic molding compound shear test, and surface roughness of the final passivation layer.

The cleanliness level reached by the final passivation layer surface after plasma treatment, detectable by the low water droplet contact angle, improves the adhesion strength with the plastic package molding compound. That can be easily revealed through plastic molding compound shear tests.

Different final cleaning techniques are used after back-grinding is completed: spin rinse with deionized water and dry, downstream oxygen plasma cleaning, isopropyl alcohol (IPA) bath, etc. The plasma cleaning process has been proven to be quite effective in removing organic contamination. Other possible sources of contamination come from the environment itself and the packing (interleaves separating the wafers from each other and foams) for shipment from diffusion to EWS or assembly plants. In any case a plasma cleaning treatment can be always performed on the dies attached to the copper "reels" just before molding.

In order to understand the surface condition evolution of the induced contamination, at least the following actions are recommended:

(1) To study the aging effect (elapsed time) and storage conditions of the wafers, to detect the flow of crucial operations mainly responsible for contamination

(EWS sequence, contact with interleaves, travel, die attach glue, etc.) to determine which plasma treatments extend the cleanliness level of the surfaces over time maintaining as closely as possible their original condition.
(2) Evaluation of the impact of organic contamination on actual adhesion strength at package level at the die/resin interface. For this reason the Jedec moisture sensitivity level (JMSL) test is usually performed on the devices considering possible variables such as the various degrees of contamination and different final passivation materials. The JMSL test consists of a simulation of the reflow soldering process with a defined oven profile, after packages have absorbed a controlled amount of humidity.
(3) To study the effect of thermal treatments of dielectric layers on their wettability properties.
(4) Owing to the dependence of the adhesion on die surface cleanliness level and die/package relative stress, estimate how the application of polyimide at wafer level or to die coating materials can act as buffer between the two on new devices.
(5) To use Auger analysis (AES) to study the fluorine contamination level left by actual pad opening operations on the metal pads vs. different plasma cleaning conditions [12].

Special low-temperature plasma treatments have been developed to handle very thin wafers for the final cleaning operation and are now available.

The possibility of using different back-grinding techniques and the requirements to obtain ultrathin dies/devices in an NVM back end of line production environment have to be correlated with testing, assembly flows, and in-line controls.

BG + EWS1,2 + ASSY flow is utilized and is suitable for NVMs and embedded devices (logics plus memory on the same chip): that is the most commonly used flow for standard wafer thickness. It is not practical for ultrathin wafer production and management, owing to classic burn-in between the first and second EWS and for electrical testing equipment handling possible issues.

EWS1,2 testing + grinding of thin wafers + assembly replaces the former one for ultrathin or 12" wafers. It can be used with dicing before grinding, grinding in combination with polishing both wet with slurry or dry one, with thin wafer released on mount tape, and rigid frames for signal devices not requiring back-side metallization (memories).

In seeking completeness, *Grinding + one level of EWS* process flow on wafers on mount tape/frame could be followed for logics not requiring back-side metallization, but not for memories that need burn-in in between EWS1 ("writing") and EWS2 ("reading").

The best solution in terms of equipment and process cannot be easily selected as unique, because there can be different solutions for different applications: for instance, Smartcards, owing to their peculiar assembly in plastic cards, require high chip flexibility and die strength, so high stress relief is required. On the other hand, for flash stacked devices die warp can be an issue during assembly, so a minimal level of silicon removal for stress relief can be sufficient to flatten the silicon.

"Dicing by thinning" process configuration can be applied simply by adding a precut step as first process: its main advantages result in good dicing quality and lower risk of breakage, provided there are no drawbacks introduced in the stress relief step.

When removing a large amount of silicon by a plasma process, wafer thickness uniformity is a key parameter to be kept under control. One possibility to be investigated is the application of a plasma process that removes just a few microns, which has the advantages of improved cosmetics, roughness control, and thickness uniformity. If applied onto already separated dies it "heals" chip sidewalls that have already opened up, achieving a more mechanical robustness of the chip.

All of these observations deal with flash memory and Smartcard devices, but special applications such as bumped wafers, die attach tape, bonded wafer, MEMS, and back-side metallization must be customized case by case, sometimes simply by adding a few options (e.g., a special back-grinding tape) and sometimes by using a completely different process.

The market is now requiring wafer thicknesses down to 75–50 μm, so technological proposals such as edge rounding, tapeless grinding solutions, and heat shrinkable tapes are all innovations worth considering for the always new shrinking and complexity challenges.

Rigid tapes that can be used as supports during handling and shipment of thin and warped wafers are also commercially available.

To install a complete back-grinding line for ultrathin wafers, in line control equipment it is necessary to monitor:

- Thickness and TTV, especially for ultra-thin and large size wafers.
- Wafer bow.
- Microcrack detection at both the edge and the back-side.
- Surface roughness.

Front-side cleanliness level has also to be controlled in order to meet the requirements of front and back end compatibility, both in terms of die-to-package plastic molding compound coupling and device reliability during its life time.

NVM Device Degradation by Wafer Thinning

Oxygen Precipitation and Gettering

Oxygen is generally oversaturated in silicon wafers used for devices [13], and it has long been known that oxygen precipitates and that related bulk microdefects (BMDs) form during the thermal treatments involved in device processing [14, 15]. The gettering ability of such defects is also well known [16]. Usually, in the past the major concern about oxygen precipitates was to avoid the formation of defects in the device region by creating a so-called denuded zone, i.e., a region free from defects close to the wafer surface. This result was obtained by oxygen out-diffusion from the near-surface region in a high-temperature treatment. (about 1100°C). After that, a treatment at moderate temperatures (700–800°C) induced the nucleation of oxygen precipitates, with a thermal sequence called HI-LO [17]. LO-HI sequences were also proposed, with the nucleation step preceding the out-diffusion step [18]. LO-HI sequences were based on the assumption that the precipitate nuclei located close to the wafer surface dissolve completely during the out-diffusion treatment. As this condition is not guaranteed, LO-HI sequences were risky from the point of view of the near-surface quality, though they provided higher concentrations of bulk defects with respect to HI-LO treatments. Therefore, HI-LO sequences have generally been used. Thermal treatments at about 1000°C cause oxygen precipitation on the previously created nuclei. In many cases the precipitate growth can be achieved during the device process flow and no additional thermal treatment is required.

Out-diffusion annealing requires rather high temperatures, so owing to surface nitridation, it cannot be done in a pure nitrogen environment, and a mixture or another gas (argon or hydrogen) must be used. If a N_2:O_2 environment is used, the wafer surface is oxidized in the denuding treatment and the interstitial oxygen concentration at the surface can be reduced only to the solid solubility at the annealing temperature, thus limiting the denuding efficiency. Out-diffusion treatments in argon or hydrogen do not have this limitation and are therefore more effective.

However, HI-LO sequences are usually rather time consuming because they require long out-diffusion treatments at high temperatures and even longer nucleation treatments. For instance, 6-h annealing at 1100°C produces an oxygen depleted region of about 30 μm. In addition, the reduction of thermal budgets reduced the growth of bulk defects and hence the intrinsic gettering efficiency [19]. For this reason, a different approach to the formation of a denuded zone was proposed, based on the out-diffusion of point defects rather than the out-diffusion of oxygen [20].

This approach led to a technique named Magic Denuded Zone (MDZ, [21]). Oxygen precipitation injects silicon self-interstitials, and the resulting self-interstitial excess inhibits further oxygen precipitation. Therefore, oxygen precipitation can be modulated by modifying the point defect concentration. Indeed, a vacancy excess assists in reducing the interstitial excess generated by oxygen precipitation, thus allowing oxygen precipitation to proceed. The MDZ technique involves the formation of a near-surface region depleted from vacancies and a bulk region rich in vacancies. During the thermal treatments of the device process flow, oxygen precipitates nucleates and increases only in the vacancy-rich region, thus producing the desired wafer structure with a near-surface denuded zone and a defective bulk zone. The concentration of bulk defects is independent of the specific thermal treatment over a wide range of annealing temperatures. The modulation of intrinsic point defect concentration is achieved by rapid thermal treatments, so the MDZ process is much less time consuming than the conventional HI-LO sequence. However, it is worth recalling that the MDZ process produces only a vacancy-rich bulk region. The defect-rich bulk region acting as the gettering region must be produced by using the thermal budget of the device process flow.

Another approach to increase the concentration of bulk defects and hence the substrate gettering efficiency is nitrogen doping [22, 23]. Nitrogen reduces both vacancy-type defects (the so-called Crystal-Originated Particles, COPs) and interstitial type defects and also enhances oxygen precipitation [24]. However, nitrogen-doped crystals require a HI-LO treatment to form a denuded zone and a bulk defect region.

Intrinsic Gettering in the Back-End Process Flow

The final part of the device process flow is characterized by low-temperature thermal treatments (about 400°C for the definition of the metal lines and about 200–300°C for packaging). During these phases, the wafer front-side is usually protected, so silicon cannot be contaminated at the front-side. Vice versa, silicon contamination is still possible at the wafer back-side. Though the thermal budget is limited, some fast-diffusing contaminants may reach the device region even when the thermal treatments are at such low temperatures. Copper, nickel, and cobalt are the fastest-diffusing silicon contaminants [25], and retain significant diffusivity even at temperatures close to room temperature, where their solid solubility is very low. Thus, these elements can diffuse through the wafer thickness and reach the device region during the back-end process flow. Owing to the very low solubility, these elements are unstable in the solid solution and either form precipitates at existing crystal defects or segregate at the oxide-silicon interface. In the device region, this results in the destruction of device functionality or in the degradation of device reliability. The only way to prevent this phenomenon is to retain enough gettering ability through the entire device process flow. This seemingly trivial requirement can become critical after wafer thinning because most getter sites are removed by thinning, and the device region can be left exposed to fast-diffuser contamination. On the other hand, nickel and copper are widely used in packaging, so contamination by these elements is not unlikely.

Fast-Diffusing Contaminants in Silicon

It has been previously mentioned that copper is a fast diffuser and hence a potentially harmful contaminant in the final part of the process flow. In this section we report the results of some experiments about copper contamination [26], with the aim of showing the possible impact of such contamination on device properties.

TEM Analysis

Wafers with intentionally induced crystal defects were prepared to simulate the copper decoration of crystal defects in a device wafer. Slip lines were induced in these wafers by a Rapid Thermal Process (RTP), where the temperature distribution over the wafer surface was intentionally made nonuniform. Then, the samples were contaminated with copper by dipping the wafers along with a metallic copper part in

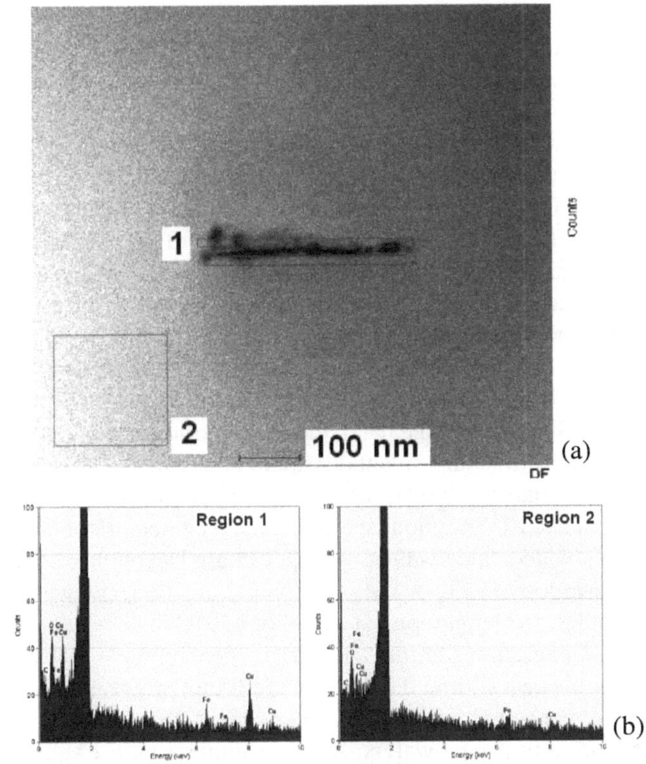

Fig. 6.13 TEM image of a dislocation in the slip-line region (**a**), EDX spectra of a wafer contaminated by HF dipping with metal copper in the dislocation region [(**b**), Region 1] and far from the dislocation [(**b**), Region 2]

a 40% HF solution. The wafers underwent thermal treatment at 250°C for 1 h and TEM samples were prepared in the slip-line region far from the wafer surfaces.

Figure 6.13 shows the TEM image of a dislocation in the slip-line region of the wafer (a) and the Energy-Dispersive X-ray (EDX) spectrum in the dislocation region (b) and in a region far from the dislocation in a wafer contaminated by HF dipping with metal copper. Copper is revealed at the dislocation, whereas no copper is found far from the dislocation. So, a thermal treatment at moderate temperature (250°C) is enough to diffuse copper through the wafer thickness as well as to make existing crystal defects decorated by copper.

Electrical Analysis

P-type, (100), 200-mm diameter, 725-μm thickness, 10-Ωcm resistivity Magnetic Czochralski (MCZ) wafers were used in this study. MCZ material was chosen because of its low oxygen content in order to prevent oxygen precipitation and consequent copper gettering at bulk defects, so in this study we show what may happen in devices when gettering is absent or insufficient. Capacitors were created by growing 1000 Å oxide with a dry O_2 oxidation cycle at 1000°C and depositing, masking, and etching an aluminum layer. The capacitor wafers were annealed at 430°C N_2/H_2 for the reduction of interface states. Then, the oxide was removed at the back-side of capacitor wafers and the wafers were ready for back-side contamination.

Two contamination methods were used: chemi-mechanical polishing (CMP) in copper-contaminated equipment and spinning of a contaminated solution on the back-side surface. The contamination obtained by CMP was in the range $1–2 \times 10^{13}\text{cm}^{-2}$, as measured by TXRF. In CMP-contaminated wafers, there was no native oxide at the back-side surface during the contamination.

Wafers to be contaminated by spinning were prepared both with a native oxide at the wafer back-side (hydrophilic surface) and with no native oxide (hydrophobic surface). The native oxide was obtained by a spray ammonia-peroxide solution (often called SC1 solution). On one hand, the hydrophilic surface could be expected to result in a more uniform contaminant distribution, but on the other hand the native oxide might act as a barrier slowing down the copper penetration into the silicon lattice, as was indicated by previous experiments. The actual contamination level was checked in a few dedicated wafers by Vapor Phase Decomposition plus Atomic Absorption Spectroscopy (VPD-AAS). After contamination, the wafers were thermally treated in single-wafer equipment at 400 or 450°C in an N_2 environment with a 2-h cool-down. Generation lifetime and surface generation velocity were extrapolated from capacitance vs. time data by the well-known Zerbst method [27, 28].

The recovery of capacitors from deep depletion was found to be very sensitive to copper contamination in p-type wafers. As an example, Fig. 6.14 shows the cumulative distributions of the transient duration in p-type wafers contaminated by CMP (copper concentration $1–2 \times 10^{13}\text{cm}^{-2}$) and in a reference wafer. The analysis of the Zerbst plot shows that this effect is essentially determined by the surface generation velocity (Fig. 6.15).

Figure 6.16 shows the capacitance transient duration as a function of the surface recombination velocity obtained from the Zerbst plots in copper-contaminated wafers. These data clearly show that the capacitance transient duration decreases in

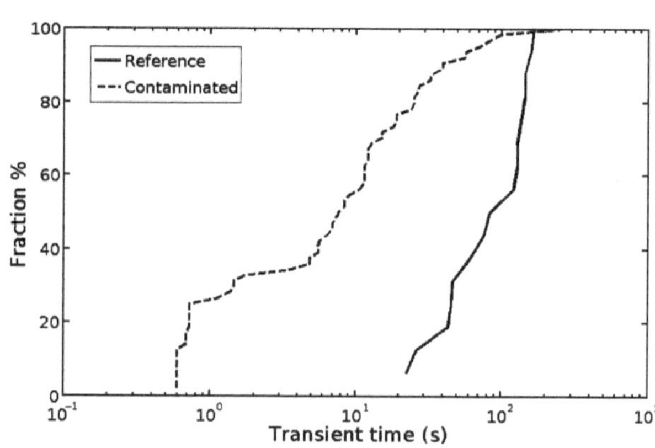

Fig. 6.14 Cumulative distribution of the capacitance transient duration in a copper-contaminated and a reference wafer

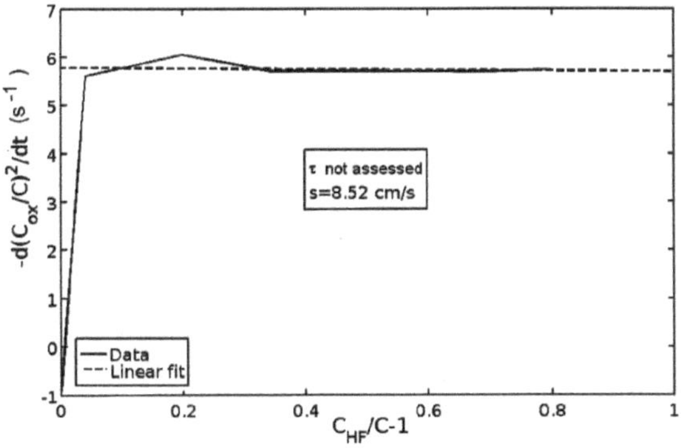

(a)

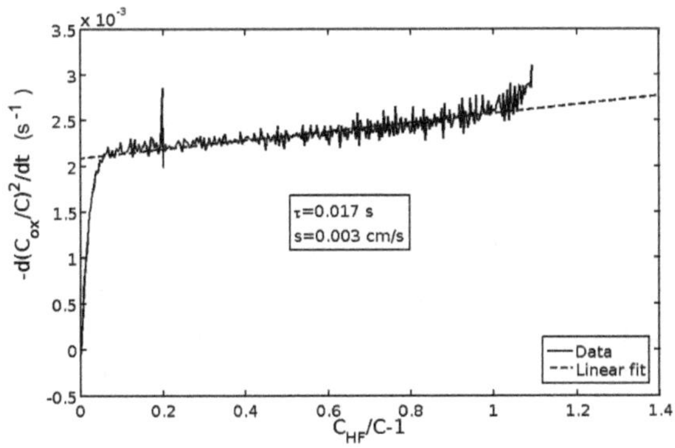

(b)

Fig. 6.15 (**a**) Zerbst plot of a capacitor in copper-contaminated wafers; (**b**) Zerbst plot of a capacitor in a reference wafer

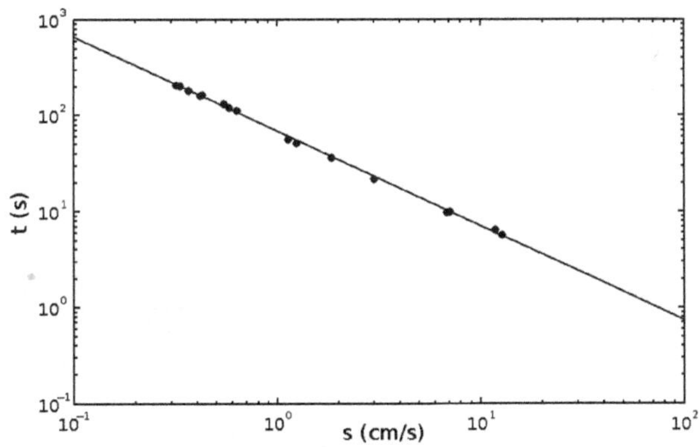

Fig. 6.16 Capacitance transient duration as a function of the surface generation velocity obtained from the Zerbst plot

proportion to the inverse surface recombination velocity, as expected when carrier generation is dominated by the surface. On the other hand, there is no correlation between the capacitance transient duration and the bulk generation lifetime.

Figure 6.17 shows the cumulative distributions of the surface generation velocity measured in p-type reference and copper-contaminated wafers treated at 400°C for 30 min. The data in Figs. 6.16 and 6.17 refer to wafers contaminated by CMP

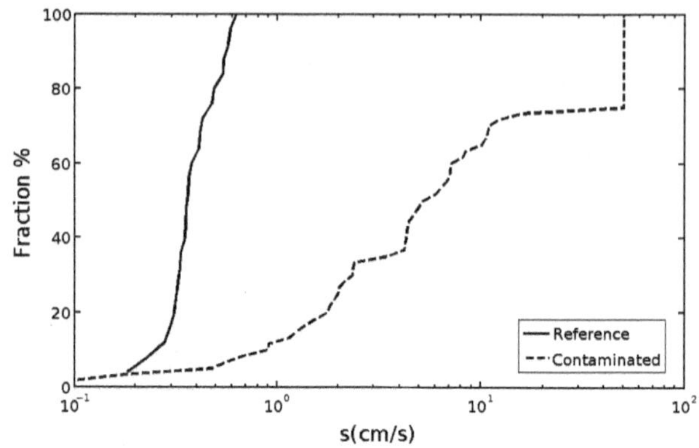

Fig. 6.17 Cumulative distributions of the surface generation velocity in copper-contaminated wafers and in reference wafers treated at 400°C for 30 min

(copper concentration 1–2 × 10^{13}cm^{-2}). It is clear that the copper contamination increases the surface generation velocity by up to two orders of magnitude.

The samples contaminated on a hydrophilic surface as well as the samples contaminated with a lower copper concentration than in Fig. 6.17 show an unexpected phenomenon, consisting of a recovery of surface generation velocity with the time spent after the thermal treatment, as shown in Fig. 6.18. It is important to note that no such effect was observed in the samples contaminated by CMP with 1–2×10^{13}cm^{-2} copper concentration. In addition, the behavior shown in Fig. 6.18 is reversible, i.e., if after recovery the wafers are heated again the same behavior as in Fig. 6.18 is reproduced.

Table 6.3 summarizes the results of surface generation velocity measurements. The surface generation velocity values at the 25th percentile ($s_{25\%}$) and at the 75% percentile ($s_{75\%}$) of the distributions are shown. Generally, when the contamination

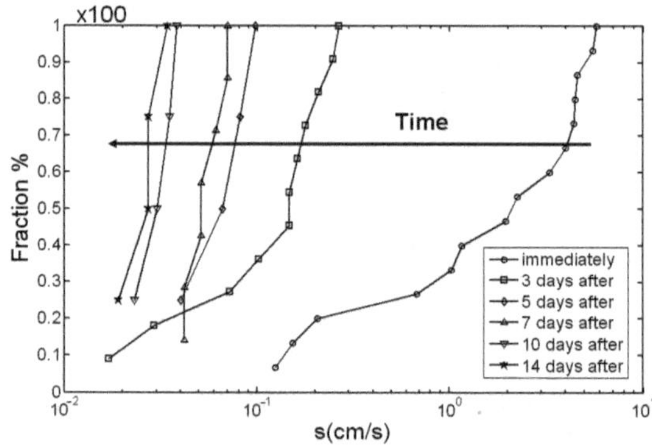

Fig. 6.18 Cumulative distributions of the surface generation velocity in copper-contaminated wafers treated at 400°C plus 450°C for 2 h, immediately after the thermal treatment and the change with time after the thermal treatment. In this example the copper concentration was 9 × 101^{2}cm^{-2}, and the back-side surface was hydrophilic

Table 6.3 Summary of the surface generation velocity results

Back-side surface preparation	[Cu] (cm^{-2})	Thermal treatment	$s_{25\%}$–$s_{75\%}$ (cm/s)	Recoverable
Reference	$\leq 10^{10}$	Any thermal treatment	0.02–0.1	–
Hydrophobic	$1\text{–}2 \times 10^{13}$	400°C, 30 min	2–60	No
Hydrophobic	2.5×10^{11}	450°C 2 h + 450°C 2 h	No effect	–
Hydrophobic	2.5×10^{12}	400°C, 30 min + 450°C 2 h	1–60	Partially
		450°C 2 h + 450°C 2 h	0.2–60	Partially
Hydrophilic	9.5×10^{11}	450°C 2 h + 450°C 2 h	No effect	–
Hydrophilic	9.5×10^{12}	400°C, 30 min 400°C,	0.17–0.19	–
		30 min + 450°C 2 h	2.4–6.5	Yes
		450°C 2 h + 450°C 2 h	10–60	Partially

was deposited on a hydrophilic surface the data show less spread, as a consequence of the better contamination uniformity on the hydrophilic surface. On the other hand, contamination deposited on a hydrophobic surface appears to be more effective even with a lower contaminant concentration, possibly because the native oxide layer may act as a surface barrier for copper penetration into the silicon lattice.

TOF-SIMS Analysis

Time-of-Flight Secondary Ion Mass Spectrometry (ToF-SIMS) data were acquired by means of an ION TOF IV dual beam ToF-SIMS with Ga^+ primary ion source and Cs/EI (Electron Impact) sputter gun dual source. TOF-SIMS measurements were used to detect copper at the oxide-silicon interface. The oxide on the silicon surface was not removed before the measurements to prevent sample contamination, as copper is a very common element. To reach the oxide-silicon interface, a depth profile with 1-KeV, 250-nA Ar sputter of the sample was obtained and used to calculate the sputter rate. Then, a new sample was sputtered under the same conditions to reach the oxide-silicon interface, and ToF-SIMS surface spectra were obtained in the center of the crater. The ToF-SIMS spectra were obtained by a 25 KeV Ga^+ primary ion source with a pulsed current of 2.4 pA. Secondary positive ions were acquired with a raster area of $30 \times 30\ \mu m^2$ and an acquisition time of 600 s. The quantification of copper concentration was based on measurements of wafers contaminated by spinning a contaminated solution [29]. As in the present case the measurement conditions were rather unusual, the copper concentration obtained in this way should be considered as an estimate within the order of magnitude.

TOF-SIMS analyses were carried out in a sample contaminated with copper by CMP and in a reference wafer. The cumulative distribution of surface generation velocity in these wafers is shown in Fig. 6.17.

As seen in Fig. 6.19, ToF-SIMS detected copper at the Si-SiO2 interface. Thus, this spectrum proves the diffusion of copper from the back-side of the wafer to the interface.

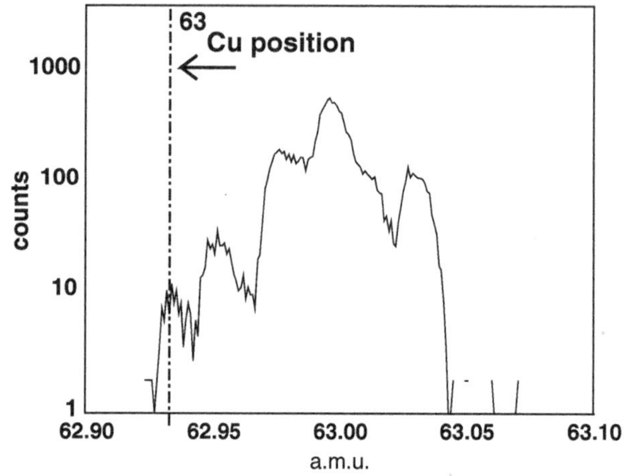

Fig. 6.19 ToF-SIMS surface analysis on the center of the crater obtained by sputtering up to the Si-SiO2 interface. The figure shows the part of the spectrum related to the main ^{63}Cu isotope

The estimated copper concentration is 5×10^9 cm^{-2}, a value that is close to the detection limit of the equipment, which is 10^9 cm^{-2}. Though these measurements provide a concentration estimate only, they clearly show that the copper concentration segregated at the Si-SiO_2 interface is a small fraction of the copper concentration deposited at the wafer back-side. This fraction is enough for a complete nonrecoverable degradation of the surface generation velocity.

Discussion

Our results indicate that the capacitor recovery from deep depletion is very sensitive to copper contamination even after low-temperature treatments (400–450°C). The phenomena involved in the deep depletion recovery are essentially the same as in determining the dynamic RAM retention time, so these results indicate that copper contamination in back-end processes may degrade the performance of dynamic devices. In addition, we have shown that copper decoration of existing crystal defects may occur even with thermal treatments typical of back-end processes. Copper decoration can enormously enhance the electrical activity of crystal defects [30], and thus cause the destruction of device functionality. To prevent the harmful effects of contamination by fast diffusers, it is necessary to ensure that enough getter sites for metal impurities are available even at the packaging level, i.e., after thinning.

Bulk Defect monitoring in Device Wafers

As previously noted, the density and the size of bulk defects in present device wafers are usually moderate. This fact is due partially to the reduction of the thermal budgets in device processes and partially to the requirement of a defect-free near-surface region. In addition, the precipitation threshold for effective gettering is estimated to be quite low ($\approx 3 \times 10^{16}$cm^{-3} precipitated oxygen atoms [31]); on the other hand heavy oxygen precipitation can be detrimental from the point of view of defect generation at the wafer surface [32] because of a decrease in the silicon yield stress. Because of these reasons, precipitates can be barely detected by conventional chemical etching techniques [33], so direct observation of the denuded and the precipitate-rich zones is problematic. This being the case, a method was set up [19] for the quantitative evaluation of denuded zone depth based upon carrier recombination lifetime data obtained from Surface Photovoltage (SPV) measurements.

Denuded Zone Data from SPV Measurements

Oxygen precipitates are effective generation-recombination centers [34], so they can be easily detected by lifetime measurement techniques [35–37]. Usually these techniques do not provide information about the in-depth distribution of recombination centers, but an estimate of the denuded zone depth can be obtained by elaborating SPV data. In this section this method is reviewed briefly.

Czochralski grown p-type wafers with oxygen concentration in the range 6–7.5 $\times$ 10^{17}cm^{-3} (ASTM 83 units) were used. CZ wafers that were subjected to a conventional denuding treatment (1100°C, 4 h) were compared to wafers with no denuding and to wafers treated with the MDZ process. All the wafers underwent the thermal treatments of a device flow using the LOCOS isolation scheme. Table 6.4 summarizes the main thermal treatments in this process flow.

Table 6.4 Thermal treatments of the process flow used to prepare the wafers to test the SPV-based method to estimate the denuded zone depth

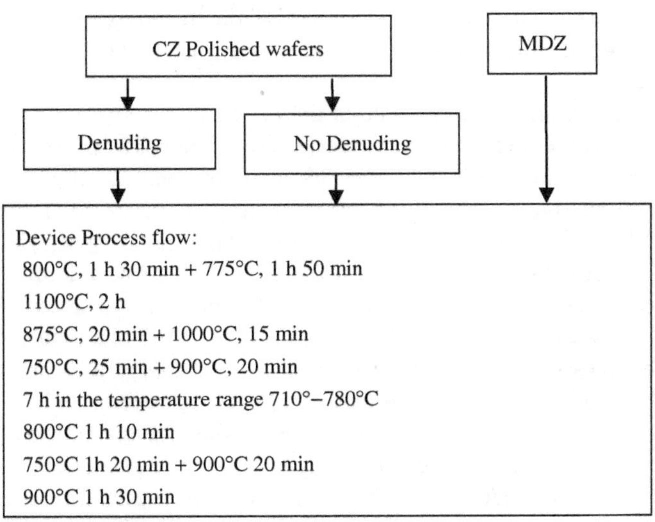

A CSM III-A SPV system by SDI [38] was used for SPV measurements. In the standard elaboration of SPV data [39] a uniform distribution of recombination centers is assumed. In wafers with a denuded zone, this procedure yields an effective diffusion length L_{eff}, based on the average between the denuded zone and the defective bulk zone.

In order to interpret surface photovoltage data of wafers with a denuded zone at the surface and a bulk zone with oxygen precipitates and related defects, this assumption was changed into the following: $L_{diff} = L_{DZ}$ if $x \leq t_{DZ}$, and $L_{diff} =$ L_{def} if $x > t_{DZ}$, where L_{DZ} is the diffusion length corresponding to the density of recombination centers in the DZ, L_{def} is the diffusion length in the defective bulk zone ($L_{DZ} >> L_{def}$), x is the depth coordinate, and t_{DZ} is DZ thickness. To get a direct measure of L_{def}, diffusion length, measurements were carried out at wafer back-side after DZ removal, thus obtaining a measurement of the carrier diffusion length in the defective bulk region. In all samples the measurement taken in the defective bulk region was lower than the that at the wafer front-side L_{eff}, thus indicating the presence of a defect-free zone at the wafer front-side (the denuded zone). Of course, no such difference was found in wafers with a uniform distribution of recombination centers (e.g., iron-contaminated samples).

For a quantitative evaluation of DZ depth, however, some data elaboration is required. The minority carrier diffusion equation was solved by assuming the in-depth distribution of diffusion length described above and by imposing the continuity of the excess minority carrier density function $\delta n(x)$ and that of its first derivative at $x = t_{DZ}$. As it is done in the standard SPV technique, it was assumed that $\Delta V_{ph} \propto n(x = 0)$, where ΔV_{ph} is the surface photovoltage. The surface photovoltage signal is defined as $\varphi/\Delta V_{ph}$, where φ is the photon flux penetrating into the silicon matrix. This quantity was then written as a function of light penetration depth z with the DZ thickness, L_{def}, and L_{DZ} as the parameters. Since L_{def} was measured and L_{DZ} is rendered ineffective by assuming $L_{DZ} >> L_{def}$, the surface photovoltage signal has one fitting parameter only, i.e., the DZ depth. A best-fit procedure of this function to experimental data of the surface photovoltage signal yields an estimate of DZ depth.

As an indirect method, the estimate of the DZ depth by SPV has to be validated by comparison with conventional microscopy techniques. The samples were cross-sectioned, preferentially etched by the Secco d'Aragona solution and inspected by Scanning Electron Microscopy (SEM) or Atomic Force Microcopy (AFM). Table 6.5 collects the results of this comparison. The SEM inspection of a sample cross section after selective etching (Secco) often fails to provide a full direct image of the denuded zone because of the lack of resolution of secondary electron images at low magnification. In addition, the density and size of precipitate-related etch pits are sometimes so low that they can hardly be detected by secondary electron SEM. AFM provides the best compromise between sensitivity and statistics, as it retains the resolution required for small etch pits even at a moderate magnification. However, AFM is limited by the sensitivity of the etching solution. Furthermore, many AFM scans are required to obtain an image of the denuded zone, so this method is very time consuming. Transmission Electron Microscopy (TEM) was

Table 6.5 Summary of the SPV analysis and of microscopy observations in various CZ substrates

	SPV analysis		Microscopy observations	
Substrate	L_{def} (μm)	t_{DZ} (μm)	SEM	AFM
No denuding			Large defects deep in the bulk	Etch pits
Average	7	23	Poor uniformity	Poor uniformity
Min–Max	5–10	10–38	No DZ estimate	$t_{DZ} = 7–10$ μm
Conventional denuding			Small etch pits in the bulk	Small etch pits
Average	12	51	No DZ estimate	$t_{DZ} = 30–39$ μm
Min–Max	10–15	44–54		
MDZ			Few, very small etch pits	Many very small etch pits
Average	21	120	No DZ estimate	$t_{DZ} = 120$ μm
Min–Max	17–24	100–130		

also tried, but was found to be unsuitable for this sort of investigation because of the very limited silicon volume explored by TEM analysis. Indeed, a few bulk defects could be found in TEM lamellas only in samples with very low bulk diffusion length (≈2–3 μm).

SPV estimates of denuded zone depth are plotted vs. the results given by AFM inspections in Fig. 6.20. AFM and SPV data agree very well with each other, although SPV estimates are somewhat larger than AFM data. SPV has very good sensitivity to oxygen precipitation and detects strong lifetime degradation even in samples where very few defects are detected by microscopy techniques. However, the SPV analysis is very rough in terms of in-depth defect distribution. For instance, SPV cannot give information about the width of the transition region between the

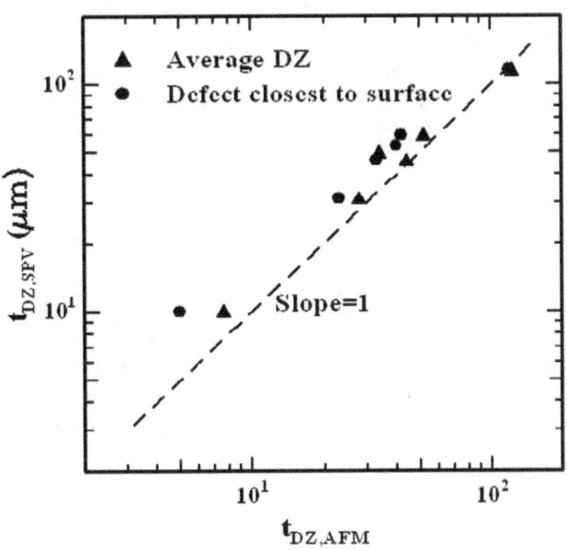

Fig. 6.20 Comparison between SPV and AFM data for DZ depth

denuded and the defective zone, nor about the uniformity of defect distribution in the bulk. The comparison with AFM inspections suggests that the SPV estimate of the DZ depth includes the transition region, at least partially.

Finally, the diffusion length in the defective bulk region is related to the density of bulk defects. Indeed, this parameter can be considered as a measurement of the bulk defect density under the hypothesis that the contamination level is comparable in the samples under study. This hypothesis is reasonable since these wafers were processed together.

Investigation of Bulk Defects in a Current Device Process

The method described in the previous section was used to study the formation of bulk defects in a current device process, using a Shallow Trench Isolation. With respect to the process flow studied in the previous section, this process is characterized by a lower initial interstitial oxygen concentration ([O$i \leq 6.75 \times 10^{17}$cm^{-3}) and by a very limited thermal budget, consisting essentially of 50 min at 800°C and 30 min at 1000°C. Under these conditions the formation of bulk defects can be highly critical. In our test we compared nitrogen-doped wafers from different suppliers with a furnace denuding treatment in an argon environment. In some cases, the wafer suppliers provided an optimized version of the material from the point of view of bulk defect formation, even with no change in the interstitial oxygen concentration. Table 6.6 reports the results of the SPV and SEM characterization of these wafers after the thermal treatments of the process flow. By comparing the results in Tables 6.5 and 6.6 one can see how the evolution of substrates and process flows impacted on the formation of bulk defects. The data in Table 6.6 were obtained in conventionally annealed wafers, so they can be compared to the data on conventionally annealed wafers in Table 6.5. First, the denuded zones (as estimated by SPV) became deeper, and this is probably due to the presence of wide transition regions, as it is difficult to reach such deep denuded zones by oxygen out-diffusion. In wafers from supplier B with no optimization, the estimate of the denuded zone depth is unreasonably high, indicating that a bulk defective region and a denuded zone can hardly be defined in these wafers. In some cases (suppliers B and C with no optimization) the parameter L_{def} is quite high, so the bulk defect density is expected

Table 6.6 Summary of the SPV analysis and of SEM observations in wafers from different suppliers

	SPV Analysis		SEM Inspection	
Supplier	DZ depth (μm)	L_{def} (μm)	BMD density (cm^{-3})	DZ depth (μm)
A	80–120	10–20	3–6 × 10^7	80–110
B	150–400	30–70	Small etch pits in the bulk	–
B, optimized	70–100	9–14	Small etch pits in the bulk	–
C	90–120	20–70	No defects revealed by SEM	–
C, optimized	50–70	5–8	Small etch pits in the bulk	–

to be low and possibly critical for gettering. In all substrates the optimization of the crystal growth and of the denuding cycle resulted in acceptable behavior from the point of view of bulk defect formation, so cooperation with wafer suppliers is important for obtaining the required wafer characteristics. However, we note that if wafers are thinned below 100 μm they are potentially critical in terms of contamination after thinning, though some improvement was obtained by optimization for factors concerning the denuded zone depth.

Through Silicon Via for a High-Density Memory System

Three-dimensional (3-D) integration is an emerging architecture used to reduce system size by stacking multiple devices vertically. Wire bonding, microbumps, contactless interconnects, and through silicon vias (TSV) are possible 3-D approaches. Figure 6.21 shows an example of vertical stack with TSV.

This section concerns TSV, isolated connections formed through silicon and used in chip level or wafer level packaging, as a promising solution for high-density applications.

TSV technology can improve electrical performances with respect to wire bonding, because it reduces delay, noise, and power consumption thanks to the minimized interconnection length. Figure 6.22 shows two examples of wire bonding replacement. In addition, it offers a wider range of applications with respect to flip chip packaging because allows heterogeneous integration of more than two dies. The limited footprint, volume, and weight are important features for today's IC market.

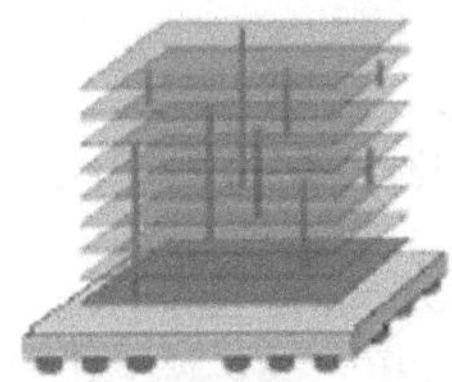

Fig. 6.21 Example of a 3-D stack with TSV

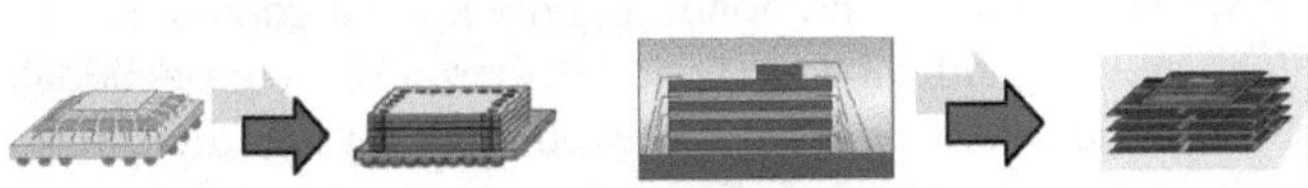

Fig. 6.22 Examples of TSV applications as a replacement for wire bonding. Low density (on the *left*) with TSV at the chip edge and high density (on the *right*) with TSV inside the chip

Players and Applications

Many research institutes and consortia have been working on TSV development for more than 10 years. ASET (Japan), Fraunhofer Institute (Germany), IMEC (Belgium), CEA-LETI (France), ITRI (Taiwan), Lincoln Labs (USA), Tohoku

Fig. 6.23 Samsung 16-Gbit NAND flash with TSV

University (Japan) and Arkansas University (USA) are among the groups most actively involved in this field.

Several semiconductor companies have recently announced TSV applications in various fields, from image sensors, to NAND, DRAM, MEMS, and power amplifiers.

In 2006 Samsung Electronics announced that it was able to stack eight 50-μm thick 2-Gbit NAND flash dies, to provide a 16-GBit flash memory for mobile applications [40].The stacked device has a 15% smaller footprint and 30% thinner package than the equivalent wire-bonded solution. The stack is shown in Fig. 6.23.

In 2007 IBM announced a breakthrough chip stacking technology [41], with its first products in 2008, including a power amplifier for wireless applications [42]. Another example of 3-D applications is the image sensors from Toshiba [43] and the DRAMs stack demonstrated by IBM [44].

Technology Trends

Different TSV technology approaches are presently under development. Some basic options are described in what follows.

Die to Wafer or Wafer to Wafer

With die to wafer alignment and bonding, only known good dies are selected for stacking, which leads to yield enhancement. Furthermore, heterogeneous integration is possible because the top die and the corresponding bottom die can have different sizes and layouts. Low throughput due to the lengthy alignment process is the main disadvantage of this option.

Direct wafer to wafer bonding dramatically increases throughput, but the total yield can be much lower because of blind assembly of not selected dies (Fig. 6.24). In addition, stringent topography requirements and same chip size stack limit the fields of application of this option.

Via First or Via Last

TSV can be formed during front-end process flow by adding specific technology steps or after full integration as the last step before assembly. These two options, called respectively the Via first and the Via last process, are illustrated in Fig. 6.25.

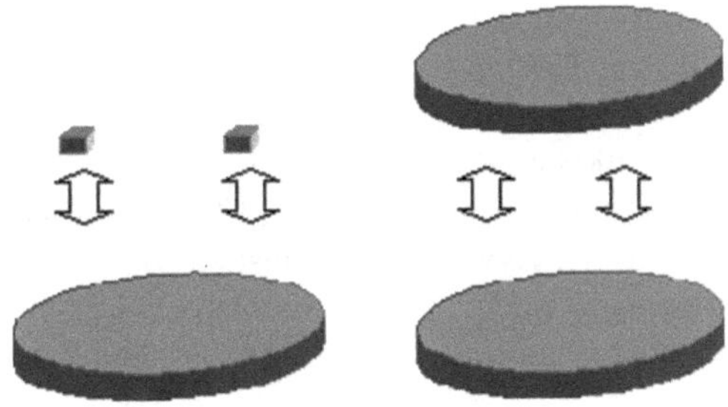

Fig. 6.24 Die to wafer (on the *left*) vs. wafer to wafer (on the *right*)

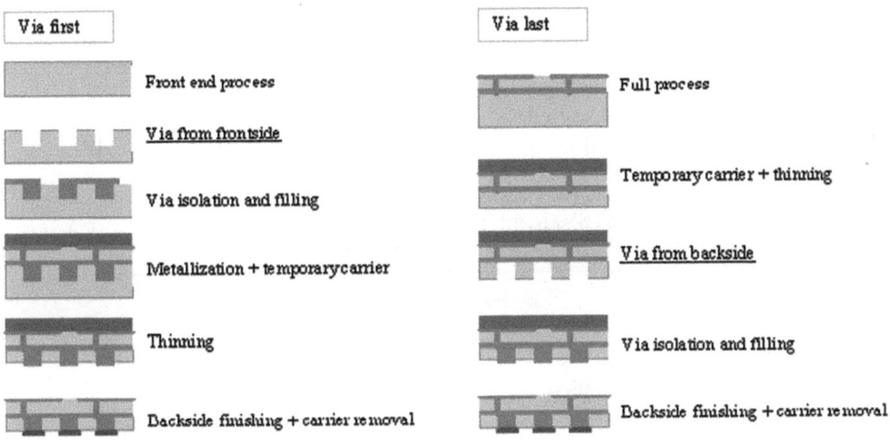

Fig. 6.25 Via first on the *left* (vias opened before BEOL) and Via last on the *right* (vias opened after BEOL)

Via first is not generally Front End of the Line (FEOL) compatible unless the via filling is limited to polysilicon, and it can be more precisely defined as Via middle if TSV are integrated after CMOS before metallization, and is hence compatible with any kind of conductor. This approach has to be chosen for high-density applications, because it provides TSV with scalable pitch and requires more advanced technology for via opening and filling. Via last, on the other hand, can be selected if vertical stacking is managed as part of a back-end process without involving the IC manufacturer; it requires back-side lithography and a low-temperature conformal insulation process.

Size and Pitch

TSV can range from large sizes, such as a via diameter of 10–100-μm down to less than a diameter of 1–10 μm and a silicon thickness of about 50–300 μm, down to 1 μm, as is used in silicon-on-insulator (SOI) technologies [45]. For example,

Takahaski et al. [46] gave a TSV technical presentation on 10-μm copper conductors utilizing TSVs for electrical interconnection at a 20-μm pitch.

TSV densities achievable using wafer to wafer integration are suggested by the ITRS 2008 roadmap, as reported in Table 6.7, as referred to Fig. 6.26.

Simulations of Cu TSV [47] show that the von Mises stress decreases with decreasing TSV pitch at constant via size and increases with decreasing via size at constant pitch, suggesting that plastic deformation of Cu vias has to be taken into account as an additional concern in defining TSV specifications.

Table 6.7 High-density through silicon vias specification according to ITRS 2008

	2010	2011	2012
High-density TSV diameter Ø[μm]	1.4	1.3	1.3
Minimum interlayer pitch [μm]	2.8	2.6	2.6
Minimum layer thickness [μm]	10	10	10
Bonding overlay accuracy (Δ) [μm]	1	1	1
Minimum size bonding pad (Ø + 2Δ)[μm]	3.4	3.3	3.3
Minimum pad spacing (STSV) [μm]	1	0.5	0.5
Minimum pitch ($Ø_{pad}$ + STSV) [μm]	4.4	3.8	3.8

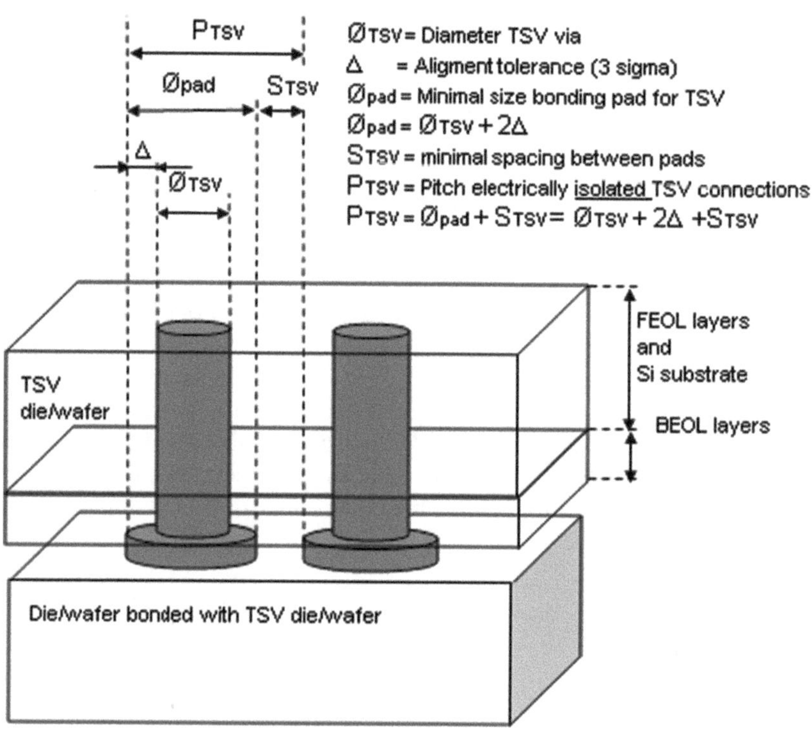

Fig. 6.26 High-density through silicon vias specifications. See Table 6.7

TSV Process Flow

One possible way to fabricate TSV consists of dry etching, liner deposition (isolation layer, barrier and seed layer), and via filling (e.g., Cu deposition and CMP).

Figure 6.27 shows a possible sequence of process steps:

- Photolithography.
- Si dry etching, with a SiO_2 hard mask, if any.
- Dielectric deposition (SiO_2 for isolation and SiN as a barrier against Cu diffusion, if required).
- Cu seed deposition.
- Cu ECD.
- Cu CMP.

The most commonly used dry etching technology is the so-called Bosch approach, a patented process developed by Robert Bosch GmbH in 1996 [48], which consists of alternating etch and passivation steps with SF_6 and C_4F_8. Via depth (D) and via size (CD top and CD bottom) are the target specifications that have be fulfilled with the proper etch rate, proper wafer uniformity, and minimum defectivity, such as undercut (U) and sidewall scallops, defined by W and L as shown in Fig. 6.28.

A tapered via can be chosen to optimize the subsequent deposition and filling. However, this option can limit the interconnection density owing to the larger TSV top size. A typical slope angle between 80° and 88° is expected to be a good trade-off with respect to seed layer deposition and top area opening.

Sidewall insulation can be considered one of the most critical steps in TSV technology [49] because of thermal budget versus conformality constraints. In addition, thermomechanical stress issues can have a detrimental impact on the final yield.

As for via filling material, CVD doped polysilicon, CVD tungsten or plated copper are possible conductor options. Doped CVD polysilicon can be integrated in front-end process modules, but it has relatively high resistance (500 times higher

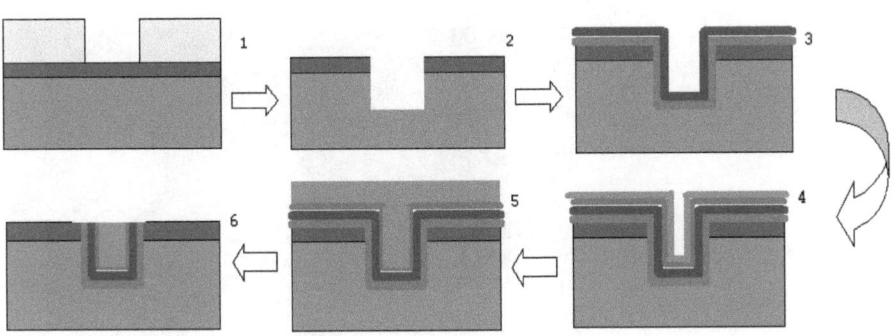

Fig. 6.27 Possible process flow for TSV formation

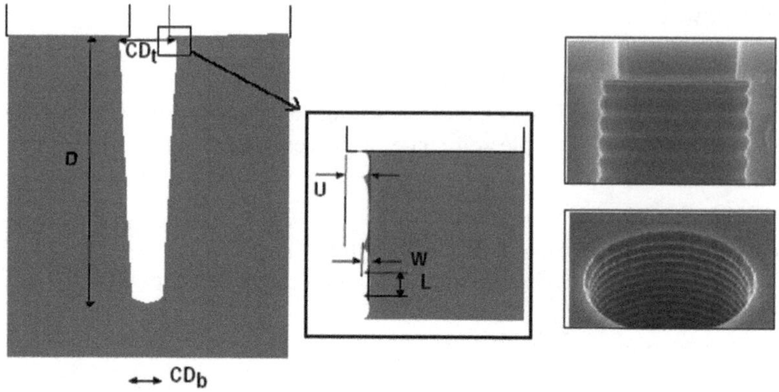

Fig. 6.28 TSV dry etching specifications on the *left*; undercut and scallops on the *right*

than copper) and it is limited to about 1 μm in thickness. CVD tungsten is more limited in regard to temperature than polysilicon, and is a better choice for Via last or Via middle integration approaches. This material is limited to about 1 μm in thickness by stress and to about 1.5 μm due to conformality constraints. An approach using tungsten-filled, multifinger, bar-shaped TSV, integrated with Via middle has been described in a recent paper by researchers IBM [50].

Plated copper is the conductor with the lowest resistivity and can be used for both high- and low-density TSV, with either conformal or bottom up filling technology, both for Via first and Via last options. However, the challenge of this metallization is in its high coefficient of thermal expansion (CTE) with respect to silicon and in its limited filling capability, not trivial at all for aspect ratios higher than 3:1 [50]. Experimental results indicate that vias with aspect ratios higher than of 3 prove difficult to plate conformally, resulting in formation of voids after Cu electrodeposition. See, for instance, Fig. 6.29. Such voids entrap the plating solution and cause subsequent failures when the device is heated above the boiling point of the trapped liquid [51] Nevertheless Cu filling can be successfully achieved by using specific

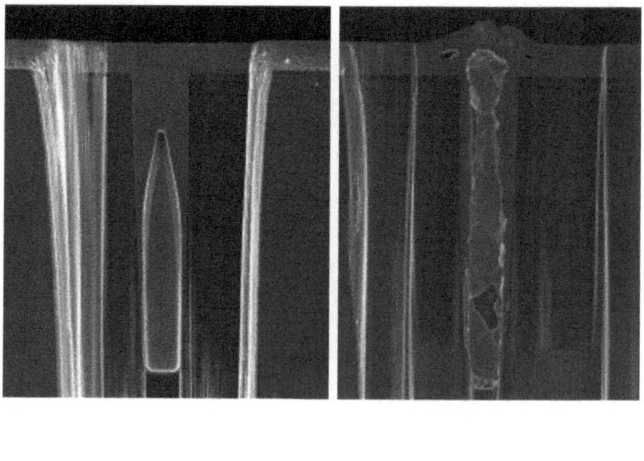

Fig. 6.29 Large void after Cu electrodeposition and its possible evolution after annealing

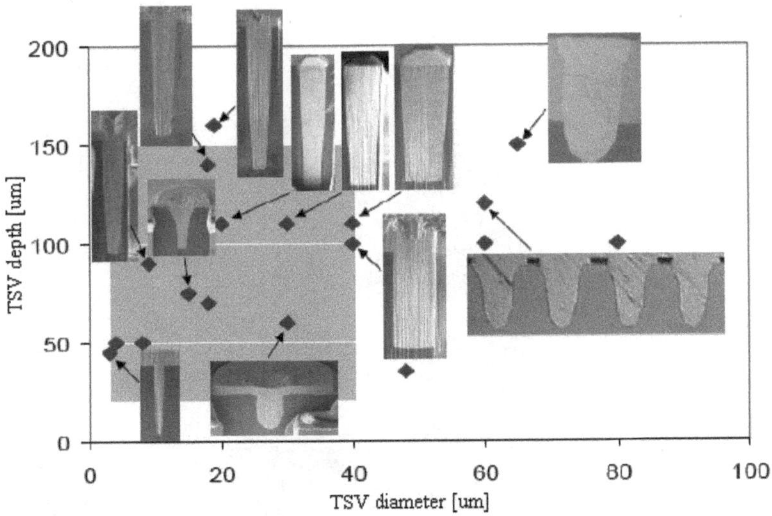

Fig. 6.30 Examples of TSV Cu void-free filling [53]

technology developed for TSV applications, as shown in Fig. 6.30, e.g., by using careful control of bath chemistry, pulse current, time, and frequency [52, 53].

Electrical Characterization

A 3- vertical stack with TSV technology can be obtained by completing the process with metallization levels, pad formation, wafer thinning, back-side finishing, and bonding, as previously mentioned and shown in Fig. 6.24. Once the multiple device stack is achieved, the electrical continuity has to be demonstrated because various weak points can be introduced by several parts of this long process. Two examples of devices connected with TSV and bonded on substrate for testing are shown in Fig. 6.31.

Proper daisy chain structures can be designed in order to evaluate specific electrical parameters, such as single-via resistance, via capacity (to substrate, to via, to bottom RDL), via chain quality, via isolation, and breakdown voltage, and also to study reliability issues associated with TSV technology. A possible schema for daisy chain stacking and assembly is shown in Fig. 6.32.

A daisy chain electrical characterization has been published by several TSV research groups. For example, scientists at the University of Arkansas performed resistance measurements on daisy chains of 48 and 72 Cu filled TSV [52]; IBM designed a via array module containing nearly 52000 annular W filled TSV [54]; and CEA-LETI and STMicroelectronics, studied annular poly filled TSV, using interdigitated chains with more than 250 TSVs per chain [55]. In all these cases, the experimental data showed that overall resistance increased with the number of vias,

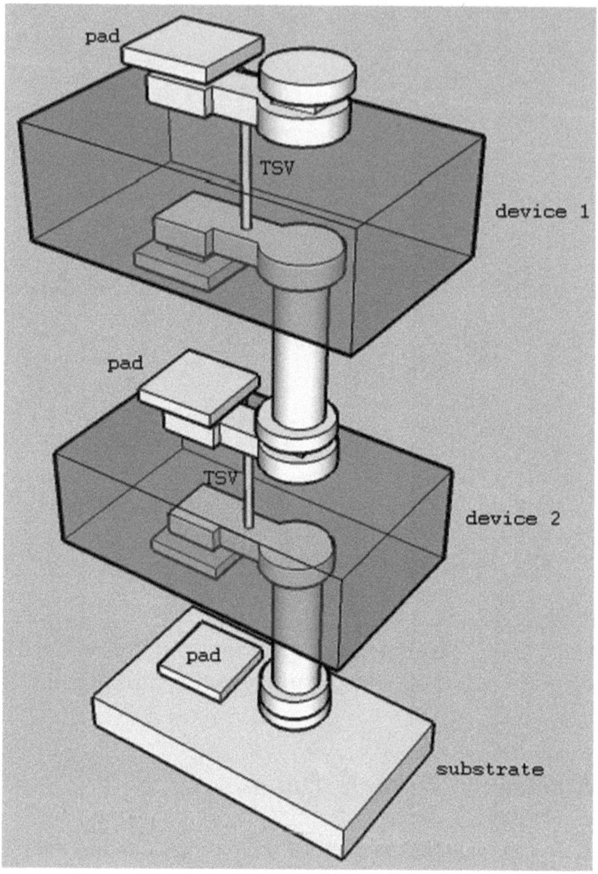

Fig. 6.31 A 3-D stack with TSV and possible architecture for testing

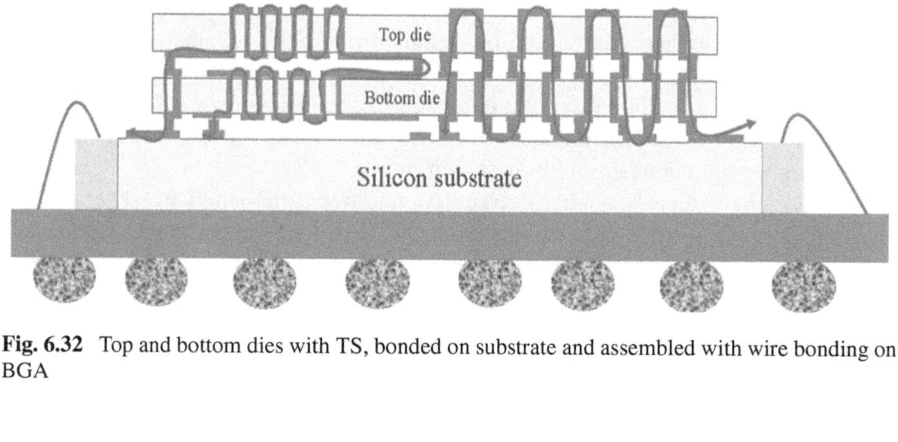

Fig. 6.32 Top and bottom dies with TS, bonded on substrate and assembled with wire bonding on BGA

as expected, with a connection yield higher than 99.9%. A recent study presented by TSMC [56] highlighted a possible yield loss (probability plots of TSV chain resistance with percentage <95%), due to nonoptimized back-side metallization.

In terms of reliability, ASET published some results about temperature cycling (–40, 125°C), push and bonding tests on stack with Cu filled TSV (CD 10 μm, depth 50 μm), showing an improvement in chip to chip joint reliability, resulting from relaxing the compressive strain of bumps [57, 58, 59].

Electromigration properties have been studied by TSMC [56] and IBM [50], which, in particular, demonstrated solid TSV performance and identified the root cause of the primary failure (resistance increase after stress) as voiding located at the expected TSV to metal interface. In addition, back-side metallization, which provides the electrical coupling between the silicon chip and the package through conductive epoxy, was considered to be another key element for overall reliability. Some sample TSMC electrical results are shown in Fig. 6.33.

Electrical functionality of active devices stacked with TSV has been investigated by Intel, among others: Cu filled TSV, obtained using a Via last approach, were integrated in 65-nm MOSFET and 4-MB SRAM without detecting any electrical degradation from the 3-D stacking technology [60]. The Fraunhofer Institute studied W filled TSV, obtained using Via first integrated with suitable testing devices containing CMOS transistors: none of the electrical measurements, including wafer maps of threshold voltage and drain saturation current, showed any impact from the TSV technology process on the transistor behavior [61]. More recently Imec showed a possible TSV impact on CMOS functionality: Cu filled vias obtained using Via first proved to modify some device performances (V_{sat} shift) if connected to the gate electrode. See the experimental data shown in Fig. 6.34. The degradation effect is supposed to be related to an antenna effect that is the result of back-side damage on TSV during wafer thinning [62].

Vertical integration improves the package density of the system by integrating a larger number of elements while reducing the overall footprint. Furthermore, architectures making use of vertical interconnections have lower power dissipation and

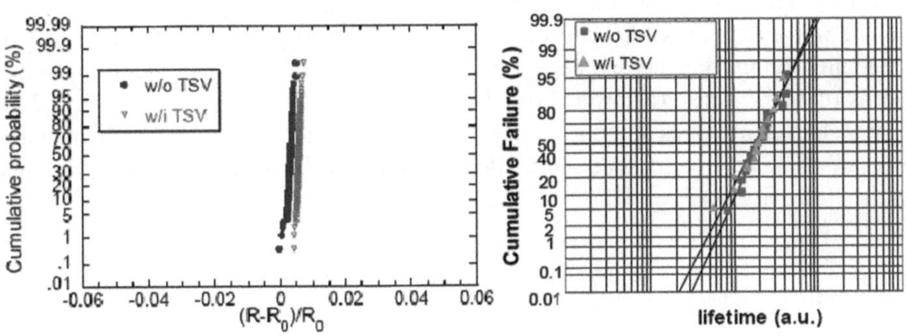

Fig. 6.33 Electromigration and stress migration results for CMOS devices with and without TSV [55]; there are no relevant differences between the data groups

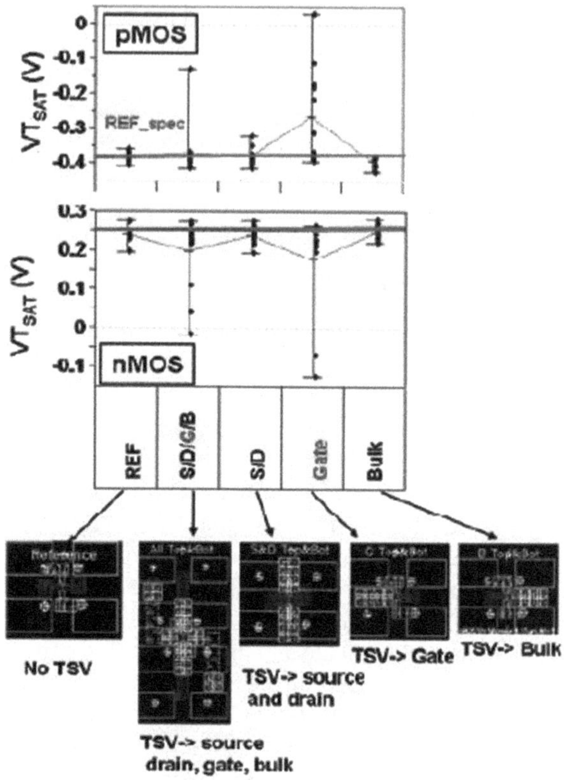

Fig. 6.34 Impact of TSV on MOSFET parameters [62]

better disturb immunity compared to the standard planar approach. This is due mainly to the reduction of interconnect length and coupling capacitance.

On the other hand, more chips in the same smaller footprint correspond to increased power densities with eventual impact on device performance: unless methods can be found to effectively remove heat from the stack, any improvement from wire-length reduction could be lost in degraded device performance.

The move to 3-D is likely to be limited by heat and yield. In any case, cost is the main driving force behind accepting any new technology into production. In the words of Mark Tuttle (Micron) "smaller, cheaper, faster and more reliable: a 3-D approach that improves any one of these attributes has the potential to influence the market" [63]. TSV technology, in particular, is one of the most promising solutions for high-density applications.

References

1. Gaulhofer E, Oyrer H (2000) Semicon Singapore 2000, May 09–11
2. Landesberger C, Klink G, Schwinn G, Aschenbrenner R (March 2001) Advanced packaging materials, processes, properties and interfaces. In Proceedings of IEEE 01TH8562, ISBN 0-930815-64-5, pp 92–97

3. Liu JH, Pei ZJ, Fisher GR (2007) Int J Mach Tools Manufact 47(1):1–13
4. McGuire K, Danyluk S, Baker TL, Rupnow JW, McLaughlin D (1997) J Mater Sci 32: 1017–1024
5. Chen J, De I (2003) Wolf Semicond Sci Technol 18:261–268
6. Landesberger C, Scherbaum S, Schwinn G, Spöhrle H (2001) In Proceedings of microsystems technologies. Mesago, Stuttgart, pp 431–436
7. Sibailly O, Wagner F, Richerzhagen B (2004) Future. Fab Intl (16):146–147
8. Perrottet D, Buchilly JM, Wagner F, Richerzhagen B (August 2004) Semiconductor manufacturing. pp 46–50
9. Schmitz TL, Davies A, Evans CJ (2003) American Society for Precision Engineering (ASPE). Winter Topical Meet 28:98–103
10. Ali YM, Zhang LC (1999) J Mater Process Technol 89–90:561–568
11. Alberici S, Dellafiore A, Manzo G, Santospirito G, Villa CM, Zanotti L (2004) Microelectronic Eng 76:227–234
12. Alberici S, Coulon D, Joubin P, Mignot Y, Oggioni L, Petruzza P, Piumi D, Zanotti L (2003) Microelectronic Eng 70:558–565
13. Zulehner W, Huber D (1982) Czochralsky-grown silicon. Crystals 8:1–143
14. Wilkes J G (1983) The precipitation of oxygen in silicon. J Crystal Growth 65:214–230
15. Borghesi A, Pivac B, Sassella A, Stella S (1995) Oxygen precipitation in silicon. Appl Phys Rev 77:4169–4244
16. Tan TY, Gardner EE, Tice WK (1977) Intrinsic gettering by oxygen precipitate induced dislocations in Czochralski Silicon. Appl Phys Lett 30:175–176
17. Murray EM (1984) Denuded zone formation in p silicon. J Appl Phys 55:536–541
18. Yamamoto K, Kishino S, Matsushita Y, Iizuka T (1980) Lifetime improvement in Czochralski-grown silicon wafers by the use of a two-step annealing. Appl Phys Lett 6:195–197
19. Polignano ML, Brambilla M, Cazzaniga F, Pavia G, Zanderigo F (1998) Denuded zone thickness from surface photovoltage measurements: comparison with microscopy techniques. J Electrochem Soc 145:1632–1639
20. Falster R, Gambaro D, Olmo M, Cornara M, Korb H (1998) The engineering of silicon wafer material properties through vacancy concentration profile control and the achievement of ideal oxygen precipitation behaviour. Mat Res Soc Symp Proc 510:27–35
21. Falster R, Cornara M, Gambaro D, Olmo M (1999) Ideal oxygen precipitating silicon wafers and oxygen out-diffusion-less process. US patent US 5994761
22. Nakai K, Inoue Y, Yokota H, Ikari A, Takahashi J, Tachikawa A, Kitahara K, Ohta Y, Ohashi W (2001) Oxygen precipitation in nitrogen-doped Czochralski-grown silicon crystals. J Appl Phys 89:4301–4309
23. von Ammon W, Dreier P, Hensel W, Lambert U, Köster L (1996) Influence of oxygen and nitrogen on point defect aggregation in silicon single crystals. Mat Sci Eng B 36:33–41
24. von Ammon W, Hölzl R, Virbulis J, Dornberger E, Schmolke R, Gräf D (2001) The impact of nitrogen on the defect aggregation in silicon. J Crystal Growth 226:19–30
25. Graf K (2000) Metal impurities in silicon device fabrication. Springer, Berlin
26. Polignano ML, Brivio J, Codegoni D, Grasso S, Altmann R, Nutsch A (2009) Revealing copper contamination in silicon after low temperature treatments. Electrochem Soc Trans 25:337–348
27. Zerbst M (1966) Relaxation effects at semiconductor-insulator interfaces. Z Angew Phys 22:30–33
28. Kang JS, Schröder DK (1985) The pulsed MIS capacitor – a critical review. Phys Stat Sol (a) 89:13–43
29. Beckhoff B, Nutsch A, Altmann R, Borionetti G, Pello C, Polignano ML, Codegoni D, Grasso S, Cazzini E, Bersani M, Lazzeri P, Gennaro S, Kolbe M, Muller M (2009) Highly sensitive detection of inorganic contamination. Solid State Phenomena 145–146:101–104
30. Dishman JM, Haszko SE, Marcus RB, Murarka SP, Sheng TT (1979) Electrically active stacking faults in CMOS integrated circuits. J Appl Phys 50:2689–2696
31. Falster R (1989) Gettering in silicon by oxygen related defects, stacking faults and thin polycrystalline films. Solid State Phenomena 6–7:13–20

32. Vanhellemont J, Claeys C (1988) Intrinsic gettering: sense or nonsense? In: Soncini G, Calzolari PU (eds) Proceedings of ESSDERC '87. North Holland, Amsterdam, pp 451–454
33. Falster R, Bergholz W (1990) The gettering of transition metals by oxygen-related defects in silicon. J Electrochem Soc 137:1548–1559
34. Hwang JM, Schroder DK (1986) Recombination properties of oxygen precipitated silicon. J Appl Phys 59, 2476–2487
35. Porrini M, Gambaro D, Gerenzani P, Falster R (1996) Influence of oxygen and oxygen-related defects on the minority carrier lifetime of high purity CZ and MCZ silicon. Electrochem Soc Proc 96–13:170–179
36. Pang SK, Rohatgi A, Sopori BL, Fiegl G (1990) A comparison of minority carrier lifetime in as-grown and oxidized float-zone, magnetic Czochralski and Czochralski silicon. J Electrochem Soc 137:1977–1981
37. Jastrzebski L (1989) Heavy metal contamination during integrated-circuit processing: measurements of contamination level and internal gettering efficiency by surface photovoltage. Mat Sci Eng B4:113–121
38. Jastrzebski L, Henley W, Nuese CJ (1992) Surface photovoltage monitoring of heavy metal contamination in IC manufacturing. Solid-State Technol 35:27–33
39. Goodman AM (1961) A method for the measurement of short minority carrier diffusion lengths in semiconductors. J Appl Phys 32 :2550–2552
40. Clendenin M (November 2006) Samsung wraps up 16 NAND die in multi-chip package. EE Times Europe. http://www.eetimes.eu/193500880
41. IBM Press Release (April 2007) http://www-03.ibm.com/press/us/en/index.wss
42. Schmidt RR, Notohardjono BD (2002) High-end server low-temperature cooling. IBM J Res Dev 46(6):739–752
43. Joseph J, Gillis JD, Doherty M, Lindgren PJ, Previti-Kelly RA, Malladi RM, Wang P-C et al (2008) Through-silicon vias enable next-generation SiGe power amplifiers for wireless communications. IBM J Res Dev 52(6):635–648
44. Andry P, Tsang C, Sprogis E, Patel C, Wright S, Webb B (May 30–June 2, 2006) A CMOS-compatible process for fabricating electrical through-vias in silicon. In Proceedings of the 56th Electronic Components and Technology Conference, pp 831–837
45. Takahaski K, Taguchi Y, Tomisaka M, Yonemara H, Hoshino M, Ueno M, Egawa Y, et al (June 1–4, 2004) Process integration of 3D chip stack with vertical interconnection. In Proceedings of the 54th Electronic Components and Technology Conference, pp 601–609
46. Knickerbocher et al (November 2008) IBM J Res Dev 52(6):553
47. Zhang J et al (November 2006) IEEE Trans Semicond Manufact 19(4):437
48. Läermer F, Schilp P, Bosch Gmbh R (1996) Method of anisotropically etching silicon. U.S. Patent 5501893
49. Henry D et al (May 2008) Through silicon vias technology for CMOS image sensors packaging. In Proceedings of the 58th Electronic Components and Technology Conference, pp 556–562
50. Joseph et al (November 2008) IBM J Res Dev 52(6):635
51. Shaper et al (August 2005) IEEE Trans Adv Packag 28(3):356
52. Abhulimen et al (Nov/Dec 2008) J Vac Sci Technol B 26(6)
53. Kim B (2007) Mater Res Soc Symp Proc 970 © 2007 Mater Res Soc 0970-Y06-02
54. Andry PS et al (November 2008) IBM J Res Dev 52(6):571
55. Laviron C et al (2009) In Proceedings of the 59th Electronic Components and Technology Conference, pp 14–19
56. Chen DY (2009) Enabling 3D-IC foundry technologies for 28 nm node and beyond: through-silicon-via integration with high throughput die-to-wafer stacking, IEDM 09, Maryland, USA
57. Tanaka (2002) In Proceedings of the 52nd Electronic Component and Technology Conference
58. Tanaka (2003) In Proceedings of the 53rd Electronic Component and Technology Conference
59. Takahashi (2003) Microelctr Reliabil 43:1267–1279
60. Morrow P (May 2006) IEEE Electr Dev Lett 27(5):335

61. Wieland R (December 2005) Microelectronic engineering in Proceedings of the 9th European Workshop on Materials for Advanced Metallization 82(3–4):529–533.
62. Katti G 3D stacked ICs using Cu TSVs and die to wafer hybrid collective bonding, IEDM09
63. (2008) Handbook of 3D integration. Wiley-VCH, p 688

Chapter 7
High-Capacity NAND Flash Memories: XLC Storage and Single-Die 3D

Rino Micheloni, Luca Crippa, Alessandro Grossi, and Paolo Tessariol

Abstract NAND Flash memory has become the preferred nonvolatile choice for portable consumer electronic devices. Features such as high density, low cost, and fast write times make NAND perfectly suited for media applications where large files of sequential data need to be loaded into the memory quickly and repeatedly. This chapter starts with an overview of the basic functionalities of the NAND Flash storage. Popular devices like Flash cards are then described as an example of NAND-based systems. The last part of this chapter deals with all the state-of-the-art technologies used to reduce the equivalent bit size. In particular, the reader will find an overview of the multilevel storage and of the 3D solutions from single-die 3D architectures to packages containing multiple dies.

Keywords NAND Flash memory · SLC · MLC · Flash card · SSD · Die stacking · 3D array · Charge trap memory

Introduction

The purpose of NAND Flash memories as nonvolatile memories is to store the user data for years without requiring a supply voltage. The state-of-the-art NAND memory cell is the 1T floating gate cell. In contrast to the 1T1C DRAM cell, which consists of an access transistor and a separate capacitor as charge storage node, the 1T floating gate cell is a MOSFET whose gate is split, having a charge storage node (floating gate) in between. This charge storage node, usually polysilicon, is completely surrounded by oxide and, therefore, electrically isolated. The electrical charges stored within the floating gate represent the nonvolatile information; the program operation adds electrons to the floating gate according to the user data.

R. Micheloni (✉)
Integrated Device Technology, Agrate Brianza, Italy
e-mail: rino.micheloni@ieee.org

G. Campardo et al. (eds.), *Memory Mass Storage*,
DOI 10.1007/978-3-642-14752-4_7,

There are a lot of different floating gate cell concepts, and corresponding program and erase methods, in use today, each optimized for a specific field of applications. In the following, we focus on the state-of-the-art floating gate cell for mainstream data Flash applications.

Flash memory is a type of nonvolatile memory that can be erased in large blocks (erase blocks) and reprogrammed in either byte/word or pages. Different Flash implementations exist: NOR and NAND are the most common, but AND and NROM exist as well [1]. This kind of device is used in memory cards and USB Flash drives for storage and data transfer between computers and digital systems. Other applications include PDAs, notebooks, MP3 players, digital cameras, and cellular phones. In the past years, Flash gained some popularity also in the game console market. In the future it is expected that Flash-based systems like Solid-State Disks (SSD) will increasingly replace conventional Hard Disk Drives (HDD).

A memory system, e.g., a Flash card like an SD card, is a small "system in package" built around the Flash memory and combined with a microcontroller, featuring a technology-independent interface.

The popularity of the Flash memory is mainly based on its distinctive characteristic of being nonvolatile (i.e., stored information is retained even when not powered, differently from other kinds of memories, like DRAM). Flash memory also offers fast access times in read (although not as fast as DRAM) and better mechanical shock resistance compared to hard disks. The high storage capacity of a Flash-based system is typically achieved by means of advanced packaging techniques. Furthermore, the smart memory management carried out by the microcontroller allows a high endurance and an appealing reliability of the resulting system.

As manufacturers increase the density of data storage in Flash devices, the size of individual memory cells becomes smaller and the number of electrons stored in the cell decreases. Moreover, coupling between adjacent cells and quantum effects can change the write characteristics of cells, making it more difficult to design devices able to guarantee reasonable data integrity. As we will see in the remaining sections of this chapter, on top of the technology shrink there are other solutions aimed at increasing the system memory capacity: multilevel storage, multichip stacking, and 3D arrays.

NAND Flash Memory

Flash memory contains an array of floating gate transistors: each of them acts as memory cell. In Single Level Cell (SLC) devices, each memory cell stores one bit of information; in Multi-Level Cell (MLC) devices, more than one bit per cell can be stored [2]. Figure 7.1 shows a schematic cross section of a floating gate memory cell. The floating gate is used to store electrons: changing the number of electrons results in a different threshold voltage V_{TH} of the associated transistor and, therefore, in a different current sunk by the cell under fixed biasing conditions.

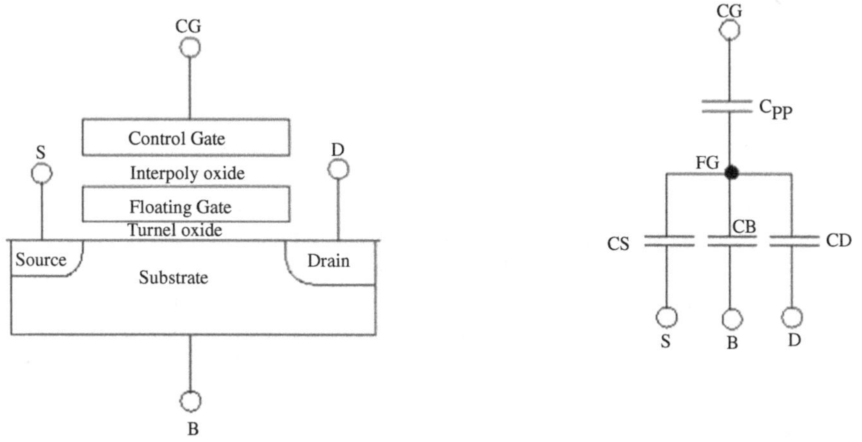

Fig. 7.1 Floating gate memory cell (*left*) and the corresponding capacitive model (*right*)

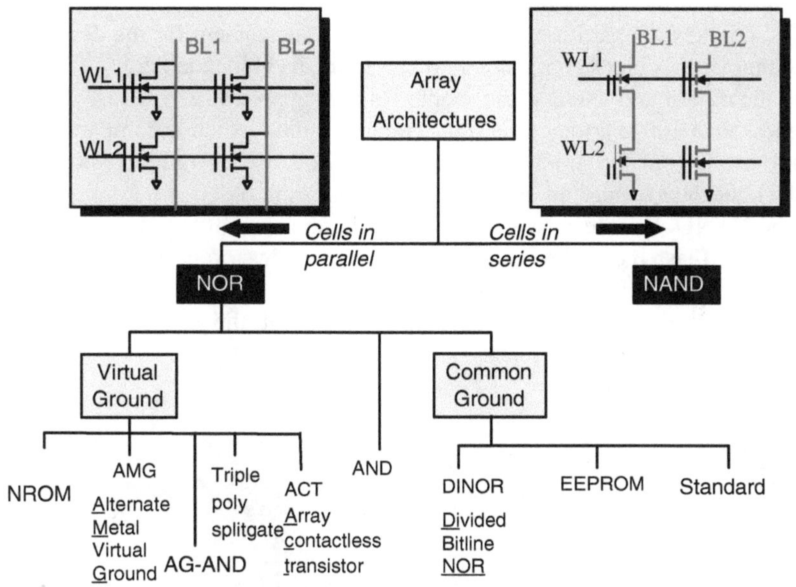

Fig. 7.2 Flash memory array architectures. *Source*: Forward Insights

There are two dominant kinds of Flash memories, namely NAND and NOR. The name is related to the topology used for the array of cells, as it is for the corresponding logic gates (Fig. 7.2). NAND Flash memory is the core of the removable USB interface storage devices known as USB drives, as well as of most Flash card formats available today on the market.

NAND Flash uses Fowler–Nordheim tunnel injection for writing and Fowler–Nordheim tunnel release for erasing.

Due to their construction principles, NAND Flash memories cannot provide "execute in place." These memories are in fact accessed like hard disks and therefore are very suitable for use in mass storage devices such as memory cards.

While programming is performed on a page basis, erase can only be performed on a block basis (i.e., a group of pages). Pages are typically 2,048 or 4,096 bytes in size. Associated with each page there are a few bytes that can be used for storage of error detection and correction checksum as well as for administration as requested by the external microcontroller. Figure 7.3 shows a schematic representation of the memory organization of a NAND Flash memory.

Another limitation is the finite number of write–erase cycles (manufacturers usually guarantee 100,000 write–erase cycles for SLC NAND). Furthermore, in order to increase manufacturing yield and, therefore, reduce NAND memory cost, NAND devices are shipped from the factory with some bad blocks (i.e., blocks where some locations do not guarantee the standard level of reliability when used) which are identified and marked according to a specified bad block marking strategy.

The first physical block of a NAND memory (block 0) is always guaranteed to be readable and free from errors when device is shipped to customers. Hence, code, all vital pointers for partitioning, and bad block management for the device can be located inside this block (typically a pointer to the bad block tables).

On the other hand, NOR Flash memories are capable of a very fast read access time (less than 100 ns); they offer a programming time which is comparable to that of the NAND (but the amount of programmed bit per operation is considerably smaller) but they feature an erase time which is some order of magnitudes higher than the NAND. For these reasons, and due to the capability to "execute in place" (XIP), NOR Flash memories are more suitable for code storage and they will not be considered in the following sections.

Table 7.1 shows the performance parameters of the two above-mentioned memories.

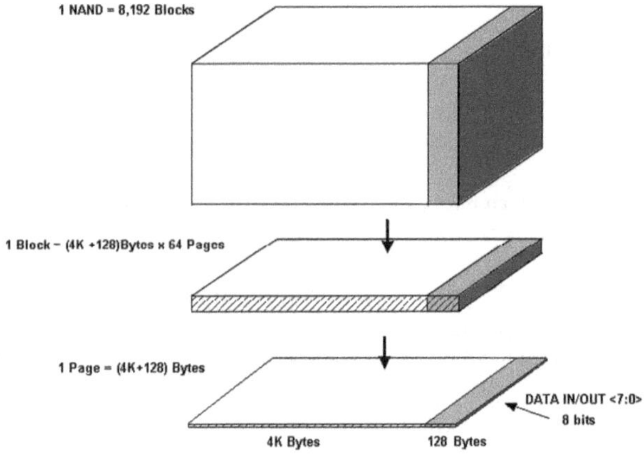

Fig. 7.3 Schematic representation of the memory organization of a NAND Flash memory

Table 7.1 NOR and NAND performances

	NOR 1 bit/cell	NOR 2 bits/cell	NAND 1 bit/cell	NAND 2 bits/cell
Page size	32 B	1 KB	4 KB	4 KB
Read access time	< 70 ns	< 80 ns	25 μs	50 μs
Program time	9 μs/word	4.2 μs/word	150 μs/page	600 μs/page
Erase time	1 s/sector	1 s/sector	2 ms/sector	2 ms/sector

The NAND Array

Modern nonvolatile memories contain billions of memory cells, organized in a memory array. Row and column decoders address the cells in the memory array, while write and read circuits control the data input and output operations.

The core components of the NAND Flash memory are shown in Fig. 7.4. In the NAND array the cells are organized in pages, blocks, and planes. A sense amplifier (page buffer) is connected to a pair of bitlines which, in a shielded bitline scheme [3], are addressed alternately. The bitlines are connected through the string select transistor to a number of memory transistors in series, which form the NAND string (stack). The row decoder controls the string select lines and the wordlines of each block.

The page size is proportional to the number of cells per wordline and is given by the wordline length (limited by its *RC* delay) and the bitline pitch (limited by lithography). The block size is proportional to the number of cells per NAND string and is limited by the circuit ability of handling its series resistance during the read operation (see section "Read"). By increasing the page size, and therefore the number of cells addressed in parallel, the read and write performance can be increased. By increasing the block size, the cell array efficiency is increased since more cells share both the necessary select gate and contact area.

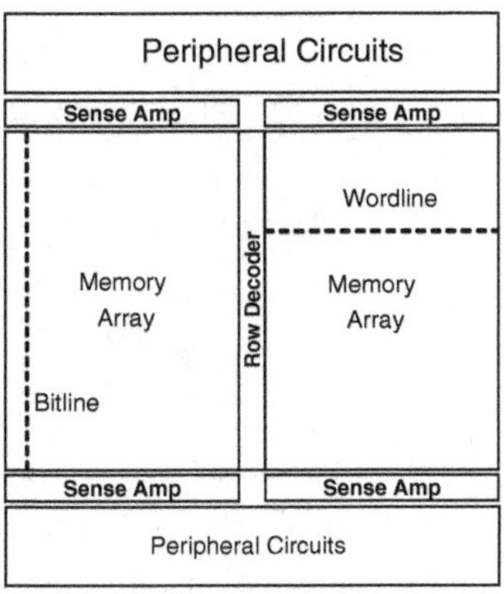

Fig. 7.4 NAND Flash memory floorplan

Read

The reading operation is designed to address specific memory cells within the array and extrapolate the information stored therein. In the case of a NAND-type Flash memory, the memory cells are connected in series, in groups (strings) of 2^k cells, up to 64. In Fig. 7.5, the string of a NAND Flash memory is illustrated: M_{BLS} and M_{SLS} NMOS selector transistors connect the string to bitline BL and to source line SL, respectively.

As in other types of Flash memories, the stored information is associated with the cell's threshold voltage V_{TH}; in Fig. 7.6, the threshold voltage distributions of cells containing one logic bit are shown.

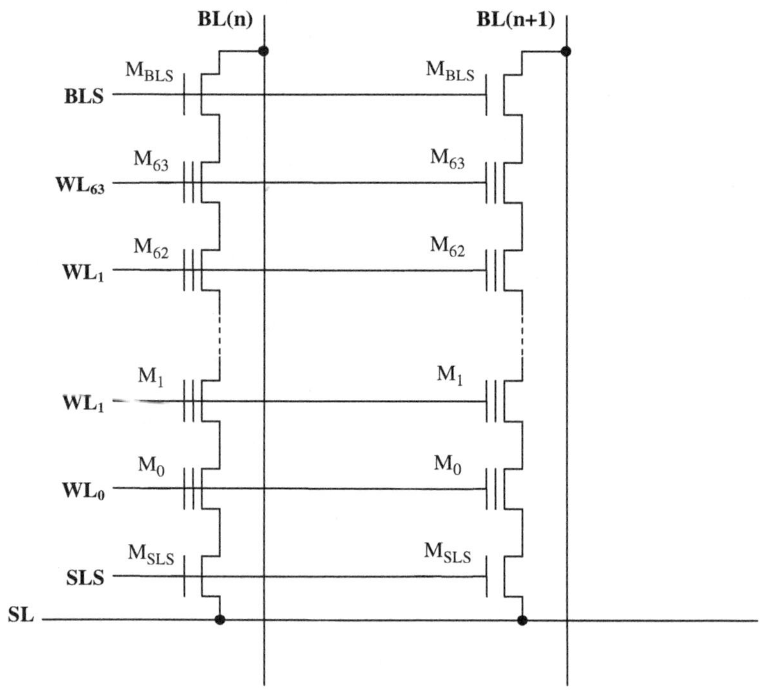

Fig. 7.5 Two strings in a NAND architecture

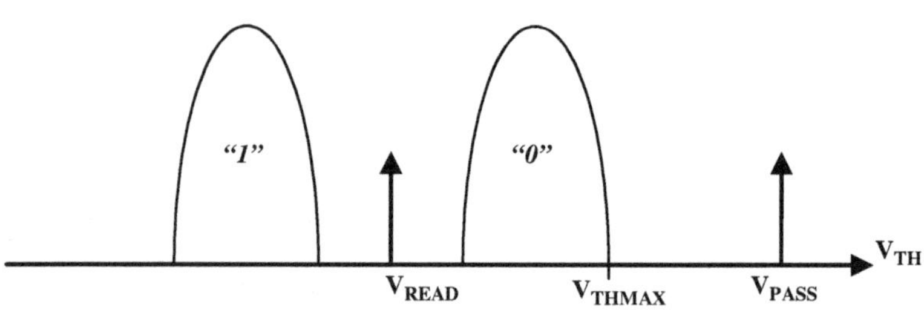

Fig. 7.6 Threshold voltage distributions for erased and programmed cells

If the cell has a V_{TH} belonging to the erased distribution, it contains a logic "1"; otherwise, if it belongs to the written distribution, it contains a logic "0." Cells containing n bits of information have 2^n different levels of V_{TH}.

Flash cells act like usual MOS transistors. Given a fixed gate voltage, the cell current is a function of its threshold voltage. Therefore, through a current measure, it is possible to understand which V_{TH} distribution the memory cell belongs to.

The fact that a memory cell belongs to a string made up of other cells has some issues. First of all, the unselected memory cells must be biased in a way that their threshold voltages do not affect the current of the addressed cell. In other words, the unselected cells must behave as pass transistors. As a result, their gate must be driven to a voltage (commonly known as V_{PASS}) higher than the maximum possible V_{TH}. In Fig. 7.6, V_{PASS} has to be higher than V_{THMAX}.

However, the presence of 2^n–1 transistors in series has a limiting effect (saturation) on the current's maximum value; this maximum current is, therefore, much lower than the one available in NOR-type Flash memories.

Figure 7.7 shows the *I–V* (current–voltage) characteristic of a NAND cell (string): V_{READ} is applied to the selected gate while V_{PASS} biases the unselected gates. V_{PASS} is a fixed voltage. Three main string working regions can be highlighted:

1. Region A: the addressed cell is not in a conductive state.
2. Region B: V_{READ} makes the addressed cell more and more conductive.
3. Region C: the cell is completely on, but the series resistance of the pass transistors (unselected cells) limits the current to I_{SSAT}.

The string current in region C can be estimated as

$$I_{SSAT} = \frac{V_{BL}}{(n-1)R_{ON}}, \tag{7.1}$$

where R_{ON} is the series resistance of a single memory cell, V_{BL} is the voltage applied to the bitline, and n is the number of cells in the string. R_{ON}, at a first approximation, is the resistance of a transistor working in the ohmic region.

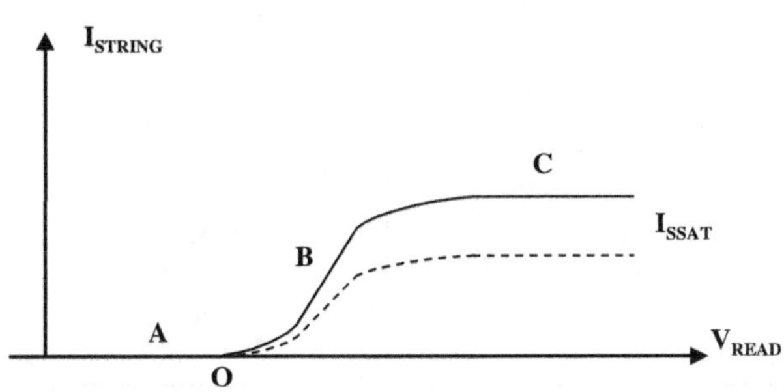

Fig. 7.7 Cell current characteristics vs. gate voltage

For a MOS transistor in ohmic region the following equation holds true:

$$I_{\mathrm{D}} = k \cdot \left[(V_{\mathrm{GS}} - V_{\mathrm{TH}}) \cdot V_{\mathrm{DS}} - \frac{V_{\mathrm{DS}}^2}{2} \right]. \tag{7.2}$$

For small V_{DS} values, as in our case, (7.2) may be simplified as

$$I_{\mathrm{D}} = k \left[(V_{\mathrm{GS}} - V_{\mathrm{TH}}) \cdot V_{\mathrm{DS}} \right]. \tag{7.3}$$

Therefore, R_{ON} is equivalent to

$$R_{\mathrm{ON}} = \frac{V_{\mathrm{DS}}}{I_{\mathrm{D}}} = \frac{1}{k\,(V_{\mathrm{GS}} - V_{\mathrm{TH}})}. \tag{7.4}$$

Equation (7.4) shows that R_{ON} is a function of V_{TH}. In other words, I_{SSAT} depends on the V_{TH} values of the n cells in series. When all the cells are programmed to V_{THMAX}, R_{ON} takes its maximum value (dashed line in Fig. 7.7). R_{ON} influences the I–V characteristic also in region B but in a more negligible way. In order to reduce the dependency from R_{ON}, the cell has to be read in region B as near as possible to point O.

The order of magnitude of the saturation current, in the state-of-the-art NAND technologies, is a few hundreds of nanoamperes, which means a reading current of some tens of nanoamperes. It is very hard to sense such small currents with the standard techniques used in NOR-type Flash memories, where the reading current is, at least, in the order of some microamperes. Moreover, in NAND devices, tens of thousands of strings are read in parallel. Therefore, tens of thousands of reading circuits are needed. Due to the multiplicity, a single reading circuit has to guarantee a full functionality with a very low area impact. As a matter of fact, the first memory NAND prototypes used traditional sensing methods, since the said currents were in the order of tens of microamperes [4].

The reading method of the Flash NAND memories consists in integrating the cell current on a capacitor in a fixed time (Fig. 7.8). The voltage ΔV_{C} across a

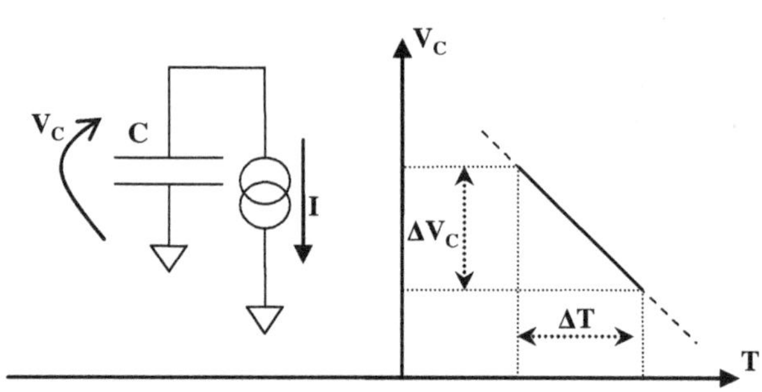

Fig. 7.8 Capacitor discharge through a constant current source

capacitor C, charged by a constant current I for a time period ΔT, is described by the following equation:

$$\Delta V_C = \frac{I}{C}\Delta T. \tag{7.5}$$

Since the cell current is related to its V_{TH}, the final voltage on the capacitor (ΔV) is a function of V_{TH} too.

There are different reading techniques, starting from the one using the bitline parasitic capacitor, ending with the most recent sensing technique which integrates the current on a little dedicated capacitor. The above-mentioned techniques can be used both in SLC and in MLC NAND memories. In the MLC case, multiple basic reading operations are performed at different gate voltages.

Historically, the first reading technique used the parasitic capacitor of the bitline as the element of the cell current integration [5–7].

In Fig. 7.9, the basic scheme is shown. V_{PRE} is a constant voltage. At the beginning, C_{BL} is charged up to V_{PRE} and then it is left floating (T_0). At T_1 the string is enabled to sink current (I_{CELL}) from the bitline capacitor. The cell gate is biased at V_{READ}. If the cell is erased, the sunk current is higher than (or equal to) I_{ERAMIN}. A programmed cell sinks a current lower than I_{ERAMIN} (it can also be equal to zero). C_{BL} is connected to a sensing element (comparator) with a trigger voltage V_{THC} equal to V_{SEN}. Since I_{ERAMIN}, C_{BL}, V_{PRE}, and V_{SEN} are known, it follows that the

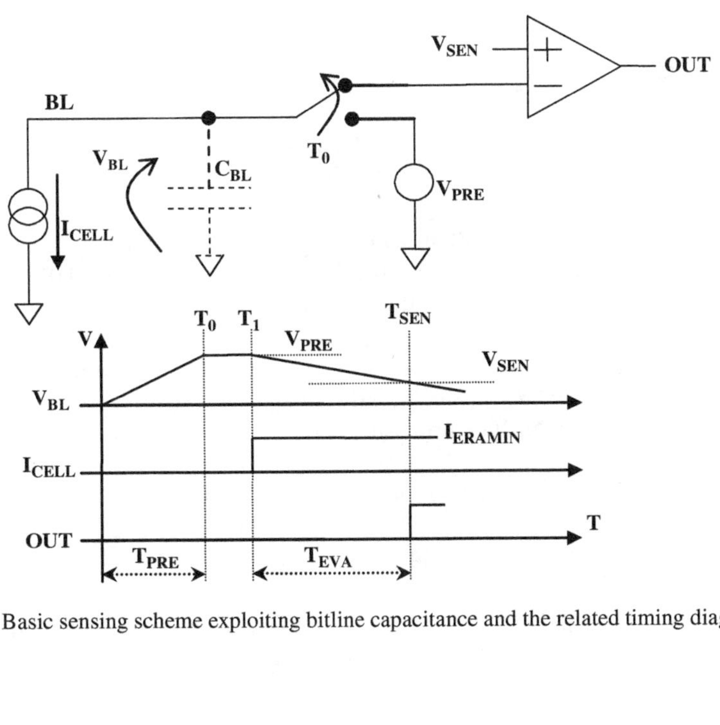

Fig. 7.9 Basic sensing scheme exploiting bitline capacitance and the related timing diagram

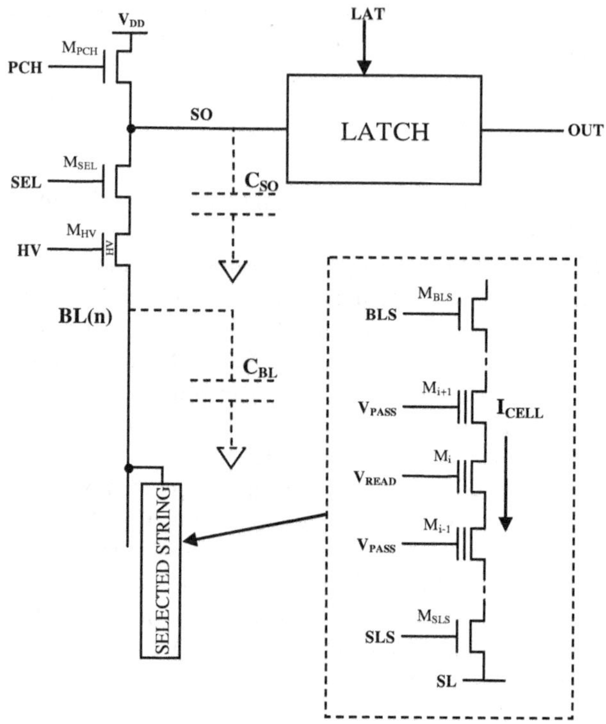

Fig. 7.10 Basic elements of the page buffer (sense amplifier)

shortest time (T_{EVAL}) to discharge the bitline capacitor is equal to

$$T_{EVAL} = C_{BL}\frac{V_{PRE} - V_{SEN}}{I_{ERAMIN}}. \tag{7.6}$$

If the cell belongs to the written distribution, the bitline capacitor will not discharge below V_{SEN} during T_{EVAL}. As a result, the output node (OUT) of the voltage comparator remains at 0. Otherwise, if the cell is erased, V_{BL} drops below V_{SEN} and the OUT signal is set to 1.

The basic page buffer structure is sketched in Fig. 7.10. During the precharge phase, T_{PRE}, M_{SEL}, and M_{PCH} are biased to V_{PRE} and $V_{DD} + V_{THN}$. V_{THN} is the threshold voltage of an NMOS transistor and V_{DD} is the device's power supply voltage.

As a consequence, C_{BL} is charged to the following value:

$$V_{BL} = V_{PRE} - V_{THN}. \tag{7.7}$$

During this phase, the SO node charges up to V_{DD}. Since V_{GS} and V_{DS} can be higher than 20–22 V, M_{HV} has to be a high-voltage (HV) transistor. In fact, during the erase phase, the bitlines are at about 20 V and M_{HV} acts as a protection element for the page buffer's low-voltage components. Instead, during the reading phase, M_{HV} is biased at a voltage that makes it behave as pass transistor. Moreover, during the precharge phase, the appropriate V_{READ} and V_{PASS} are applied to the string. M_{BLS} is biased to a voltage (generally V_{DD}) that makes it work as pass transistor. Instead, M_{SLS} is turned off in order to avoid cross-current consumption through the string.

Typically, V_{BL} is around 1 V. From (7.7), V_{PRE} values are approximately 1.4–1.9 V, depending on V_{THN} (NMOS threshold voltage). The bitline precharge phase usually lasts 5–10 μs and depends on many factors, above all the value of the distributed bitline parasitic RC.

Sometimes this precharge phase is intentionally slowed down to avoid high current peaks from V_{DD}. In order to achieve this, the M_{PCH} gate could be biased with a voltage ramp from GND to $V_{DD} + V_{THN}$.

At the end of the precharge phase, PCH and SEL are switched to 0. As a consequence, the bitline and the SO node parasitic capacitor are left floating at a voltage of $V_{PRE} - V_{THN}$ and V_{DD}, respectively. M_{SL} is then biased in order to behave as pass transistor. In this way the string is enabled to sink (or not) current from the bitline capacitor.

At this point, the evaluation phase starts. If the cell has a V_{TH} higher than V_{READ}, no current flows and the bitline capacitor maintains its precharged value. Otherwise, if the cell has a V_{TH} lower than V_{READ}, the current flows and the bitline discharges.

Disturb effects alter the memory transistor's threshold voltage unintentionally during memory access operations. Since it is essential to optimize area consumption, cells are arranged in a matrix, sharing voltages. During the read and program operations, positive voltages are applied to the gate nodes of the memory transistors, wherein the channel is at a lower potential or even grounded. Therefore, this condition is called gate disturb and is the most common disturb mechanism in NAND Flash memory arrays. The positive voltage is able to cause Fowler–Nordheim tunneling of electrons to the floating gate.

Read disturb occurs when reading many times the same cell (wordline) without any erase operation in between. All the cells belonging to the same string of the cell to be read must be driven in an ON state, independently from their stored charge. The relatively high V_{pass} bias applied on the control gate, and the sequence of V_{pass} pulses applied during successive read operations, may trigger SILC effects in some cells that, therefore, may gain charge. These cells suffer a positive shift of their threshold voltage that may lead to read errors (when addressed). Since the SILC effect is not symmetrical, the cells that may be affected by SILC effects induced by read disturbs are not necessarily the same that exhibit data retention problems. Figure 7.11 shows the typical read disturb configuration.

The probability of suffering a read disturb increases with the number of write/erase cycles (i.e., toward the end of the memory useful lifetime) and it is higher in damaged cells. Read disturbs do not provoke permanent oxide damages: erase operation restores the initial conditions.

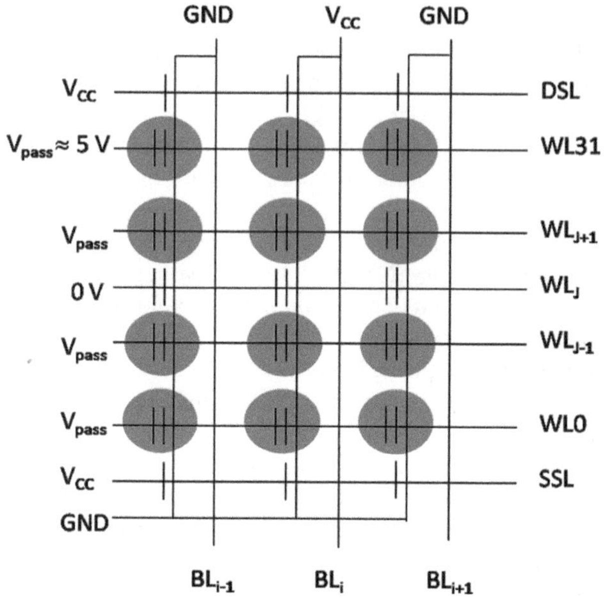

Fig. 7.11 Read disturb

Program

V_{TH} is modified by means of the *Incremental Step Pulse Programming* (ISPP) algorithm (Fig. 7.12): a voltage step (whose amplitude and duration are predefined) is applied to the gate of the cell. Afterward, a verify operation is performed, in order to check whether V_{THR} has exceeded a predefined voltage value (V_{VFY}). If the verify operation is successful, the cell has reached the desired state and it is excluded from

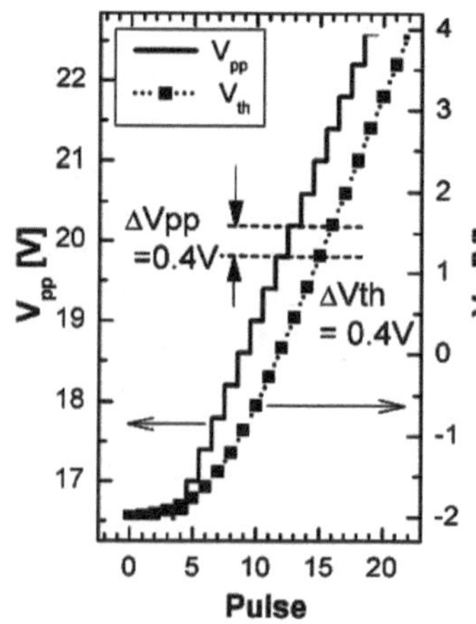

Fig. 7.12 Incremental step pulse programming (ISPP): constant V_{TH} shift

the following program pulses. Otherwise, another cycle of ISPP is applied to the cell; this time the program voltage is incremented by ΔV_{pp}.

During the program operation, the cells share the high programming voltage on the selected wordline but the program operation has to be bit-selective. Therefore, a high channel potential is needed to reduce the voltage drop across the tunneling dielectric and prevent the electrons tunneling from the channel to the floating gate, as indicated by Fig. 7.13a. In the first NAND devices, the channel was charged by applying 8 V to the bitlines of the program inhibited NAND strings. This method suffers from several disadvantages [8], especially power consumption and high stress on the oxide between adjacent bitlines.

The self-boost program-inhibit scheme consumes less power. By charging the string select lines and the bitlines connected to inhibited cells to V_{cc}, the select transistors are diode connected (Fig. 7.13b). When the wordline potentials rise (selected wordline to V_{pp} and unselected wordlines to V_{ppass}), the channel potential is boosted by the parasitic series capacitance of control gate, floating gate, channel, and bulk.

In fact, when the voltage of the channel exceeds $V_{cc} - V_{TH,SSL}$, then SSL transistors are reverse-biased and the channel of the NAND string becomes a floating node.

Two important typologies of disturbs are related to the program operation: the *Pass disturb* and the *Program disturb,* which are shown in Figs. 7.14 and 7.15, respectively.

The former is similar to the read disturb and affects cells belonging to the same string of a cell to be programmed.

With respect to the read disturb, the Pass one is characterized by the higher V_{pass} voltage applied to cells that are not to be programmed (thus enhancing both the

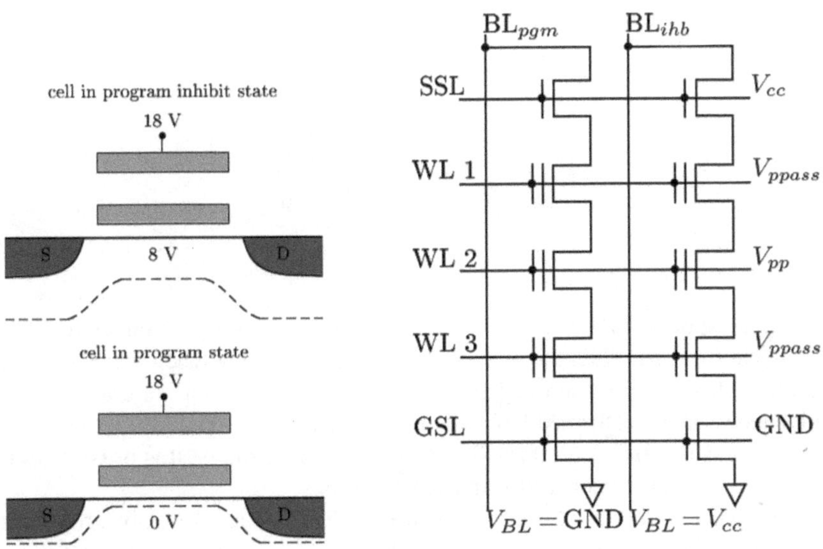

Fig. 7.13 Self-boosted program inhibit scheme. (**a**) Conditions of cells selected for program and program inhibit. (**b**) Biased conditions during the self-boosted program inhibit

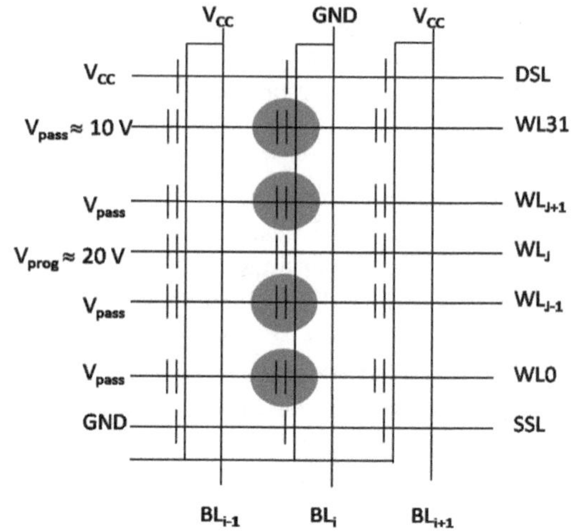

Fig. 7.14 Pass disturb during programming: the cells affected by the disturb are highlighted in gray

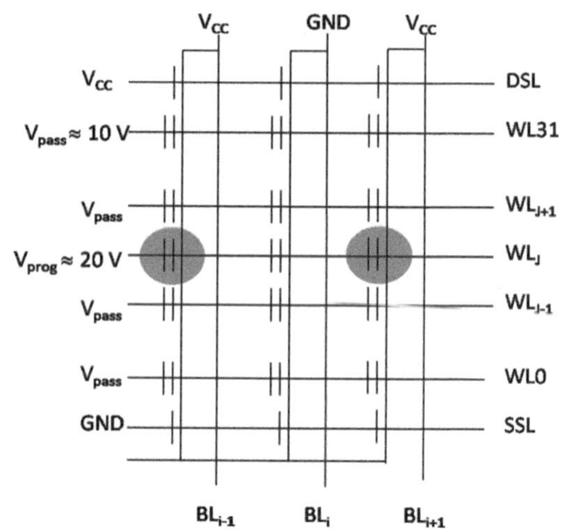

Fig. 7.15 Program disturb during programming: the cells affected by the disturb are highlighted in *gray*

electric field applied to the tunnel oxides and the probability of undesired charge transfer). On the other side, the worst case for Pass disturb is when there is only one remaining cell to be programmed in the NAND string. In fact, an erase operation must be necessarily performed before any other reprogram. Therefore, the Pass disturb duration is much shorter than the stress read time: usually, it is possible to read one memory block between 100 k and 1,000 k times.

The Program disturb, on the contrary, affects cells that are not to be programmed and belong to the same wordline of those that are to be programmed. In this case, the Program disturb is strongly related to the voltages and pulse sequences used for the self-boosting techniques.

These soft programming mechanisms disturb the memory transistor's state. Since the tunneling current has an exponential dependency from the voltage drop across the tunneling dielectric, the soft programming time required to cause the same V_{TH} shift as during the standard programming pulse is much longer. However, this fact limits the *Number of Programs* (NOP) which can be realized without influencing the stored memory state. Since the floating gate potential and, therefore, the electric field across the tunneling dielectric are dependent on the cells' V_{TH}, erased cells are most vulnerable to soft programming disturb effects.

Erase

The erase operation simultaneously resets the information of all the cells belonging to one block.

Tables 7.2 and 7.3 summarize the erase voltages. During the erase pulse, all the wordlines belonging to the selected block are kept at ground, the matrix ip-well must rise (through a staircase) to 23 V, and all the other nodes are floating. This phase lasts almost a millisecond and it is the phase when the actual electrical erase takes place.

Since the matrix ip-well (as well as the surrounding n-well) is common to all the blocks, it reaches high voltages also for the unselected blocks. In order to prevent an unintentional erase on those blocks, wordlines are left floating; in this way, their voltage can rise thanks to the capacitive coupling between the wordline layer and

Table 7.2 Electrical erase pulse voltages for the selected block

	T_0	T_1	T_2	T_3	T_4
BLeven	Float	Float	Float	Float	Float
BLodd	Float	Float	Float	Float	Float
DSL	Float	Float	Float	Float	Float
WLs	0 V	0 V	0 V	0 V	0 V
SSL	Float	Float	Float	Float	Float
SL	Float	Float	Float	Float	Float
ip-well	0 V	V_{ERASE}	V_{ERASE}	0 V	0 V

Table 7.3 Electrical erase pulse voltages for unselected blocks

	T_0	T_1	T_2	T_3	T_4
BLeven	Float	Float	Float	Float	Float
BLodd	Float	Float	Float	Float	Float
DSL	Float	Float	Float	Float	Float
WLs	Float	Float	Float	Float	Float
SSL	Float	Float	Float	Float	Float
SL	Float	Float	Float	Float	Float
ip-well	0 V	V_{ERASE}	V_{ERASE}	0 V	0 V

the underneath matrix layer. Of course, the voltage difference between wordlines and ip-well should be low enough to avoid Fowler–Nordheim tunneling.

After each erase pulse, an erase verify (EV) follows. During this phase, all the wordlines are kept at ground. The purpose is verifying if there are some cells that have a V_{TH} higher than 0 V, so that another erase pulse can be applied. If EV is not successful for some columns of the block, it means that there are some columns programmed too. If the maximum number of erase pulses is reached (typically 4), then the erase exits with a fail. Otherwise, the voltage applied to the matrix ip-well is incremented by ΔV_E and another erase pulse follows.

MLC and XLC Storage

The obvious advantage of a 2 bit/cell implementation (MLC) with respect to a 1 bit/cell device (SLC) is that the area occupation of the matrix is half as much; on the other hand, the area of the periphery circuits, both analog and digital, increases. This is mainly due to the fact that the multilevel approach requires higher voltages for program (and therefore bigger charge pumps), higher precision and better performance in the generation of both the analog signals and the timings, and an increase in the complexity of the algorithms.

Figure 7.16 shows an example of how 2 bits are associated with the four read threshold distributions stored in the cell and how the set of programmed distributions

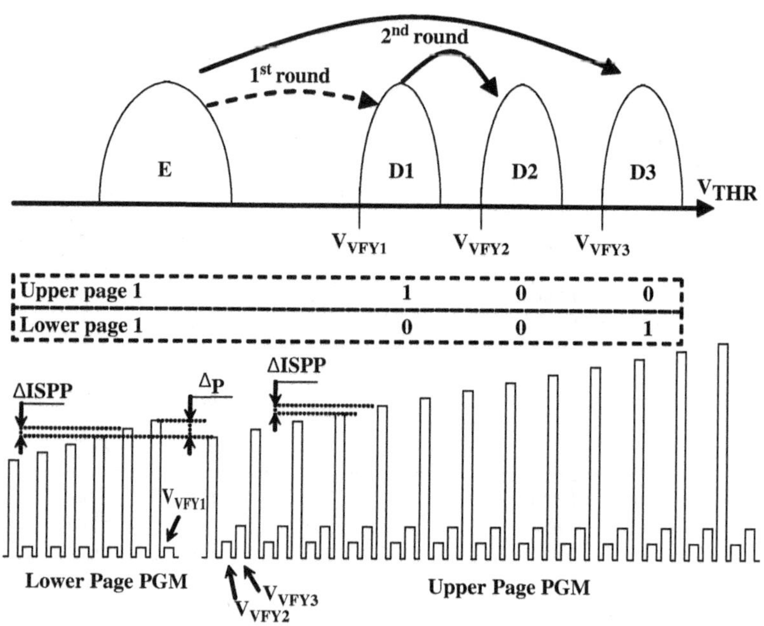

Fig. 7.16 Two rounds of MLC program operation

is built starting from the erased state "E." In this case the multilevel is achieved in two distinct rounds, one for each bit to be stored [9–11].

In the first round, the so-called lower-page (associated with the *Least Significant Bit* – LSB) is programmed. If the bit is "1," the read threshold of the cell V_{TH} does not change and, therefore, the cell remains in the erased state, E. If the bit is "0," V_{TH} is increased until it reaches the D1 state.

In the second round, the upper-page (associated with the *Most Significant Bit* – MSB) is programmed. If the bit is "1," V_{TH} does not change and, therefore, the cell remains either in the erased state, E, or in the D1 state, depending on the value of the lower-page.

When MSB is "0," V_{TH} is programmed as follows:

– if, during the first round, the cell remained in E state, then V_{TH} is incremented to D3;
– if, during the first round, the cell was programmed to D1, then, in the second round, V_{TH} reaches D2.

Even in this case, the program operation uses ISPP, and the verify voltages are V_{VFY2} and V_{VFY3}. Lower-page programming only needs the information related to LSB, while for the upper-page it is necessary to know both the starting distribution (LSB) and the MSB.

Because of technological variations, V_{TH} is not perfectly related to the amplitude of the program pulse (during ISPP): there are "fast" cells which reach the desired distribution with few ISPP pulses, while other "slow" cells require more pulses.

The amplitude of the first program pulse ($V_{PGMLSB0}$) of the lower-page should not allow the threshold V_{THR} of the "fastest" cell to exceed V_{VFY1}. Should it happen, an undesired widening of distribution D2 occurs or, in the worst case scenario, V_{THR} might reach D2 distribution at once.

Typical $V_{PGMLSB0}$ is around 16 V. In case of program of "slow" cells from E to D1, the last programming step needs values as high as 19 V. Assuming ΔISPP equal to 250 mV, it takes 12 steps to move from 16 to 19 V.

Similarly, the starting pulse of the upper-page $V_{PGMMSB0}$ should have an amplitude such that the "fastest" cell does not go beyond V_{VFY2}.

$$V_{PGMMSB0} = V_{PGMLSB0} + (V_{VFY2} - V_{VFY1}). \tag{7.8}$$

The value of $V_{VFY2} - V_{VFY1}$ is typically around 1 V and, therefore, the initial voltage is about 17 V. As shown in Fig. 7.16, the upper-page ISPP does not start from the last voltage used for the lower-page programming, but it begins at $V_{PGMLSB0} - \Delta_P$. For example, instead of starting at 19 V, it could start at 17 V, eight steps below.

Driven by cost, Flash manufacturers are now developing 3 bit/cell (8 V_{TH}) distributions and 4 bit/cell (16 V_{TH}) distributions [12–14]. Three and 4 bits/cell are usually referred to as XLC (8LC and 16LC, respectively). Unfortunately, due to reliability reasons, the V_{TH} window remains the MLC one; in fact, the highest verification level must be low enough to prevent bit failures caused by program disturb

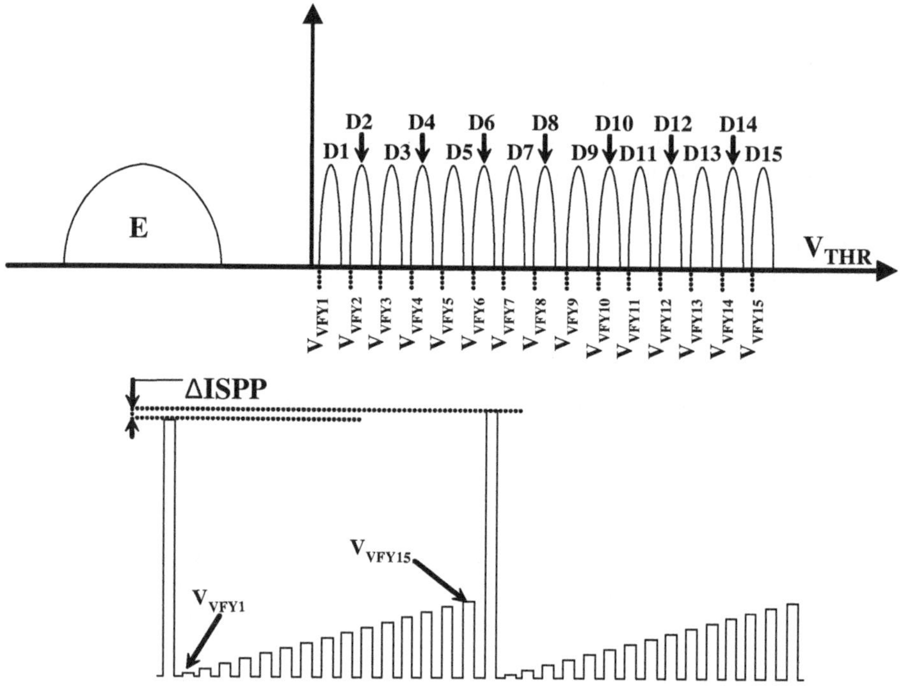

Fig. 7.17 A 4 bits/cell programming algorithm

and read disturb. The more states a memory cell is made to store, the more finely divided is its V_{TH} window.

Of course, the main drawback is a slow program time. As the distribution width needs to be tighter, ISSP program step is smaller and the number of verify operations increases, as depicted in Fig. 7.17.

NAND-Based Systems

Flash cards, USB sticks, and *Solid-State Disks* (SSDs) are definitely the most known examples of electronic systems based on NAND Flash.

Several types of memory cards are available on the market [15–17], with different user interfaces and form factors, depending on the needs of the target application, e.g., mobile phones need very small-sized removable media like μSD. On the other hand, digital cameras can accept larger sizes as memory capacity is more important (CF, SD, MMC). Figure 7.18 shows different types of Flash cards.

The interfaces of the Flash cards (including USB sticks) support several protocols: parallel or serial, synchronous or asynchronous. Moreover, the Flash cards support hot insertion and hot extraction procedures, which requires the ability to manage sudden loss of power supply while guaranteeing the validity of stored data.

Fig. 7.18 Flash card form factors: (from the *left-hand side*) μSD, Mini-SD, SD, MMC, CF, and a miniaturized version of a USB device

Fig. 7.19 Block diagram of a typical memory card (or SSD)

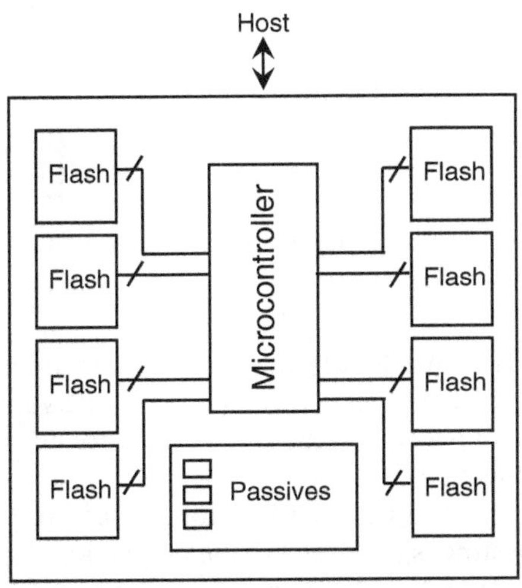

For the larger form factors, the card is a complete, small system where every component is soldered on a PCB and is independently packaged (Fig. 7.19). For example, the NAND Flash memories are tested devices in TSOP packages. It is also possible to include some additional components: for instance, an external DC–DC converter can be added in order to derive an internal power supply (CompactFlash cards can work at either 5 or 3.3 V), or a quartz can be used for a better clock precision. Usually, reasonable filter capacitors are inserted for stabilizing the power supply. Same considerations apply to SSDs.

For small form factors like μSD, the size of the card is comparable to that of the NAND die. Therefore, the memory chip is mounted as bare die on a small substrate. Moreover, the die thickness has to be reduced in order to comply with the thickness of μSD, considering that several dies are stacked, i.e., mounted one on top of each

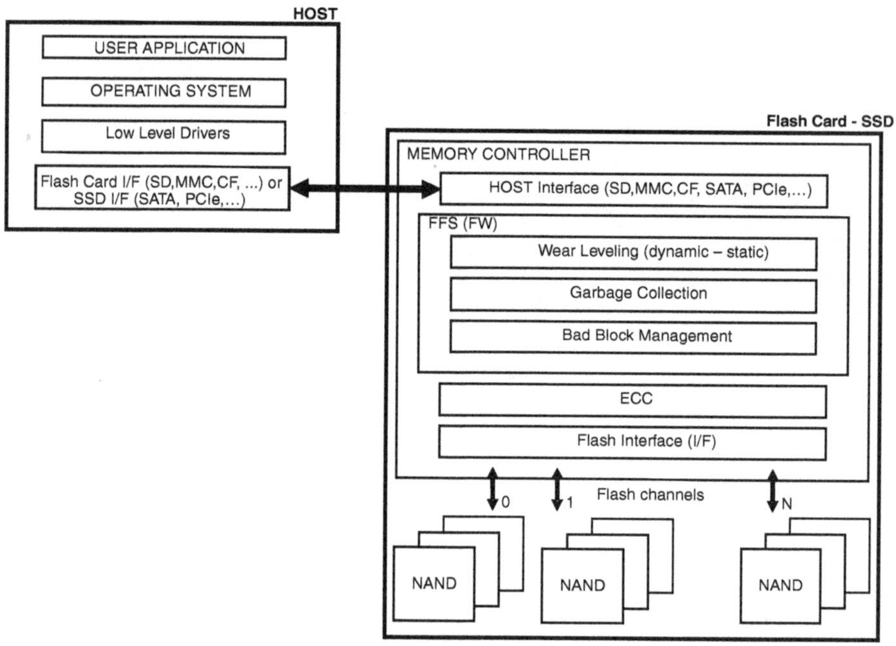

Fig. 7.20 Functional representation of a Flash card (or SSD)

other. All these issues cause a severe limitation to the maximum capacity of the card; moreover external components, like voltage regulators and quartz, cannot be used. In other words, the memory controller of the card has to implement all the required functions.

The assembly stress for small form factors is quite high and, therefore, system testing is at the end of the production. Hence, production cost is higher.

Figure 7.20 shows a schematic representation of a memory card or SSD; two types of components can be identified: the memory controller and the Flash memory components. Actual implementation may vary, but the functions described in the next sections are always present.

Memory Controller

The aim of the memory controller is twofold:

1. to provide the most suitable interface and protocol toward both the host and the Flash memories;
2. to efficiently handle data, maximizing transfer speed, data integrity, and information retention.

In order to carry out such tasks, an application-specific device is designed, embedding a standard processor – usually 8/16 bits – together with dedicated hardware to handle timing-critical tasks.

For the sake of discussion, the memory controller can be divided into four parts, which are implemented either in hardware or in firmware. Proceeding from the host to the Flash, the first part is the host interface, which implements the required industry-standard protocol (MMC, SD, CF, etc.), thus ensuring both logical and electrical interoperability between Flash cards and hosts. This block is a mix of hardware – buffers, drivers, etc. – and firmware – command decoding performed by the embedded processor – which decodes the command sequence invoked by the host and handles the data flow to/from the Flash memories.

The second part is the Flash File System (FFS) [18], that is, the file system which enables the use of Flash cards, SSDs, and USB sticks like magnetic disks. For instance, sequential memory access on a multitude of subsectors which constitute a file is organized by linked lists (stored on the Flash card itself) which are used by the host to build the File Allocation Table (FAT). Just like the case with HDD, defragmentation can be invoked by the host in order to optimize access speed and data organization.

The FFS is implemented in firmware and manages (at NAND Flash level) all data accesses to/from the host with a minimum granularity of 512 bytes (one subsector). This block is of utmost importance during data transfer operations. As already outlined in the previous section, Flash memories have intrinsic limitations, some of which can be overcome by performing erase operations, while some others lead to unrecoverable situations and require specific management.

The FFS is usually implemented in the form of firmware inside the controller, each sublayer performing a specific function. The main functions are Wear Leveling Management, Garbage Collection, and Bad Block Management. For all these functions, tables are widely used in order to map sectors and pages from logical to physical (Flash Translation Layer, FTL) [19, 20]. From the host perspective, data are transparently written and overwritten inside a given logical sector: due to Flash limitations, overwrite on the same page is not possible, therefore a new page (sector) must be allocated in the physical block and the previous one is marked as invalid. It is clear that, at some point in time, the current physical block becomes full and therefore a second one (Buffer) is assigned to the same logical block.

The required translation tables are always stored on the memory card itself, thus reducing the overall card capacity.

Wear Leveling

Usually, not all the information stored within the same memory location changes with the same frequency: some data are often updated while others remain always the same for a very long time – in the extreme case, for the whole life of the device. It is clear that the blocks containing frequently updated information are stressed with a large number of write/erase cycles, while the blocks containing information updated very rarely are much less stressed.

In order to mitigate disturbs, it is important to keep the aging of each page/block as minimum and as uniform as possible, that is, the number of both read and program cycles applied to each page must be monitored. Furthermore, the maximum number of allowed program/erase cycles for a block (i.e., its endurance) should be considered: in case SLC NAND memories are used, this number is in the order of 100 k cycles, which is reduced to 10 k when MLC NAND memories are used. Wear leveling techniques rely on the concept of logical to physical translation, that is, each time the host application requires updates to the same (logical) sector, the memory controller dynamically maps the sector onto a different (physical) sector, keeping track of the mapping either in a specific table or with pointers. The out-of-date copy of the sector is tagged as both invalid and eligible for erase. In this way, all the physical sectors are evenly used, thus keeping the aging under a reasonable value. Two kinds of approaches are possible: Dynamic Wear Leveling is normally used to follow up a user's request of update for a sector; Static Wear Leveling can also be implemented, where every sector, even the least modified, is eligible for remapping as soon as its aging deviates from the average value.

Garbage Collection

Both wear leveling techniques rely on the availability of free sectors that can be filled up with the updates: as soon as the number of free sectors falls below a given threshold, sectors are "compacted" and multiple, obsolete copies are deleted. This operation is performed by the Garbage Collection module, which selects the blocks containing the invalid sectors, copies the latest valid copy into free sectors, and erases such blocks.

In order to minimize the impact on performance, garbage collection can be performed in background. The equilibrium generated by the wear leveling distributes wear-out stress over the array rather than on single hot spots. Hence, the bigger the memory density, the lower the wear-out per cell is.

Bad Block Management

No matter how smart the wear leveling algorithm is, an intrinsic limitation of NAND Flash memories is represented by the presence of so-called Bad Blocks (BB), i.e., blocks which contain one or more locations whose reliability is not guaranteed. The *Bad Block Management* (BBM) module creates and maintains a map of bad blocks: this map is created during factory initialization of the memory card, thus containing the list of the bad blocks already present during the factory testing of the NAND Flash memory modules. Then it is updated during device lifetime whenever a block becomes bad.

ECC

This task is typically executed by a specific hardware inside the memory controller. Examples of memories with embedded ECC are also reported [21–23]. Most popular ECC codes, correcting more than one error, are Reed–Solomon and BCH [24]. While the encoding takes few controller cycles of latency, the decoding phase can take a large number of cycles and visibly reduce read performance as well as the memory response time at random access.

There are different reasons why the read operation may fail (with a certain probability):

- noise (e.g., at the power rails);
- V_{TH} disturbances (read/write of neighbor cells);
- retention (leakage problems).

The allowed probability of failed reads after correction is dependent on the use case of the application. Price-sensitive consumer application, with a relative low number of read accesses during the product lifetime, can tolerate a higher probability of read failures as compared to high-end applications with a high number of memory accesses. The most demanding applications are cache modules for processors.

The reliability that a memory can offer is its intrinsic error probability. This probability could not be the one that the user wishes. Through ECC it is possible to fill the discrepancy between the desired error probability and the error probability offered by the memory; the latter probability can be written as

$$p = \frac{\text{Number of bit errors}}{\text{Total number of bits}}, \tag{7.9}$$

while the *Chip Error Probability* (CEP) is defined as

$$p = \frac{\text{Number of chip errors (p)}}{\text{Total number of chips}}. \tag{7.10}$$

Figure 7.21 shows a typical system composed of a memory array and an ECC block. CEP is usually calculated before (CEP_{in}) and after (CEP_{out}) the ECC block.

Figure 7.22 shows the graphs of CEP_{in} (indicated in the legend as "no ECC") and CEP_{out} as a function of p for a system with 512 Mbit memory, 512 byte block, and ECC able to correct one, two, three, or four errors. The page size of 512 bytes is the typical sector size of host operating systems.

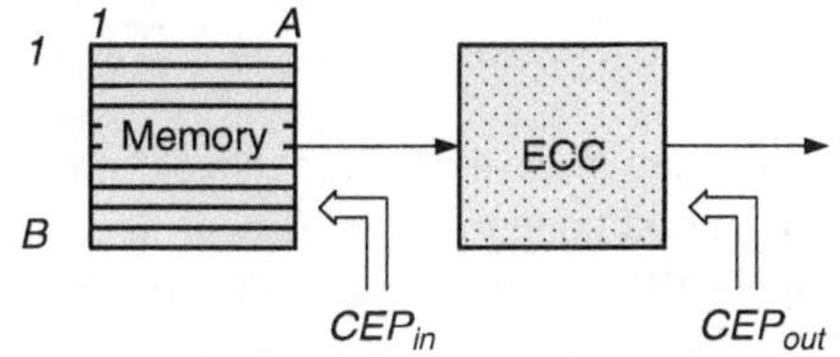

Fig. 7.21 Simplified block diagram of a memory system

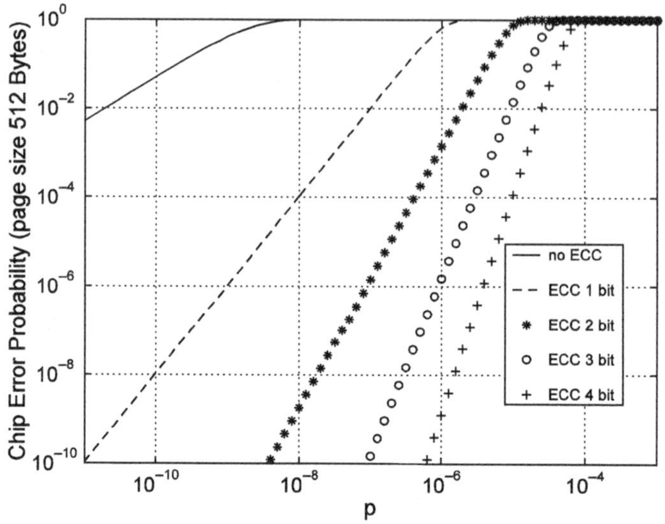

Fig. 7.22 Chip Error Probability as a function of the error probability of the cell

If, for example, the memory is able to guarantee a probability p of 10^{-8}, the use of an ECC correcting two errors allows a CEP of 0.001 ppm (defective parts per million) with respect to 100 ppm, guaranteed by an ECC correcting only one error. On the other hand, if a CEP of 100 ppm is required, the use of an ECC correcting a single error leads to a probability p of 10^{-8}, while an ECC correcting four errors leads to the higher probability of 10^{-5}.

The object of the theory of error correction codes is the addition of redundant terms to the message, such that, on reading, it is possible to detect the errors and to recover the message that has most probably been written.

Methods of error correction are applied for the purpose of data restoration at read access. Block code error correction is applied on subsectors of data. Depending on the used error correcting schemes, different amounts of redundant bits called parity bits are needed.

Between the length n of the code words, the number k of information bits, and the number t of correctable errors, a relationship known as Hamming inequality exists, from which it is possible to compute the minimum number of parity bits:

$$\sum_{i=0}^{t} \binom{n}{i} \leq 2^{n-k}. \tag{7.11}$$

It is not always possible to reach this minimum number: the number of parity bits for a good code must be as near as possible to this number. On the other hand, the bigger the size of the subsector is, the lower the relative amount of spare area (for parity bits) is. Hence, there is an impact in Flash die size.

Table 7.4 Comparison of the number of parity bits required by the different ECC algorithms vs. data size

Data bits	Min ($t = 3$)	BCH ($t = 3$)	RS ($t = 3$)	Min ($t = 4$)	BCH ($t = 4$)	RS ($t = 4$)
2,048	31	36	54	40	48	72
4,096	34	39	54	44	52	72
8,192	37	42	60	48	56	80
16,384	40	45	66	52	60	88
32,768	43	48	72	56	64	96

BCH and Reed–Solomon codes have a very similar structure, but BCH codes require less parity bits and this is one of the reasons why they were preferred for an ECC embedded in the NAND memory [23].

Table 7.4 shows the number of parity bits required by BCH and RS algorithms compared to the minimum number (Min) calculated with the Hamming inequality. Different data sizes and number of correctable errors are considered.

Die Stacking

Reduced form factor has been one of the main drivers for the success of the memory cards; on the other hand, capacity requirement has grown dramatically to the extent that standard packaging (and design) techniques are no longer able to sustain the pace. In order to solve this issue, two approaches are possible: advanced die stacking and 3D technologies (see section "3D NAND Arrays").

The classic way to increase capacity is to implement a multichip solution, where several dies are stacked together. The advantage of this approach is that it can be applied to existing bare die, as shown in Fig. 7.23: dies are separated by means of a so-called interposer, so that there is enough space for the bonding wires to be connected to the pads. On the other hand, the use of the interposer has the immediate drawback of increasing the height of the multichip, and height is one of the most relevant limiting factors for memory cards.

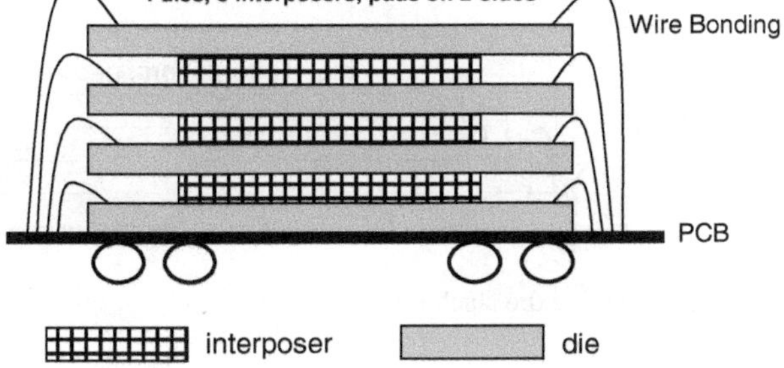

Fig. 7.23 Classic die stacking

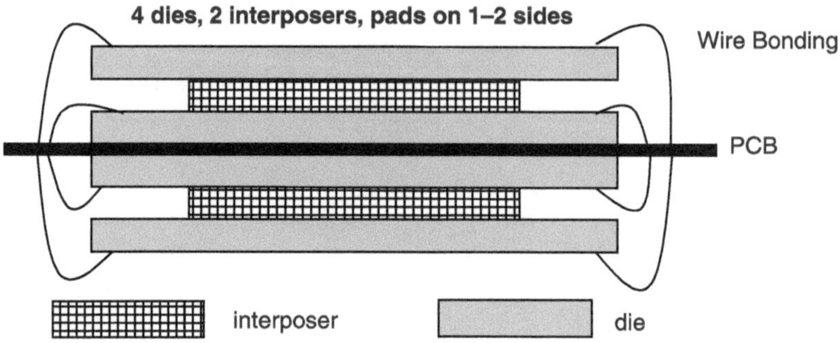

Fig. 7.24 Flipped die stacking

One way to overcome this issue is to exploit both sides of the PCB, as shown in Fig. 7.24: in this way, the PCB acts as interposer, and components are evenly flipped on the two sides of the PCB. Height is reduced, but there is an additional constraint on design: in fact, since the lower die is flipped, its pads are no longer matching those of the upper die. The only way to have corresponding pads facing one another is to design the pad section in such a way that pad to signal correspondence can be scrambled: that is, when a die is used as the bottom one, it is configured with mirrored pads. Such a solution is achievable, but chip design is more complex (signals must be multiplexed in order to perform the scramble) and chip area is increased, since it might be necessary to have additional pads to ensure symmetry when flipping.

The real breakthrough is achieved by completely removing the interposer, thus using all the available height for silicon (apart from a minimum overhead due to the die-to-die glue). Figure 7.25 shows an implementation, where a staircase arrangement of the dies is used: any overhead is reduced to the minimum, bonding does not pose any particular issue, and chip mechanical reliability is maintained (the disoverlap between dies is small compared to the die length, and therefore the overall stability is not compromised, since the upmost die does not go beyond the overall center of mass).

The drawback is that such a solution has a heavy impact on chip design, since all the pads must be located on the same side of the die. In a traditional memory component, pads are arranged along two sides of the device: circuitry is then evenly

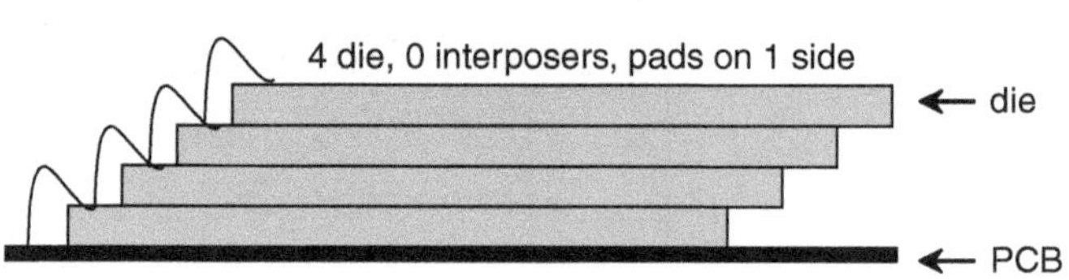

Fig. 7.25 Staircase die stacking

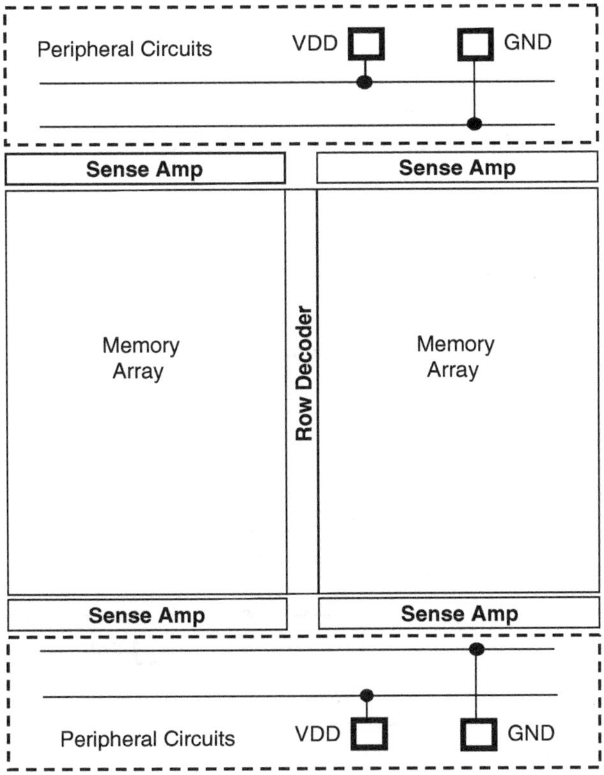

Fig. 7.26 Memory device with pads along opposite sides

located next to the two pad rows and the array occupies the majority of the central area. Figure 7.26 shows the floorplan of a memory device whose pads lie on two opposite sides.

If all pads lie on one side, as shown in Fig. 7.27, chip floorplan is heavily impacted [25]: most of the circuits are moved next to the pads in order to minimize the length of the connection and to optimize circuit placement. But some of the circuits still reside on the opposite side of the die (for instance, part of the decoding logic of the array and part of the page buffers, i.e., the latches where data are stored, either to be written to the memory or when they are read from the memory to be provided to the external world).

Of course, such circuits must be connected to the rest of the chip, both from a functional and from a power supply point of view. Since all pads are on the opposite side, including power supply ones, it is necessary to redesign the power rail distribution inside the chip, making sure that the size and geometry of the rails is designed properly, in order to avoid IR drops issues (i.e., the voltage at the end of the rail is reduced due to the resistive nature of the metal line).

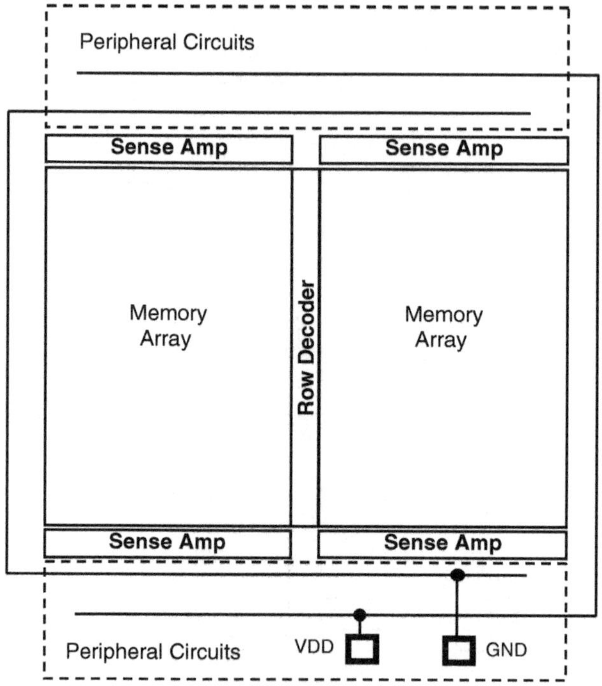

Fig. 7.27 Memory device with pads along one side

Table 7.5 Die stacking options: pros and cons

Die stacking method	Height efficiency	Bonding complexity	Design complexity
Classic	Poor	Average	None
Flipped	Good	Average	Medium
Staircase	Best	Average	High

Table 7.5 summarizes the main features of the suggested solutions.

Recently, another stacking option came into the game: *Through Silicon Via* (TSV) [26, 27]. With this technology, dies are directly connected without asking for a wire bonding, as depicted in Fig. 7.28. One of the main advantages is the reduction of the interconnection length and the associated parasitic *RC*. As a result, data transfer rate can be definitely improved, as well as power consumption. In fact, DRAMs are the main drivers for TSV technology as, with the standard bonding technology, they cannot stack more than two dies in the same package.

The solutions presented so far exploit advances in stacking techniques, eventually requiring changes in chip design floorplan. Quite recently, advanced design and manufacturing solutions have been presented, where the 3D integration is performed directly at chip level.

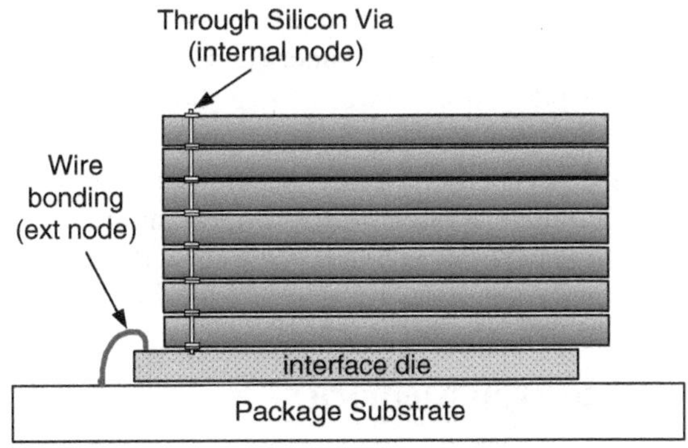

Fig. 7.28 Multiple stacking with TSV technology

3D NAND Arrays

Market request for bigger and cheaper NAND Flash memories triggers continuous research activity for cell size shrinkage. Up to now there is no advice of scaling path slowdown and workarounds have been found to many issues that could limit scalability. Some examples of these solutions are the improved algorithms introduced to control electrostatic interference between adjacent cells and the double patterning techniques to overcome lithography restrictions.

Unfortunately, other physical phenomena are expected as drawbacks of further cell size reduction. Number of electrons stored in floating gate is now extremely low: only few hundreds of electrons differentiate two different levels in a technology node around 30 nm. As reported in the literature, channel doping spread [28] and random telegraph noise [29] can induce large native threshold distributions, and electron injection statistics [30] can cause additional variability after program, impacting both cell endurance and retention. Scaling string dimension increases electric field between programming wordline and adjacent ones, enhancing failure rate during cycling. In the next decade, these and other limitations are expected to cause a drastic delay in the introduction of new planar technologies.

Three-dimensional array organizations represent a promising opportunity to realize new NAND products overcoming the bounds of actual planar devices. In recent years, research activity has generated new interesting architecture proposals in order to try to match several requirements that must be satisfied by the ideal postfloating gate technology: manufacturable and cheap process flow, reliable cell compatibility with multilevel cell approach, capability to realize high-density products, and compliance with current NAND device specification.

A fundamental process change related to the evolution from planar to 3D memories is the transition of baseline flow from conventional floating gate (FG) to charge trap (CT) cells [31]. Almost all NAND technologies in production use planar arrays with polysilicon floating gate as storage element. Charge trap cells that use a dielectric material to store data represent only an option for device scaling and cannot be considered the baseline process of planar arrays. On the contrary, almost all

3D architectures presented up to now use charge trap cells that have a simpler integration due to their thinner stack.

The next section is devoted to a short introduction about basic concepts of charge trap cells; then a review of most promising 3D array is presented with some comments about economic effectiveness. The remaining part of this chapter reports some considerations about the most important architectural choices that characterize 3D array organizations and their impact on device performances.

Charge Trap Memories

Like in the floating gate cell, charge trap threshold voltage shift is determined by the number of elementary charges stored, but as explained in the previous section, the basic difference between FG and CT devices is the storage material: a semiconductor is used by FG cells, whereas in CT case elementary charges are trapped in a dielectric layer (evenly in a wide gap semiconductor) using localized trapping states in the band gap.

Also, program and erase operations are impacted by storage layer characteristics. In CT case more mechanisms are present and devices are often characterized by a small working window due to unavoidable backward currents across blocking layer that cause unwanted saturation during write operations.

As sketched in Fig. 7.29, charge trap system has an intrinsic asymmetry with respect to the floating gate one. CT cells are characterized by a different energy barrier for charge injection and extraction; furthermore, we can assume that write operations in FG cells are always due to quantum mechanical tunneling of free carriers (e.g., from conduction band to conduction band) without energy exchange.

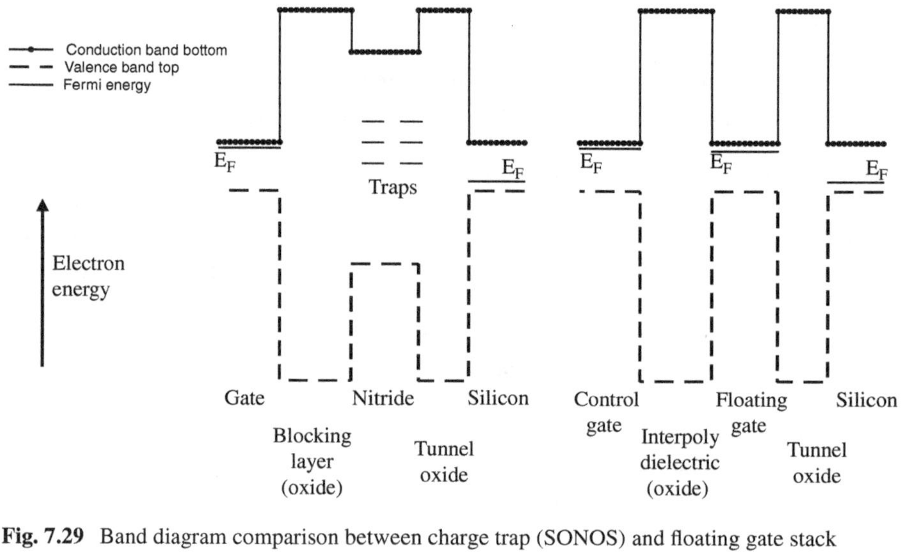

Fig. 7.29 Band diagram comparison between charge trap (SONOS) and floating gate stack

Instead, CT cells use bounded states whose charge and discharge require accumulation and relaxation of lattice energy [32]: these asymmetries induce different speeds for electron and hole injection to or from charge trap storage elements [33].

Being quantum mechanical tunneling more probable for light particles, it is generally convenient to write using electrons since holes have a bigger effective mass in the silicon lattice [34].

Consequently, the ideal best CT cell moves electrons into trapping layer to achieve fast program and removes those electrons from the trapping layer with a relaxed erase timing. Addition of significant hole current by introduction of engineered tunnel barrier has been proposed to overcome poor erase performances of CT cell [35].

Even if 3D cell stack is, in general, thinner to simplify process integration, its composition is aligned to typical planar devices. As shown in Fig. 7.30, it is possible to list the following basic elements:

- gate;
- blocking layer;
- storage layer;
- tunnel layer;
- substrate.

Gate requirements change according to the architecture, but in any case it is important that the material in contact with the blocking dielectric has a high work function to improve erase speed, limiting parasitic back-tunneling current.

Silicon oxide or alumina is usually proposed as blocking layer. Silicon oxide is preferred for architectures that need a limited thermal budget or a very thin stack as, for example, in cells adopting wrap-around-gate approach (see section "Impact of the Wrap-Around-Gate"), whereas alumina is used for cells that suffer for early erase saturation (e.g., in case of not wrap-around-gate cells).

Even if alternatives are available, silicon nitride is probably the best storage material since it is characterized by a high trap density and by a giant lifetime of the

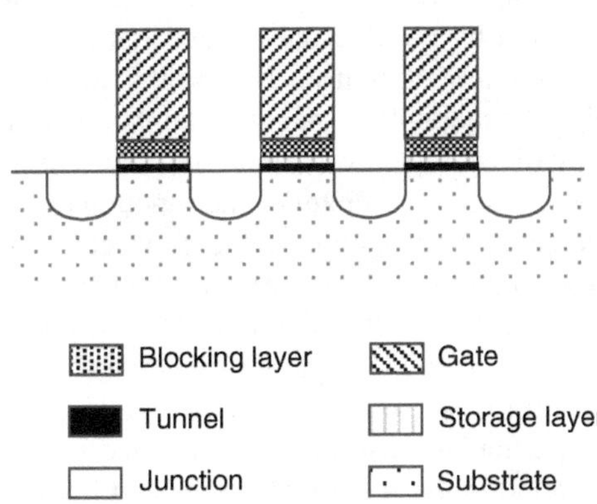

Fig. 7.30 Cross section of a typical CT cell

charged state that ensure large threshold windows and excellent data retention in memory applications [36].

Like in planar NAND, silicon oxide is the most common tunnel material, and only for architectures slow in erase, band engineered tunnel has been proposed [37].

Three-dimensional arrays have polycrystalline silicon substrates and this difference with planar devices, which use silicon, impacts cell performances limiting electron mobility and increasing diode leakage.

3D Array Organizations

Different criteria can be adopted to sort the different 3D memories proposed but, probably, the classification based on topological characteristics is the most effective since the choice of a specific topological organization has a direct impact on cost, electrical performances, and process integration. The following cases can be considered:

- horizontal channel and gate;
- vertical channel and horizontal gate;
- horizontal channel and vertical gate.

Array with Horizontal Channel and Gate

As shown in Fig. 7.31, this array is obtained by stacking many planar memories. Drain contacts and bitlines are common for strings at different levels whereas all the other terminals (source, source selector, wordlines, and drain selector) can be decoded separately, stack by stack. Since this 3D organization is the natural evolution of the conventional planar array, this architecture is the first proposed, and many considerations about costs, process, and electrical performances can be derived by flat memories.

No special requirements are needed by the typical process; the only issue is represented by the thermal budget that must be limited as much as possible to avoid degradation of bottom layers and to have a similar behavior among cells. The big advantage of horizontal channel and gate architecture is flexibility: each level is realized separately, removing many issues that are present with the remaining approaches (e.g., channel and junction doping). Process can be easily changed to let the cells work in enhancement or depletion mode, but typically enhanced mode is preferred to reuse more easily the know-how developed on planar memories.

From the electrical point of view, the biggest difference compared to conventional memories is floating substrate. As shown in Fig. 7.33, with this organization it is not possible to contact the body and this constraint impacts device operations (in particular erase). A more detailed analysis of floating body management will be reported in the section "Options and Important Mechanisms for 3D Cells."

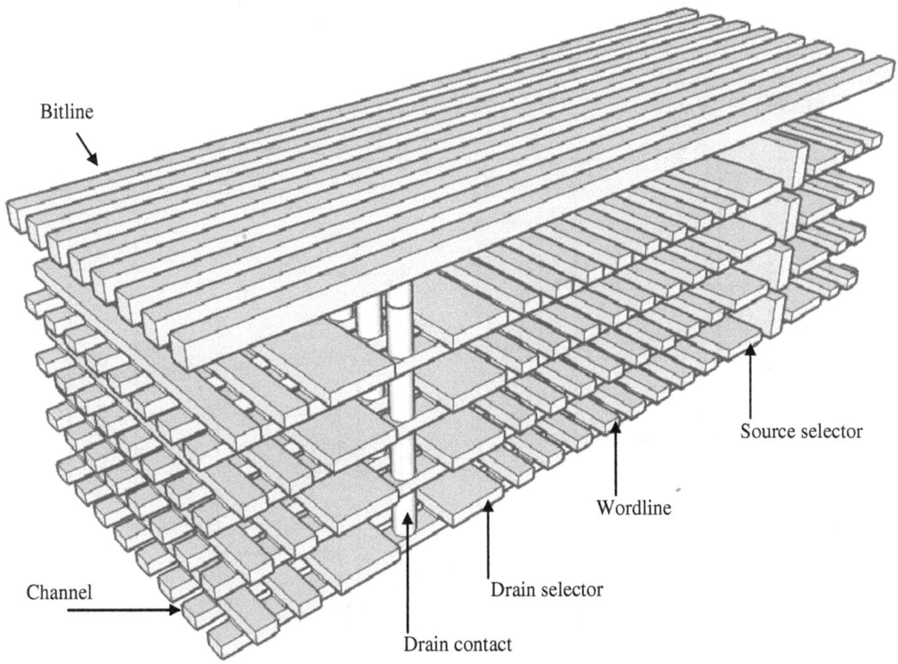

Fig. 7.31 Cross section of a general array with horizontal channel and gate

From an economic point of view, this approach is not very effective since it multiplies the costs to realize a planar array by the number of levels. The only improvement compared to conventional flat memories is represented by circuitry and metal interconnections that are realized only once. In order to limit wafer cost, the number of vertical levels must be as low as possible and, to compensate this limitation, it is fundamental to use small single cells in this specific architecture.

Many publications using this array organization have been presented [38, 39]. Flexibility and easier reuse of know-how developed in planar CT cell investigation are probably the reasons to explain the remarkable activity in this area.

Figure 7.32 shows a schematic description of two NAND layers [40]: on the first one both matrix (MAT1) and peripheral circuits are formed; on the second layer only matrix MAT2 is present. The main peripheral circuits are the sensing circuitry, source line (SL) and P-WELL voltage generators, and two NAND string decoders, one for each layer. Bitlines (BL) are only on MAT1 and they are connected to MAT2 through the contacts shown in Fig. 7.32. Metal BLs are not present on MAT2; this is true also for SL and PWELL networks.

Sensing circuits can access both MAT1 and MAT2 and, due to the fact that the bitline is shared, the capacitive load of a BL is comparable to that of a conventional planar device.

Therefore, there is no penalty on power consumption and timings. Only the load of the vias is added (less than 5% of the total BL capacitance).

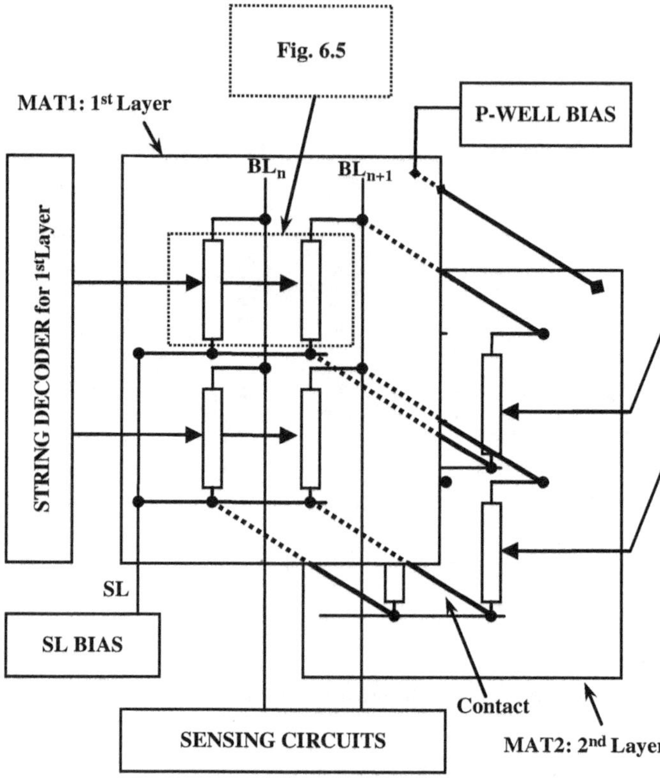

Fig. 7.32 Two NAND layers of Fig. 7.31

Thanks to the two independent string (row) decoders, WL load is in the same range of a planar device. Furthermore, there are no additional program and read disturbs, because only one layer at a time is accessed. Just the doubling of the PWELL parasitic load is a penalty, but it can be negligible as the charge time of the PWELL capacitor is not the dominant term in the overall erase time.

Table 7.6 shows NAND string biasing conditions. During read and program operations, the required biasing voltages are applied only to the strings of the selected layer MAT1. Strings of MAT2 have WLs floating and BSL and SLS biased at 0 V. During Erase, since PWELL is common, WLs of unselected MAT2 layer are floating, like in the unselected blocks of MAT1. In this way, erasing of the blocks in MAT2 is avoided.

MAT1 and MAT2 are defined in different steps; as a consequence, the memory cell's V_{TH} distribution may be different. Figure 7.33 shows a typical V_{TH} distribution after a single program pulse ΔISPP1. Two different distributions D_{MAT1} and D_{MAT2} are generated. Using a conventional program method, this results in a degradation of program performances; in fact, the starting voltage $V_{STARTPGM}$ of the ISPP algorithm is determined by the fastest cell, which is located at the rightmost side of the

Table 7.6 NAND string biasing conditions for MAT1 and MAT2

		Read	Program	Erase
	BLn	V_{PRE} (0.5–1 V)	0/V_{DD}	Floating
	BLS1	V_{PASS}	V_{DD}	Floating
Selected	Selected WL1	V_{READ}	V_{PROG}	0 V
Layer1	Unselected WL1	V_{PASS}	$V_{PASSPGM}$	0 V
	SLS1	V_{PASS}	0 V	Floating
	BLS2	0 V	0 V	Floating
Unselected	WL2	Floating	Floating	Floating
Layer2	SLS2	0 V	0 V	Floating
	SL	0 V	V_{DD}	Floating
	P-WELL	0 V	0 V	18–20 V

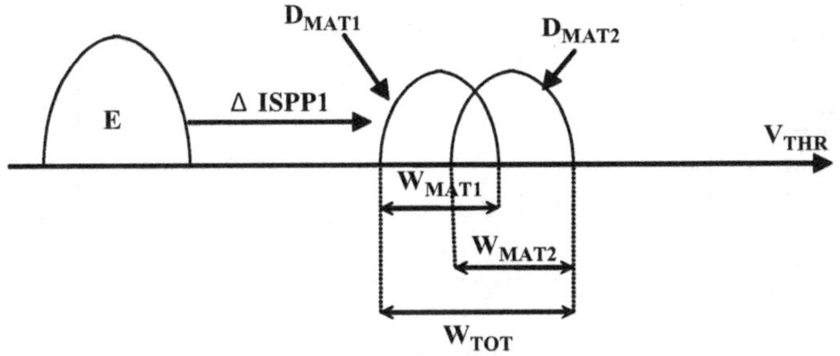

Fig. 7.33 MAT1 and MAT2 distributions after a single ISPP step

V_{TH} distribution. Indeed, it causes an increase in the required number of program steps, due to the enlargement of W_{TOT}.

Increasing the number of ISPP steps means reducing the program speed. The proposed solution in [40] is a dedicated programming scheme for each MAT layer. Depending on the specific MAT layer, program parameters such as $V_{PGMSTART}$, ΔISPP, and maximum number of ISPP steps are properly chosen (Fig 7.34b).

A dedicated control scheme can also be used for erasing: with both PWELLs at the same erase voltage, slightly different voltages are applied to the wordlines of the different layers. Having two row decoders, it is possible to randomly erase two blocks, one for each MAT layer.

Array with Vertical Channel and Horizontal Gate

In the latest years, some interesting architectures with vertical channel and horizontal gate have been presented. In this case, the number of critical masks is low, since the entire stack is etched at the same time.

Limited dependence of wafer cost from the level number grants a positive economic evaluation of these arrays, even if the typical cell size is relatively large and

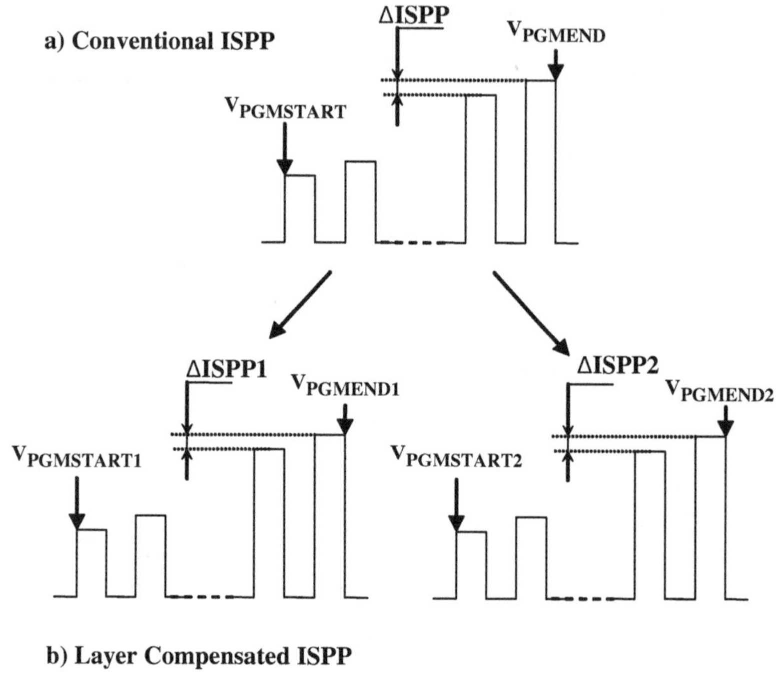

Fig. 7.34 Conventional (**a**) and layer compensated (**b**) ISPP algorithm

many levels are necessary to reach a small equivalent cell area (i.e., single cell area divided by the number of levels).

Typical cross section is reported in Fig. 7.35: the quantity of cells inside a string is defined by the number of vertical wordline levels stacked in the array. Bitlines and drain selector lines run horizontally and are used to select string in the two directions.

The three most important architectures with vertical channel and horizontal gate are BiCS [41], VRAT [42], and TCAT [43]. As shown in Table 7.7, few commonalities are present among proposed flows to realize the arrays and different process options have a remarkable impact on cell characteristics.

BiCS was proposed for the first time in 2007 and an improved version named P-BiCS [44] was presented in 2009 to improve retention, source selector performances, and source line resistance.

The most important characteristics of BiCS are depletion mode cell operation and a channel wrapped around by gate. Depletion cell approach is forced by the decision to adopt a light n-doped or undoped channel in order to avoid process issues related to p-type doping of a vertical polysilicon, whereas use of a cell with gate all around the channel improves electrical performances.

VRAT is the only vertical channel cell with alumina as blocking layer. As reported in the literature [45], cell performances are increased by the use of materials

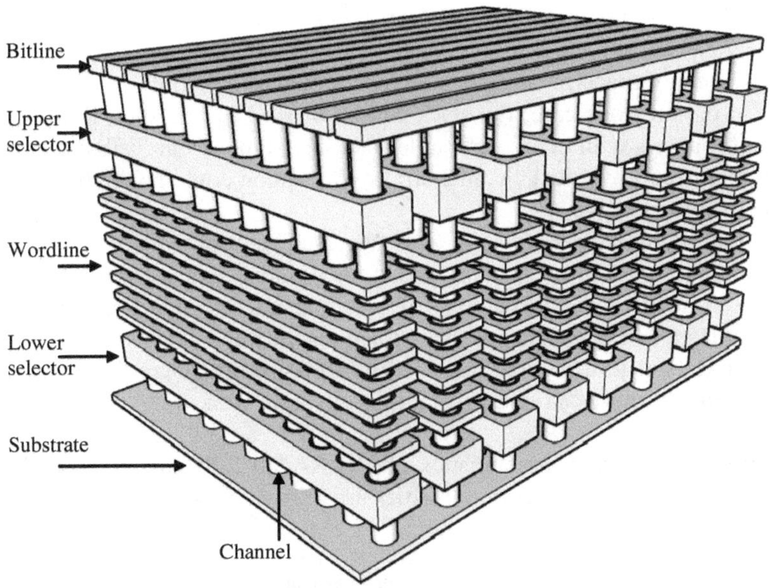

Fig. 7.35 Cross section of a general array with vertical channel and horizontal gate

Table 7.7 Summary of most important characteristics of 3D vertical channel arrays

	BiCS	VRAT	TCAT
Cell operation mode	Depletion	–	Enhancement
Gate all around	Yes	No	Yes
Gate/channel first	Gate first	Channel first	Channel first
Gate	p+ polysilicon	n+ polysilicon	Tungsten
Blocking/storage/tunnel	Oxide/nitride/oxide	Alumina/nitride/oxide	Oxide/nitride/oxide
Channel	n polysilicon	Polysilicon	p polysilicon

with high dielectric constant over storage layer and this improvement balances the lack of a gate all around the channel.

In 2009, an evolution of VRAT, named VSAT [46], was proposed. VSAT is characterized by a gate-first process to simplify integration, whereas all other features reported in Table 7.7 seem unchanged, even if no details about doping of polysilicon gate and choice of high-K material for blocking layer are reported.

TCAT is characterized by many interesting features that get better cell performance and reliability: enhancement mode operation, gate all around the channel, and possibility to integrate a metal as first gate layer. As in the VRAT case, the channel is deposited before the gate and this choice could have beneficial effects on active dielectric quality, but the proposed flows need an extremely high conformity of active layers.

Array with Horizontal Channel and Vertical Gate

A cross section of this array organization is reported in Fig. 7.36.

Like in the previous case, the stack can be defined in one shot, limiting wafer cost also for arrays with many vertical levels. Since typical unit cell size should be smaller compared to vertical channel arrays, this option could be the best one from an economic point of view.

Nevertheless, only in 2009, the first two architectures with horizontal channel and vertical gate have been proposed. A possible explanation of this delay could be ascribed to the difficulties of realizing an effective decoding for this array organization. Since channels are defined together and also wordlines are common to all vertical levels, drain contacts must be used to select the level but this decoding scheme causes a remarkable string increase, worsening cost effectiveness evaluation.

The most important characteristics of VG-NAND [47] and Φ-Flash [48] are reported in Table 7.8. Many commonalities are present between these architectures

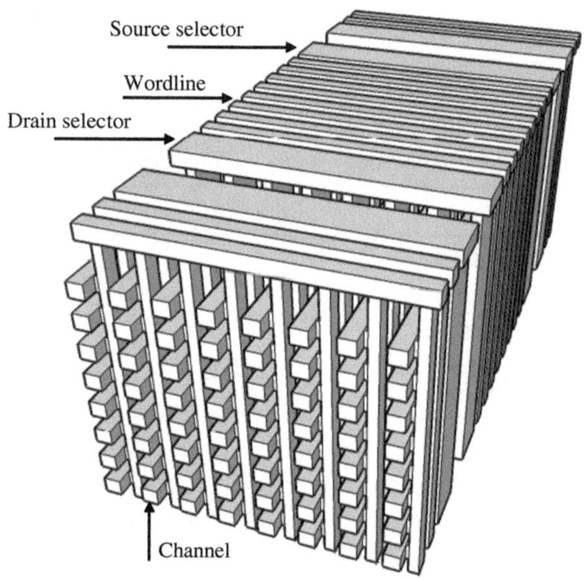

Fig. 7.36 Cross section of a general array with horizontal channel and vertical gate

Table 7.8 Summary of most important features of horizontal channel and vertical gate arrays

	VG-NAND	Φ-Flash
Cell operation mode	Enhancement	–
Gate all around	No	No*
Gate/channel first	Channel first	Channel first
Gate	n+ polysilicon	n+ polysilicon
Blocking/storage/tunnel	Oxide/nitride/oxide	Oxide/nitride/oxide
Channel	p polysilicon	Polysilicon

*An option with gate all around has been proposed in the same paper

if analysis is limited to cell structure, but proposed solutions to decode vertical levels are completely different.

VG-NAND uses a multiple selector scheme with enhancement and depletion mode transistors, whereas Φ-Flash uses a double-decoding scheme in which drain selectors address column and drain contacts identifying the vertical level.

Options and Important Mechanisms for 3D Cells

As already anticipated in the section "3D NAND Arrays," all considered 3D architectures have in common the necessity to adopt a CT cell as storage memory element and require a deep revision of the cell-associated MOSFET structure and operative mode compared to the planar floating gate.

In the following sections we will touch some specific aspects of CT cell inside 3D arrays that characterize this kind of memories. Due to the relevant number of proposed 3D options, it is not possible to detail how these topics impact each mentioned architecture since a complete treatment should require an accurate analysis of NAND string structure of each proposal. Nevertheless this review could be useful to understand advantages and drawbacks of all presented array organizations.

Operative Mode

In Fig. 7.37, a schematization of the two possible operative modes of the cell-associated MOSFET is presented. Due to the fact that the depletion device has the same doping flavor for junctions and channel, its conduction is always granted by

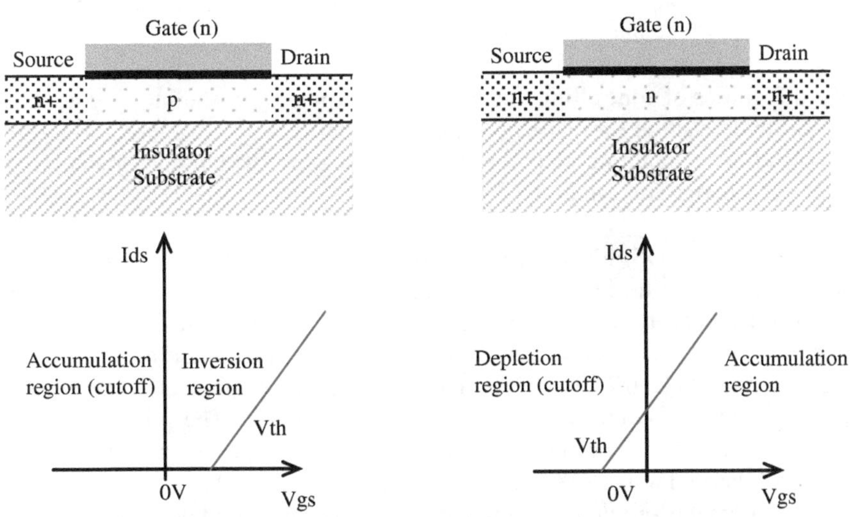

Fig. 7.37 Enhancement (*left*) and depletion (*right*) n MOSFET: structure and transcharacteristic comparison

majority carriers. In an enhancement device, source and drain junction doping is different with respect to the channel flavor and the conduction is granted by channel minority carriers [49].

Since the channel in 3D arrays is done by polycrystalline silicon strips or pillars, these cells have many similarities with planar SOI (Substrate On Insulator) devices in which the MOSFET channel is not formed over a bulky semiconductor substrate but it is a semiconductor layer with limited thickness.

This fact has different consequences for each operative mode; the most important are

- the requirement to find a method to generate minority carriers in the channel of a depletion mode device;
- no trivial solution to grant an ohmic contact to cell substrate in case of an enhancement mode device.

Depletion Operative Mode

Depletion mode MOSFET has a conductive channel between source and drain with 0 V at the gate and no charge into the storage element. This channel increases its driving capability if the gate voltage accumulates majority substrate carriers (e.g., applying positive voltage for n-channel devices and negative voltage for p-channel devices) or if charges of the same sign of majority substrate carriers are removed from the storage element (e.g., electrons in an n-channel device). Channel is switched off applying at gate an opposite voltage bias or equivalently injecting into the storage element a charge of the same sign of majority substrate carriers. Channel cutoff happens when a complete depletion of majority carriers beneath the gate is induced by electrostatic repulsion [49].

Major motivations to select n-channel depletion mode option in 3D architectures (see Table 7.7) are related to

- the difficulties of controlling the doping of p-type polycrystalline silicon in case of in situ deposition for realizing the cell channel;
- the need of maintaining electron injection (program) as the fast operation (see section "Charge Trap Memories").

The depletion MOSFET has a major problem in NAND Flash memory applications: the capability to drive electric field across tunnel dielectric is granted only for a specific polarity, the accumulation one. For the opposite polarity, in fact, the depleted silicon acts as an insulator, strongly reducing the active electric field over the tunnel dielectric. Then, the associated operation (typically, the erase for n-channel devices) becomes slow and poorly controllable. The problem is therefore to provide minority carriers to the cell substrate in order to sustain the reverse polarization needed to remove majority carriers from the storage element.

To overcome this problem a solution has been first suggested in the original BiCS paper [41]. The proposed mechanism for a faster and uniform erase is the raising up of cell substrate potential by means of holes generated by GIDL at the string selector n+/n homojunction [50].

Enhancement Operative Mode

For almost the totality of Flash memory technologies in production today, the operative mode of the cell-associated MOSFET is the enhancement one; this means that at the flat band voltage, without charge into the storage element, the conduction between source and drain is inhibited by the reverse bias of the p/n junction. The raise of the gate voltage over the threshold voltage induces the formation of a layer of minority carriers at the silicon surface beneath the gate, connecting the source to the drain.

The only reported 3D architecture with direct access to the cell body is TCAT, in which each pillar is connected to a common p-type substrate. Enhancement architectures with the option to directly access the substrate do not need any special algorithm for erasing.

For the remaining 3D proposals with the enhancement option, SOI-like matrix structure makes impossible the direct access to the cell substrate. For these architectures, therefore, there is the problem of anchoring the majority channel carrier accumulation layer to an external potential (e.g., the erase voltage). This fact could seriously impact the cell operations and could require special corrections to device working algorithms.

For a whole class of 3D architectures, as for example the vertical channel and horizontal gate (Fig. 7.35), gate self-aligned junction implant is almost impossible, but in case of NAND, especially for highly scaled string length, there is the possibility of skipping the local junctions between cells and use string selector junctions to furnish the minority carriers in order to generate the inversion layer [51, 52].

The comparison between NAND strings with standard local junctions (upper side) and the above-mentioned "junctionless" approach (lower side) is shown in Fig. 7.38. In junctionless string the minority carriers coming from selector junctions

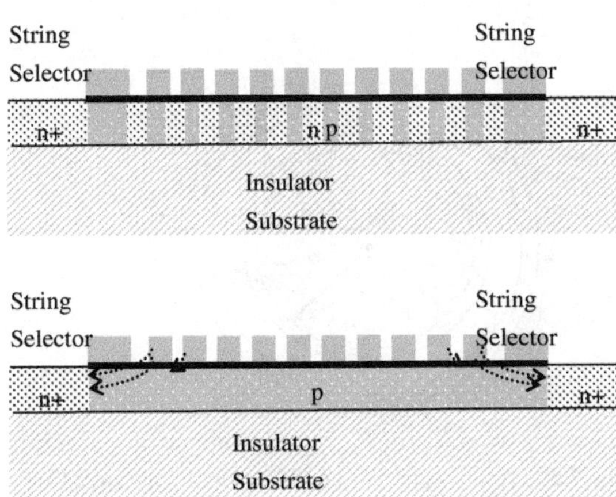

Fig. 7.38 Standard local junction (*upward*) vs. "junctionless" option (*downward*) NAND string

invert the entire string channel thanks to the fringing capacitance that each gate has toward the neighbor cell channels (sketched by the black arrows in the figure).

Junctionless approach has some advantages: simpler integration process, superior string leakage characteristic (no short channel effect), and major immunity to Program/Pass disturb but it has some issues in terms of driving capability of NAND string and mainly it has still undemonstrated capability to fulfill the reliability requirements.

Impact of the Wrap-Around-Gate

Reduction of charge trap stack to scale elementary cell and thermal budget constraints force, in many cases, to avoid high *k* materials as blocking layer; in fact, those materials usually have a higher thickness with respect to the standard silicon dioxide and they need a high-temperature crystallization annealing [53]. The stack choice for many 3D arrays is therefore SONOS or MONOS.

Unfortunately, planar SONOS or MONOS cells have bad erase performances. In particular, they suffer from early erase saturation due to the high electric field across the blocking oxide. The simplest solution to overcome this effect is to adopt one of the so-called wrap-around-gate cell architectures.

These solutions, abundantly explored for standard shrink path of Flash memories [54, 55], can be categorized into

- tri gate-FET, also called fin-FET or omega-FET: in this class the cell has a gate that wraps the channel along three sides;
- gate all around (GAA) FET: in this class the cell channel is completely surrounded by the gate.

For both categories the basic principle is to enhance the electric field across the tunnel oxide and to relax the electric field across the blocking oxide. This is possible thanks to the curvature effect. In Fig. 7.39, two-dimensional section of an

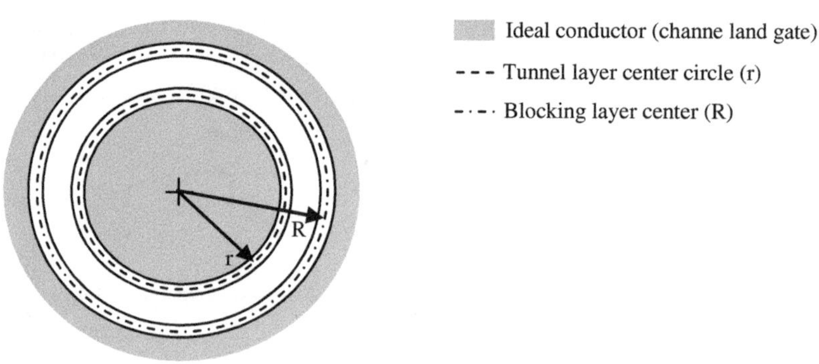

Fig. 7.39 A 2D section of an ideal cylindrical GAA-FET

ideal perfectly cylindrical GAA-FET is sketched. Using Gauss' law for this simple 2D system, an analytic solution to calculate the ratio between the electric field in the center of the tunnel layer (at radius r) and the center of the blocking layer (at radius R) is presented hereafter with assumptions of no charge stored into charge trap medium and homogeneous, nondispersive, and linear dielectric material with permittivity ϵ for tunnel and blocking layers. **D** being the electric displacement field and Q the total charge of inner conductor, we can write

$$\varepsilon \cdot \mathbf{E} = \mathbf{D}, \tag{7.12}$$

$$\oint_{S_r} \mathbf{D} \cdot \mathrm{ds} = Q = \oint_{S_R} \mathbf{D} \cdot \mathrm{ds} \Rightarrow \frac{|\mathbf{E}(r)|}{|\mathbf{E}(R)|} = \frac{R}{r} \tag{7.13}$$

From (7.13) we can observe that the magnification of electric field across the tunnel layer (or equivalently the reduction of electric field across the blocking layer) is tunable changing the curvature radius of the stack.

The need of using a high k material for the blocking layer is, in principle, avoided thanks to the curvature effect: this enables the usage of SONOS and MONOS stacks for charge trap 3D cells that exploit this effect.

Polycrystalline Channel/Substrate Impact

The presented 3D architectures are based on polycrystalline channel devices; even if the possibility of substituting polycrystals with monocrystals can be, in principle, always claimed, it should then be considered as the ultimate solution, being generally cost expensive and still not mature [38].

Use of polycrystalline channel/substrate has detrimental effects on different aspects of the MOSFET operations: it degrades the lifetime of minority carriers and induces shallow states that could impact the ability of cutting off the conduction between source and drain, inducing a leakage not controllable by means of the gate polarization.

NAND array is rather tolerant with respect to the leakage of a single cell but is very sensitive to the selector leakage worsening. NAND standard column decoding and cell sensing are basically based on the ratio between on and off currents; therefore, if the leakage of string selectors becomes too high, it is necessary to limit the number of strings tied to a single column in order to avoid erroneous reading of the programmed cells. This constraint induces an unacceptable reduction of the array efficiency.

High selector leakage worsens program inhibit efficiency as well: this fact could impact the specification of the sustainable maximum number of successive program operations and/or the page size, with the same above-mentioned difficulties to organize an efficient array.

Fortunately, leakage properties of the polycrystalline channel MOS seem to improve as the critical dimension of the channel itself scales down; this fact is due to the reduced probability of intercepting grain boundary. Recent work [56] shows that subthreshold slope of a polycrystalline device approaches monocrystalline device value if the channel width is less than 100 nm, resulting in a similar control of the leakage current.

Acknowledgements The authors would like to thank Emilio Camerlenghi (Numonyx, R&D Technology Development) for the useful suggestions.

References

1. Cappelletti P et al (1999) Flash memories. Kluwer, Norwood, MA
2. Campardo G, Micheloni R, Novosel D (2005) VLSI-design of non-volatile memories. Springer, Berlin
3. Tanaka T et al (Nov 1994) A quick intelligent page-programming architecture and a shielded bitline sensing method for 3 V-only NAND flash memory. Solid State Circuits IEEE J 29(11):1366–1373
4. Itoh Y et al (Feb 1989) An Experimental 4 Mb CMOS EEPROM with a NAND Structured Cell Solid-State Circuits Conference, 1989. Digest of Technical Papers. 36th ISSCC, IEEE International, pp 134–135
5. Suh K-D et al (Nov 1995) A 3.3 V 32 Mb NAND flash memory with incremental step pulse programming scheme. IEEE J Solid Stare Circuits 30(11):1149-1156.
6. Iwata Y et al (Nov 1995) A 35 ns cycle time 3.3 V Only32 Mb NAND flash EEPROM.IEEE J Solid State Circuits 30(11):1157–1164.
7. Kim J-K et al (May 1997) A 120-mm 64-Mb NAND flash memory achieving 180 ns/Byte effective program speed. IEEE J Solid State Circuits 32(5):670–680
8. Suh K-D et al (Nov 1995) A 3.3 V 32 Mb NAND flash memory with incremental step pulse programming scheme. Solid State Circuits IEEE J 30(11):1149–1156
9. Lee S et al. (Feb. 2004) A 3.3 V 4 Gb four-level NAND flash memory with 90 nm CMOS technology. IEEE Int Solid State Circuits Conf ISSCC, Digest of Technical Papers 1:52–53, 513
10. Byeon D-S et al. (Feb. 2005) An 8 Gb multi-level NAND flash memory with 63 nm STI CMOS process technology. Solid-State Circuits Conference, ISSCC, Digest of Technical Papers, vol 1, pp 46–47
11. Micheloni R, Crippa L, Marelli A (2010) Inside NAND flash memories, chap. 5. Springer
12. Li Y et al (Feb 2008) A 16 Gb 3b/ Cell NAND Flash Memory in 56 nm with 8 MB/s Write Rate. Solid-State Circuits Conference, 2008. Digest of Technical Papers, ISSCC 2008 IEEE International, pp 506–507, 632
13. Shibata N et al (April 2008) A 70 nm 16 Gb 16-Level-Cell NAND flash Memory. IEEE J Solid State Circuits 43(4):929–937
14. Trinh C et al (Feb 2009) A 5.6 MB/s 64 Gb 4b/Cell NAND flash memory in 43 nm CMOS. Solid-State Circuits Conference, 2009. Digest of Technical Papers, ISSCC 2009 IEEE International February 2009, pp 246–247
15. www.mmca.org
16. www.compactflash.org
17. www.sdcard.com
18. Kawaguchi A, Nishioka S, Motoda H (1995) A flash-memory based file system. Proceedings of the USENIX Winter Technical Conference, pp 155–164
19. Kim J, Kim JM, Noh S, Min SL, Cho Y (May 2002) A space-efficient flash translation layer for compactflash systems. IEEE Trans Consumer Electron 48(2):366–375

20. Lee S-W, Park D-J, Chung T-S, Lee D-H, Park S-W, Songe H-J (August 2005) FAST: A log-buffer based ftl scheme with fully associative sector translation. 2005 US-Korea Conference on Science, Technology, & Entrepreneurship
21. Tanzawa T, Tanaka T, Takekuchi K, Shirota R, Aritome S, Watanabe H, Hemink G, Shimizu K, Sato S, Takekuchi Y, Ohuchi K (May 1997) A compact on-chip ECC for low cost Flash memories. IEEE J Solid State Circuits 32:662–669
22. Campardo G, Micheloni R et al (Nov 2000) 40-mm^2 3-V-only 50-MHz 64-Mb 2-b/cell CHE NOR Flash memory. IEEE J Solid State Circuits 35(11):1655–1667
23. Micheloni R et al (Feb. 2006) A 4 Gb 2b/cell NAND flash memory with embedded 5b BCH ECC for 36 MB/s system read throughput. IEEE International Solid-State Circuits Conference Dig. Tech. Papers, pp 142–143
24. Micheloni R, Marelli A, Ravasio R (2008) Error correction codes for non-volatile memories. Springer, Berlin
25. Kanda K et al (Feb. 2008) A 120 mm^2 16 Gb 4-MLC NAND flash memory with 43 nm CMOS technology. IEEE International Solid-State Circuits Conference Dig. Tech. Papers, pp 430–431
26. Hwang CG (2006) New Paradigms in the Silicon Industry, International Electron Device Meeting (IEDM). pp 1–8
27. Kawano M et al (2006) A 3D Packaging Technology for a 4Gbit Stacked DRAM with 3Gbps Data Transfer. Int Electron Dev Meeting (IEDM) 1–4
28. Mizuno T et al (November 1994) Experimental study of threshold voltage fluctuation due to statistical variation of channel dopant number in MOSFET's. IEEE Trans Electron Devices 41(11)
29. Kurata H et al (2006) The impact of Random Telegraph Signals on the scaling of multilevel Flash memories. Symposium on VLSI technology
30. Compagnoni CM et al (October 2008) Ultimate accuracy for the NAND Flash program algorithm due to the electron injection statistics. IEEE Trans Electron Devices 55(10)
31. Micheloni R, Crippa L, Marelli A (2010) Inside NAND flash memories. Springer, Berlin
32. Nasyrov KA et al (2003) Multiphonon ionization of deep centers in amorphous silicon nitride: experimental and numerical simulations. JETP Lett 77(7):385–388
33. Bouchet M et al (2009) An in-depth investigation of physical mechanisms governing SANOS memories characteristics. IEEE IMW, May 10–14, Monterey CA, USA
34. Sze SM (1981) Physics of semiconductor devices, 2nd ed. Wiley (WIE), New York, NY
35. Likharev KK (1998) Layered tunnel barriers for nonvolatile memory devices. Appl Phys Lett 73(15):2137
36. Eitan B et al (2000) NROM: a novel localized trapping, 2-bit nonvolatile memory cell. IEEE EDL 21(11)
37. Lue HT et al (2005) BE-SONOS: a Bandgap Engineered SONOS with excellent performance and reliability. IEDM Tech Digest
38. Jung SM et al (2006) Three dimensionally stacked NAND Flash memory technology using stacking single crystal Si layers on ILD and TANOS structure for beyond 30 nm node. IEDM Tech Digest
39. Lai EK et al (2006) A multi-layer stackable thin-film transistor (TFT) NAND-type flash memory. IEDM Tech Digest
40. Park K-T et al (Jan 2009) A fully performance compatible 45 nm 4-gigabit three dimensional double-stacked multi-level NAND flash memory with shared bit-line structure. Solid-State Circuits IEEE J 44(1):208–216
41. Tanaka H et al (2007) Bit Cost Scalable technology with punch and plug process for ultra high density Flash memory. Symposium on VLSI technology, June 12–16, Kyoto, Japan
42. Kim J et al (2008) Novel 3-D structure for ultra high density Flash memory with VRAT (Vertical-Recess-Array-Transistor) and PIPE (Planarized Integration on the same PlanE). Symposium on VLSI technology, June 17–20, Honolulu, USA

43. Jang J et al (2009) Vertical cell array using TCAT (Terabit Cell Array Transistor) technology for ultra high density NAND Flash memory. Symposium on VLSI technology, June 15–17, Kyoto, Japan
44. Ishiduki M et al (2009) Optimal device structure for Pipe-shaped BiCS Flash memory for ultra high density storage device with excellent performance and reliability. IEDM Tech Digest
45. Lee CH et al (2003) A novel SONOS structure of $SiO_2/SiN/Al_2O_3$ with TaN metal gate for multi-Giga bit Flash memories. IEDM Tech Digest
46. Kim J et al (2009) Novel vertical-stacked-array-transistor (VSAT) for ultra-high-density and cost-effective NAND Flash memory devices and SSD (Solid State Drive). Symposium on VLSI technology, June 15–17, Kyoto, Japan
47. Kim W et al (2009) Multi-layered Vertical Gate NAND Flash overcoming stacking limit for terabit density storage. Symposium on VLSI technology, June 15–17, Kyoto, Japan
48. Hubert A et al (2009) A stacked SONOS technology, up to 4 levels and 6 nm crystalline nanowires, with gate-all-around or independent gates (Φ-Flash), suitable for full 3D integration. IEDM Tech Digest
49. Tsividis Y (2003) Operation and modeling of the MOS transistor. Oxford University Press, Oxford
50. Fukuzumi Y et al (2007) Optimal integration and characteristics of vertical array devices for ultra-high density, bit-cost scalable flash memory. IEDM Tech Digest
51. Park KT et al (2006) A 64-Cell NAND flash memory with asymmetric S/D structure for sub-40 nm technology and beyond. Symposium on VLSI technology, June 13–17, Honolulu, USA
52. Lee CH et al (2008) Highly scalable NAND flash memory with robust immunity to program disturbance using symmetric inversion-type source and drain structure. Symposium on VLSI technology, June 17–20, Honolulu, USA
53. Scozzari C et al (2008) Al2O3 optimization for charge trap memory application. ULIS 191–194
54. Xuan P et al (2003) FinFET SONOS flash memory for embedded applications. IEDM Tech Digest
55. Specht M et al (2004) 20 nm tri-gate SONOS memory cells with multi-level operation. IEDM Tech Digest
56. Tzu TH et al (2009) Study of Sub-30 nm thin film transistor (TFT) charge-trapping (CT) devices for 3D NAND flash application. IEDM Tech Digest

Chapter 8
Optical Data Storage

Yang Wang, Yiqun Wu, Haifeng Wang, Mingju Huang, and Yan Wang

Abstract In recent 30 years, optical data storage has undergone persistent development in response to the ever-growing information storage demands as a result of technological and market competition from magnetic and semiconductor memory technology. This chapter reviews basic principles and some important R&D progress on popular optical disk technology and other high-density optical storage technologies, such as superresolution, near-field, and three-dimensional optical storage technologies.

Keywords Optical data storage · Optical disk · Super-resolution optical storage · Near-field optical storage · Holographic optical storage · Two-photon absorption optical storage

Introduction

With the rapid development of the computer, multimedia and network applications, the volume of information has increased to a point where global volume reaches more than zetta bytes. Hence the storage and transfer of mass information requires the superhigh density and superhigh transfer rates.

Basically, optical storage refers to the storage of data on an optically readable medium. Data are recorded by making marks in a pattern that can be read back with the aid of light (generally laser). Thus optical memory has some innate technological advantages over conventional magnetic and flash memory technologies, including such features as high data stability, long media life, and data unrewritable protection. Optical memory is less susceptible than magnetic disks and semiconductor memory to corruption by environmental influences such as electromagnetic fields. Furthermore, the possibility of a read/write head crash is extremely remote

Y. Wang (✉)
Shanghai Institute of Optics and Fine Mechanics, Chinese Academy of Sciences,
P.O. Box 800-211, Shanghai 201800, China
e-mail: ywang@siom.ac.cn

G. Campardo et al. (eds.), *Memory Mass Storage*,
DOI 10.1007/978-3-642-14752-4_8,

because of the noncontact read/write manner of optical memory. Optical memory is unique in being able to support the write-once-read-many (WORM) type of recording media. When data are written to WORM optical media, a permanent, virtually incorruptible mark is made. Thereafter, when the data are updated to a WORM optical memory, they are written to a new location and the original locations cannot be reused. Thus the risk to integrity from modifying data can be greatly reduced.

In recent years, optical storage technology has undergone rapid development in response to the ever-growing demands as a result of technological and market competition from magnetic and semiconductor storage technology. Optical disk drives, optical disk towers, and optical disk libraries can be used as separate units or as parts of network-based mass storage systems for any data archive application. Optical mass storage devices and systems can provide absolute data authenticity for a wide variety of markets and applications, including medical, financial, government, broadcast, and entertainment. This chapter reviews basic principles and some important R&D progress on popular optical disk technology and other high-density optical storage technologies.

Optical Disk Technology

Brief History

The most common example of optical memory is the optical disk, which was invented in 1958 by David Paul Gregg. In 1969, the first patent (US Patent No. 3430966) for an analog optical disk was registered (Fig. 8.1). In 1978, Philips and MCA marketed the first commercial laserdisk product. Philips and Sony successfully developed the audio compact disk in 1983. In the mid-1990s, a consortium of manufacturers developed the second generation of the optical disk, the DVD. The third generation, Blu-Ray Disk, was developed in 2000–2006 by the Blu-Ray Disk Association (BDA) [1]. Up to the end of the last century, digital audio disk replacing the audio cassette, digital video disk replacing video tape, and recordable/rewritable disk replacing the floppy disk were widely used in data distribution, for backup of computer data, and in commercial digital video/audio applications.

Basic Principle

The recording/reading principle of the optical disk is that a highly coherent and monochromatic laser beam is focused on a near-diffraction-limited micro spot, and the micro-spot region on the recording medium produces physical or chemical changes that cause a change in the micro-area optical properties (such as refractive index and reflectivity, etc.) and a detectable contrast with the surrounding medium. With the modulated laser beam and spinning media, analog or digital information can be stored dynamically.

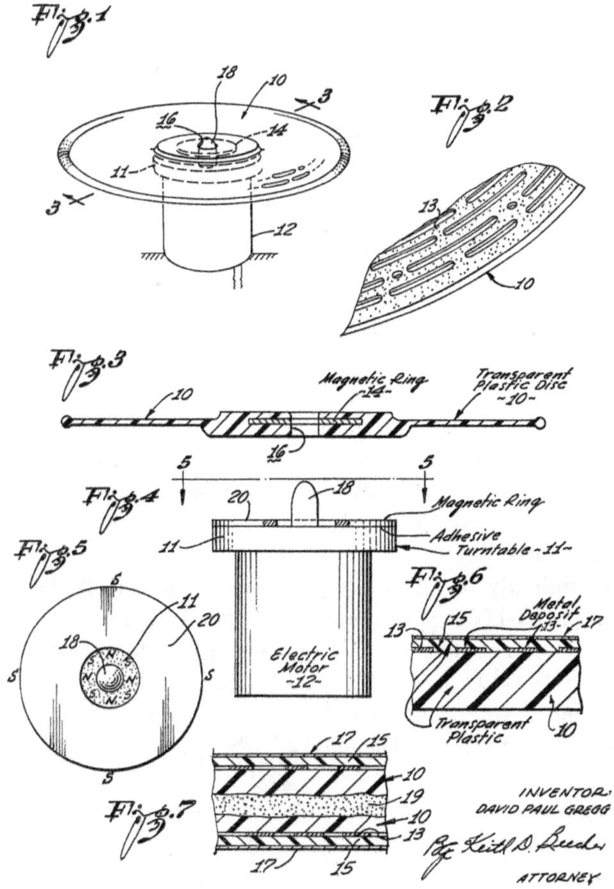

Fig. 8.1 Drawing from D.P. Gregg's optical disk patent (US Patent No. 3430966)

Classification

Currently commercialized optical disks can be divided into the following four types:

(1) Compact Disk (CD) series, operating wavelength: 780–830 nm.
(2) Digital Versatile Disk (DVD) series, operating wavelength: 635–650 nm.
(3) Blue laser disk series, operating wavelength: 405 nm, including Blu-Ray Disk (BD) and High Definition-DVD (HD-DVD).
(4) Other format optical disks in China: China red-light high-definition optical disks (operating wavelength: 650 nm), including Enhanced Versatile Disk (EVD), High-Definition Video Disk (HDV), High-Definition Versatile Disk (HVD), Forward Versatile Disk (FVD), Next-generation Versatile Disk (NVD) et al., and China Blue High-Definition Disk (CBHD) (operating wavelength: 405 nm).

Table 8.1 Some technical parameters of three mainstream optical disks

Parameters	CD	DVD	BD
Capacity/GB	0.6	4.7	23~27
Wavelength/nm	780–830	635–650	405
NA	0.45	0.6	0.85
Track pitch/μm	1.6	0.74	0.32
Minimum pit length/μm	0.83	0.39	0.13
Channel bit rate/Mb/s	4.3218	26.16	66
CLV/m/s	1.2–1.4	3.49–3.84	4.5–5.29
Modulation	EFM	8/16, RLL(2,10)	17PP, RLL(1,7)
ECC	CIRC	RSPC	LDC+BIS
User data transfer rate	150 K/s	11 Mb/s	36 Mb/s

Most of China's red-light high-definition optical disks have the physical format of DVD. Some newly developed audio and video codes (such as EAC and VP6) were used to enable the disk to store HDTV resolution, which the standard DVD format does not support.

CBHD was originally based on the physical format of HD-DVD, which was abandoned by the HD-DVD promotion group in 2008. But CBHD uses China's AVS video codec, DRA audio codec, and a new copy protection system, DKAA, as an alternative to HD-DVD's AACS [2].

Tables 8.1 and 8.2, respectively list some technical parameters of three mainstream series of optical disks and China's red-light high-definition optical disks [3].

An optical disk is designed to support one of three recording types: read-only (prerecorded), recordable (write once read many times), or rewritable (erasable).

Development Trends

The primary development goals of optical disk technology have been increasing storage density and capacity. In the region of far-field optical recording, the minimum diameter of focused spot d is determined by the diffraction limit of the optical system and is proportional to λ/NA: $d \propto \lambda/\mathrm{NA}$, where λ is writing and reading laser wavelength and NA is the numerical aperture of the focusing lens. Therefore, the practical approach to minimizing the mark length is to decrease λ and increase NA. The working wavelength is decreased from 780 to 830 nm (near infrared light, using a GaAlAs semiconductor laser) of the CD series to 635–650 nm (red light, using a GaAlInP semiconductor laser) of the DVD series, and then to 405 nm (blue light, using a GaN semiconductor laser) of blue laser optical disks. The NA is increased from 0.45 of the CD series to 0.6 of the DVD series, and then to 0.65 or 0.85 of the blue laser optical disks. The capacity is increased from 700 Mb of the CD series to 4.7 GB of the DVD series and then to 15–27 Gb of the blue laser optical disks. Several new optical disk technologies have been developed in recent years to further increase the storage density and capacity, including near-field

Table 8.2 Some technical parameters of red-light high-definition optical disks in China

	EVD	HDV	HVD	FVD	NVD
Release Time	2004	2004	2004	2004	2004
Capacities/single layer (GB)	4.7	4.7	4.7	5.4~6.0	12/dual layer
Developer	Beijing Fuguo Digital	Beijing Kaicheng	Shanghai Jingchen	ITRI	NVD Association
Wavelength (nm)/NA	635(650)/ 0.6	DVD	DVD	DVD	650/0.65
Track pitch	DVD	DVD	DVD	0.64	0.62
CLV(m/s)	DVD	DVD	DVD	DVD	3.38
Min. pit length (μm)	DVD	DVD	DVD	0.4	0.388
Modulation	DVD	DVD	DVD	8/15, RLL (2,12)	DVD
ECC	DVD	DVD	DVD	RSPC+	DVD
Channel bit rate/Channel T	DVD	DVD	DVD	DVD	DVD
User data transfer Rate	DVD	DVD	DVD	DVD	DVD
Audio code	ExAC	Dolby AC-3, DTS	Dolby AC-3, DTS	WMA-9	AVS, ExAC
Video Code	MPEG-2 MP@HL	HD12	MPEG-2 optimized	WMV-9	MPEG4, H.264, AVS
Copy Protection	Digital water mark tech.	CSS	CSS	Dual mode AES	CSS+
Max. recording time for HDTV(minutes)/ Resolution, data rate	47/1080i, 22 Mbps	150/720p	132/720p	132/720 p	132/1080i, 12Mbps

optical data storage, superresolution optical data storage, holographic optical storage, photonic multidimensional information storage, optical-magnetic hybrid recording, etc.

Transfer rate can be thought of as the speed of reading/writing data from or to an optical disk. Obtaining higher data transfer rates and shorter reading/recording times are the foci of other primary development directions in optical disk technology. The main approaches to improve the data transfer rate include increasing rotating speed, reducing the weight of pickup, and optimizing addressing methods and data processing programs. The data transfer rate of an optical disk (1X) is increased from 1.2 Mb/s of CD to 11 Mb/s and 36 Mb/s of DVD and BD, respectively. The maximum data transfer rate that can be achieved is higher than 400 Mb/s, which is almost on the scale of hard magnetic disks. Miniaturization and integration of the optical head is an important approach to reducing its size and weight and increasing its speed.

A light-emitting diode LD is often used as laser source of optical pickup, but as its beam quality is not good enough, the objective lens has to meet more demanding technical requirements. As the combined spherical glass lens is too heavy, a spherical plastic lens has been developed to decrease the weight and improve flexibility. Optimizing addressing units and then reducing addressing time can also effectively enhance the data transfer rate. The average addressing time is shortened from about 400 ms of 1X CD to below 100 ms of 40–50X CD, then to about 30 ms of DVD and below 20 ms of BD.

Data processing programs, including those governing data structure, data format, coding method, and so on, will directly affect the data transfer efficiency. The introduction of new coding schemes and multichannel or parallel reading/writing are effective ways to enhance the data transfer rate. For recordable and rewritable disks, the recording and erasing speed, which is closely related to the sensitivity of recording media, is also a key factor. Thus, stable recording media with rapid response times are of crucial importance in shortening storage time of recordable and rewritable optical disks. At present, the recording laser pulse has reached the order of tens of nanoseconds.

Read Only Memory (ROM) Optical Disk

A read only memory (ROM) optical disk carries information that is inserted at the time of manufacture and cannot be changed. In today's market, popular ROMs include CD-ROM, DVD-ROM, and BD-ROM. The data or audio/video signal after coding is recorded on the glass master disk by lithography, and then stampers are duplicated from the master. With the injection reproduction method, the information prerecorded on stampers is pressed onto plastic substrates and reflective and protective layers are coated onto the data substrate surface to create a CD-ROM. Bonding of another blank substrate is needed for DVD-ROM and BD-ROM.

Manufacturing Process for a Master Disk

The traditional manufacturing process for a master disk is shown in Fig. 8.2. At least eight steps are required [4, 5].

(1) Cleaning and polishing of glass substrates: A master disk is formed on the glass substrate, which is a carrier of the photoresist and the master. The glass substrates should be thoroughly cleaned and carefully polished. After chemical removal of the Ni and photoresist residues and multiple washings with de-ionized water, the glass substrates can be reused.
(2) Spin-coating and hardening of the photoresist: Photoresist is uniformly spin-coated on the glass substrate. Its thickness (CD: 140–150 nm, DVD: 100–110 nm) can be controlled by its viscosity and spin-coating speed. In order to increase the adhesiveness between the resist and the substrate, coating of an

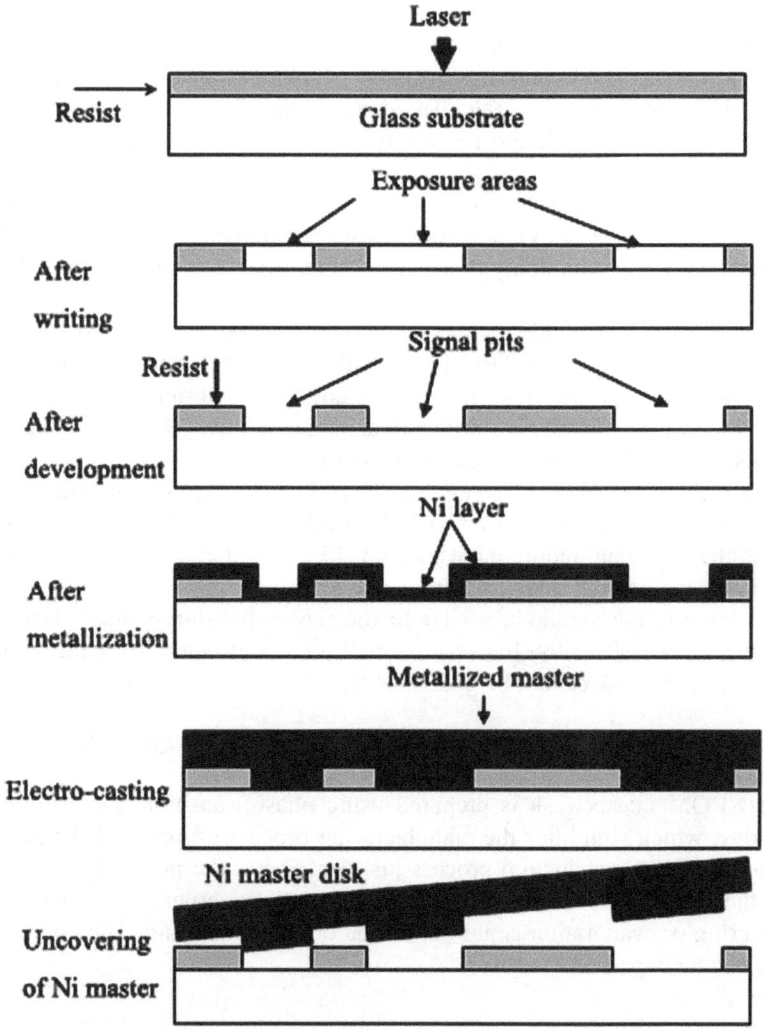

Fig. 8.2 Manufacturing process for a master disk

adhesion layer should precede the coating of resist. The sample should be baked to evaporate solvents and improve the adhesiveness.

(3) Laser writing: Blue or ultraviolet laser is often used to expose the light-sensitive photoresist. Electro-optical modulator (EOM) is often used to stabilize the laser power and reduce the high-frequency noise. Laser pulse modulation and servo focusing technology are used to guarantee the writing quality. For radial mobile production of CD-ROM masters, writing information usually begins from the inner circle to the outer one along a spiral lines. The rotating speed of the glass substrate and moving speed of the objective lens should be well matched to

maintain the required tracking pitch. The precision of spindle motor should be of the order of tens of nanometers.

(4) Development: The exposed part of the photoresist layer is dissolved to create pits on the glass substrate. The concentration of developing solution and the developing time should be strictly controlled to ensure the quality of the signal pits.
(5) Metallization: A conductive thin nickel layer should be deposited on the surface of the glass master before it is electrocasted to create a metallized master disk. A nickel layer is commonly prepared by sputtering, evaporation, or a chemical method.
(6) Electrocasting: The cathode and the anode, respectively are connected to the metallized master disk and to a nickel ball immersed in electrocasting solution. Driven by the electric field, nickel ions are deposited onto the surface of the master to form a nickel plate with a thickness of about 300 ($\pm$5) μm. Stress and distortion should be avoided during this process.
(7) Back surface polishing: The quality of back surface polishing affects the life and signal quality of a master disk. Typically, back surface polishing should meet the following requirements: $R_a < 0.15$ μm and $R_{\max} < 1.0$ μm. Polishing methods include dry and wet grinding styles.
(8) Punching: A hole should be used to fix the master disk during injection molding. If that inner hole is too big or too small, or the eccentricity is too large, the quality of the disk cannot be guaranteed.

Figure 8.3 shows the atomic force microscopy (AFM) images of CD-ROM and DVD-ROM master disks [5].

A BD-ROM master disk is prepared using phase transition mastering (PTM) technology, which simplifies the manufacturing process of master disks compared with the traditional production process just described. The phase-change material acts as the photoresist, and, owing to its conductive properties, metal layer coating by sputtering or evaporation is not needed before electrocasting. Thus step (2) and step (5) can be bypassed.

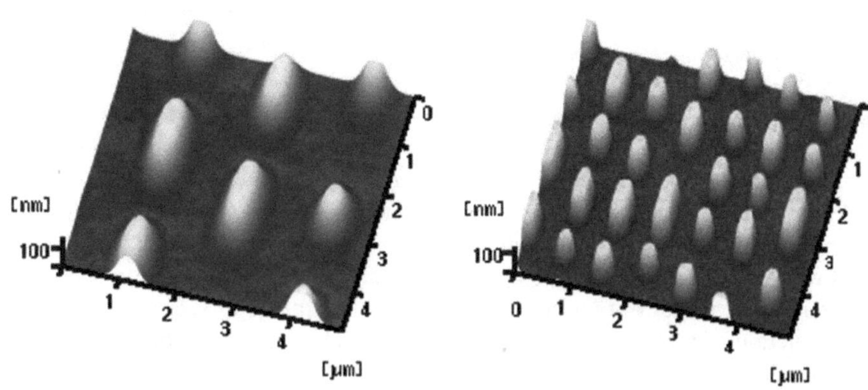

Fig. 8.3 AFM images of CD-ROM (*left*) and DVD-ROM (*right*) master disks (5 μm × 5 μm)

Structure of ROM Optical Disk

The structure of optical disk should match the corresponding optical system. The thickness of the substrate is determined from the working distance of the objective lens. CD-ROMs are made from a 1.2-mm-thick prerecorded polycarbonate (PC) substrate, a thin layer of aluminium to make a reflective surface, a protective layer, and a printing layer. The typical structure of a CD-ROM is shown in Fig. 8.4 [4].

The typical structure of a DVD-ROM is shown in Fig. 8.5 [4]. Its basic structure of a DVD-ROM (4.7 Gb) consists of two 0.6-mm-thick PC substrates, a prerecorded substrate and a blank substrate bonded together. Double layers (dual layers) or double sided DVD-ROMs were made to achieve double or more capacity.

Owing to the use of a higher NA objective lens, the thickness of the cover layer for BD is decreased to 0.1 mm, as shown in Fig. 8.6 [5]. Since the data layer of BD is closer to the surface of the disk compared to the DVD and CD standard, all BD media are required by specification to be scratch-resistant.

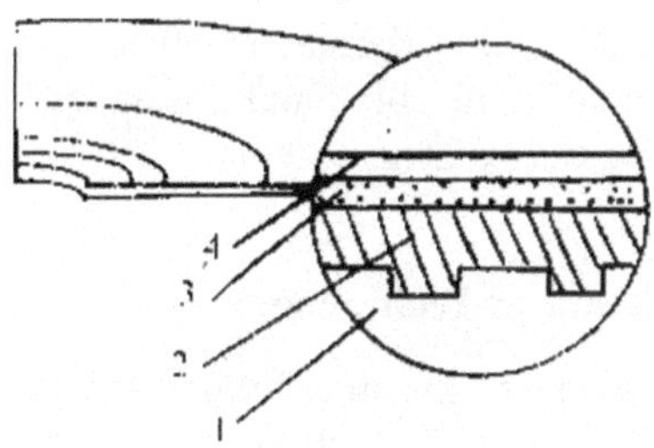

Fig. 8.4 Typical structure of a CD-ROM: (1) PC substrate; (2) reflective layer; (3) protective layer; (4) printing layer

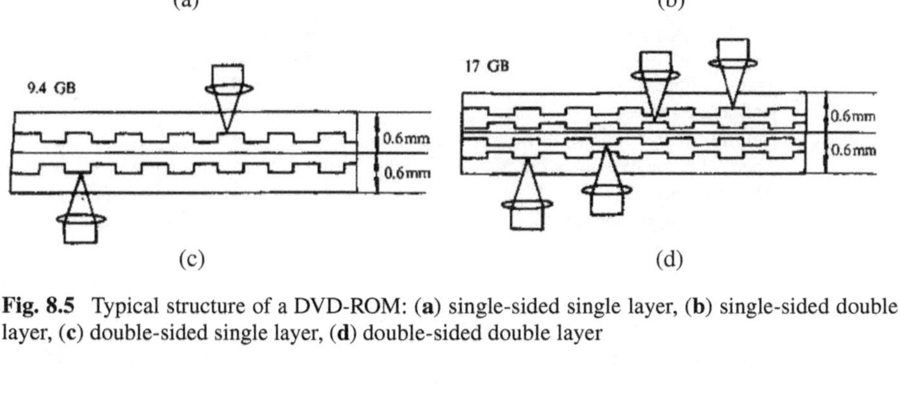

Fig. 8.5 Typical structure of a DVD-ROM: (**a**) single-sided single layer, (**b**) single-sided double layer, (**c**) double-sided single layer, (**d**) double-sided double layer

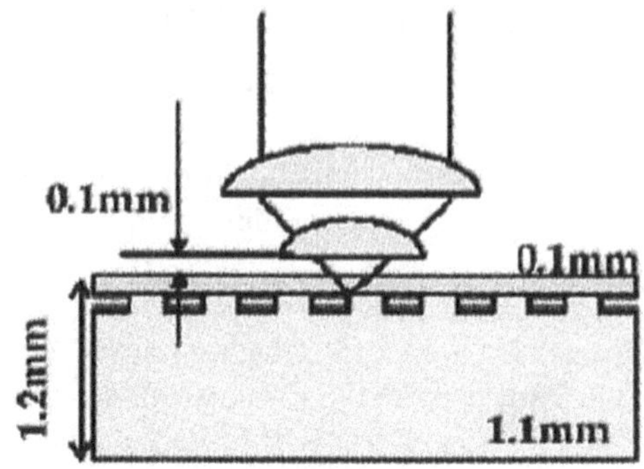

Fig. 8.6 Structure of a single-layer BD-ROM

Substrate Materials

Polycarbonate is often used to produce optical disk substrate. Pure polycarbonate is highly transparent, volume stable, and impact resistant (see Table 8.3) [4]. Optical-grade polycarbonate is baked to remove any moisture, then melted in the injection molding machine and a stamper is used to impress it into the substrate under high pressure during cooling.

Readout Technology

Owing to the interference effect, the intensity of light reflected from the pit surfaces is significantly less than that reflected from the land surface. The high-contrast signals can be distinguished. Figure 8.7 shows the readout principle of the ROM optical disk [5], where L is the pit width, I is the incident intensity, I_l, Φ_l is the land reflection intensity and phase, I_p, Φ_p is the pit reflection intensity and phase, and the phase difference $|I_l - I_p| = \lambda/2$.

The readout system of a ROM disk is shown in Fig. 8.8 [5]. The readout signals are collected by the four-quadrant detector. Focusing and tracking servo signals are collected by the four-quadrant and two-quadrant detectors, respectively.

Table 8.3 Physical and chemical properties of PC

Transmittance (%)	93
Refractive index	1.59
Minimum birefringence (nm)	<60
Thermal refractive index coefficient (°C^{-1})	-1.2×10^{-4}
Abbe number	31
Impact strength (kg·cm^{-2})	80~100
Density (g/cm^3)	1.2
Glass transition temperature (°C)	138
Tensile strength (kg-cm^2)	560
Water aborbability (%)	0.4
Linear expansibility (°C^{-1})	0.7×10^{-4}

Fig. 8.7 Schematic readout principle of a ROM optical disk

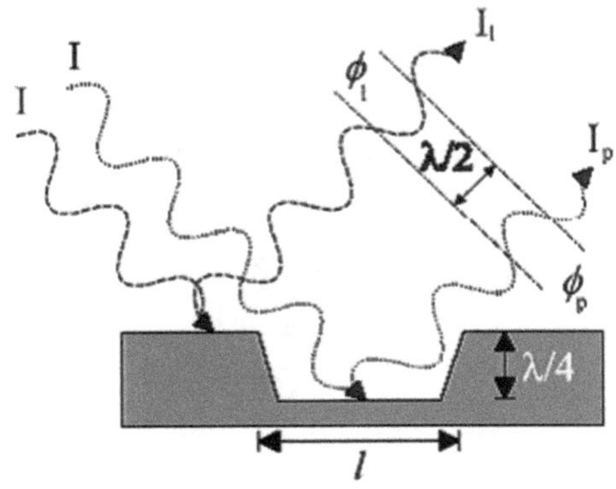

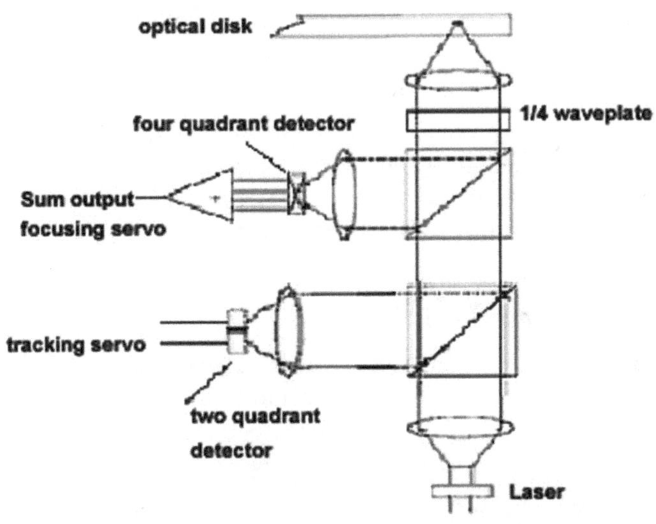

Fig. 8.8 Readout system of a ROM optical disk

Recordable Optical Disk

Recording/Readout Mechanism of a Recordable Optical Disk

A recordable optical disk is a type of write once and read many (WORM) optical disk. Information is recorded on a recordable optical disk using a photothermal method. With laser irradiation on the recording layer, material molecules in the film absorb energy from the laser pulses and undergo melting, evaporation, break, and then form bumps, pits, or bubbles, which result in changes in the physical (optical) properties at the irradiated spot (e.g., a reflectivity change). The two physically (optically) different states before and after irradiation are used for information storage [4].

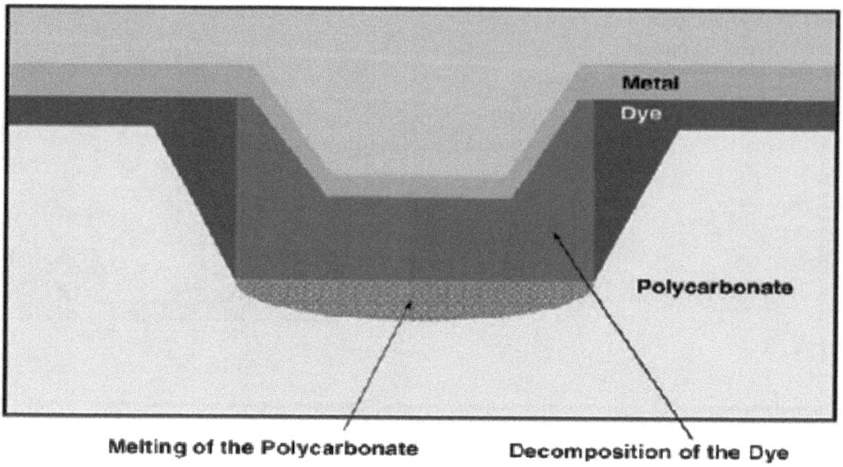

Fig. 8.9 Effect in the formation of the recorded pits in a CD-R

The recording process for recordable CD (CD-R) is based on irreversible deformation of dye in the recording layer by laser irradiation (Fig. 8.9) [6]. The deformed areas, which have less reflectivity, are called "pits"; and the unmodified areas remaining between these pits, which have higher reflectivity, are called "lands".

Structure of a Recordable Optical Disk

As shown in Fig. 8.10, injection-molded CD-R polycarbonate substrate contains a wobble pregroove for tracking, CLV speed control, and time purposes. A dye layer

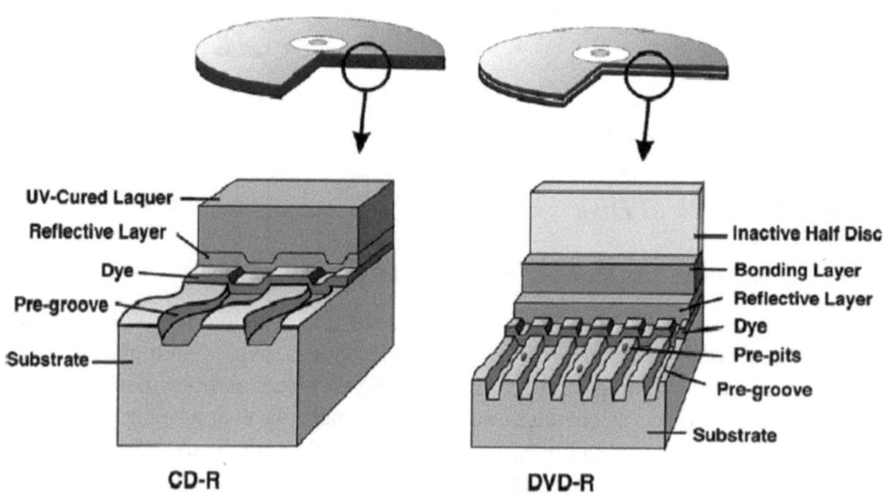

Fig. 8.10 Structure of a CD-R and a DVD-R

as the recording medium spin-coats the substrate. A silver layer, applied by argon sputtering, serves as a reflector and also acts as a heat sink during the recording of the pits. The surface of the CD-R is formed by a protective lacquer and a printing layer.

As with a recordable CD (CD-R), a recordable DVD (DVD-R), the version that uses an organic-dye as a recording layer has by far the largest share of the market. Recordable DVDs are available in two different formats: the DVD+R ("plus R") and the DVD-R ("minus R"). Although there are two significant differences in the way the data are addressed, the two differently formatted recording disks both have high compatibility with the DVD-ROM. The structure of DVD-R is shown in Fig. 8.10. Similar to the DVD-ROM, the DVD-R also has a second polycarbonate half disk without information content attached.

Recordable Blu-Ray disk (BD-R) with larger storage capacity has won out in next-generation write-once media contest. The inorganic and the dye layer are both included as recording media in the specifications (as shown in Fig. 8.11).

Recording Materials for Recordable Optical Disks

Organic dyes commonly used as optical recording media for CD-R and DVD-R are cyanine dyes (Fig. 8.12) and phthalocyanine dyes (Fig. 8.13) for CD-Rs and cyanine dyes (Fig. 8.14) and azo dyes (Fig. 8.15) for DVD-Rs [4, 5].

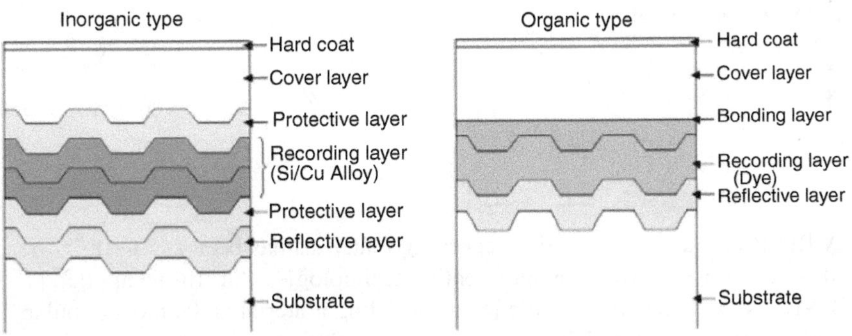

Fig. 8.11 Cross section of the two types of BD-R structures

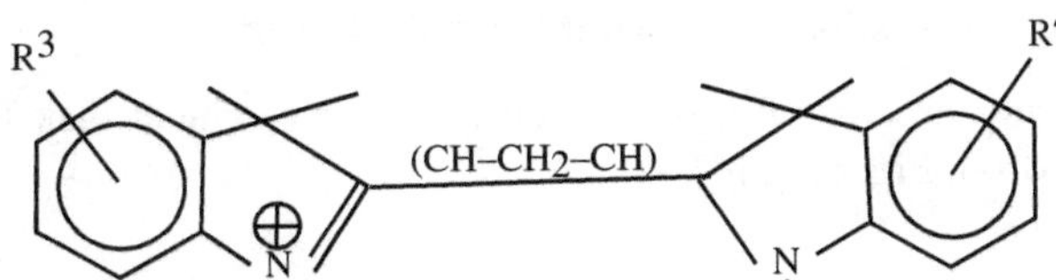

Fig. 8.12 General structural formula of CD-R cyanine dye

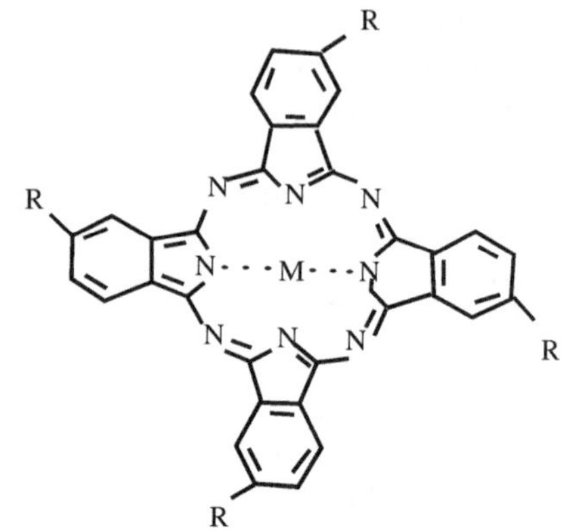

Fig. 8.13 Structural formula of phthalocyanine with a central metal ion

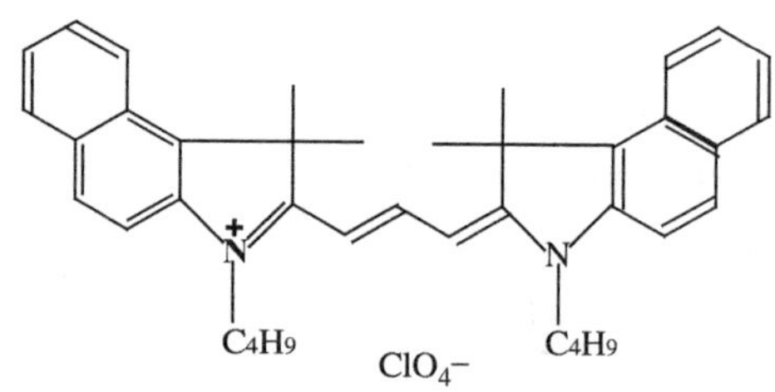

Fig. 8.14 A typical cyanine molecule for DVD-R

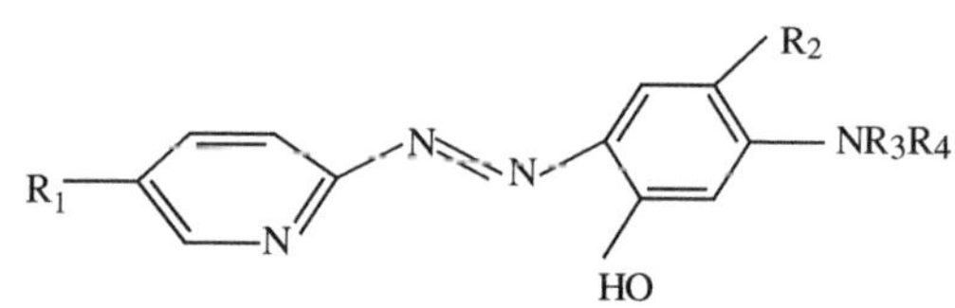

Fig. 8.15 An azo dye with an hydroxyl group for DVD-R

A BD-R based on organic dye recording materials has been successfully developed, so the existing low-cost spin-coating technologies can still be applied [7, 8]. A BD-R disk with a Cu/Si double layer recording material is the most popular [9], but other inorganic materials such as Bi-Ge nitride and Te-Pd-O have also used successfully as BD-R recording media [10, 11].

Manufacturing Process for a Recordable Optical Disk

The manufacturing process of a dye-based recordable optical disk (DVD-R), for example, is shown in Fig. 8.16 [4].

Recording/Readout Technology of Recordable Optical Disks

As shown in Fig. 8.17, the optical system of a recordable optical disk is similar with that of a ROM [5]. Recording information on an optical disk requires that

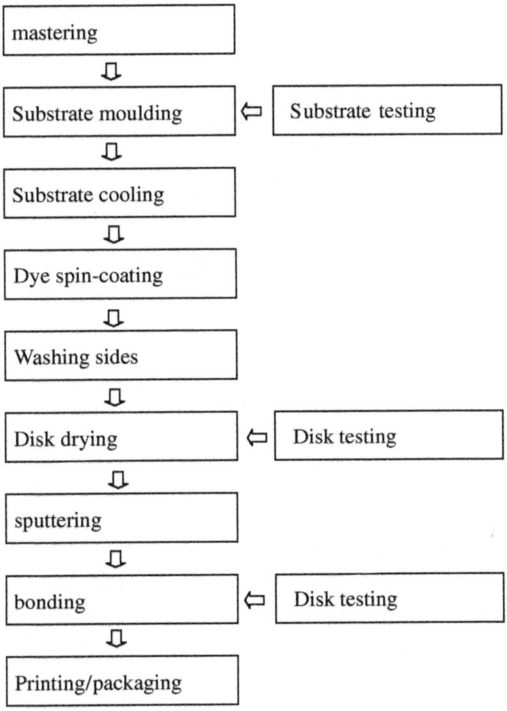

Fig. 8.16 Manufacturing process for a DVD-R

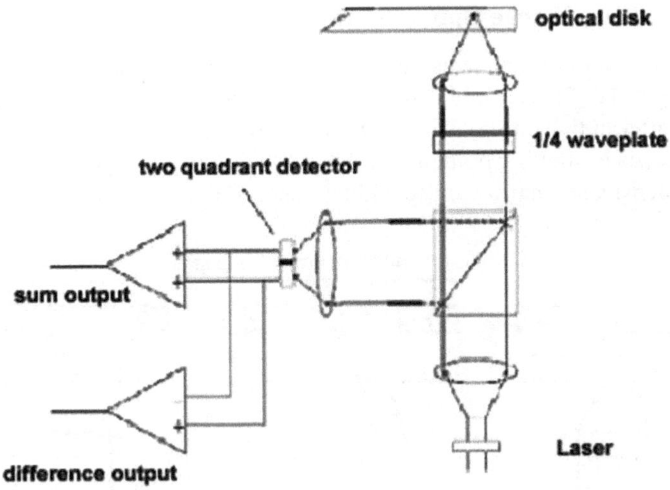

Fig. 8.17 Recording/readout system of a recordable optical disk

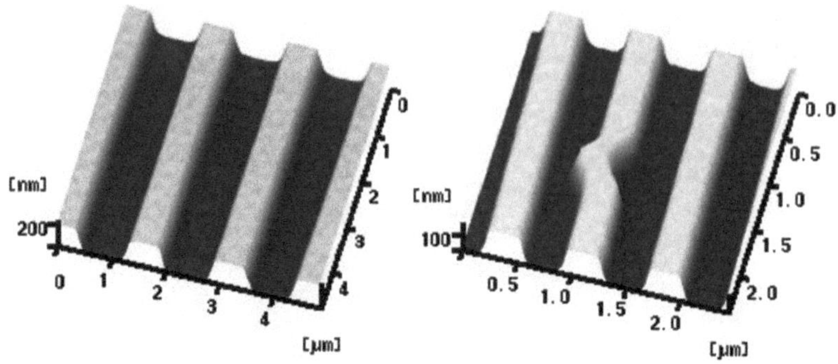

Fig. 8.18 Pre-grooved substrates of a CD-R (*left*) and DVD-R (*right*)

the maximum output power of the semiconductor laser should be relatively large. Recordable disks are tracked by prerecorded wobble grooves (Figs. 8.10 and 8.18), whereas ROM disks are tracked by prerecorded pits [5]. The prerecorded wobble grooves contain the information for absolute time in pregroove (ATIP) for CD-R and absolute address in pregroove (ADIP) for DVD-R. The readout and tracking servo signals are obtained from the sum and difference signals collected by the two-quadrant detector.

Phase-Change Rewritable Optical Disks

Principle of Phase-Change Optical Storage

The recording, reading, and erasing principles of a phase-change optical disk is shown in Fig. 8.19 [5]. The initial state of phase-change materials in such a disk is crystalline (after initialization). The micro-area temperature of crystalline recording material exceeds its melting point under the irradiation of a focused laser beam with high power density and narrow pulse width (tens to 100 ns). The melted liquid phase

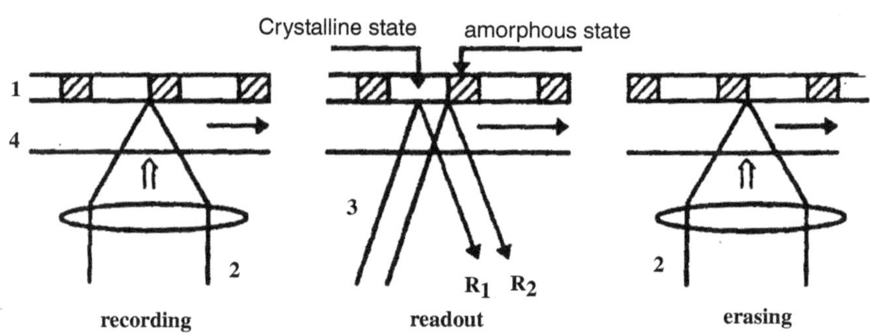

Fig. 8.19 Principle of phase-change optical storage: (1) phase-change material layer, (2) recording and erasing laser beam, (3) readout (detecting) laser beam, (4) substrate

cools and solidifies with a high cooling rate (10^9–10^{10} K/s) to an amorphous phase. Information is recorded by this amorphization process. If the amorphous region is irradiated by a focused laser beam with medium power density and a wide pulse width, the micro-area temperature will be between the glass transition temperature and melting temperature, it will be recrystallized through crystal nucleation and growth. The recorded information is erased by this recrystallization process. As the reflectivity of the amorphous state is lower then that of crystalline state, it is easy to read out the recorded information by detecting this difference in reflectivity. Reading is carried out by quasi-continuous wave laser beam with low power density. There will be no phase change during readout.

Structure of a Phase-Change Optical Disk

Typical structures of phase-change rewritable optical disks are shown in Fig. 8.20 [5].

Phase-Change Materials

The phase-change materials used in rewritable optical disks are often composed of III-IV group semiconductor elements, such as Te-based and InSb-based alloys. Te-based phase change materials include mainly GeSbTe, GaSeTe, SeSbTe, SeInTe, SeSnTe, SeCuTe, GeAsTe, GeSeTe, GeSnTe, $GeTeO_x$, $SnTeO_x$, GeSbTeN, GeSbTeS, GeSbTeSe, GeSbTePd, GeSbTeCr, SbTeAg, InSbTe, GeSnSb, GeSbTeO, GeSbTeO(N), GeSbTeAg, GeSbTeSn, GeSbTeB, GeSbTe (Fe;Zn;Bi), GeSbTeBi(B), GeSbTeIn, Te-Ge-Sn-Au, and so on. In-Sb-based phase-change materials include mainly InSb, AgInSbTe, and so on [4].

The ternary phase diagram depicting different phase-change alloys and corresponding rewritable optical disk products is shown in Fig. 8.21 [12].

The most often-used phase-change material is $Ge_2Sb_2Te_5$. Its main physical and chemical properties are listed in Table 8.4 [5].

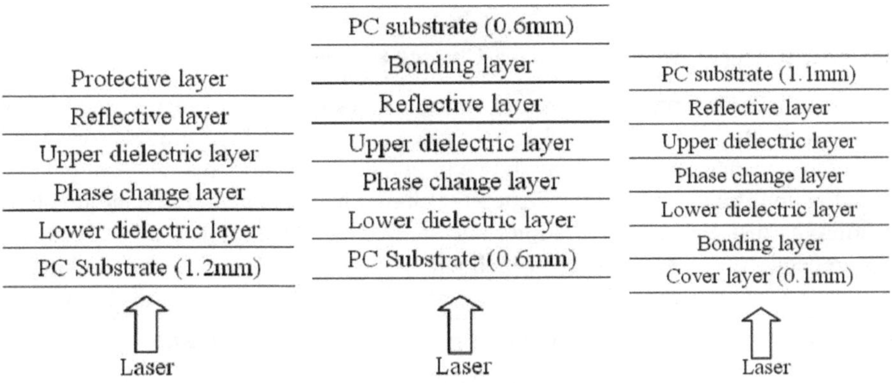

Fig. 8.20 Structure of rewritable CD (*left*), DVD (*middle*), and BD (*right*)

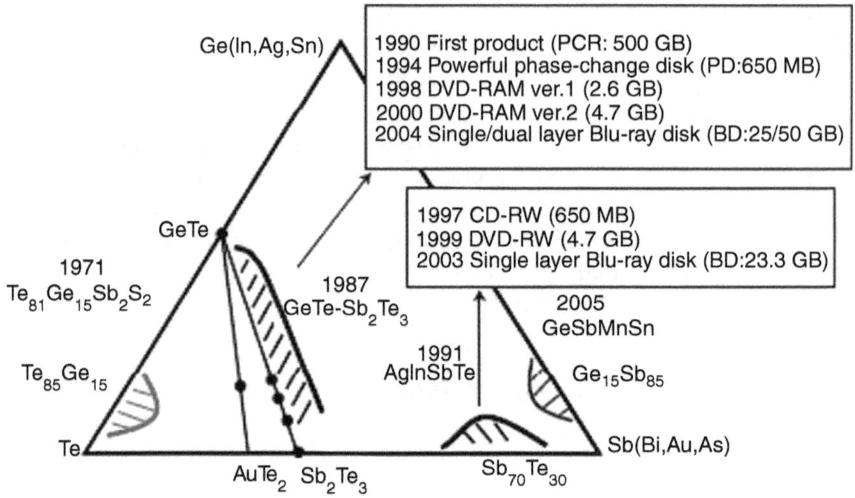

Fig. 8.21 Ternary phase diagram of different phase-change alloys and corresponding rewritable optical disk products

Table 8.4 Physical and chemical properties of $Ge_2Sb_2Te_5$

Crystallization temperature	~ 180°C
Melting temperature	~ 600°C
Crystallization time	~ 50 ns
Structural properties	As-deposited state : amorphous state; Low temperature (~ <300°C) : NaCl FCC; High temperature (~ >300°C): HCP; Laser initialization : FCC
Lattice constant	FCC : $a = 0.601$ nm; HCP : $a = 0.4216$ nm, $c = 1.7174$ nm
Activation energy	amorphous ⇒ FCC: 2.1 ~ 2.4 eV; FCC ⇒ HCP: 3.4 ~ 3.8 eV
Crystallization mechanism	Nucleation-driven crystallization
Thermal conductivity	$0.058\ J \cdot cm^{-1} \cdot K^{-1} \cdot s^{-1}$
Specific heat	$0.21\ J \cdot g^{-1} \cdot K^{-1}$
Density	$\sim 6.15\ g \cdot cm^{-3}$

Manufacturing Process for a Phase-Change Optical Disk

The manufacturing processes for the various series of phase-change optical disks are basically similar to one another, so a brief description of the process for a CD-RW phase-change optical disk, as shown in Fig. 8.22, can suffice as an example [4]. Initialization is a special technical step for phase-change optical disks and involves heating the disk by laser irradiation to crystallize the phase-change layer, so that a high reflectivity can be obtained, which will be helpful for focusing and tracking.

Recording/Readout Technology of a Phase-Change Optical Disk

A typical optical system for a phase-change rewritable optical disk is shown in Fig. 8.23 [5]. C1, C2, and C3 indicate signal output, focusing servo, and tracking servo, respectively.

Fig. 8.22 Manufacturing process for a CD-RW

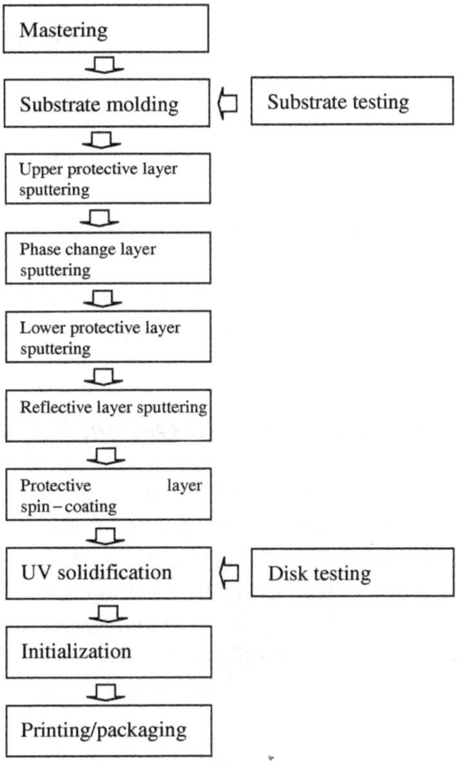

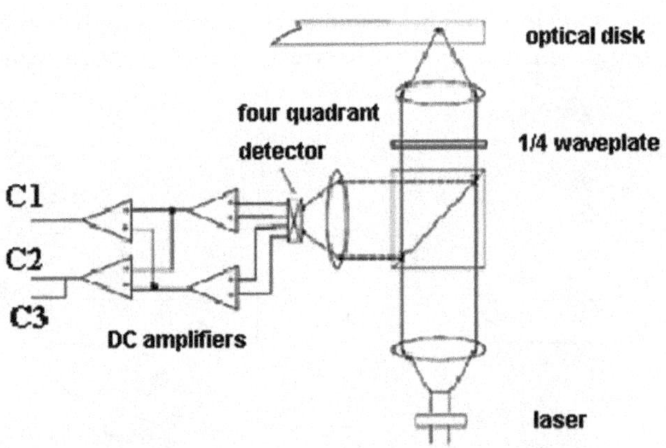

Fig. 8.23 A typical recording/readout system for a phase-change rewritable optical disk

In direct rewriting technology, the pulse duration (pulse width) of an erasing pulse is same as that of a writing pulse (about 50 ns). However, with double-pulse rewriting technology, the duration of the erasing pulse is much longer than that of the writing pulse. A comparison of double-pulse rewriting and direct rewriting

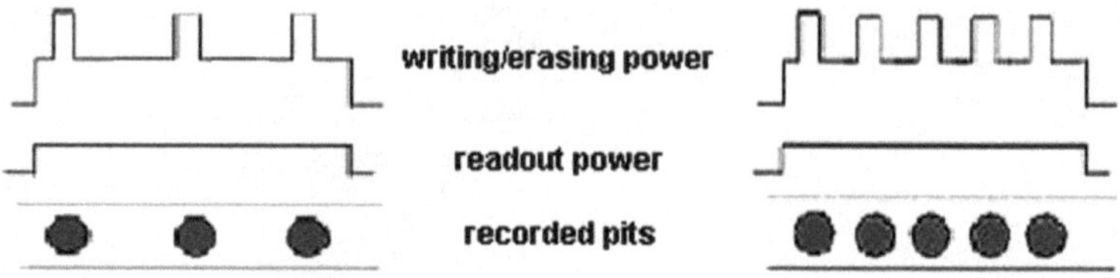

Fig. 8.24 Comparison of double-pulse rewriting (*left*) and direct rewriting (*right*)

technology is shown in Fig. 8.24. A shorter erasing laser pulse results in lower thermal diffusion, which is helpful for decreasing cross-talk and enhancing density.

Rewritable Magneto-Optical Disks

A magneto-optical (MO) disk is capable of writing and rewriting data by magneto-optical Kerr effects. The MO systems exploit basic principles of both the magnetic and the optical storage systems: MO systems write magnetically (with thermal assist) and read optically. Although more like optical disks, they appear as magnetic hard disks to the computer's operating system and do not require a special file system. Magneto-optical disks are still common in some countries such as Japan, but have fallen into disuse in other countries in recent years.

Principles of Magneto-Optical Storage

In magneto-optical recording, the information is stored in the form of thermally induced magnetic domains (Fig. 8.25) and read-out is accomplished by sensing the resulting polarization change (optical Kerr effect) in the optical beam (Fig. 8.26) [13, 14].

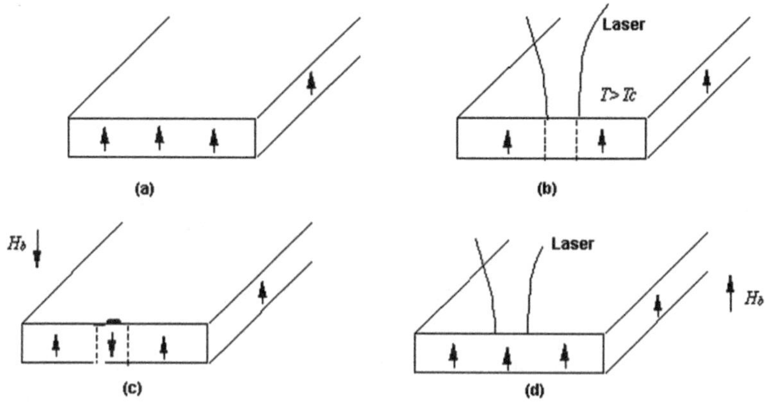

Fig. 8.25 Magneto-optical recording (**a–c**) and erasing process

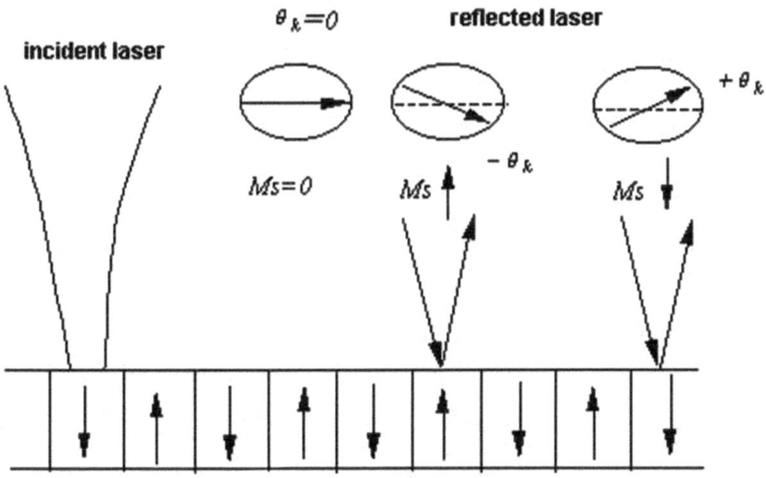

Fig. 8.26 Magneto-optical readout by the Kerr effect

A high signal-to-noise ratio (S/N) is required for magneto-optical recording. The S/N ratio can be expressed simply as:

$$\mathrm{S/N} = ARP_0 \sin^2 \theta_k, \tag{8.1}$$

where R is the reflectivity of the multilayer film, P_0 is the laser power, θ_k is the Kerr rotation angle, and A is a constant of the sensing system. It would be seen that a high S/N ratio could be obtained by increasing the laser power, but this strategy is limited because higher laser power could lead to a rise in temperature at the beam spot, and thus to a decrease in the Kerr angle. High values of θ_k (and R) are equally necessary for good MO recording performance.

Structure of an MO Disk

Quadrilayers are usually used in the design of MO disks [5]. The aluminum layer can serve as both the light reflector and the heat sink (to minimize lateral heating of the active layer). Sometimes, two disks are bonded together to achieve double capacity in a double-sided disk, as shown in Fig. 8.27.

Magneto-Optical Storage Materials

The materials for MO recording [15] should meet the following major criteria:

- Have an amorphous structure (smooth surface and domain boundaries to decrease system noise).
- Low thermal conductivity (to limit lateral heating to the recording layer).

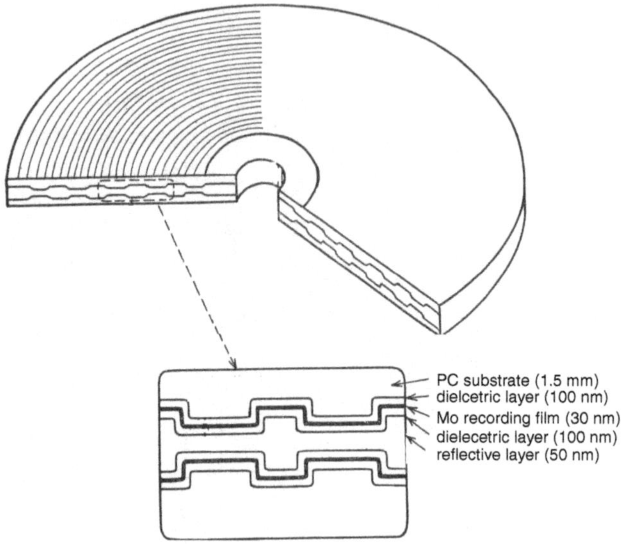

Fig. 8.27 Typical structure of an MO disk

- High melting point at about 200°–300°C (media stability, prevention of accidental loss of data).
- Rapid drop of coercivity near the Curie temperature (sharp recording threshold).
- High coercivity at room temperature (media stability, prevention of accidental loss of data).
- Vertical anisotropy (perpendicular magnetic recording).
- Chemical stability (constant material properties under repeated heating-cooling).

Amorphous rare earth-transition metal (RE-TM) alloys are typically used for the MO media. A shorter-wavelength light (~400 nm) is required to increase the recording density. An optical Kerr rotation angle (θ_k) is the most important requirement for MO media. Figure 8.28 illustrates the wavelength dependence of θ_k for amorphous binary and ternary RE-TM films. It can be seen from the Kerr rotation spectra that the value of θ_k decreases with wavelength for most of the heavy rare earth and transition metal (HRE-TM) alloys.

It is well known that TbFeCo alloy films are popularly used in MO recording, but their θ_k is small at short wavelengths. TbFeCo films are not suitable for high-density optical data storage with shorter-wavelength laser recording. Development of new media that show large Kerr rotation at short wavelengths is expected. The amorphous RE-TM alloys with light RE (LRE) elements, such as Nd, Pr, Ce, and Sm can increase θ_k at short wavelengths; however, as the magnetization of (Nd, Ce, Pr, Sm)-TM films is higher than that of Tb-Fe-Co, it is difficult to prepare films with perpendicular magnetization.

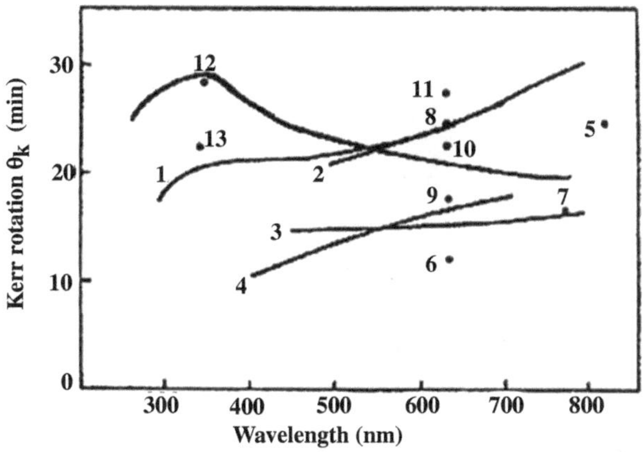

Fig. 8.28 Wavelength dependence of the Kerr rotation angle of binary and ternary amorphous RE-TM alloys: (1) $Gd_{26}(Fe_{81}Co_{19})_{74}$; (2) $Tb_{21}(Fe_{50}Co_{50})_{79}$; (3) $(Gd_{50}Tb_{50})_{28.6}Co_{71.4}$; (4) GdTbFe; (5) $Gd_7Tb_{13}Fe_{80}$; (6) TbDyFe; (7) GdTbDyFe; (8) GdFeBi; (9) $Gd_{26}Fe_{74}$; (10) $(Gd_{26}Fe_{74})_{84}Sn_{16}$; (11) $Gd_{28}(Fe_{90}Co_{10})_{68}Bi_4$; (12) $Nd_{19}Co_{81}$; (13) $Ce_{21}Co_{79}$

It has been found that artificially superstructured multilayer films (MLFs) have an axis of easy magnetization perpendicular to the plane of the film and display strong interfacial magnetic anisotropy. Examples include PrGd/FeCo and Nd/FeCo superstructure multilayer films [13].

Superresolution Technology in MO Disks

High-density MO data storage is achieved using magnetically induced superresolution (MSR) technology, which includes mainly center aperture detection (CAD), a magnetic amplifying magneto-optical system (MAMMOS), and domain wall displacements (DWDD) [13]. MSR technology can make it possible to read the magnetic domain, which is smaller than a laser spot. Figure 8.29 shows the schematic diagram of magnetic superresolution with the central aperture detection (CAD-MSR) mechanism [13]. The magnetic layer configurations are exchange-coupled double-layer (ECDL) or exchange-coupled multilayer (ECML), consisting of the readout layer and the recording layer or the intermediate (switching) layer. The function of the readout layer is to replicate the magnetic domain of the recording layer by magnetic coupling or the switching layer by exchange catena force. The RE-rich readout layer has high magnetization and low coercivity, which makes magnetic coupling easy, and the TM-rich recording layer has high coercivity to guarantee perpendicular recording.

Superresolution Optical Data Storage

The readout system for optical disk is comparable to a scanning optical microscope [16–18]. The resolving power of a coherent optical system is determined by the uncertainty principle [19], i.e., $\Delta x \geq \frac{h}{\Delta P_x}$, where $\Delta P_x = \hbar k_x$ is the momentum of photons, h and $\hbar$ are Planck's constants, and $k_x = k_0 \sin\theta = \frac{2\pi}{\lambda} \sin\theta$ is the

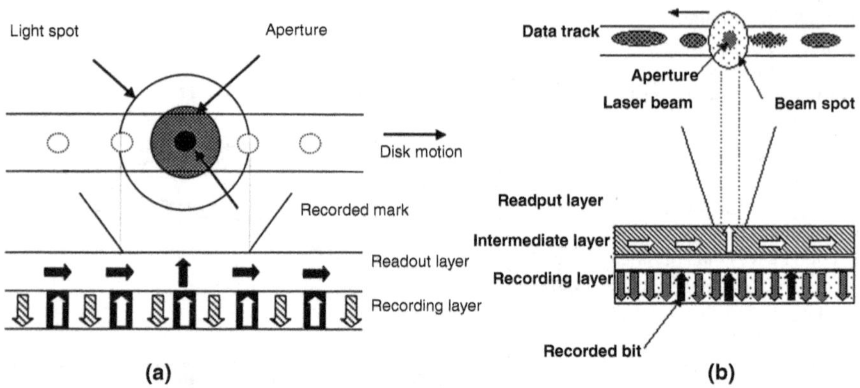

Fig. 8.29 Schematic diagram of CAD-MSR: (**a**) ECDL, (**b**) ECML

wave number in the measurement direction x. The maximum difference in photon momentum is $\Delta P_x = 2\hbar k_x$, thus the highest resolving power of a coherent optical system is $\Delta x = \frac{\lambda}{2\sin\theta} = \frac{\lambda}{2NA}$. This formula is also usually used to describe the size of the point spread function of an optical system. However, for a scanning optical system, the maximum difference in the photon momentum is $\Delta P_x = 4\hbar k_x$, which corresponds to a cut-off resolution of $\Delta x = \frac{\lambda}{4NA}$; because of finite coherence of laser and noise issues, the maximum achievable resolving power of an optical disk system is between $\Delta x = \frac{\lambda}{2NA}$ and $\Delta x = \frac{\lambda}{4NA}$ [20].

Many ways have been suggested for overcoming this resolution limit,; such as the apodization [21–29], which is achieved by manipulating the amplitude [25] or phase [21–24, 26–29] or both [30] of the field on the exit pupil of an imaging system to reduce the point spread function. A diffraction grating inserted between the imaging lens and the object can also improve the resolution of an objective lens [31], but is not suitable for an optical disk system because of the difficulty of integrating it into the system. In another technique, using an absorbing layer that shows pronounced nonlinear behavior reduces the size of the read-out spot owing to local thermal bleaching of the nonlinear auxiliary layer [32, 33]. However, practical problems such as the stability of the nonlinear response layer, power control, and photon efficiency have precluded the use of this technique.

The resolving power of an optical data storage system is usually enhanced by increasing the numerical aperture of the pickup and decreasing the laser wavelength, which results in a rapid decrease in the depth of the focus [34]. However, optical pickup has to follow the vibration of the disks, which can best be done at low vibration frequencies, as the optical pickup cannot follow the disk at high-frequency vibrations. When the vibration amplitude is larger than the depth of the focus of the pickup lens, the readout process cannot be completed because the extensive defocusing and spherical aberration degrade the resolving power of the pickup lens. A solution to the high-frequency vibration problem is to use a pure-phase

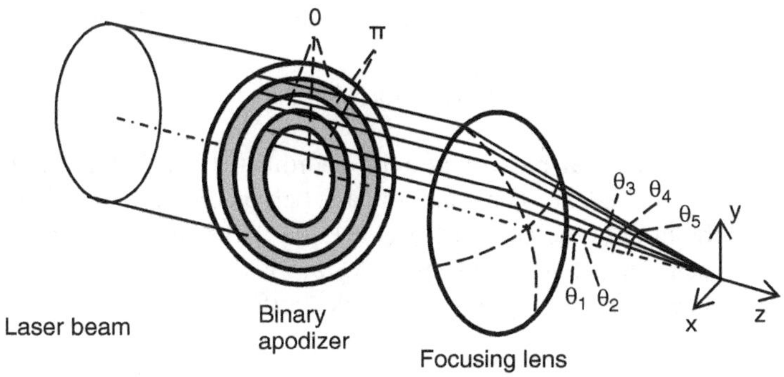

Fig. 8.30 Schematic configuration of the superresolution and nondiffraction system

apodization technique to increase both the depth of focus and the resolving power of the optical pickup by generating a localized nondiffraction beam with super-resolution [26–29].

Design of the Binary Apodizer

The structure of the binary apodizer, shown in Fig. 8.30, consists of a multiple annular ring structure; the phase difference between adjacent rings is P_i. This kind of binary apodizer can be made of transparent substrate with annular grooves or bumps with depths of $\lambda/[2(n\text{-}1)]$, where λ is the wavelength of the light and n is the refractive index of the substrate. To achieve super resolution and eliminate diffraction, the apodizer has to be placed on the exit pupil of the objective lens, as shown in the figure.

Apodizer Design Based on Scalar Focusing

When a uniform distribution of light intensity is subjected to a multibelt annular apodizer and then focused by the lens, as is shown in Fig. 8.30, the normalized amplitude distribution in the focal region can be simplified as [26, 27]:

$$G(\rho, u) = 2\sum_{j=1}^{N} \exp\left(i\phi_j\right) \int_{j-1}^{r_j} rg\left(r\right) J_0\left(\rho r\right) \exp\left[-\left(1/2\right) iur^2\right] dr \tag{8.2}$$

Here $J_0(\rho, u)$ is the Bessel function of zero order, r is the radial coordinate of the objective lens pupil plane, and $g(r)$ is the amplitude distribution in the radial direction of the pupil plane of the lens; ρ and u are simplified radial and axial coordinates, respectively.

$$\rho = (2\pi/\lambda)(\mathrm{NA})\mathrm{R}, \tag{8.3}$$

$$u = (2\pi/\lambda)(\mathrm{NA})^2\mathrm{Z}, \tag{8.4}$$

where R and Z are the real radial and axial coordinates and NA is the numerical aperture of the objective lens. The phase of each belt on the pupil plane is ϕ_i, $(i=1, 2, \ldots, n)$, and the radii of the belts are r_i, $(i=1, 2, \ldots, n)$ and $r_{i-1} < r_i < r_n = 1$.

For a three-belt pure phase apodizer, $r_1 = b$, $r_2 = a$, $r_3 = 1$, axial optimization can be achieved by solving a second-order differential equation for the axial intensity curve:

$$\frac{\partial^2 I(0,u)}{\partial u^2}|_{u=0} = 0 \tag{8.5}$$

Then the relationship between a and b is

$$-\frac{1}{2} - 2(a^2 - b^2)^4 - (1-a^2)^4 + (1-b^2)^4 + a^8 - b^8 = 0 \tag{8.6}$$

Two curves that describes the relationship between a and b can be obtained by solving equation (8.6), as is shown in Fig. 8.31. When the numerical aperture of the pickup is 0.85, the axial intensity distribution for some of a and b pairs are as plotted in Fig. 8.32. The solid curve corresponds to the system without the apodizer. However, not all the a and b pairs for optimized axial intensity can result in super-resolution of light spot, as is shown in Fig. 8.33. The solid curve corresponds to the system without the apodizer. Only two of the selected pairs can result in a smaller beam spot than that obtained without the apodizer, i.e., $b = 0.28$, $a = 0.5575$ and $b = 0.3$, $a = 0.5813$, so this optimization process also includes a beam spot size comparison.

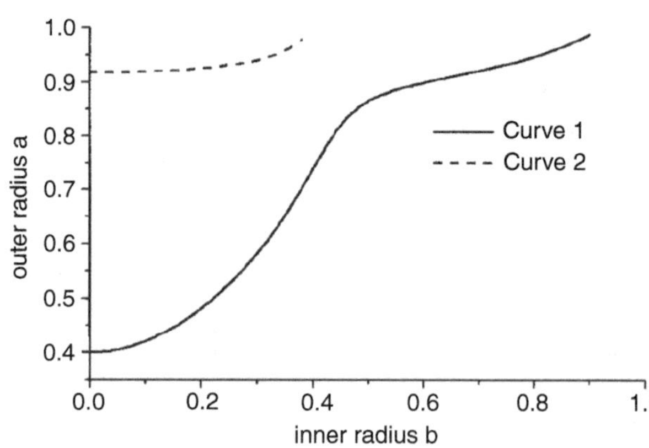

Fig. 8.31 Relationship between a and b

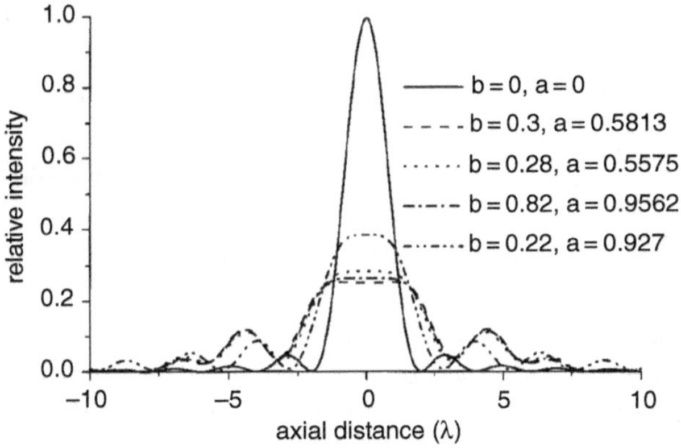

Fig. 8.32 Axial intensity for different *a* and *b* pairs

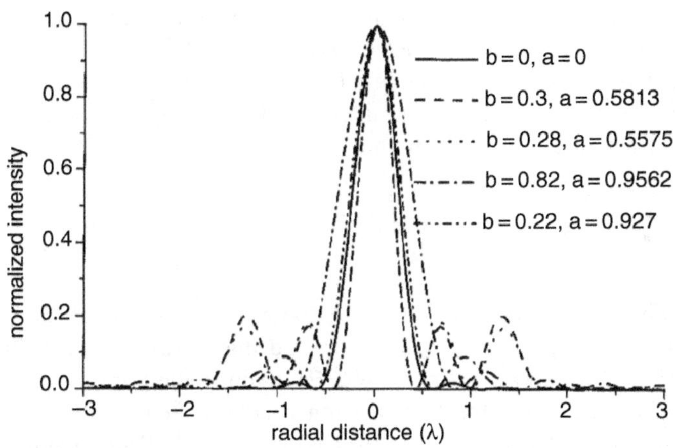

Fig. 8.33 Radial intensity for different *a* and *b* pairs

Apodizer Design Based on Vector Focusing

When the numerical aperture of the optical pickup is above 0.6, vector focusing is the preferred method in apodizer design. If the field on the exit pupil of the focusing lens is circularly polarized, the field in the focal region is given as [34]

$$E_x = -iA\left(I_0 + I_2\left(\cos(2\varphi) + i\sin(2\varphi)\right)\right), \tag{8.7}$$

$$E_y = -iA\left(iI_0 - iI_2\left(\cos(2\varphi) + i\sin(2\varphi)\right)\right), \tag{8.8}$$

$$E_z = -2AI_1\left(\cos(\varphi) + i\sin(\varphi)\right), \tag{8.9}$$

and

$$I_0 = \int_0^\alpha \sqrt{\cos(\theta)} \sin(\theta)(1 + \cos(\theta))J_0(k_0 r \sin(\theta)) \exp(ik_0 z \cos(\theta))d\theta, \quad (8.10)$$

$$I_1 = \int_0^\alpha \sqrt{\cos(\theta)} \sin^2(\theta)J_1(k_0 r \sin(\theta)) \exp(ik_0 z \cos(\theta))d\theta, \quad (8.11)$$

$$I_2 = \int_0^\alpha \sqrt{\cos(\theta)} \sin(\theta)(1 - \cos(\theta))J_2(k_0 r \sin(\theta)) \exp(ik_0 z \cos(\theta))d\theta, \quad (8.12)$$

where $\alpha = \arcsin(\mathrm{NA})$, denotes the largest focusing angle, φ is the azimuthal angle, $k_0 = 2\pi/\lambda$, and J_0, J_1, and J_2 are zero-order, first-order, and second-order Bessel functions. The constant $A = \pi l_0 f/\lambda$, with $l_0 = 1$ for uniform illumination, and f is the focal length.

The circular polarized plane wave undergoes a multibelt binary optical element and is then focused by an aplanatic lens with an NA of 0.85; the transmission of the aperture is $T(\theta)$. Optimization of the binary optics can be accomplished by making the intensity on the optical axis constant within certain range; because of the circular polarized light is rotationally symmetric and E_z is equal to zero on the optical axis, the optimization can be achieved by only look at the E_x field intensity on the optical axis towards obtaining constant axial intensity. The strength of the electric field at an axial point with a distance z from the focal plane is given as [28]:

$$F(z) = \int_0^\alpha (1 + \cos\theta)(\cos\theta)^{1/2} T(\theta) \exp(ikz\cos\theta) \sin\theta d\theta. \quad (8.13)$$

For n-belt binary optics (n =1, 2, 3, ...), the radius of each belt is r_i, (i =1, 2, ..., n), and $r_{i-1} < r_i < r_n = 1$, $r_0 = 0$, and the corresponding focusing angle is $\alpha_i = \arcsin(r_i NA)$, so $r_i = \sin(\alpha_i)/NA$. By making the transmission coefficient within each belt the same, the transmission function $T(\theta)$ can be expressed as a function of the belt order, and is given by $T(i) = (-1)^{i+1}$. Thus equation (8.13) can be further expressed as

$$\begin{aligned} F(z) &= \sum_{i=1}^{n} T(i) \int_{\alpha_{i-1}}^{\alpha_i} (1 + \cos\theta)(\cos\theta)^{1/2} \exp(ikz\cos\theta) \sin\theta d\theta \\ &= \sum_{i=1}^{n} (-1)^{i+1} \left(f(\alpha_i, z) - f(\alpha_{i-1}, z)\right), \end{aligned} \quad (8.14)$$

where

$$f(\alpha_i, z) = \Big(\exp(jkz)\,(3 - 4jkz) + j\exp(jkz\cos\alpha_i)\,(\cos\alpha_i)^{1/2}\,(3j + 2kz + 2kz\cos\alpha_i)\Big) \Big/ \left(2k^2z^2\right) + \sqrt{j\pi kz}\,(3j + 2kz)\left(Erfi\left(\sqrt{jkz}\right) - Erfi\left(\sqrt{jkz\cos\alpha_i}\right)\right) \Big/ \left(4k^3z^3\right), \quad (8.15)$$

and $j = \sqrt{-1}$, $Erfi(x)$ is the imaginary error function.

Equation (8.14) describes the field on the optical axis generated by the system using the multibelts π-phase binary optical element. With this expression and the relation $\alpha_i = \arcsin(r_i\text{NA})$, series of r_i can easily be found in order to obtain the expected axial intensity. Nondiffraction beams can be obtained through optimizing the radius (r_i) of each belt towards obtaining constant axial intensity, those with beam size smaller than diffraction limit are superresolution beams. For example, the axial intensity (as shown in Fig. 8.34) is made constant within an appreciable range by using a seven-belt ($r_1 = 0.0896$, $r_2 = 0.2852$, $r_3 = 0.4869$, $r_4 = 0.6136$, $r_5 = 0.6755$, $r_6 = 0.7688$, $r_7 = 1$) phase element, the full width half-maximum of the total electric density profile in the focal plane is 0.53λ, the diffraction limit of this objective lens is 0.59λ, and the beam size is about 9% smaller than the diffraction limit. The image of the beam in the focal region before and after using the binary apodizer is shown in Fig. 8.35; it is clear that in the original system, the beam diverges rapidly away from the focal plane, whereas for the system with the binary apodizer, the beam does not diverge within a five-wavelength range.

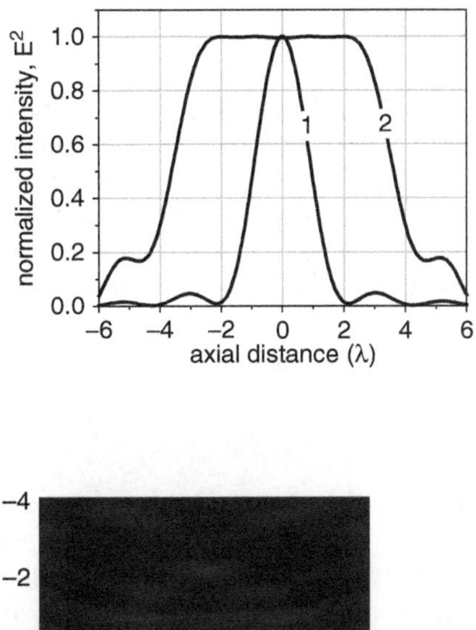

Fig. 8.34 Axial intensity in the focal region of the NA = 0.85 lens: (1) original system and (2) with the seven-belt optimized binary apodizer

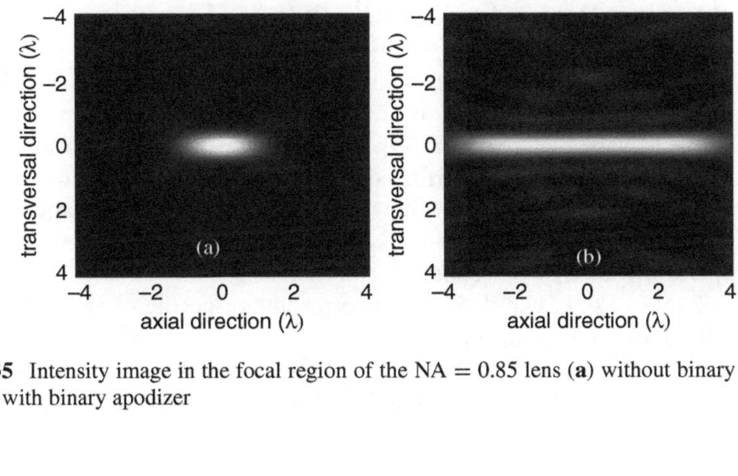

Fig. 8.35 Intensity image in the focal region of the NA = 0.85 lens (**a**) without binary apodizer, and (**b**) with binary apodizer

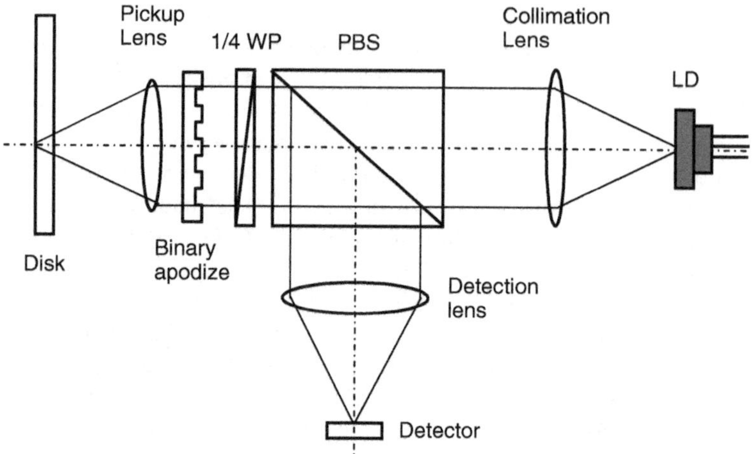

Fig. 8.36 Schematic of the system configuration for optical data storage

System Configuration

One configuration designed to achieve superresolution of the system's reading and writing is shown in Fig. 8.36; the binary apodizer is placed just in front of the pickup lens. A beam from the laser diode is collimated by the collimation lens, goes through a polarization beam splitter, and is then turned into circular polarized light by a quarter-wave plate. The phase of the beam is modulated by the binary apodizer and a superresolution and nondiffraction beam is generated at the focal region of the pickup lens. When this beam is used to scan the optical disk, a high-resolution optical signal passes through the same optics and is picked up by the detector.

Binary Apodizer Design for Radially Polarized Light

Recently, it has been demonstrated that focusing of radially polarized light using a ring aperture to obstruct the center part of the incident beam can result in the smallest light spot achievable in free space [35]. This can happen because the suppression of the radial field component contributed by the lower aperture field effectively reduces the cross-polarization effect; however, such a beam also diverges when it is out of focus and the obstruction results in very low optical efficiency. Replacing the ring aperture with a binary apodizer and increasing the numerical aperture of the focusing lens to 0.95 can deliver a superresolution needle of (nondiffracting) longitudinally polarized light [29].

The electric fields in the focal region for illumination of the high aperture lens with a radially polarized Bessel-Gaussian beam is expressed as [29, 36],

$$E_r(r,z) = A\int_0^{\alpha}\cos^{1/2}\theta\sin(2\theta)\ell(\theta)J_1(kr\sin\theta)e^{ikz\cos\theta}d\theta, \tag{8.16}$$

$$E_z(r,z) = 2iA\int_0^{\alpha}\cos^{1/2}\theta\sin^2\theta\,\ell(\theta)J_0(kr\sin\theta)\,e^{ikz\cos\theta}d\theta. \tag{8.17}$$

where α = arcsin(NA), NA is the numerical aperture, $J_0(x)$ and $J_1(x)$ denote Bessel functions. The function $\ell(\theta)$ describes the amplitude distribution of the Bessel-Gaussian beam, which is given by

$$\ell(\theta) = \exp\left[-\beta^2\left(\frac{\sin\theta}{\sin\alpha}\right)^2\right]J_1\left(2\gamma\frac{\sin\theta}{\sin\alpha}\right), \tag{8.18}$$

where β and γ are parameters that are taken as unity in this configuration and the numerical aperture of the focusing lens is NA = 0.95 ($\alpha \approx 71.8'$). The corresponding field distribution is shown in Fig. 8.37.

As shown in Fig. 8.37a, the radial component of the electric density E_r^2 is about 30% that of the longitudinal electric density E_z^2, and this strong cross-polarization effect makes the beam size as big as FWHM = 0.68λ, which is larger than the diffraction limit for this focusing lens λ/(2NA) = 0.526λ. Furthermore, this beam diverges rapidly away from the focus, as shown in Fig. 8.37b. Applying a binary apodizer to the exit pupil of the focusing lens can also achieve a superresolution and nondiffraction beam.

For example, when a five-belt binary apodizer (r_1 = 0.091, r_2 = 0.391, r_3 = 0.592, r_4 = 0.768, r_5 = 1) is applied, the radial electric density E_r^2 can be suppressed to around 8% of the longitudinal electric density E_z^2, and the beam size, i.e.,

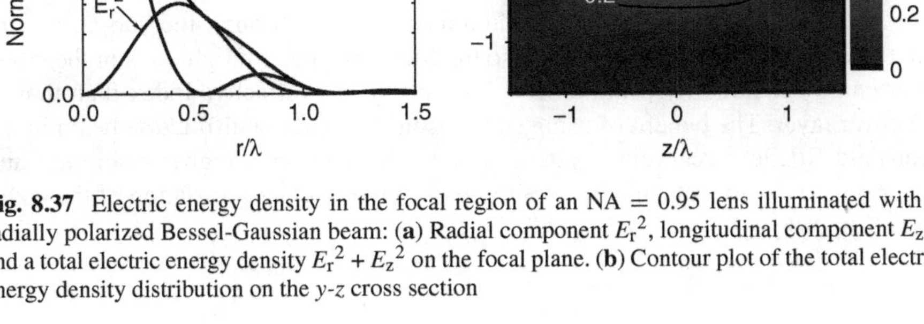

Fig. 8.37 Electric energy density in the focal region of an NA = 0.95 lens illuminated with a radially polarized Bessel-Gaussian beam: (**a**) Radial component E_r^2, longitudinal component E_z^2, and a total electric energy density $E_r^2 + E_z^2$ on the focal plane. (**b**) Contour plot of the total electric energy density distribution on the y-z cross section

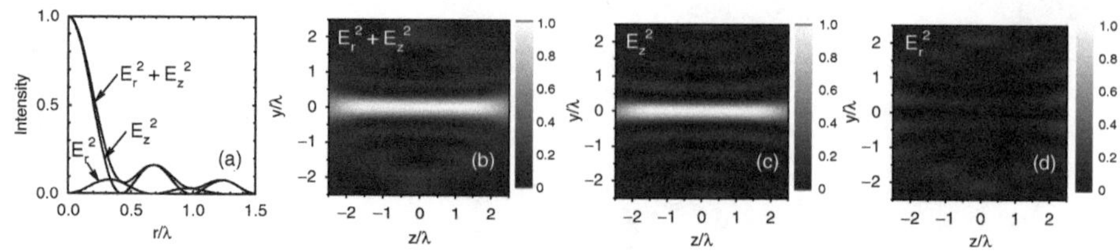

Fig. 8.38 Electric density profile on the focal plane and contour plots for the electric density distributions in y-z plane after using the binary apodizer: (**a**) electric density profile on focal plane; (**b**) the total energy density distribution; (**c**) longitudinal components; (**d**) radial component

the FWHM of the $E_r^2 + E_z^2$ profile is only 0.43λ, which is about 18% smaller than the diffraction limit of the optical system; the nondiffraction effect is also achieved, as shown in Fig. 8.38b. The total electric density beam size is constant within four-wavelength range and the longitudinally polarized electric density dominates, as can be seen in Fig. 8.38c. The radially polarized electric density is quite low, as indicated in Fig. 8.38(d); this beam is substantially longitudinally polarized. The binary apodizer acts like a polarization filter, scattering the radial polarized light away from the center of the beam.

Discussion and Conclusion

Although the superresolution effect obtained for circular polarized light is only 9% when the NA of the lens is 0.85, which corresponds to a decrease of 18% in beam area or a 22% increase in disk capacity; this superresolution ratio may increase to around 20% for lower numerical apertures such as NA = 0.65, which corresponds to a 36% decrease in beam area, or around a 50% increase in disk capacity. The binary optics design is suitable for both far- and near-field optical data storage systems. However, the design of the binary apodizer for radially polarized light is only suitable for near-field storage. This is the case because when focused radially polarized light goes from air to disk substrate, the focusing angle becomes smaller, which results in an increase in the radially polarized component in the focal region; a large radially polarized component makes the beam size larger, which reduces the resolution of the beam. This problem can be overcome in the near-field SIL lens system [37, 38], because higher-resolution photons can jump through the sub-50-nm air gap to the cover layer of optical disk, so the focusing angle of light rays in the cover layer can be ensured by using a SIL lens with higher refractive index than that of the cover layer. The benefit of using superresolution and nondiffraction beam in the near-field SIL lens recording system is that it may provide higher resolution and lower sensitivity to a change in the air gap or in the thickness variation of the cover layer [38, 39].

Holographic Storage

The concept of holographic storage was first proposed by Pieter van Heerden, a researcher at Polaroid in 1963, who suggested using an optical method to store data in three dimensions. In theory, it is a great idea. Existing media store data in only two dimensions and adding another dimension would make storage devices far more efficient, but a complete, general-purpose commercial system has eluded both industrial and academic researchers for over four decades.

Such a system is still elusive, but it is getting closer. The first commercial holographic memory should be on the market in the relatively near future, and more designs are expected to follow. When that happens, there may be a ballooning of computer storage capacity that will make existing disks look like leaflets compared with the *Encyclopaedia Britannica*.

Principle of Digital Holographic Storage

Principle

Digital holographic storage is a method of optical information recording by optical holography. Holography breaks through the density limits of conventional storage by going beyond recording only on the surface, to recording through the full depth of the medium. Unlike other technologies that record one data bit at a time, holography records and reads over a million bits of data with a single flash of light. This enables transfer rates significantly higher than those of the current optical storage devices. Combining high storage densities and fast transfer rates with durable, reliable, low-cost media, make holography poised to become a compelling choice for next-generation storage and content distribution needs.

Hologram Writing and Reading Process

- **Recording data**
  In holographic data storage, light from a coherent laser beam is split in two, a signal beam (data-carrying) and a reference beam [40]. Encoding data onto the signal beam is accomplished by a spatial light modulator (SLM), which translates the electronic data of 0's and 1's into an optical "checkerboard" pattern of light and dark pixels (see Fig. 8.39). The data are arranged in an array or page of over one million bits. The exact number of bits is determined by the pixel count of the SLM. The hologram is formed where the reference beam and the data-carrying signal beam intersect in the recording medium. At the spot where these two beams intersects, a chemical or physical change occurs. By varying the reference beam angle or the position of the medium hundreds of unique holograms are recorded in the same volume of material.

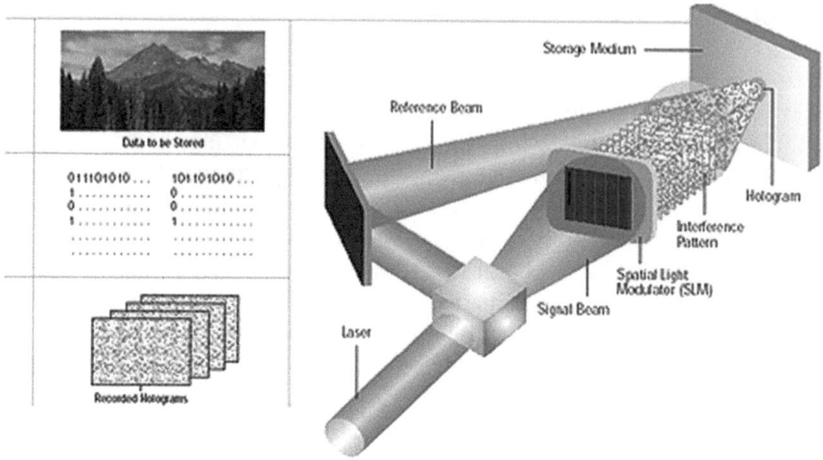

Fig. 8.39 Hologram writing process

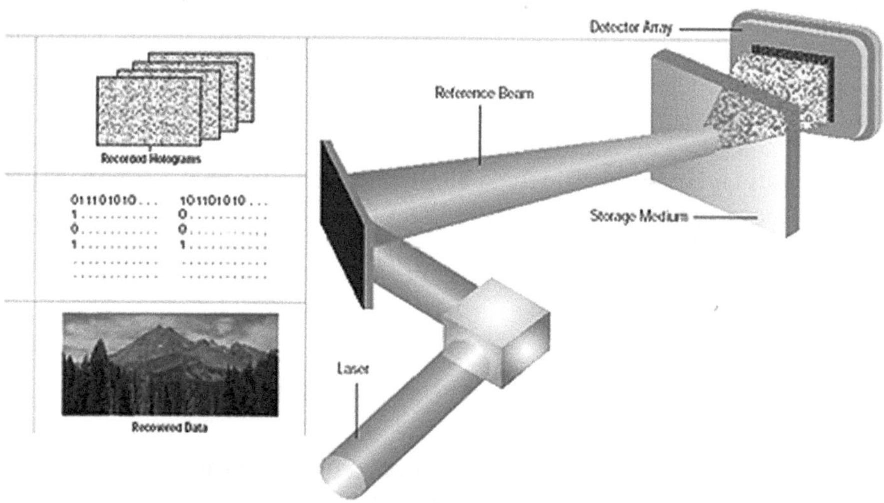

Fig. 8.40 Hologram reading process

- **Reading data**
  The interference pattern induces the modulations in the refractive index of the recording material yielding diffractive volume gratings. In order to read the data, the reference beam deflects off the grating, thus reconstructing the stored information (see Fig. 8.40). The reconstructed array is projected onto a pixilated detector that reads the data in parallel (such as "charge couple device," CCD). This parallel readout provides holography with its fast data transfer rates (10's to 100's of Mb/s). The readout of data depends sensitively on the characteristics of the reference beam. By varying the reference beam, for instance, by changing its

angle or the wavelength of the incidence light, many different data pages can be recorded in the same volume of material and read out by the reference beam that was used during writing. This process of multiplexing data yields the enormous storage capacity of holography.

Types of Holograms

- **Reflection Hologram**
  The reflection hologram, in which a truly three-dimensional image is seen near its surface, is the most common. The hologram is illuminated by the light source at a specific angle and distance and located on the viewer's side of the hologram. The image consists of light reflected by the hologram. Recently, these holograms have been created and displayed in color, which provides optical images that are indistinguishable from the original objects.
- **Transmission holograms**
  The typical transmission hologram is viewed with laser, usually of the same type used to make the recording. Transmission holograms are viewed with the light source on the opposite side of the hologram from the viewer, so that the light is transmitted through the hologram. The virtual image can be very sharp and deep. For example, through a small hologram, a full-size room with people in it can be seen as if the hologram were a window. If this hologram is broken into small pieces, one can still see the entire scene through each piece. Depending on the location of the piece (hole), a different perspective can be observed. Furthermore, if an undiverged laser beam is directed backward through the hologram, a real image can be projected onto a screen located at the object's original position.
- **Hybrid holograms**
  There are many variations of both reflection and transmission holograms:
- **Embossed holograms**
  To mass produce cheap holograms for security applications such as the eagle on the VISA cards, a two-dimensional interference pattern is pressed onto thin plastic foils, which is viewed with reflected light; this is actually a transmission holograms "mirrorized" with a layer of aluminum on the back. The original hologram is usually recorded on photosensitive material called photoresist. When developed, the hologram consists of grooves on the surface. A layer of nickel is deposited on this hologram and then peeled off, resulting in a metallic "shim."
- **Multiplex holograms**
  A transmission or reflection hologram can be generated from a series of photographs of an object – which can be a live person, an outdoor scene, a computer graphic, or an X-ray image. Usually, the object is "scanned" by a camera, which records many discrete views. Each view is shown on an LCD screen illuminated with laser and is used as the object beam to record a hologram on a narrow vertical strip of holographic plate. The next view is similarly recorded on an adjacent strip and so on. When viewing the finished composite hologram, the left and right eyes see images from different narrow holograms and a stereoscopic image is observed.

- **Holographic interferometry**
  Making two exposures of a changing object can measure and record the microscopic changes of an object quantitatively. The two images interfere with each other and fringes can be seen on the object that reveal the vector displacement. In real-time holographic interferometry, the virtual image of the object is compared directly with the real object.
- **Multichannel holograms**
  As the angle of the viewing light on the same hologram is changed, absolutely different sights can be observed, and this has immense potential for massive computer memories.
- **Computer-generated holograms**
  There are three basic elements in holography: the light source, the hologram, and the image. If any two of the elements are known, the third can be computed. For example, if there is parallel beam of light of a certain wavelength and a “double-slit” system (a simple “hologram”), the diffraction pattern can be calculated. Also, from the diffraction pattern and the details of the double-slit system, the wavelength of the light can be calculated [41].
- **Rainbow holograms**
  In transfer transmission holograms the original "master" hologram is masked off through a horizontal slit. This reduces the vertical parallax, but enables the hologram to be viewed by white light without causing a rainbow smear. The object appears to change color as the viewing point moves vertically.

Multiplexing Methods

Optical memory, which enables the storage of large amounts of information, is one of the most desired targets for computing technology, since society continues to demand better tools to transmit more data at higher rates [42]. Owing to its large capacity and high recording speed, holographic optical recording is regarded as one of the very promising approaches for increasing recording density [43].

If one can store a page of bits in a hologram, one can create and interface to a computer. The problem is that storing only one page of bits is not beneficial, but fortunately, the properties of holograms provide a unique solution. Unlike a magnetic storage mechanism, which stores data on its surface, holographic memories store information throughout their whole volume. After a page of data is recorded in the hologram, a small modification to the source beam before it reenters the hologram records another page of data in the same volume. This method of storing multiple pages of data is called multiplexing. The thicker the volume becomes, the smaller the modifications to the source beam can be.

Angular Multiplexing

When a reference beam recreates the source beam, it has to be at the same angle that it was during recording because even a very small change in this angle will make the regenerated source beam disappear. Owing to this property, angular multiplexing changes the angle of the source beam by an infinitesimal amount after each page of data is recorded. By harnessing the sensitivity of the recording material, thousands of pages of data can be stored in the same point of laser beam entry [44]. Unlike conventional data access systems that move mechanical matter to obtain data, the angle of entry of the source beam can be deflected by high-frequency sound waves in solids [45]. The elimination of mechanical access methods reduces access times from milliseconds to microseconds.

Wavelength Multiplexing

Used mainly in conjunction with other multiplexing methods, wavelength multiplexing alters the wavelength of source and reference beams between recordings. Sending beams to the same point of origin in the recording medium at different wavelengths allows multiple pages of data to be recorded, but this form of multiplexing is limited because lasers have such a small tuning range.

Spatial Multiplexing

Spatial multiplexing is the method of changing the point of entry of source and reference beams into the recording medium. This form tends to break away from the nonmechanical paradigm because either the medium or the recording beams must be moved physically. Like wavelength multiplexing, this is combined with other forms of multiplexing to maximize the amount of data stored in the holographic volume. Two commonly used forms of spatial multiplexing are peristrophic multiplexing and shift multiplexing.

Peristrophic Multiplexing

This form of spatial multiplexing rotates the recording medium as the source beams remain in fixed positions [46]. For instance, a holographic cube could be rotated so each of its six sides could take in a source beam. This would provide six times the number of pages that could be stored in the volume.

However, certain problems arise when implementing this method of multiplexing. The rotational axes have to be positioned in a way that does not interfere with the laser beams. As with all spatial multiplexing, bringing the recording medium back to its original position for data retrieval would have to be done very precisely, and this is much easier to accomplish when the medium remains static.

Shift Multiplexing

Shift multiplexing alters the point of entry on one surface of the recording media. The recording optics or medium could be repositioned to allow the source beam to enter multiple positions on a surface. Depending on the size of the laser beam and the beam takes into it can be immense [46]. This form of multiplexing combined with peristrophic multiplexing could cover a very large percentage of the hologram.

Phase-Encoded Multiplexing

Rather than manipulating the angle of entry of a laser beam or rotating/translating the recording medium, phase-encoded multiplexing changes the phases of individual parts of a reference beam. The main reference beam is split into many smaller partial beams, which cover the same area as the original reference beam. These smaller beamlets vary by phase, which changes the state of the reference beam as a whole. The reference beam intersects the source beam and records the diffraction relative to the different phases of the beamlets. The phases of the beamlets can be changed by nonmechanical means, thus speeding up the access times [44].

Combining Multiplexing Methods

No single multiplexing method by itself is the best way to pack a hologram full of information. The true power of multiplexing is brought out in the combination of two or more methods. Hybrid wavelength and angular multiplexing systems have been tested and the results are promising. Recent tests have also been performed using spatial multiplexing methods to create a hologram that is the size of a compact disk can hold 500 times more data [47].

Holographic Recording Media

Photopolymer

Traditional holographic recording media includes silver halide, dichromated gelatin, and photoresist, etc. Silver-halide-based materials are quite suitable for making both hologram originals and hologram copies. The disadvantages include that its' original material is expensive, the processing method is complicated, and it must be processed in solution, etc. Dichromated gelatin and conventional photoresist materials, which are less expensive, have also been tried, but they require wet processing steps to develop and fix the image and thus delay availability. Relatively inexpensive copies can be made by a diazo film process, but, the exposure time is relatively long and the spatial frequency capability and the image intensity are low compared to the original. Electrostatic imaging processes offer high speed and quick access, but are restricted to low spatial frequencies because of the size of the toner particles.

Photopolymer materials can be used for recording phase holograms, where applications in the mass production of display holograms and optical elements are of primary interest. Normally, the material has a short shelf life and a rather limited refractive index change. The exposures for transmission holograms and reflection holograms are, respectively, about 5 and 30 mJ/cm^2. Diffraction efficiency can be as high as 60% for a transmission hologram and 85% for a reflection hologram, and the signal-to-noise ratio is about 90:1 for exposures that give the highest diffraction efficiency.

Spatially nonuniform illumination during holographic recording produces free radicals by initiator decomposition. The subsequent reaction of free radicals with monomer molecules leads to vinyl polymerization of the monomers in the bright regions. This polymerization process lowers the chemical potential of monomers in these regions, leading to their migration (diffusion) from the dark to the bright regions. Photopolymer materials are practical choices for digital holographic recording media, as they are inexpensive and self-processing (dry processed). The photochemical mechanisms that act during recording in these materials must be understood if there is to be further development [48]. There are many reports on various monomers, hosts, and techniques employed for holographic recording [49–52]. During holographic recording, the optical interference pattern initiates polymerization in the photoreactive system; polymerization occurs in the light intensity maxima of the interference pattern, whereas there is no polymerization in the nulls.

Reports on photopolymer-based holographic recording were available during early 1970s [53]. A photopolymer layer containing a polymeric binder, an additional polymerizable monomer, a photoinitiator system, a chain transfer agent, and a plasticizer was used. Exposure of the layer to the radiation bearing holographic information results in polymerization with accompanying changes in various physical properties in the exposed areas. The use of polymerizable compositions for holograms enables the recording of very high spatial frequencies free of any particle size limitation, since the photosensitive compositions used do not contain any particulate material. The single step of exposure alone produces a useful and stable finished hologram. The inventors reported recording in the range of 200 to 2000 lines/mm spatial frequencies. Reports of a similar composition indicated a refractive index modulation of at least 0.005 and a spatial line frequency of about 1000 lines per mm [53].

Photorefractive Crystal

Photorefractive effect is the short term for the effect of light-induced refractive index change; i.e., the phenomenon of the refractive index varying with the spatial distribution of the light intensity under the light irradiation in certain photoelectric materials. Just as a complex electro-optical process takes place within the electro-optic materials. Under the light irradiation, inside the electro-optical crystal that has certain impurities or defects formed the charge spatial distribution corresponding to the space distribution of radiation intensity, and the resulting matching space-charge field. As the linear electro-optic effect, ultimately, the refractive index spatial modulation is formed inside the crystal, and the refractive index modulating phase grating is written into the crystal. As light scattering is real-time read out owing to the effect of self-written phase grating, the grating in the photorefractive crystal is dynamic grating, so that photorefractive materials are suitable for real-time holographic recording. Moreover, the hologram memorized in the photorefractive material can be fixed through a series of technologies. It can be written, read, erased, and fixed, with a performance level that makes the photorefractive crystal the first choice for holographic storage.

The physical processes involved in the photorefractive effect can be summarized as follows: (1) The carrier process is excited owing to the nonuniform distribution of light. (2) The process in which the space-charge field is produced by light-excited carrier mobility and trapped. (3) The process of refractive index modulation by a space-charge field through the linear photoelectric effect.

Types of photorefractive crystals:

(1) Ferroelectric crystal, such as lithium niobate ($LiNbO_3$), lithium tantalite ($LiTaO_3$), barium titanate ($BaTiO_3$), potassium niobate ($KNbO_3$) etc. These crystals have high electro-optic coefficients and can be used to store holographic grating with high diffraction efficiency, but response speed is slow.
(2) Bismuth silicon groups oxide, such as bismuth silicate ($Bi_{12}SiO_{20}$), bismuth germinate ($Bi_{12}GeO_{20}$), bismuth titanate ($Bi_{12}TiO_{20}$), etc. These crystals are characterized by a fast response speed and high sensitivity, but have smaller optical coefficients.
(3) Compound semiconductor, such as indium phosphide (InP), gallium arsenide (GaAs), gallium phosphide (GaP), cadmium telluride (CdTe), cadmium selenide (CdSe), zinc sulfide (ZnS), etc. These materials have a higher response speed and more photorefractive sensitivity than the first two groups, but have small optical coefficients spectral responses of $0.95 \sim 1.35\ \mu m$ in near-infrared region.

Nanoparticle Dispersed Polymer

Holographic dry photopolymers are useful for data storage applications if both the refractive index modulation (Δn) and the dimensional stability are high enough [54]. However, this is difficult to achieve in conventional all-organic photopolymers owing to the limited refractive-index range of organic materials. Methods for holographic manipulation of nanoparticle assemblies in photopolymer syrups and its use in photonics applications have been described by Suzuki and Tomita et al. [55–57]. Inorganic materials have a wide range of refractive indices. Thus, using inorganic nanoparticles to provide an optically mobile component in a photopolymer syrup offers the opportunity to attain larger Δn compared with conventional all-organic photopolymers. The inclusion of nanoparticles also ensures the appreciable suppression of polymerization related shrinkage and so to high dimensional stability.

Photoinsensitive nanoparticles are not consumed and undergo counter-diffusion from the bright to the dark regions. Their chemical potential increases in the bright regions owing to consumption of the monomer, so the mutual diffusion process continues until the photopolymerization is complete. This leads to redistribution of nanoparticles under holographic exposure, which, results in compositional and density differences between the bright and the dark regions, thus creating a refractive index grating (see Fig. 8.41). Tests have demonstrated that there is very little scattering loss in volume holographic recording in ZrO_2 nanoparticle dispersed acrylate

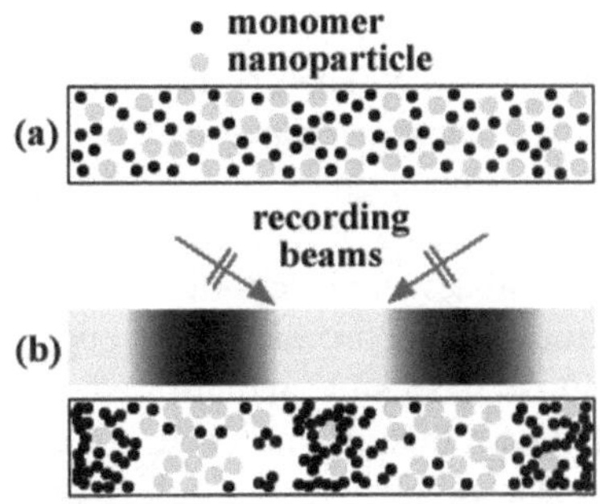

Fig. 8.41 Schematic diagram of grating formation by the components (**a**) before and (**b**) during holographic exposure to describe the holographic recording process in a nanoparticle-dispersed photopolymer

photopolymer films [58], and that a scattering coefficient more than thirtyfold smaller than those of previously reported TiO_2 nanoparticle dispersed photopolymers can be achieved. It has been shown that a refractive index modulation as high as 5.3×10^{-3}, together with substantive photopolymerization shrinkage suppression, is obtained at a nanoparticle concentration of 15 vol.%. The dependence of nanoparticle concentration and grating spacing on the refractive index modulation has also been investigated, and photopolymerization of aromatic methacrylate in organic-inorganic nanocomposite films for holographic writing has been studied as well [59]. High photoconversion leading to diffraction efficiency greater than 95% has been achieved.

Reports have shown that the nanoparticle dispersed polymers used for storing optical information as a spatial variation of the refractive index require: (a) a compound that has at least one polymerizable functional group; (b) a photopolymerization initiator; and (c) colloidal silica particles having an average particle diameter of from 4 to 30 nm [53]. Another invention involves an optical storage medium that includes nanoparticles. This medium has two or more recording layers with nanoparticles, each layer being made of a dielectric material, where the dielectric materials of at least two of the layers have different dielectric permittivity [53].

Photoactive Liquid Crystalline Polymer

Azobenzene-containing polymers are among the most well-researched photoactive polymers [60–62]. The reversible photoisomerization between the rod like trans and cis isomers of the chromophore has been effectively utilized in all studies. However, when azobenzene polymers are used for holographic recording, a simple grating formation owing to a periodic modulation of the refractive index between disordered trans- and cis-azobenzene only creates a small change in the refractive index. On the other hand, liquid-crystalline azobenzene polymers have shown great promise [63]. The large modulation of the refractive index arises from a photoinitiated phase transition between the nematic state and the isotropic state. The nematic state is compatible with the rodlike trans-azobenzene and the disordered state is created due to disorganization of the LC state by bend cis-azobenzene. Holographically formed polymer dispersed liquid crystals (H-PDLC) are phase-segregated liquid crystal/polymer composites that make electrically switchable holographic recording possible. The crystals are formed using a holographic exposure apparatus to create an interference pattern that is recorded through polymerization to produce Bragg-mode gratings. Application of an electric field eliminates the Bragg grating, and the material then appears to be optically transparent.

Shibaev et al. reported a series of cholesteric copolymers containing combined chiral-photochromic side-groups, which are characterized by selective light reflection in the visible and infrared spectral regions [64]. The polymers have been shown to be promising candidates for color data recording.

The action of light leads to the isomerization of both chiral photochromic groups and the photochromic group alone. The helical twisting power of the chiral moiety is affected through a process that results in sharp changes in the helical structure and optical properties of the polymer. By properly selecting the copolymer composition or preparing blends of LC polymers with low-molar-mass chiral and photochromic dopants, it is possible to obtain materials with different kinds of light-induced transformations [65]. The photochromic LC copolymers are considered to be promising materials for reversible and irreversible black/white and color data recording for use in, e.g., optoelectronics, data storage (in optical memory systems), holography, and color projection techniques.

A few inventions that utilize the isomerization phenomenon have been patented, It has been claimed that one device for reversible optical information storage uses polymeric liquid crystals as the storage medium [53]. The device has a film comprised of a liquid crystal main chain polymer and stores the information by means of a local variation of the molecular ordering. Hence, this reversible optical information storage does not depend on the polymer material being in a viscous state and storage is reliable and not subject to the influence of spurious fields. Information can be repeatedly stored and read without decomposition of the device. All the recorded information can be erased and the original state can be restored by increasing the temperature of the storage medium above T_i (isotropization temperature) and cooling in an electric or magnetic field.

One suggestion to avoid the difficulties in high temperature recording related to degradation is the use of certain polymeric side-chain liquid crystals, which would make it unnecessary to keep the temperature below the glass temperature (T_g) to store the device. Stable storage should be possible for many years at temperatures above T_g and below a temperature (T_i) at which the polymeric material begins to become fluid. A device for reversible optical information storage without maintaining the polymer in a viscous state has been described [53]. Polymeric liquid crystals, which contain photochromic groups as side chains, were used for storage applications. Storage and reading can be done on the solid film at room temperature. The information is erased by heating the sample above T_g.

The prealignment procedures, as followed in earlier embodiments, may not be suitable for commercial storage devices as they tend to increase the complexity of the fabrication processes, thus leading to increased chances of failure of the final product. Furthermore, all the liquid-crystalline polymers used earlier appear to be of relatively low molecular mass, which results in the poor mechanical properties. A macroscopically isotropic side-chain liquid crystal polymer containing photochromic mesogenic groups has been synthesized [53]. By irradiation with light it can be permanently or substantially permanently converted into an optically anisotropic phase without any preorientation. The polymer is usually a polyester having a mesogenic group. Holographic recording has been demonstrated with a stored spatial frequency of the holographic grating at about 5000–6000 lines/mm.

The films used for holographic recording are cast from an organic solvent and are irreversibly cross-linked. The invention that addresses this problem is a reversibly cross-linked, orientable liquid crystalline (LC) polymer film, comprising an LC poly (methacrylate) PP and/or a mixture of PP with low molecular weight cross-linking components [53]. LC methacrylates form one comonomer, whereas the cross-linking groups residing on the other comonomer are COOH or –CONH. A macroscopically isotropic side-chain liquid crystal polymer containing photochromic mesogenic groups is used for permanent or substantially permanent conversion into an optically anisotropic phase using suitable radiation without preorienting the LC. Grating strength increases with time of illumination and a 15% grating efficiency has been reported. Oligomeric siloxane compounds having liquid crystal properties (smectic liquid crystal phase) have also been used for optical data recording and storage applications [53].

Studies of the effect of the molecular weight of the polymer on storage properties have also been carried out [53]. Side-group polymers (mol. wt. 5000 to 2,000,000) having permanently shape-anisotropic side groups are used for optical components that can be employed in optical data storage and optical transfer. Side-group polymers contain photochromic side groups (azo) and permanently shape-anisotropic side groups (benzoates) having high anisotropy of the molecular polarizability, which can be easily converted into an optically isotropic, transparent, amorphous state that does not scatter light.

Holographic optical elements are formed from a polymer dispersed liquid crystal (PDLC) material comprising a monomer, a dispersed liquid crystal, a cross-linking monomer, a co-initiator, and a photoinitiator dye [53]. These PDLC materials exhibit clear and orderly separation of the liquid crystal and cured polymer, whereby the PDLC material advantageously provides high-quality optical elements. The PDLC materials used in the holographic optical elements can be formed in a single step. The holographic optical elements can also use a unique photopolymerizable prepolymer material that permits in situ control over characteristics of resulting gratings, such as domain size, shape, density, ordering, and the like.

A holographic recording material, including a photoresponsive molecule, a reactive molecule having an intrinsic birefringence, and a photopolymerization initiator that accelerates polymerization and/or cross-linking of the molecule has been reported [53]. The concentration of the photopolymerization initiator is maintained in a range of less than about 0.1 wt% relative to the holographic recording material, and that of the reactive molecule having an intrinsic birefringence is in a range of about 30–80 wt % relative to the material.

Holographic Storage System and Holographic Property

Holographic Storage System

- **Laser**
  The pulsed laser or cw laser can also be used as the read/write source in the digital holographic recording system [66–70]. With two-dimensional (2D), information storage, storage density is inversely proportional to the second power of the laser wavelength, but with holographic data storage via the volume of the medium, storage density is inversely proportional to the third power of the wavelength. Owing to the inverse relationship between the density of holographic storage and wavelength, as far as possible one should select a shorter wavelength. However, the range of sensitive wavelength of materials and detectors must be considered in that choice. The single transverse mode TEM00 laser can be used in the storage program except the wavelength multiplexing; but in contrary, the tunable lasers would be used. For the time being, helium-neon lasers, argon ion lasers, krypton ion lasers, ruby lasers, and neodymium-doped yttrium aluminum garnet (Nd: YAG) lasers are commonly used in holographic storage; moreover, a diode-pumped solid-doubled laser will gradually be applied with the development of a small and integrated memory device.
- **Spatial light modulation**
  A spatial light modulator (SLM) is used for creating binary information out of laser [71, 72]. The SLM is a 2D plane, consisting of pixels that can be turned on and off to create binary 1's and 0's. An illustration of this is a window and a window shade. It is possible to pull the shade down over the window to block incoming light. If light is desired again, the shade can be opened. A spatial light modulator contains a two-dimensional array of windows that is only microns wide. These windows block some parts of the incoming laser light and let other parts go through. The resulting cross section of the laser beam is a two-dimensional array of binary data, exactly the same as what was represented in the SLM. After the laser beam is manipulated, it is sent into the hologram to be recorded. The data is written into the hologram as page form, called so owing to its representation as a two-dimensional plane, or page, of data [43].
  SLM is employed to convert electronic information into light ray variation so that the information can be saved as a hologram in an optically sensitive medium. One type of SLM is a liquid crystal display (LCD). With holographic information storage, the imprinted images are generally angle multiplexed. Angle multiplexing includes superimposing holograms inside the medium so that they can be accessed independently by changing the reference beam angle.
- **Photodetector**
  A photodectector is used to receive the optical signal of reconstructed information from holographic grating, which is typically achieved by a CCD(charge-coupled device array) [73]. When read a CCD receives optical signals diffracted by the medium; the reproduction of the object optical signal clearly contains the

SLM modulation of information; the light with information data in the CCD induces data information containing the charge or current signals, which can exist in the CCD and can also be entered into the computer processing or storage. The main factors determining the quality of a CCD include quantum efficiency, data transfer rate, and single-detection (also known as CCD pixels) number and size. Quantum efficiency, which is defined as the average number of electrons induced by each photon hitting the detector's photosensitive surface, is often a strongly related to the wavelength. In most commercially available CCDs quantum efficiency can be low in the long-wavelength region. Thus they are somewhat insensitive when wavelength is larger than 700 nm, which directly affects the choice of light source. Data transfer rate should be about 1 Gb/s, the current approach to increasing this is to divide the CCD array into many small areas (or channels) and then read these out in parallel.

A more difficult problem to solve is the matching of the number and the size of the pixels of SLM and the CCD; the ideal situation would be that the number and the size of SLM pixels are the same as CCD pixels, but CCD pixel size is getting smaller and smaller (4–5 μm), whereas the SLM pixel size is limited to around 10 μm due to the constraints of the LCD, thereby leading to a mismatch between the two. The solution to this problem is being sought now, but it will certainly affect the cost of the system and the data transfer rate.

- **Addressing apparatus**

An addressing device is used to quickly and accurately locate the address of the corresponding hologram (data page) in the data reading, writing, and rubbing processes in the holographic recording system [73]. Addressing devices are different which are lied on storage solutions selected for different multiplexing methods, such as wavelength multiplexing addressing device is a laser wavelength tuner, the phase addressing device is a phase modulator, and the angle multiplexing addressing device is a beam deflector. At present the most commonly used is angle multiplexing, so addressing device refers mainly to the beam scanning or the deflection device, which are often divided into devices with and without mechanical motion.

Mechanical addressing devices usually can be power-controlled galvanometers, with a deflection angle near 30°, angular resolution up to 0.001°, and a minimum time for random addressing up to 10^{-4} s. Addressing devices with no mechanical movement are commonly acousto-optic deflectors (AOD) or electro-optic deflectors (EOD). The latter are seldom used because there is usually a need to add a high voltage in order to achieve a large deflection angle and high precision of angle recovery. The former use sound waves to modulate the sound and light medium pressure and perturbation, so that a refractive index of dielectric grating is formed; the grating diffraction's properties of light direction are controlled by the frequency of the sound wave. By changing the sound waves frequency, there are no mechanical moving parts in Acousto-optic deflector addressing devices, so the system is relatively stable, simple, and the random access time is 10^{-5} s.

Holographic Properties

- **Hologram Capacity**
From the optical diffractive limit and the theoretical limit of information theory, it is easy to understand that the capacity of high-density digital holographic storage system is limited not only by the methods used for the optical settings, but also by the limitations of the materials: —the former is limited to V/λ^3 and the latter is limited to the material's space bandwidth product (SBP), where V is the storage volume, λ is a record wavelength. However, the actual capacity of holographic storage depends not only on the methods used for optical settings and material and optical resolution limitations, but also to the signal-to-noise ratio required by system and the signal intensity, namely the diffraction efficiency (DE). Moreover, the system capacity with the requirements of the relationship between the SNR and DE are also different for different methods of multiplexing and optical setting, but basically the capacity is inversely proportional to the DE and SNR of nth power, where n is a certain value from 1/2 to 3.The storage capacity is also proportional to the thickness of the material.
- **Diffraction Efficiency (DE)**
DE is a value that expresses the extent to which energy can be obtained from diffracted light with respect to the energy of the incident light. In the general case, it is determined by the ratio of the power of the diffracted light beam P_{diff} to the incident power of the beam P_{inc} given by the relation:

$$\eta = \frac{P_{diff}}{P_{inc}}. \tag{8.19}$$

DE is an important parameter in holographic memory. especially in connection with volume holographic storage, as it not only affects the brightness of information page reconstruction, but also determines the number of pages that can be stored in the uniform volume.

The chemical composition of the materials, the recording light intensity, and the ratio of reference beam intensity to the object beam intensity, etc., are all factors that affect the DE of the digital holographic storage system.
- **Noise analysis**
Commonly, the noise of holographic storage is divided into the system noise and the hologram noise. System noise include the phase difference between the optical components, SLM defects, detector noise, electronic noise in the circuit, light scattering multiple reflections in a variety of optical components, the nonuniformity and fluctuations of the laser beam, laser speckle noise, and SLM and CCD pixel mismatch noise, etc. Hologram noise includes intrapage and interpage cross-talk noise, which both relate to the multiplexing method, multiplexing parameters, optical settings, and the pixel number of SLM, etc.
- **Signal-to-noise ratio and bit error rate**
Binary image noise affects the reproduction error rate in the reconstruction image of volume holographic storage; the signal-to-noise ratio (SNR) reflects the amount of noise, so it is often used to evaluate the quality of the reconstructed

binary image in holographic storage [74–84]. The signal-to-noise ratio is usually defined as

$$\mathrm{SNR} = \frac{I_s}{I_n}, \tag{8.20}$$

where I_s is the mathematical expectation of measure value; I_n is the noise intensity, i.e., the disturbance intensity compared with the signal and is commonly expressed as the standard deviation of the measured values.

Bit error rate (BER) is the final standard for evaluating the quality of binary imaging in holographic storage, which reflects the fidelity of digital data storage. BER is defined as the wrong odds of the determining of a bit by detector circuit. To calculate the BER, we have to know the statistical distribution characteristics of the signal and noise.

The relationship between the SNR and BER of digital holographic data storage is

$$\mathrm{SNR} = \frac{I_1 - I_d}{\sigma_1} a \tag{8.21}$$

When $\sigma_0 \cong \sigma_1$,

$$I_d = \frac{\sigma_0 I_1 + \sigma_1 I_0}{\sigma_0 + \sigma_1} \tag{8.22}$$

$$\mathrm{SNR} = \frac{I_1 - I_0}{\sigma_1 + \sigma_0} \tag{8.23}$$

$$\mathrm{BER} = \frac{\exp[-(\mathrm{SNR})^2/2]}{\mathrm{SNR}\sqrt{2\pi}} \tag{8.24}$$

where I_1 and I_0 are the average intensity of bright and dark pixels, σ_0 and σ_1 are the standard deviations of bright and dark pixels, and I_d is the judging threshold intensity of bright and dark pixels in reconstructed data page.

- **Data transfer rate**

  An important indicator of the performance of computer storage devices is the data transfer rate, which is also a measure by which the performance of holographic memory can be evaluated. The data transfer rata is determined by the access time of memory. Holographic memory access time is very asymmetric. Recording time contains the electronic device transmission time, the addressing positioning time, the filling time of SLM, and the holographic exposure time in each data page recorded, etc. Reading time includes the addressing time, the integrating time of the diffraction light in the detector and the detector response time, etc. The recording time is longer than the reading time, so the recording speed is slower than the reading speed of holographic storage.

In a general way, for getting the equal diffraction efficiency in the multiplexing holographic storage, it will take different exposure time for recording different data pages. As a result, the holographic recording process does not proceed at the same rate. Thence, the average recording rate is used to examine its record speed in usual. In addition, it is related to the sensitivity of recording materials and the intensity of light.

Outlook

Much progress has been made in recent years on new and improved recording materials for holographic data storage, as well as on improved components, especially large-format CCD, complementary metal-oxide semiconductor (CMOS), and SLM arrays based on ferroelectric liquid-crystal media and microelectronic and mechanical systems. Lasers are smaller and more powerful than ever, and improved algorithms have been developed and implemented. The outlook for holographic data storage has never been brighter.

Commercialization, however, still depends mainly on the longevity of other optical storage technologies. For example, it is highly likely that DVD technology will require some kind of volumetric storage scheme, either in the form of a multilayer or a holographic approach to increase data-storage capacity. But multilayer technologies lack the advantage of the high data transfer rates available through holographic systems.

It is therefore quite possible that holographic data storage could be commercialized as a follow-up to blue-laser DVD technology within 10 years. But holographic data storage will require a new——and costly——manufacturing infrastructure to produce competitively priced products in sufficient quantity and quality to satisfy the storage market, which already has many high-performance options.

As potential mass-market products, polymer-based systems will most likely be the first to be commercialized, because they match the existing manufacturing infrastructure better than the crystal based all-solid-state devices. For instance, systems using polymer materials allow 100–1000 times faster write speeds than those using inorganic photorefractive crystals like lithium niobate.

Near-Field Optical Storage Technology

Introduction

For the optical disk, the storage density is determined by the size of a recording bit (the density is inversely proportional to the square of mark size). Thus, reducing the recording mark size is the main approach to increasing the storage capacity and density.

An optical disk driver has an objective lens to read, write, and erase the recorded bit and obtain the servo signal for focusing and tracking. Because the distance between the objective lens and the recording medium (or prerecorded pits in a ROM disk) is far longer than the laser wavelength, the conventional optical disk technology is classified as far-field optical storage.

For far-field optical storage, the mark size is determined by the size of the focused laser beam. According to the Abbe principle (or Rayleigh criterion), the minimum diameter of the focused laser spot, d, is defined by

$$d = 1.22\lambda / (2\mathrm{NA}), \tag{8.25}$$

where λ is the wavelength and NA is the numerical aperture of the objective lens; d is the minimum resolvable distance of two objects in the conventional optical system. The minimum resolvable size depends on the diffraction of the light wave, so d is usually called the optical diffraction or resolution limit. From this equation, it can be estimated that the optical system cannot resolve an object smaller than $\lambda/2$ even using a lens with a large numerical aperture.

Under the optical diffraction limit condition, the practical approach to minimize the mark size is to decrease λ and increase NA. Two main methods for improving storage density and capacity are reducing the laser wavelength and increasing the numerical aperture of the objective lens. At present, a GaN laser (wavelength: 405 nm) and a high numerical aperture lens (NA: 0.85) are used in the Blu-ray disk system to achieve a storage capacity of 25 GB (12-cm diameter, single side). However, it is difficult to increase the numerical aperture and shorten the laser wavelength any further, as to do so will cause serious technical problems in connection with the substrate materials, the recording medium, optical signal detection, and so on. Thus obtaining higher data density will depend on overcoming the diffraction limit.

As we know, near-field microscopy can overcome the diffraction limit and possesses nanometer resolution. The near-field microscopy techniques have laid a foundation for high-density optical storage applications [85].

Near-field optical microscopy is based on the following concept. Although the beam cannot be focused to a spot smaller than $\lambda/2$ by traditional optical systems, the small spot can be obtained using a small enough hole (aperture). In the near-field region, the size of the light spot is determined by the size of the hole and is not related to the wavelength. If the sample is scanned in two-dimensions by the near-field spot, the surface image can be reconstructed. The image will have subwavelength resolution if the hole is smaller than the wavelength. With near-field optical microscopy, data pits smaller than the diffraction limit can be recorded, read out, and erased by the near-field light spot.

The principle of breaking through the diffraction limit by near-field optics is the generation of evanescent fields (or evanescent wave) and the interaction between evanescent fields and objects. Generally, recording or reading information by an evanescent field can be called near-field optical storage regardless of the existence of a small hole.

Technical Approaches of Near-Field Optical Storage

Near-field optical storage can be achieved by different technical approaches. Some typical methods are described in what follows.

Scanning Near-Field Optical Microscopy (SNOM)

Scanning near-field optical microscopy (SNOM), also called near field-scanning optical microscopy (NSOM), is one of the important devices to achieve near-field optical storage. The apertured fiber-tip generating evanescent field, which is called a probe, is an important part of SNOM, and it is a main factor in determining the resolution of the whole system. Figure 8.42 shows some typical probe designs [86]. Figure 8.43 shows the schematic structure of a SNOM, which is made up of a laser, a probe, an illuminating system, a focusing system, photodetectors, scanning units, and a control/display system [86]. The probe is often fabricated by a metal film coated fiber tip and controlled by PZT for modulation and scanning. The transmitted and/or reflected signals are collected by photodetectors, such as PMT and CCD. The optical signals and image information that are obtained are processed by a computer. The probe and its coupling efficiency to the evanescent wave are key features of SNOM. The resolution and sensitivity of SNOM can be improved by optimizing fabrication and design of the probe and accurately controlling the distance between the probe and the recording medium. Optical recording with nanometer-scale recording marks can be achieved with SNOM.

Near-field optical recording was first achieved by E. Betzig and his colleagues (Bell Labs) on magneto-optical thin films using a SNOM in 1992 [87]. and they obtained minimum recorded domains of 60 nm. Thereafter, SNOM-based near-field optical storage was carried out on phase-change and organic photochromic materials by scientists from the Hitachi Laboratory and the Tokyo Institute of Technology, respectively [88–90]. Although the recording density of 250 Gb/in.^{2} can be achieved

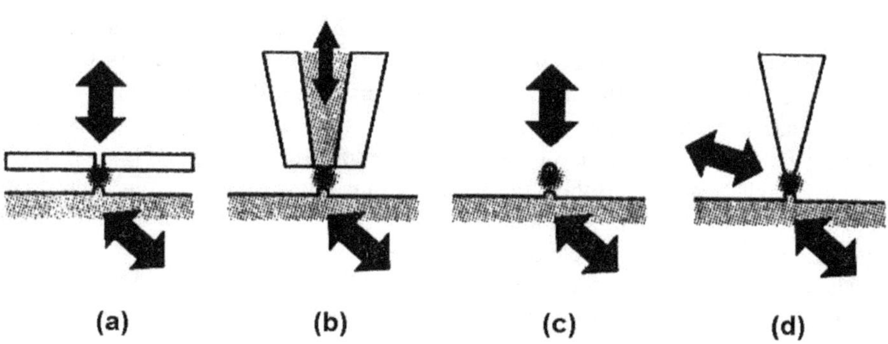

Fig. 8.42 Some typical near field probes: (**a**) aperture, (**b**) aperture-type fiber probe, (**c**) small ball scatter, (**d**) scattering-type fiber probe

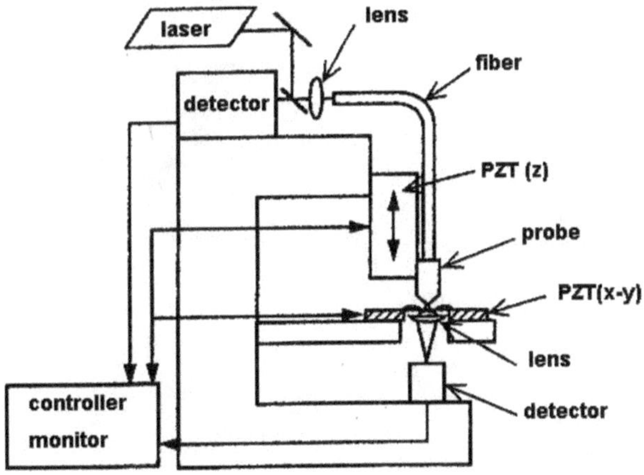

Fig. 8.43 Schematic structure of SNOM

by using SNOM, there are some well-known shortcomings for practical applications: (1) The laser energy is seriously attenuated through the nano-apertured fiber probe, so the readout signal is greatly weakened. (2) The distance between the probe tip and recording medium is hard to control and the response speed of the feedback system is low, so the enhancement of writing/readout speed is limited. (3) Preparation of the fiber probe is difficult and the fiber tip can be easily destroyed when scanning a rough surface. (4) The SNOM system is complex, expensive, and hard to miniaturize.

Some new techniques have been tried to improve SNOM's practicability. Borrowing slider technology from hard disk systems allows the near-field distance between the rapid rotating medium and probe to be controlled without feedback. The generation efficiency of near-field light and recording/readout speed can be improved by using the a silicon probe array. The schematic structure of SNOM with a silicon probe array on a slider is shown in Fig. 8.44 [86]. By using this improved SNOM, scientists at the Tokyo Institute of Technology achieved high-speed near-field optical recording. A recording mark size of 110 nm, a readout speed of 2 Mb/s (corresponding to 200 Mb/s of 10 × 10 tip array), and a CNR of 10 dB were obtained [91].

Solid Immersion Lens (SIL)

A solid immersion lens (SIL) is a hemispherical or superhemispherical aplanatic lens [92] (Fig. 8.45). The working mechanism of SIL is similar to that of an oil immersion lens, which can increase the effective numerical aperture of an objective lens by filling the object space with a high-refractive-index material. But an SIL is more suitable for optical storage applications because it does not directly contact the recording medium, so it can be scanned without friction. In the SIL optical storage system, the SIL is placed below the optical pickup head and the laser beam is focused onto its bottom surface. The numerical aperture of the pickup head has

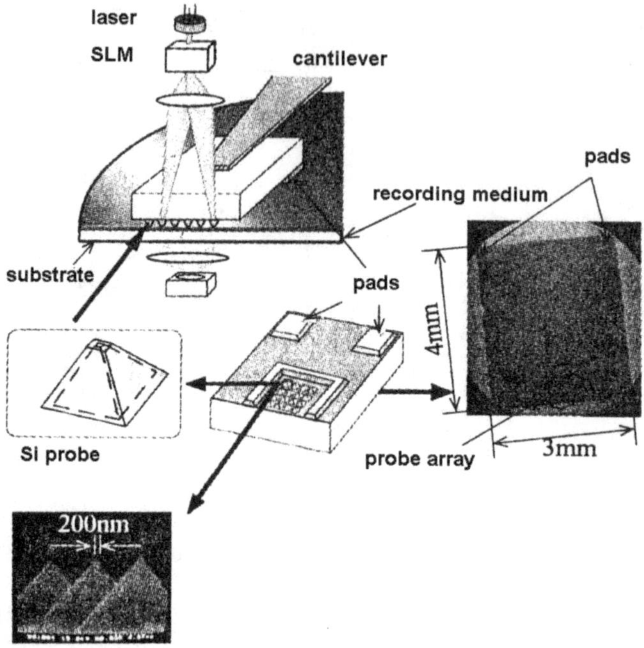

Fig. 8.44 SNOM with a silicon probe array on a slider

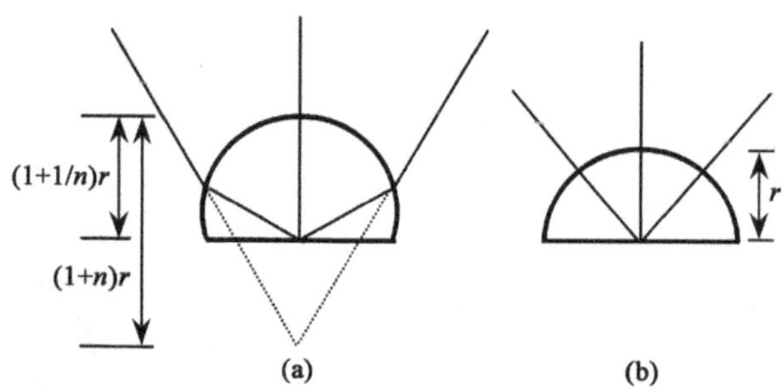

Fig. 8.45 Schematic structure of SIL: (**a**) superhemispherical structure, (**b**) hemispherical structure. (n: refractive index of SIL, r: radius of SIL)

been improved and the laser spot size correspondingly reduced. For hemispherical and superhemispherical SIL, the effective numerical aperture of the optical head is increased by a factor of n (where n is refractive index of SIL) and n^2 respectively. The numerical aperture for optical systems with SIL is often larger than 1. Evanescent waves will be generated at the bottom surface of the SIL when the laser beam passes through it. In order to make the SIL work better, the gap between its bottom surface and the recording medium should be kept at the near-field scale. SIL near-field optical recording can achieve storage density similar to that of SNOM, and the SIL system can overcome some of the technical shortcomings of SNOM, such as low optical throughput and low scanning speed.

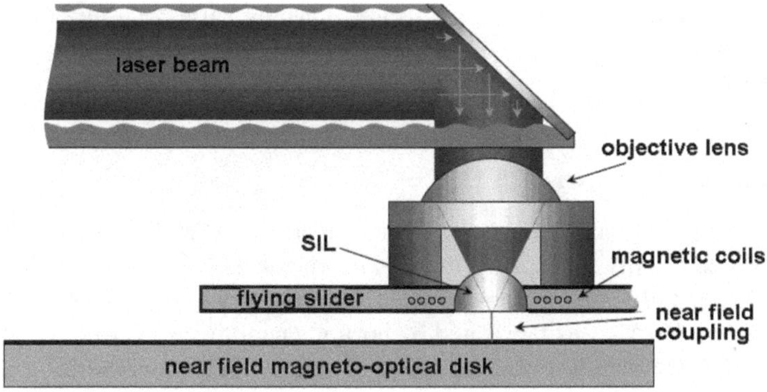

Fig. 8.46 A design of a flying head with SIL for magneto-optical recording

The first application of SIL in information storage can be traced back to the US patent (No.5125750) taken out by T. R. Corel et al. (Stanford University) in 1991. In 1994, static near-field optical storage experiments using SIL were carried out by the IBM research center and Stanford University. Minimum mark sizes of about 350 nm (laser wavelength: 780) were obtained [93]. Subsequently, they achieved dynamic near-field recording using SIL on the magneto-optical medium and obtained a domain size of about 360 nm (laser wavelength: 780 nm) [94]. Thereafter, the Sony laboratory achieved a domain size of 330 nm on a magneto-optical medium (laser wavelength: 532 nm) [95]. A minimum pit size of 150 nm on magneto-optical and phase-change materials were obtained at the Toyota Institute of Technology (laser wavelength: 680 nm) [96] and the Sony laboratory (laser wavelength: 657 nm) [97], respectively.

High-index SILs have some advantages, such as high NA, small size, and light weight and are suitable for practical applications in miniaturized flying pickup heads.

Recently, SIL technology has become a candidate for the next generation of optical disks. Sony has realized a 100-GB-capacity Si-etched read only memory (ROM) Blu-ray disk readout using a SIL with a 2.05 NA [98] and 112-GB-capacity phase-change disk recording on a 12-cm-diameter disk using a SIL with a 1.84 NA [99].

The SIL technology can also be applied to high-density magneto-optical recording and hybrid recording (heat assisted magnetic recording) systems. Figure 8.46 shows a SIL pickup design for magneto-optical recording (Terastor Co.) [100].

At present, it is not that difficult to process a SIL with the development of precious machining technique. Achieving a higher NA and integrating it with other devices (such as GMR head) are more challenging. As a higher refractive index enables us to obtain a higher NA, new SIL materials have been developed, such as S-LAH79 glass [98], Bi4Ge3O12 monocrystal [99], diamond [101]. and so on. We can get to a refractive index of single-crystal diamond at a wavelength of 405 nm of 2.458 and to a corresponding effective numerical aperture of a diamond SIL of 2.34. Using the diamond SIL, a 150-GB-capacity (104.3 Gb/in.^{2}) disk with a track pitch of 130 nm and a bit length of 47.6 nm can be read out with a blue laser system [101].

Very Small Aperture Lens (VSAL)

Near-field optical storage with a very small aperture laser (VSAL) has been proposed in order to solve the problem of low optical throughput from the fiber tip for SNOM near-field optical storage. The schematic structure of VSAL is shown in Fig. 8.47 [5]. A VSAL was fabricated by boring a nanometer aperture on the metallic coating surface of a laser diode (LD). The near-field illumination from VSAL records a pit on the recording medium to achieve near-field optical storage. At the same time, the reflected signal from the disk is coupled back into the laser cavity and the recorded information can be read out via the feedback effect.

As a near-field light source, VSAL has many advantages, such as: (1) High optical throughput. The laser power of VSAL can be enhanced by 10^4 times compared with that of a common fiber probe and the signal-to-noise ratio and data transfer rate can be correspondingly enhanced. (2) High integration. Owing to its small size (typically: $750 \times 300 \times 150\ \mu m$), VSAL may be developed into a new type of integrated pickup head. (3) The near-field gap can be controlled through some mature technology used in hard disk flying heads and near-field optics. (4) It will be easy to achieve parallel recording/readout with VSAL arrays, which will result in higher data transfer rates.

VSAL near-field recording was first demonstrated in 1999 by Bell Labs. In order to control the nanometer scale gap between VSAL and the recording medium in order to couple the evanescent wave to the recording layer, the VSAL was placed on the slider of a hard disk. A rectangular aperture with a side length of 250 nm was used for this VSAL, whose working wavelength was 980 nm. The recording pits with a diameter of 250 nm (corresponding to 7.5 Gb/in.^{2} storage density) was obtained on the GeSbTe phase-change thin film. A data transfer rate and a carrier-to-noise ratio (CNR) of 24 Mb/s and 45 dB, respectively, can be achieved [102].

In order to improve the storage density and data transfer rate, a vertical cavity surface emitting laser (VCSEL) was introduced to near-field storage by K. Goto (Tokai University, Japan) [103]. VCSEL arrays were used as the light source for multibeam

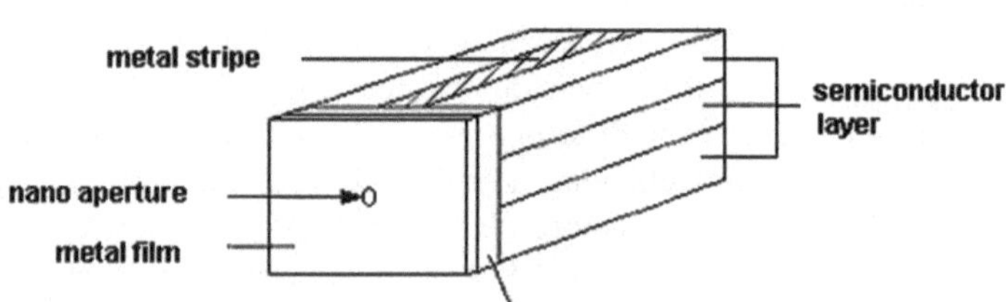

Fig. 8.47 Schematic structure of VSAL

parallel recording/readout. The improvement in aperture design and nanofabrication technology should make it possible to develop an integrated near-field parallel optical storage system based on VCSEL.

The optical disk matching with the VSAL is very complex owing to changes in addressing and recording/readout methods, which will be a serious obstacle in the rapid development of a VSAL-based near-field optical storage technology.

Superresolution Near-Field Structure (Super-RENS)

Superresolution near-field structure (Super-RENS), which is based on the eclipse effect and near-field storage techniques, was firstly proposed by J. Tominaga (National Institute of Advanced Industrial Science and Technology, Japan) in 1998. A typical Super-RENS optical disk is shown in Fig. 8.48. A dielectric layer/nonlinear mask layer/dielectric layer sandwich structure is inserted into the phase-change optical disk to achieve near-field optical storage [104].

The working mechanism of a Super-RENS is similar to the fiber probe used for near-field optical storage, as shown in Fig. 8.49. The mask layer plays the role of a near-field aperture or a light scattering center in the Super-RENS. A thin protective layer (SiN or $ZnS\text{-}SiO_2$), which separates the mask and the recording layers, functions as a space layer like that between the fiber probe head in NSOM and the recording medium. This is an advantage of Super-RENS because a smooth solid layer with uniform thickness over a wide area can be easily prepared by vacuum deposition techniques. It is no longer necessary to control the near-field gap using complex feedback equipment, since the distance between the aperture and the recording layer is always fixed at a constant height over the entire disk surface.

The formation of aperture in the mask layer is a transient process, as it is in eclipse superresolution optical disk [105]. The principle of eclipse aperture formation in the mask layer is shown in Fig. 8.50. During laser pulse irradiation, the temperature increase in the rear part of the spot is higher than that in the front part owing to the rotation of the disk. When the local temperature in the mask layer exceeds the melting temperature, the layer changes from a solid to liquid (melted) state. Because the reflectivity of the material at the melted state is often

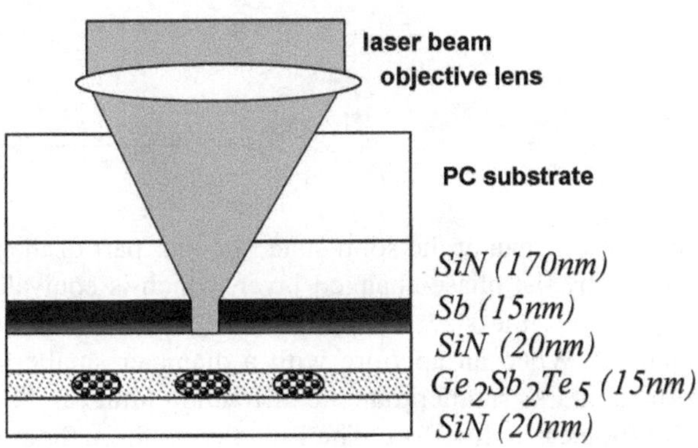

Fig. 8.48 Structure of a typical Super-RENS disk

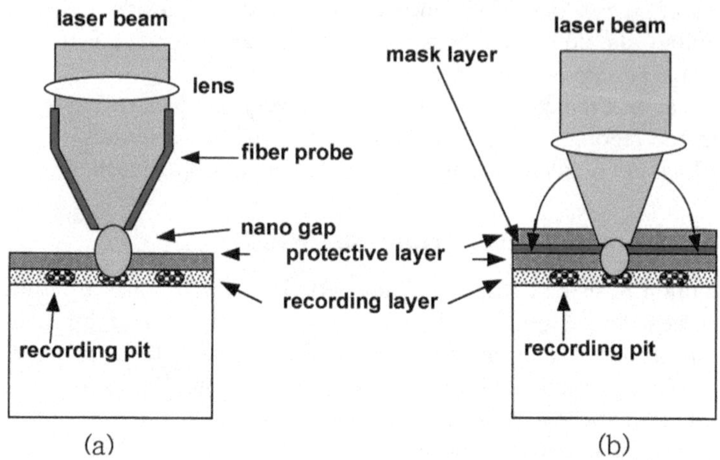

Fig. 8.49 Comparison of Super-RENS and SNOM: (**a**) SNOM, (**b**) Super-RENS

Fig. 8.50 Working mechanism of the eclipse mask layer

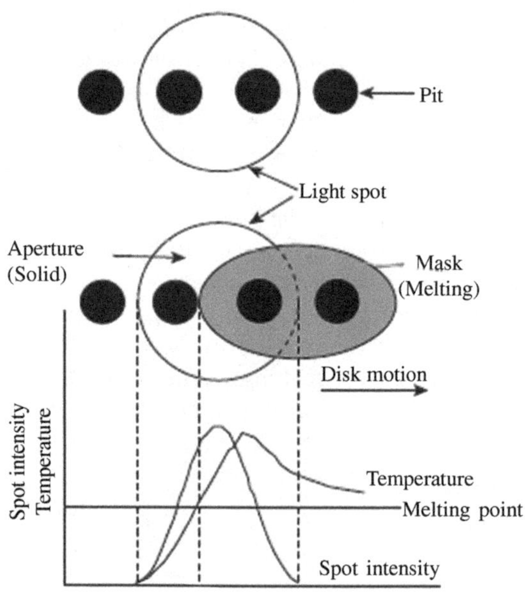

much lower than at the solid state, the rear part of the laser spot can be effectively masked by the phase-changed layer, which is equivalent to a spot size reduction. In principle, the size of the aperture can be controlled by modulating the laser intensity. When an aperture with a diameter smaller than the diffraction limit is formed, marks smaller than the diffraction limit can be recorded/read out combined with the near-field effect. The low optical throughput still seems to be a technical shortcoming of this approach because at least half of the laser energy is masked by the functional material layer, which often has high extinction coefficient. A high throughput is another principal characteristic of Super-RENS compared with the traditional eclipse self-masking approach.

The scattering-type Super-RENS was developed soon after with a working mechanism similar to the scattering-type SNOM [106]. Its structure is similar to the aperture-type Super-RENS, but a noble metal oxide (such as AgO_x, PtO_x, PdO_x) layer forms the mask layer or the recording layer [107–111]. The focused laser beam decomposes the metal oxides into metal nanoparticles and oxygen in a local area. The metal nanoparticles in a bubble pit can be seen as scattering centers and the scattered light in the near field enhances the optical throughput and signals at the same time, so superresolution readout and recording can be realized. The scattering-type Super-RENS disk shows better resolution than the aperture-type structure. A CNR of 33 dB can be achieved for 37.5-nm recorded marks on a modified PtO_x-based scattering-type Super-RENS optical disk, which corresponds to the recording capacity of 100 GB for a 5.25-in. optical disk. Such a high-density recording is obtained by stack structure optimization and by adopting the write strategy of forming elliptic bubbles [112].

In fact, the working mechanism of nonlinear material and nanofilm structure in Super-RENS has not been well understood until now. The nonlinear optical properties [113] and local surface plasmon effects [114, 115] seem to afford reasonable explanations for transient near-field enhancement at current levels of awareness. Table 8.5 lists some existing or potential Super-RENS materials and their corresponding properties that may be related to near field superresolution.

Optical Transducers (OT)

Optical transducers (OTs) that concentrate optical energy in the near field to dimensions much smaller than the standard diffraction limit are often classified as apertures or antennas. Such transducers may have applications in optical or heat-assisted magnetic data storage (hybrid recording) [144].

Optical apertures or antennas can be used as near-field generators in surface plasmon optical storage devices [145–147]. Figure 8.51 shows a schematic structure of a so-called interference light enhanced surface plasmon head [148]. The metallic nanorod acts as an optical transducer, and it has been confirmed by FDTD simulation that the head has sufficient resolution and efficiency for 1 Tbit/in.^{2} hybrid recording density.

The apertures in VSAL can also be considered as optical transducers. Various modifications to the simple rectangular or circular aperture have been considered in order to improve the near-field efficiency. The shape of the aperture can be changed to elliptical [149] or more complex shapes, including the "C" aperture [150], the "I" or the "H" aperture [151, 152], the "L" aperture [153], and the bow tie aperture [153, 154]. In fact, even simple apertures when optimized are able to couple about 1.5% of the incident power into optical spots with a full width at half-maximum diameter of a tenth of a wavelength [144].

Table 8.5 Some superresolution materials and their corresponding properties

Mask Layer Materials		Properties that may be related to near field super-resolution	References
Chalcogenides	As_2S_3	Large nonlinear refractive index	[116–118]
	AsSe	Ultrafast optical Kerr effect	
	GeAsSe	Photoinduced darkening	
	GeAsSSe		
	CdSSe	Nonlinear saturation absorption	[119]
	GeSbTe	Photothermal nonlinearity Phase transition (from NaCl FCC to hexagonal)	[105, 120]
	AgInSbTe	Photothermal nonlinearity Phase transition (from hexagonal to rhombohedral)	[120, 121]
Silicon	Si	Laser-induced refractive index distribution Photothermal nonlinearity	[122–124]
Semimetals	Sb	Thermal lens effects Third-order nonlinearity	[104, 125, 126]
	Bi	Third-order nonlinearity	[127–129]
	Te	Photothermal nonlinearity	[130]
Noble metal oxides	AgO_x	Nanoparticles	[106–111, 131, 132]
	PtO_x	Local surface plasmon	
	PdO_x	Large nonlinear refractive index	
Photochromic and thermochromic compounds	Diarylethene	Photochromism Thermochromism Reversible hysteresis	[133–137]
	Thermochromic Dyes		
	TeO_x		
	VO2		
	CoO_x		
Nonsilicon semiconductors	$Zn_{1-x}Cd_xSe$	Band gap adjustment	[138, 139]
	ZnO		
	CeO2		
	TiO2		
Micro- and nanostructured materials	Photonic crystals Surface plasmon subwavelength structure Random nanostructures	Negative refractive index Superlens Super transmission Focusing effect	[140–143]

Near-Field Optical Storage Materials

The near-field optical storage methods discussed above involve optical recording and readout. In fact, other readout methods can be used such as optical-assisted magnetic recording and magnetic head readout, laser-induced phase-change recording, and electrical or thermal conductivity readout, as well as other recording methods and optical readouts (such as charge injection recording and fluorescence readout).

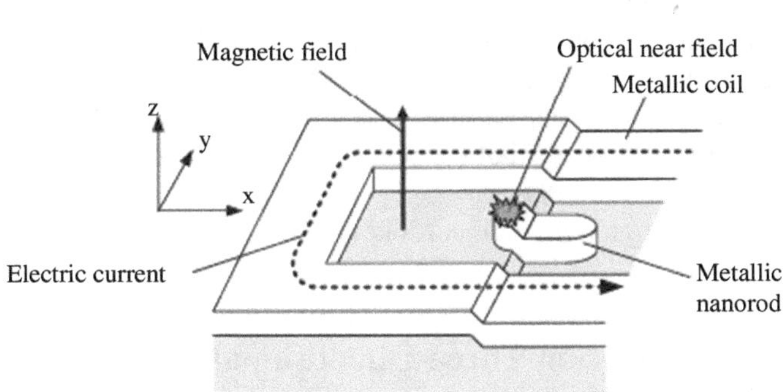

Fig. 8.51 Schematic illustration of a lensless surface plasmon head for hybrid recording

Table 8.6 Typical near-field optical recording materials

Recording Materials	Functions	Technical Approaches	References
Phthalocyanine ($C_{32}H_{18}N_8$)	Write once	Super-RENS	[155]
AgO_x	Write once	Super-RENS	[156]
PtO_x	Write once	Super-RENS	[110, 111]
Pt/Co	Erasable	SNOM	[87]
	Erasable	SIL	[157]
TbFeCo	Erasable	SIL	[93, 94, 96]
	Erasable	Super-RENS	[158]
Co	Erasable	OT	[148]
GeSbTe	Erasable	SNOM	[88, 89]
	Erasable	SIL	[97]
	Erasable	VSAL	[102]
	Erasable/write once	Super-RENS	[104, 129, 159]
Azo polymer	Erasable	SNOM	[90, 160]
	Erasable	SIL	[161]
Diarylethene	Erasable	SNOM	[162–164]
	Erasable	SIL	[165]
Bacteriorhodopsin	Erasable	SNOM	[166]
Perinaphthothioindigo	Erasable	SNOM	[167]
Polycarbonate (PC)	Read only	Super-RENS	[168]
	Read only	SIL	[101]

Near-field optical storage can be utilized to support recordable (write once), rewritable (erasable), or read only (ROM) functions. Different recording materials should be adopted depending on recording/readout method and function.

Apart from the general requirements for optical storage materials (such as optical and thermal properties), near-field optical storage has some special requirements for recording media: (1) High sensitivity, so recording pits can be formed even if the laser power is greatly lowered through the fiber probe. (2) Easily prepared homogeneous thin films with smooth surfaces, so the near-field distance between the probe tip and the recording medium can be maintained during scanning. (3) Matching of the lubrication layer materials; a lubrication cover layer is needed for most near-field systems with flying heads, but the properties of the recording layer should not be affected by the lubrication layer.

Some typical near-field optical recording materials are listed in Table 8.6.

Two-Photon Absorption Three-Dimensional Optical Storage

Introduction

As a result of rapid development in multimedia technology, there is increasing demand for optical storage devices with high density and large capacities. Present optical data storage devices such as the compact disk (CD), the digital video disk (DVD), and the Blu-ray disk (BD), which are quite essential as audio and visual storage media and very popular, involve one-photon processes to write and retrieve data or information on a two-dimensional (2D) surface, and have nearly achieved an upper limit with commercially available disks owing to the diffraction of electromagnetic waves. Hence, advanced optical storage technology is being exploited to obtain more efficient optical storage devices with higher density and larger capacities.

Two-photon absorption (2PA) storage is a three-dimensional (3D) optical information storage technology where the storage density is dependent on the reciprocal of the wavelength (λ) to the power of the dimension used to store the information [169]. This relationship implies that storage density of 2PA three-dimensional optical storage can be theoretically as high as 10^{14} bits/cm^3, which is 10^3–10^5 times higher than that of two-dimensional storage using the same laser wavelength. Thus, 2PA three-dimensional optical storage has become one of the most promising techniques. In the following section, the main principle, the storage materials, and the recording and readout technology of 2PA three-dimensional optical storage are described.

Main Principle of Two-Photon Absorption Storage

The theoretical aspects on which two-photon absorption is based were presented by Göppert-Mayer in 1931 [170] and were experimentally verified by Kaiset and Garrett in the early 1960s [171] when the pulsed laser, which can provide high-intensity light, was invented. Two-photon absorption is a nonlinear process in which two photons can be absorbed simultaneously (see Fig. 8.52). When laser light irradiates 2PA material, the virtual state may form and be sustained for a very short time, on the order of a few femtoseconds. Two-photon absorption can be

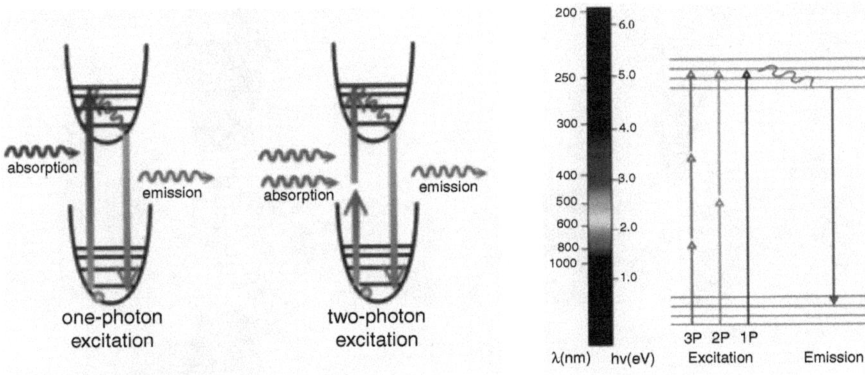

Fig. 8.52 Two-photon absorption and one-photon absorption processes

produced if a second photon arrives before virtual state decays , with the probability of 2PA scaling with the square of the light intensity. Two-photon absorption involves the concerted interaction of two photons that combine their energies to produce an electronic excitation analogous to that conventionally caused by a single photon of a correspondingly shorter wavelength. If the two photons are of the same energy (wavelength), the process is referred to as degenerate 2PA, whereas if the two photons are of different energy (wavelength), the process is called nondegenerate 2PA [172].

Unlike single-photon absorption, whose probability is linearly proportional to the incident intensity [see (Eq. 8.26)] [173], the 2PA process depends on both a spatial and temporal overlap of the incident photons and takes on a quadratic or nonlinear dependence on the incident intensity (see Eq. (8.27)) [173], giving rise to highly localized photoexcitation with a focused beam.

$$n^{(1)} = \sigma_{(\acute{i})} \bullet N_g \bullet \left(I/h_{\acute{i}}\right) \tag{8.26}$$

Here $n^{(1)}$ is the number of molecules excited by 1PA per unit time and unit volume in the material, $\sigma_{(\acute{\imath})}$ is the cross section of the absorption process at frequency $\acute{\imath}$, N_g is the density of molecules in the ground state g, I is the intensity of the excitation source (in energy per unit time and area), and $h_{\acute{\imath}}$ is the photon energy.

$$n^{(2)} = 1/2 \bullet \delta_{(\acute{i})} \bullet N_g \bullet \left(I/h_{\acute{i}}\right)^2, \tag{8.27}$$

where $n^{(2)}$ is the number of molecules excited by 2PA in a unit volume per time unit in the material, $\delta_{(\acute{\imath})}$ is the 2PA cross section for a photon of energy $h_{\acute{\imath}}$. The prefactor of 1/2 reflects the fact that two photons are needed to excite one molecule.

In other words, this is the reason that the 2PA process is feasible for exciting molecules inside a volume rather than on the surface (see Fig. 8.53) [174].

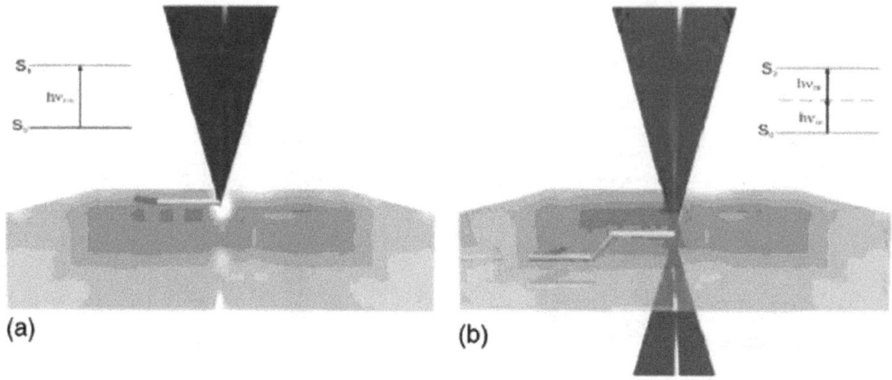

Fig. 8.53 (**a**) Single-photon absorption with UV light. Light is absorbed on the surface of the photosensitive material. (**b**) Two-photon absorption with near-infrared light. The photosensitive reaction is confined in a focused volume

Therefore, 2PA three-dimensional optical data storage systems have taken advantage of 2PA features to achieve high-density memory with high spatial resolution in 3D optical storage.

In fact, 2PA is a process for activating photochemical, photophysical, or photoresponse phenomena in 3D optical data storage systems made up of two distinct components. Currently, audio and visual storage media with computer information is stored in the form of binary code, 0's and 1's. Consequently, compared to above mentioned concept, with the photoresponse changes (such as making it isomerism or fluorescent), two distinct molecular components are similar meaning with binary code as the storage form. The structure of typical 2PA materials now exist and have been utilized and achieved in data storage application such as Anthracene derivatives whose molecular structure are changed via photochemical after the simultaneous absorption of two photons to produce two distinct forms: the write, closed form, and the read, open form (see Fig. 8.54). It is necessary for the write a data bit of information to correspond to a binary format, namely the preliminary closed form indicated by 0 and the novel open form indicated by 1. Subsequently, to read this data, the stored bits are excited or irradiated with a reading laser diode that induces fluorescence or other physical and chemical property changes from the written bits, which are collected by a detecting device such as a photomultiplier tube or a photodiode array in order to retrieve information in a storage disk.

Materials for Two-Photon Absorption Storage

Many kinds of materials, especially some organic materials, are used as the recording media for 2PA optical storage. Depending on the storage mechanism, we introduce a few series of typical two-photon absorption storage materials that evidence photochromism (photoisomer), photo-oxidation, photocycloreversion-photocycloaddition, or photopolymerization.

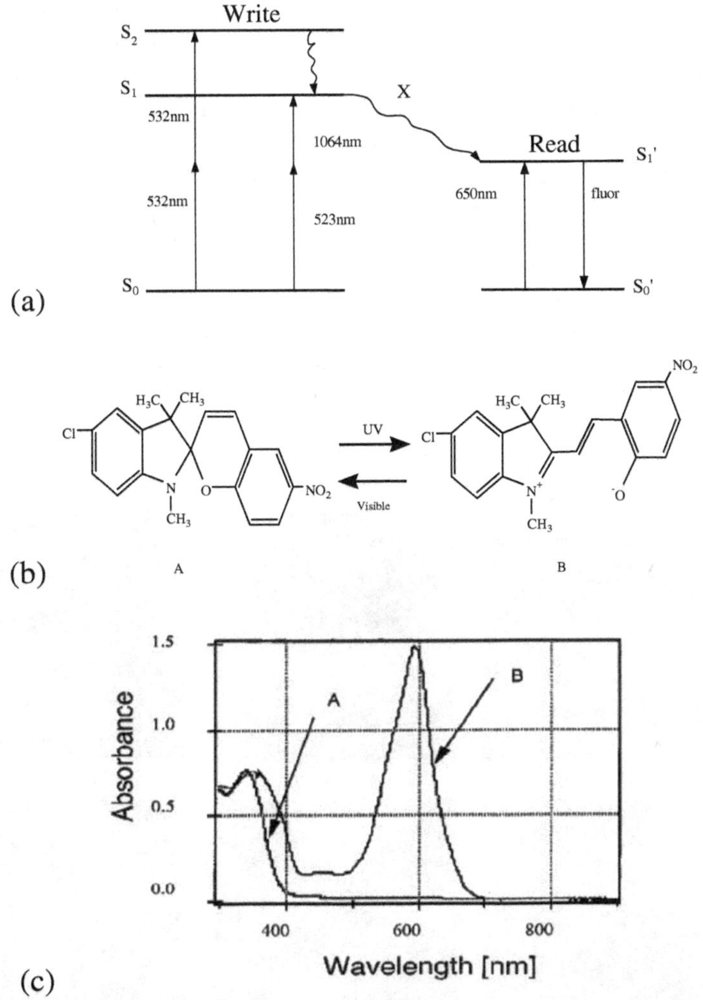

Fig. 8.54 (**a**) Two-photon process energy level diagram. (**b**) Photochromic reaction, A and B are spirobenzopyan isomers. (**c**), Absorption spectra of a polystyrene film including spirobenzopyan irradiated by 442-nm light both before (A) and after (B)

- **2PA photochromism materials**

  For optical memory media photochromic materials are characterized by the ability to alternate between two different chemical forms that have different absorption spectra and properties: (1) thermal stability of two isomers, (2) fast response time, (3) high sensitivity, (4) resistance to fatigue during cyclic write and erase processes, and (5) nondestructive readout capability [175].

  Photochromic spirobenzopyran was first demonstrated as recording material for 3D optical memory systems by Rentzepis et al. [176, 177]. For writing data, spirobenzopyran molecule structure A converts to isomer B by absorbing two photons (either a 1064-nm photon and a 532-nm photon or two 532-nm photons). After the transition from the S_0 to the S_1 state, optical ionization; then through intersystem crossing, reach the lowest triplet excited state energy levels S_1, is reached. For reading stored data, a 650-nm laser beam was used to excite molecule issued fluorescence with back to S_0 state (Fig. 8.54).

Diarylethene derivatives were also developed as 2PA photochromism materials [178, 179] and Toriumi et al. have successfully achieved 26-consecutive-layer storages via photochromism memory [180] (Fig. 8.55).

- **2PA photo-oxidation materials**

  A. D. Xia et al. [181, 182] found that photo-oxidation enhances the fluorescence intensity of the C60 molecule by two-photon excitation in a wide wavelength region (780–910 nm) with a 100 fs Ti:sapphire laser. Thus, the possibility of

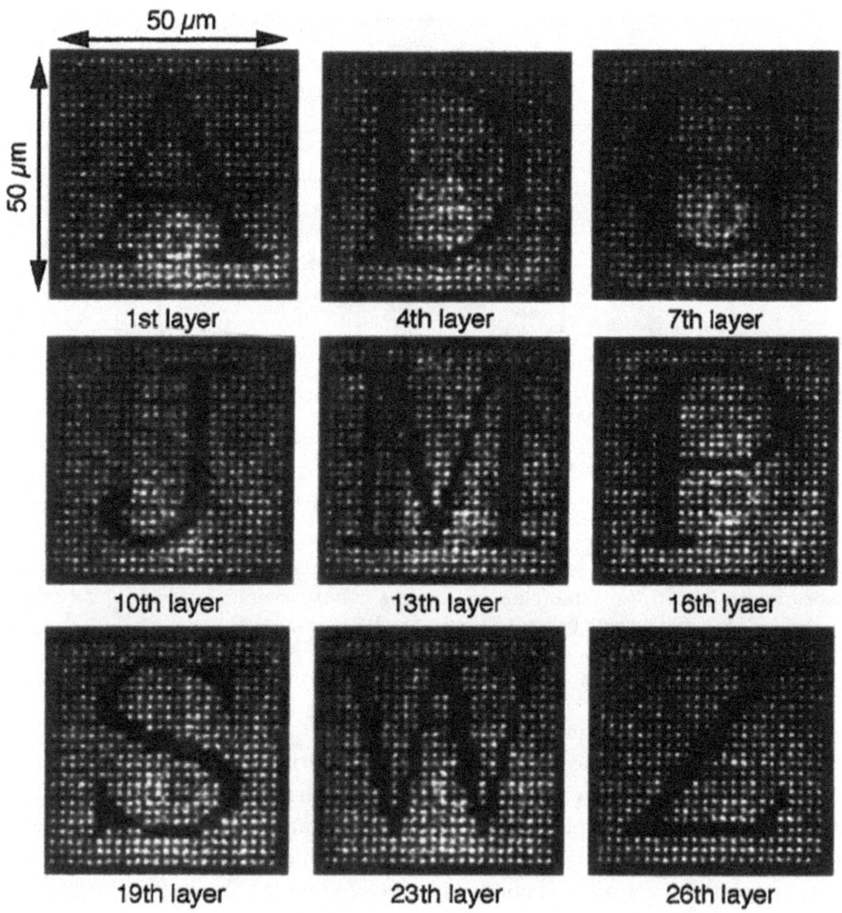

Fig. 8.55 Images of the bits written in 26 layers with a diarylethene derivative

modulating the fluorescence intensity by photo-oxidation makes C60 an interesting candidate for an optical data storage material with the high chemical stability of C60. In this system, C60 is changed into the photo-oxidation product, such as C60 by two photon absorption process. When the oxidized areas in C60-doped polystyrene films are exposed to a reading laser beam, the fluorescence is much stronger than in the nonoxidation areas, and the high fluorescence signals can easily be examined as bit 1 and the low-fluorescence signals from nonoxidation areas as bit 0. Figure 8.56 shows the pattern of three layers of bit data written in a C60-doped polystyrene film, where the letters A, B, and C are in the first, second, and third layers, respectively. The separation between two adjacent layers is about 25 mm, the distance between two adjacent dots in each layer is about 10 mm, and the diameter of one dot is about 2 mm. The intensities for both writing and reading are about 1.0 W/cm^2 at 488 nm. Scale bars are 10 mm. This change in the fluorescence intensity on C60 could be used to encode information for read-only memory.

- **2PA photocycloreversion-photocycloaddition materials**
  In 2008, Fengyu Li et al. [183] designed and synthesized a diarylethene and tetraarylcyclobutane reversible system, in which a novel thermally stable photoresponsive fluorescent switches molecules with good photo trans–cis isomerization

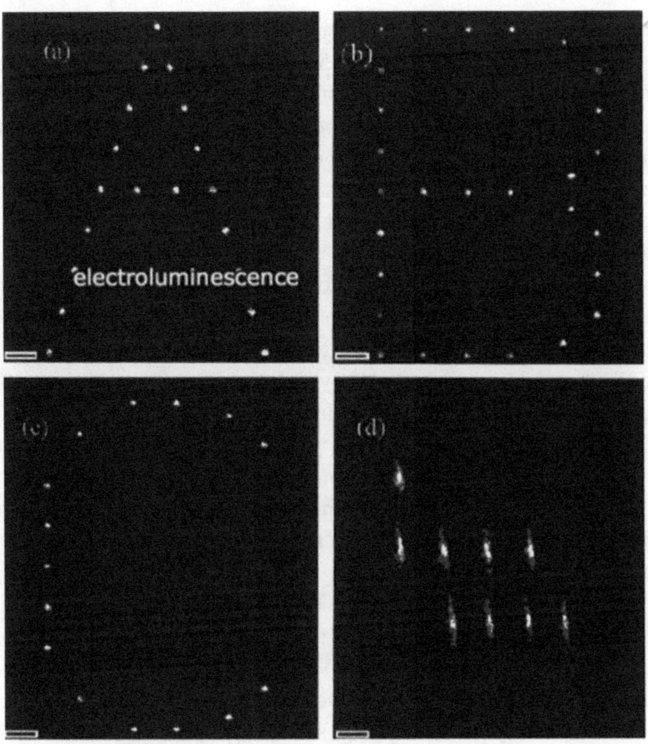

Fig. 8.56 Imagine of three layers (A, B, C) of bit data in a C60 doped polystyrene film

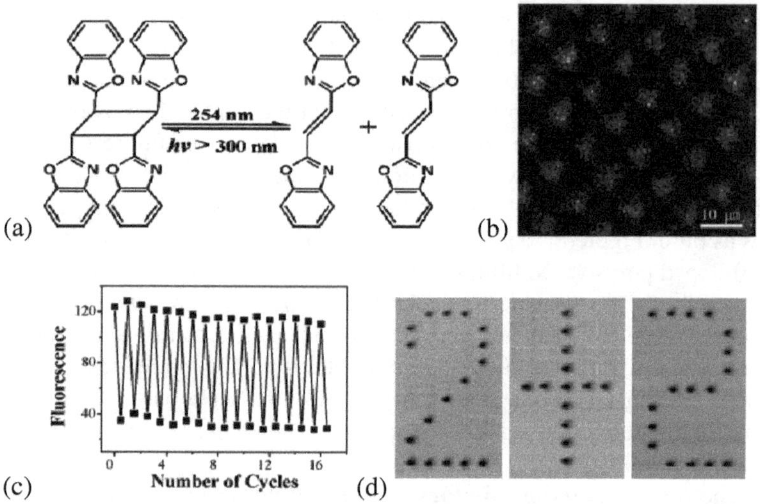

Fig. 8.57 (**a**) Photoreaction between TBC (*left*) and DBE (*right*); (**b**) Fluorescence recording image of the TBC-PMMA film; (**c**) Cycles with the photoreaction reversibility of the DNBC-PMMA film (alternating irradiation of 254 and 393 nm light); (**d**) Recording and readout images for three-layered optical memory in the DNBC-PMMA film

and electrocyclic interconversion properties and demonstrated its application in 3D optical data storage. Figure 8.57a shows the reversal of diarylethylene and tetraarylcyclobutane by introducing fluorescent groups into the system. Figure 8.57b illustrates a typical recording image in which the areas exposed to 254-nm light (5.0×10^2 μW/cm^2 for 2 min) are luminescent and the masked areas are dark. The size of each dot is about 6 μm. Sixteen cycles (Fig. 8.57c) and three-layer storages were achieved (Fig. 8.57d).

- **2PA photopolymerization materials**
  Two-photon photopolymerization, which transforms the submicron structure of a photopolymerizable resin, is a modern method in 3D optical data storage. With two-photon photopolymerization the resin is illuminated by a tightly focused ultrafast laser beam at a wavelength beyond the single-photon absorption band of the resin, which is different from that in the single-photon case. Where is highly localized focused by the high photon density initiates the non-linear effect of the resin and occasions the two- photon absorption simultaneously, in order that provides sufficient energy to trigger the photochemical reaction, namely, interacting with the surrounding monomers to trigger a polymerization chain reaction as shown in reaction (8.28). Multifunctional monomers can form a cross-linked network within the focal volume of the laser beam, which cannot be dissolved by any solvent. By moving the laser focus three-dimensionally, the resin is solidified along the pathway of the light beam. The non-cross-linked monomers maintain good solubility in certain solvents and can be washed away after the two-photon photopolymerization process, leaving 3D storage data that is an exact replica of the pathway of the laser focus (Fig. 8.58).

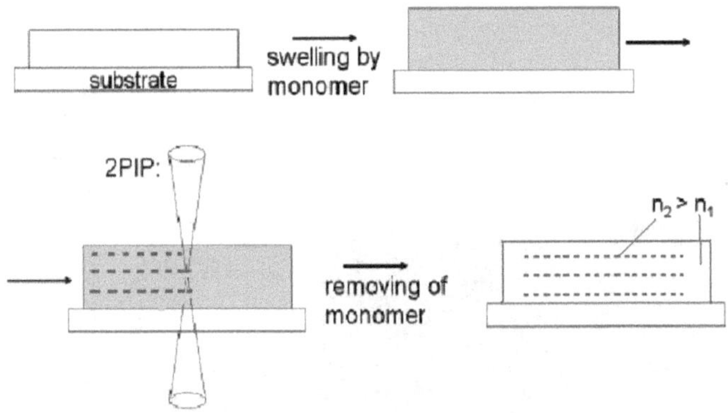

Fig. 8.58 Principle of waveguide writing by 2PA

$$2h\nu_{\mathrm{NIR}} + initiator \rightarrow R^{\bullet}\ or\ R^{+} \quad (8.28)$$

According to the above process, two-photon polymerization offers two obvious benefits over other photon absorption processes [184, 185]: (1) Achievable resolution is about one order of magnitude better than with other 2D data storage methods. Owing to a quadratic dependence on the laser intensity and the threshold effect of two-photon polymerization, only the resin closed to the laser focal spot can be polymerized, and converted from the monomer to the polymer, which results in a high spatial resolution of 100 nm or less thickness available [186, 187]. (2) It is possible to directly write inside a given volume to enhance the data storage dimensions.

Kevin D. Belfield et al. [188] have developed a photosensitive polymeric system for 2PA WORM optical data storage, consisting of 5 wt % PAG, 1 wt% 2PA dye, and 94 wt% host polymer, which was proved to be resilient to overexposure. Furthermore, the inherent nonlinearity and sensitivity of the system enabled multilayer recording and readout of cross-talk-free 3D optical data with subdiffraction-limited voxel sizes. The advantages of two-photon writing and readout were clearly demonstrated, providing a storage density capacity of 1.8×10^{13} bits/cm^3. Two-photon writing was performed at 730 nm (2.4 mW), 200 fs, 60 ms exposure/voxel with a 60×, 1.4 NA oil immersion objective. Two-photon readout was performed, layer-by-layer (0.4 μm/layer), at 860 nm (7 mW), 200 fs, with the same objective as that used for writing (Figs. 8.59 and 8.60).

Recording and Reading System of 2PA Optical Storage

Two-Beam Recording and Reading System

A two-beam recording and reading system was first proposed to demonstrated a bit-oriented two-photon absorption 3D optical storage optical memory with a photochromic spirobenzopyran by Rentzepis et al. [176]. Two beams were used to access a point in a volumetric recording medium. The experimental system [189,

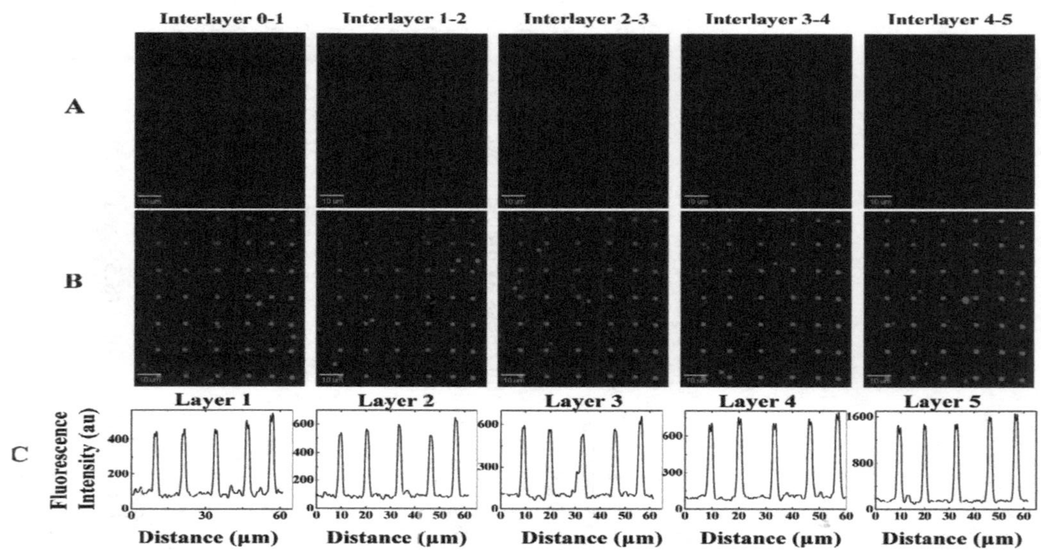

Fig. 8.59 Photosensitive polymeric system used for 3D, two-photon ODS and two-photon readout. (**a**) Blank interlayers (unrecorded volume between voxel layers). (**b**) Two-photon readout of five layers of recorded voxels. (**c**) Fluorescence intensity scan of each layer showing consistently good signal-to-noise ratios throughout all five recorded layers of data

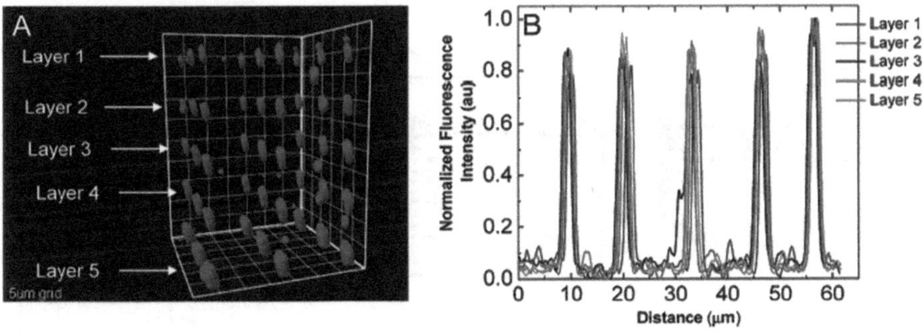

Fig. 8.60 (**a**) Three-dimensional image reconstruction done by overlaying all readout layers using SlideBook 4.1 (surface mode). (**b**) Normalized fluorescence intensity scan of all layers showing excellent signal-to-noise ratios throughout the entire polymer matrix. Note that because there is virtually no fluorescence signal between the layers, the system is cross-talk-free. Scale: 5 μm grid

190] is shown in Fig. 8.61 [191]. As was seen in Fig. 8.54, the absorption band is shorter than 450 nm for isomer A and around 600 nm for isomer B. Isomer B fluoresces around 700 nm. For writing data, isomer A absorbs two photons simultaneously (equivalent of a 355 nm pulse) at the intersection of the two beams of 1064 and 532 nm and photoisomerizes to isomer B (written form). For reading data, isomer B molecules are excited by the absorption of two 1064-nm photons, but the excitation causes only the written molecules to fluoresce.

The 1064-nm information-carrying beam passes through a spatial light modulator that contains the information to be stored. The 532-nm addressing beam is focused by a spherical lens onto the information-carrying beam. The optical path lengths are adjusted to synchronize the arrival of both the information-carrying and the addressing beam at the same time at a preselected place. It is possible to store multiple data pages in the volume of the memory device. However, of the two-beam recording and reading system has disadvantages: (1) It is difficult to get two beams to intersect at the same point. (2) The working distance of the objective lenses is limited. A single-beam recording and reading system that addresses these problems is described below.

Single-Beam Recording and Reading System

A typical single-beam recording and reading system for 2PA 3D optical storage is shown in Fig. 8.62 [191].

In this system only one beam is used and as spatial and temporal alignment of the short pulses is not required, the optical alignment is relatively simple. The recording location inside the medium is controlled by the laser beam irradiance profile and records one bit at a time, rather than a complete plane. A readout path and a detector for evaluation of the recorded data are also included. Fluorescence of the recorded spots can be excited by single- or two-photon absorption. With low power, low cost, and high efficiency, single-photon excitation could be used for readout.

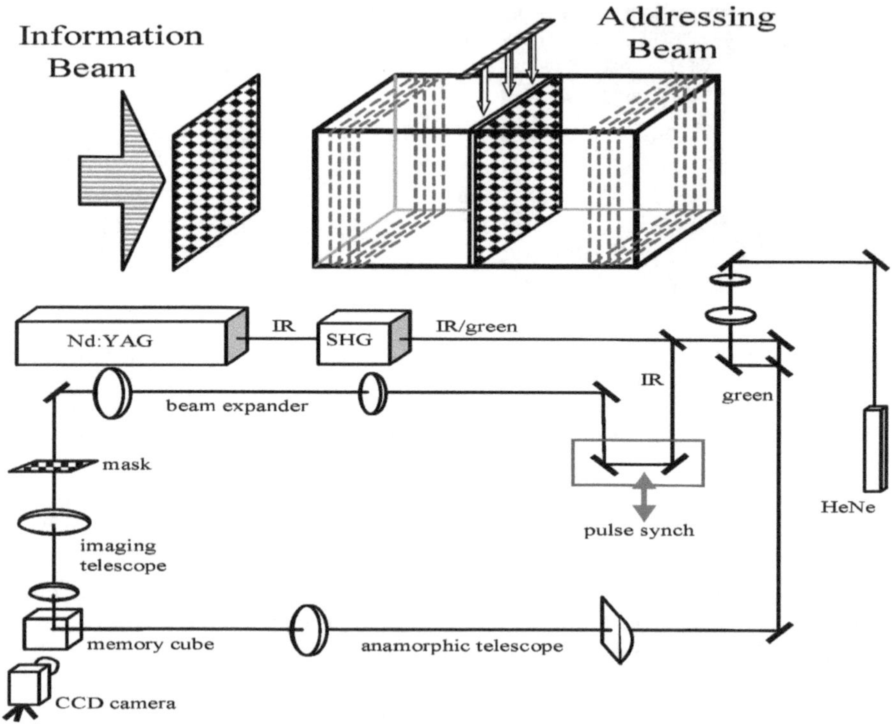

Fig. 8.61 Two-beam storage system (1064-nm and 532-nm beams were emitted by a 35-ps, 20-Hz Nd:YAG laser)

Two-photon absorption probability is proportional to the square of the intensity, so a proper laser system, such as Ti:sapphire laser, should be selected to provide efficient means for achieving high power [192, 193]. Other systems have been developed using a mode-locked Ti:sapphire laser [194–197] and a mode- locked dye laser [192].

2PA Optical Disk System

With the development of two-photon absorption 3D optical storage, a two-photon 3D optical disk with 1 Tb of data in 200 layers was prepared by Ed Walker and Peter M. Rentzepis et al. in 2008 [198] (Fig. 8.63). Each layer contains 5 Gb of data similar to the capacity of a single-layer DVD. Very-high-sensitivity materials can be recorded with bit energies as low as 250 pJ/bit. Figure 8.64 [199] shows a single-beam two-photon recording and a one-photon readout system. In this system, the high Q laser system Picotrain series 532-nm Nd:vanadate laser, which emits pulses with a 6.5-ps pulse width, 7 nJ/pulse energy at a repetition rate of 75 MHz, was used. A liquid immersion singlet (LIS) with an NA of 1.0 and a working distance of 1.2 mm was utilized in the recording system, where the liquid was contained in

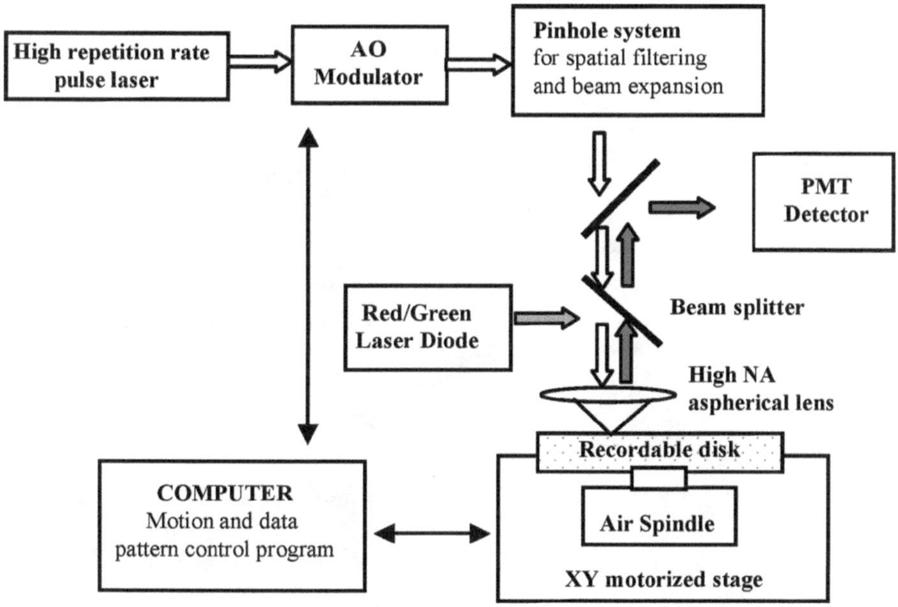

Fig. 8.62 Single-beam storage system

Fig. 8.63 Photograph of a 120-mm-diameter 1 Tb 3D disk before (*left*) and after (*right*) recording

a thin polyurethane membrane that was in contact with the lens on one side and with the spinning disk on the other (Fig. 8.65) [199]. The volume of the liquid was adjusted to allow it to access inner layers of the disk volume without any change in spherical aberration.

To access the 1 Tb of data after recording, a 635-nm CW BlueSky circulaser diode operating at less than 0.5 mW is used to induce fluorescence via a one-photon process, from the recorded data bits. The fluorescence is then picked up by the same objective lens and focused onto a detector such as a Hamamatsu R7400U PMT, with a 25-μm confocal pinhole that is used to decrease layer and tracking cross-talk from adjacent tracks and layers (Fig. 8.66) [191]. A 10-μm pinhole placed at a 1:2 image

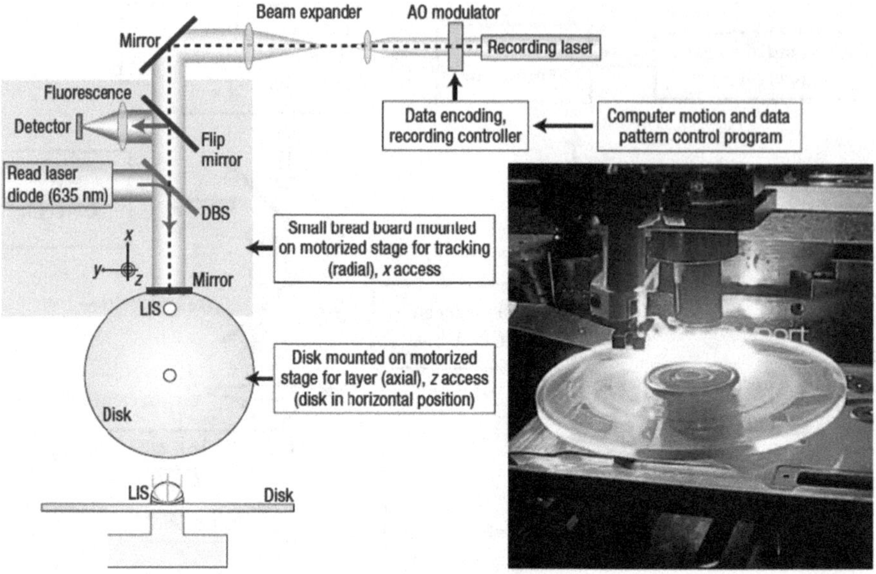

Fig. 8.64 Experimental system used for recording 1 Tb in a DVD size disk. (DBS: dichroic beam splitter; LIS: liquid immersion system; AO: acousto-optic system)

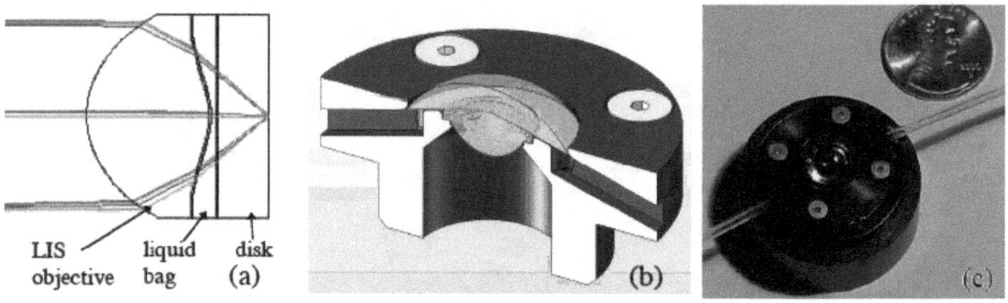

Fig. 8.65 (**a**) Zemax layout of liquid immersion singlet (LIS) objective lens. (**b**) SolidWorks design of prototype package for testing. (**c**) Assembled LIS in a prototype optomechanical package

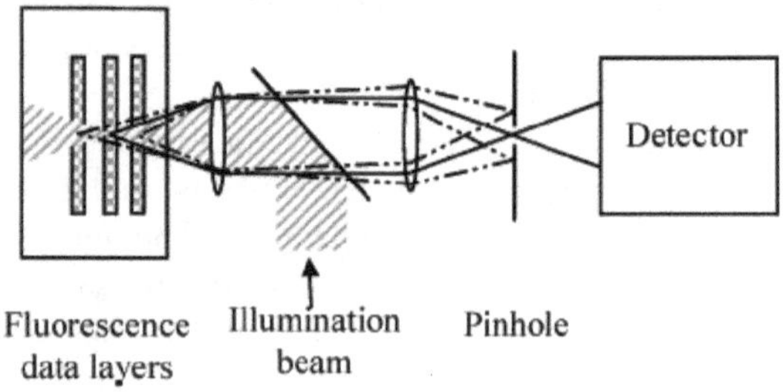

Fig. 8.66 Confocal pinhole in the collinear readout system used to decrease layer cross-talk

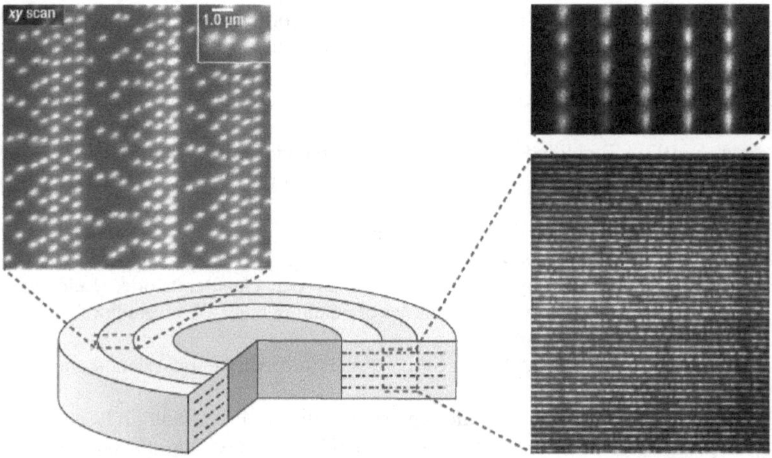

Fig. 8.67 Multiple layers of data stored in the 3D memory disk

system can block over 90% of the fluorescence coming from adjacent data layers in addition to preventing spurious background noise from impinging on the detector. Figure 8.67 [200] shows the typical confocal scanned images of data cross-talk-free tracks within one layer and multiple layers along the depth.

References

1. http://en.wikipedia.org/wiki/Optical_disc
2. http://www.chinahda.org.cn/en/index.html
3. He H (2009) Review on the optical disc formats. Proc SPIE 7125:712528
4. Gan F (ed) (1998) Digital optical disk technology. Science Press, Beijing (in Chinese)
5. Gan F et al (2006) Optical storage technology. In: Liu S (ed) Photonics technolgy and application. pp 1227–1346. Guangdong Science and Technology Press, Guangzhou (in Chinese)
6. Mustroph H, Stollenwerk M, Bressau V (2006) Current development in optical data storage with organic dye. Angew Chem Int Ed 45:2016–2035
7. Kubo H, Shibata M, Katayama K et al (2004) Progress in organic write-once technology for blu-ray disc-recordable and DVD-R media. Proc SPIE 5380:128
8. Usami Y, Kakuta T, Ishida T et al (2003) Blue-violet laser write-once optical disc with spin-coated dye-based recording layer. Proc SPIE 5069:182
9. Inoue H, Mishima K, Aoshima M et al (2003) Inorganic write-once disc for high speed recording. Jpn J Appl Phys 42:1059–1061
10. Hosoda Y, Mitsumori A, Megumi et al (2004) Recording mechanism of high-density write-once disks using inorganic recording material. Jpn J Appl Phys 43:4997–5000
11. Miyagawa Naoyasu. Kitaura Hideki. Katsuyuki Takahashi et al (2006) Over 500 years lifetime dual-layer Blu-ray disc recordable based on Te-O-Pd recording material. Proc SPIE 6282:F1–5
12. Wuttig M, Yamada N (2007) Phase-change materials for rewriteable data storage. Nat Mater 6:824–832
13. Gan F, Xu L (ed) (2006) Photonics glass. World Scientific, Singapore

14. Gan F (1996) Rare earth alloy and oxide thin films for optical storage. In: Kumar Das VG (ed) (1996) Main group elements and their compounds. Narosa Publishing House, New Delhi, pp 63–76
15. Gambino RJ, Takao Suzuki T (ed) (1999) Magneto-Optical Recording Materials. Wiley-IEEE Press
16. Korpel A (1978) Simplified diffraction theory of the video disk. Appl Opt 17:2037–2042
17. Hopkins HH (1979) Diffraction theory of laser read-out systems for optical video disks. J Opt Soc Am 69:4–24
18. Wilson T, Sheppard CJR (1984) Theory and practice of the scanning optical microscope. Cambridge University Press, Cambridge
19. Heisenberg W (1930) Physikalische Prinzipien der Quantentheorie (Leipzig: Hirzel). English translation. The physical principles of quantum theory. University of Chicago Press, Chicago
20. van de Nes AS, Braat JJM, Pereira SF (2006) High-density optical data storage. Rep Prog Phys 69:2323–2363
21. di Francia GT (1952) Nuovo pupille superresolventi. Atti Fond. Goirgio Ronchi 7:366–372
22. di Francia GT (1952) Super-gain antennas and optical resolving power. Nuovo Cimento Suppl 9:426–435
23. Ando H (1992) Phase-shifting apodizer of three or more portions. Jpn J Appl Phys 31:557
24. Ando H, Yokota T, Tanoue K (1993) Optical head with annular phase-shifting apodizer. Jpn J Appl Phys 32:5269
25. Wang H, Gan F (2001) New approach to superresolution. Opt Eng 40(5):851–855
26. Wang H, Gan F (2002) Phase-shifting apodizer for increasing the focal depth. Appl Opt 41(25):5263–5266
27. Wang H, Gan F (2001) High focal depth with pure-phase apodizer. Appl Opt 40(31): 5658–5662
28. Wang H, Shi L, Yuan G, Miao XS, Tan W, Chong T (2006) Subwavelength and super-resolution nondiffraction beam. Appl Phys Lett 89:171102
29. Wang H, Shi L, Luk'yanchuk B, Sheppard C, Chong CT (2008) Creation of a needle of longitudinal polarized light in vacuum using binary optics. Nat Photonics 2:501–505
30. Luo H, Zhou C (2004) Comparison of superresolution effects with annular phase and amplitude filters. Appl Opt 43(34):6242–6247
31. Lokosz W (1966) Optical system with resolving powers exceeding the classical limit. J Opt Soc Am 56:1463–1472
32. Bouwhuis G, Spruit J (1990) Optical storage read-out of nonlinear disks. Appl Opt 29: 3766–3768
33. Tominaga J, Fuji H, Sato A, Nakano T, Fukaya T, Atoda NJ (1998) The near-field super-resolution properties of an antimony thin film. Jpn J Appl Phys Part 2—Lett 37: L1323–L1325
34. Wang H, Yuan G, Tan W, Shi L, Chong T (2007) Spot size and depth of focus in optical data storage system. Opt Eng 46:065201
35. Dorn R, Quabis S, Leuchs G (2003) Sharper focus for a radially polarized light beam. Phys Rev Lett 91:233901
36. Visser TD, Foley JT (2005) On the wavefront spacing of focused, radially polarized beams. J Opt Soc Am A 22:2527–2531
37. van de Nes S, Billy L, Pereira SF (2004) Calculation of the vectorial field distribution in a stratified focal region of a high numerical aperture imaging system. Opt Express 12: 1281–1293
38. Park N-C, Yang1 H-S, Rhim1 Y-C, Park1 Y-P (2008) A study on enhancing data storage capacity and mechanical reliability of solid immersion lens-based near-field recording system. Jpn J Appl Phys 47:6646–6654
39. Yoon Y-J, Kim W-C, Park K-S, Park N-C, Park Y-P (2009) Cover-layer-protected solid immersion lens-based near-field recording with an annular aperture. J Opt Soc Am A 26:1882–1888

40. Lisa D, Curtis K, Facke T (2008) Holographic data storage: coming of age. Nat Photonics 2:403–405
41. Leith EN (1966) Holographic data storage in three-dimensional media. Appl Opt 5: 1303–1311
42. Bianco A (2005) A theoretical study to understand the effect on the IR spectrum and a simple way to read optical memory in the mid-IR. Chem Mater 17:869–874
43. Chen Y (2004) Photochromic fulgide for holographic recording. Opt Mater 26:75–77
44. Kann JL, Canfield BW, Jamberdino AA, Clarke BJ, Daniszewski Ed, Sunada G (1996) Mass storage and retrieval at Rome laboratory. Proceedings of the 5th NASA Goddard mass storage systems and technologies conference (College Park, Maryland)
45. Demetri P (1995) Holographic memories. Scient Am 273:70–76
46. Coufal HJ, Sincerbox GT, Psaltis D (2000) Holographic data storage. Springer, New York, NY
47. Hardin RW (July 1999) Optical storage stacks the deck. OE reports No.187 (http://spie.org/x19548.xml?ArticleID=x19548)
48. Michael RG et al (2006) Effects of absorption and inhibition during grating formation in photopolymer materials. J Opt Soc Am B 23:2079–2088
49. Bloom A et al (2004) The effect of polymer host on volume phase holographic recording properties. Polym Eng Sci 17:356–358
50. Jeong YC et al (2007) Holographic diffraction gratings with enhanced sensitivity based on epoxy-resin photopolymers. Opt Exp 15:1497–1504
51. Gong Q et al (2005) Humidity-resistant highly sensitive holographic photopolymerizable dry film. Mater Lett 59:2969–2972
52. Blaya S et al (2003) Optimization of a photopolymerizable holographic recording material based on polyvinyl- alcohol using angular responses. Opt Mater 23:529–538
53. Samui AB (2008) Holographic recording medium Haugh. Recent Pat Mater Sci 1:74–94
54. Barachevskii V (2006) Photopolymerizable recording media for three dimensional holographic optical memory. High Energ Chem 40:165–176
55. Suzuki N et al (2002) Holographic recording in TiO2 nanoparticle-dispersed methacrylate photopolymer films. Appl Phy Lett 81:4121–4123
56. Suzuki N, Tomita Y (2004) Silica-nanoparticle-dispersed methacrylate photopolymers with net diffraction efficiency near 100%. Appl Opt 43:2125–2129
57. Tomita Y et al (2005) Holographic manipulation of nanoparticle distribution morphology in nanoparticle-dispersed photopolymers. Opt Lett 30:839–841
58. Suzuki N et al (2006) Highly transparent ZrO2 nanoparticle-dispersed acrylate photopolymers for volume holographic recording. Opt Exp 14:12712– 12719
59. Park J, Kim E (2005) Preparation and characterization of organic inorganic nanocomposite films for holographic recording. Key Eng Mater 277–279:1039–1043
60. Yamamoto T et al (2000) Holographic grating and holographic image storage via photochemical phase transitions of polymer azobenzene liquid-crystal films. J Mater Chem 10:337–342
61. Ichimura K (2000) Photoalignment of liquid-crystal systems. Chem Rev 100:1847–1873
62. Corvazier L (1999) Induction of liquid crystal orientation through azobenzene-containing polymer networks. Macromolecules 32:3195–3200
63. Yoneyama S (2002) High-performance material for holographic gratings by means of a photoresponsive polymer liquid crystal containing a tolane moiety with high birefringence. Macromolecules 35:8751–8758
64. Shibaev V et al (2000) Fieldresponsive chiral-photochromic side-chain liquid-crystalline polymers. Polym Int 49:931–936
65. Shibaev V et al (2003) Photoactive liquid crystalline polymer systems with light-controllable structure and optical properties. Prog Polym Sci 28:729–836
66. Yin S (1993) Wavelength multiplexed holographic storage in a sensitive photorefractive crystal using a visible-light tunable diode laser. Opt Commun 101: 317–321

67. Randolph G (1995) Green microlaser technology promises new applications. Laser Focus World 31:121–125
68. Goertzen B (1996) Volume holographic storage for large relational databases. Opt Eng 335:1847–1853
69. Huo Y (1996) Miniature green lasers for optical recording and storage. Proc SPIE 2890: 49–52
70. Sefler G (1997) Radio-frequency matched filtering by photorefractive optical holography. Appl Opt 36:7415–7421
71. Coy J (1996) Characterization of a liquid crystal television as a programmable spatial light modulator. Opt Eng 35:15–19
72. Ward C (1991) Operating modes of the microchannel spatial light modulator. Opt Eng 30:1442–1427
73. Shiquan T, Dayong W, Zhuqing J (1998) Optical holographic storage. Beijing Institute of Technology Press, Beijing (in Chinese)
74. Kulich H (1991) Reconstructing volume holograms without image field loss. Appl Opt 30:2850–2857
75. Peter A (1994) Noise-free holographic storage in iron-doped lithiume niobate crystals. Opt Lett 19:1583–1585
76. Yariv A (1993) Interpage and interpixel cross talk in orthogonal holograms. Opt Lett 18: 652–654
77. Yi X (1994) Statistical analysis of cross-talk noise and storage capacity in volume holographic memory. Opt Lett 19:1580–1582
78. Gu C (1996) Bit-error rate and statistics of complex amplitude noise in holographic data storage. Opt Lett 21:1070–1072
79. Neifeld M (1996) Technique for controlling cross-talk noise in volume holography. Opt Lett 21:1298–1300
80. Bashaw M (1994) Cross-talk in multiplexed holograms using phase-encoded multiplexing in volume holography. J Opt Soc Am B 11:1820–1836
81. Yi X (1995) Cross-talk noise in volume holographic memory with spherical reference beams. Opt Lett 20:1812–1814
82. Yi X (1995) Statistical analysis of cross-talk noise and storage capacity in volume holographic memory: image plane holograms. Opt Lett 20:779–781
83. Dai F (1997) Statistical analysis on extended reference method for volume holographic data storage. Opt Eng 36:1691–1699
84. Steckman G (2001) Storage density of shift-multiplexed holographic memory. Appl Opt 40:3387–3394
85. Courjon D, Bainier C (1994) Near field microscopy and near field optics. Rep Prog Phys 57:989–1028
86. Nanba S et al (ed) (2003) Nanotechnology handbook. Ohmsha, Tokyo
87. Betzig E, Trautman JK, Wolfe R, et al (1992) Near-field magneto-optics and high density data storage. Appl Phys Lett 61(2):142–144
88. Hosaka S, Shintani T, Miyamoto M, et al (1996) Nanometer-sized phase-change recording using a scanning near-field optical microscope with a laser diode. Jpn J Appl Phys 35(1B):443–447
89. Kim MR, Park JH, Jhe W (2000) Near field optical recording by reflection-mode near-field scanning optical microscopy: submicron-sized marks and their thermodynamic stability. Jpn J Appl Phys 39(2B):984–985
90. Jiang S, Ichihashi J, Monobe H, et al (1994) Highly localized photochemical processes in LB films of photochromic material by using a photon scanning tunneling microscope. Opt Commun 106:173–177
91. Yatsui T, Kourogi M, Tsutsui K, et al (2000) High-density-speed optical near-field recording-reading with a pyramidal silicon probe on a contact slide. Opt Lett 25(17): 1279–1281
92. Mansfield SM, Kino GS (1990) Solid immersion microscope. Appl Phys Lett 57(24): 2615–2616

93. Terris BD, Mamin HJ, Rugar D, et al (1994) Near field storage using a solid immersion lens. Appl Phys Lett 65(4):388–390
94. Terris BD, Mamin HJ, Rugar D, et al (1996) Near field optical data storage. Appl Phys Lett 68(2):141–143
95. Ichimura I, Hayashi S, Kino GS (1997) High-density optical recording using a solid immersion lens. Appl Opt 36(19):4339–4348
96. Chekanov A, Birukawa M, Itoh Y, et al (1999) "Contact" solid immersion lens near-field optical recording in magneto-optical TbFeCo media. J Appl Phys 85(8):5324–5326
97. Ichimura I, Kishima K, Saito K, et al (2001) Near-field optical recording on a pre-grooved phase-change disk in the blue-violet. Jpn J Appl Phys 40(3B):1821–1826
98. Shinoda M, Saito K, Ishimoto T, Kondo T, Nakaoki A, Furuki M, Takeda M, Yamamoto M (2003) Proc SPIE 5069:306
99. Shinoda M, Saito K, Ishimoto T, Kondo T, Nakaoki A, Furuki M, Takeda M, Akiyama Y, Shimouma T, Yamamoto M (2004) Proc SPIE 5380:224
100. www.thic.org/pdf/Jul98/terastor.gknight.pdf
101. Shinoda M, Saito K, Kondo T et al (2006) High-density near-field readout using diamond solid immersion lens. Jpn J Appl Phys 45:1311
102. Partovi A, Peale D, Wuttig M et al (1999) High-power laser light source for near-field optics and its application to high-density optical data storage. Appl Phys Lett 75(11):1515–1517
103. Goto K (1998) Proposal of ultrahigh density optical disk system using a vertical cavity surface emitting laser array. Jpn J Appl Phys 37(4B):2274–2278
104. Tominaga J, Nakano T et al (1998) An approach for recording and readout beyond the diffraction limit with a Sb thin film. Appl Phy Lett 73(15):2078–2080
105. Yasuda K, Ono M, Aratani K et al (1993) Premastered optical disk by superresolution. Jpn J Appl Phys 32(11B):5210–5213
106. Tominaga J, Fuji H, Sato A et al (2000) The characteristics and the potential of super resolution near-field structure. Jpn J Appl Phys 39(2B):957–961
107. He YC, Lan YC, Hsu WC et.al (2004) Effect of constituent phases of reactively sputtered AgOx film on recording and readout mechanisms of super-resolution near-field structure disk structure disk with a silver oxide mask layer. J Appl Phys 96(3):1283–1285
108. Kolobov AV, Rogalev A, Wilhelm F et al (2004) Thermal decomposition of a thin AgOx layer generating optical near-field. Appl Phys Lett 84(4):1641–1643
109. Shima T, Tominaga J (2003) Optical and structural property change by the thermal decomposition of amorphous platinum oxide film. Jpn J Appl Phys 42(6A):3479–3480
110. Kim J, Hwang I, Yoon D, Park I, Shin D, Kikukawa T, Shima T, Tominaga J (2003) Super-resolution by elliptical bubble formation with PtOx and AgInSbTe layer. Appl Phys Lett 83(9):1701–1703
111. Kim J, Hwang I, Kim H, et al (2004) Signal characteristics of super-resolution near-field structure disk in blue laser system. Jpn J Appl Phys 43(7B):4921–4924
112. Kim J, Hwang I, Kim H, Bae J, Jung M, Park I, Tominaga J (2005) Proceedings of ISPS 2005, Awaji Yumebutai ICC, p 46
113. Seo M, Im S, Lee J. (2009) Nonlinear modeling of super-resolution near-field structure. Jpn J Appl Phys 48:03A051
114. Tsai DP, Lin WC (2000) Probing the near fields of the super-resolution near-field optical structure. Appl Phys Lett 77:1413
115. Liu WC, Wen CY, Chen KH, Lin WC, Tsai DP (2001) Near-field images of the AgO_x-type super-resolution near-field structure. Appl Phys Lett 78:685
116. Song KB, Lee J, Kim JH et al (2000) Direct observation of self-focusing with subdiffraction limited resolution using near-field scanning optical microscope. Phys Rev Lett 85: 3842–3845
117. Song KB, Kim J, Park KH (2002) Technique to enhance the throughput on a near-field aperture by the use of self-focusing effect. Appl Phys Lett 80:2827–2829

118. Gan FX, Liu QM (2002) Nonlinear optical effects in chalcogenide glasses (invited paper). In: 13th International Symposium on "Non-Oxide Glasses and New Optical Glasses", Pardubice, Czech Republic
119. Nagase T, Ashida S, Ichihara K (1999) Super-resolution effect of semiconductor-doped glass. Jpn J Appl Phys 38:1665–1668
120. Tominaga J et al (2004) Ferroelectric catastrophe: beyongd nanometer-scale optical resolution. Nanotechnology 15:411–415
121. Zhang F, Wang Y, Xu WD et al (2004) High-density read-only memory disk with $Ag_{11}In_{12}Sb_{51}Te_{26}$ super-resolution mask layer. Chin Phys Lett 21(10):1973–1975
122. Avrutsky I et al (2004) Super-resolution in laser annealing and ablation. Appl Phys Lett 84(13):2391–2393
123. Cohn K, Simanovskii D, Smith T et al (2002) Transient photoinduced diffractive solid immersion lens for infrared microscopy. Appl Phys Lett 81:3678–3680
124. Wei JS, Gan FX (2003) Study on readout of super-resolution pits with Si films. Proc SPIE 5060:167–170
125. Ou DR, Zhu J, Zhao JH (2003) Approach for imaging optical super-resolution based on Sb films. Appl Phys Lett 82(10):1521–1523
126. Wei JS, Gan FX (2003) Thermal lens model of Sb thin films in super-resolution near-field structure. Appl Phys Lett 82(16):2607–2609
127. Pan Z et al (1995) Linear and nonlinear optical response of bismuth and antimony implanted fused silica: annealing effects. Opt Mater 4:675–684
128. Liu DR et al (2002) Giant nonlinear optical properties of bismuth thin films grown by pulsed laser deposition. Opt Lett 27(17):1549–1551
129. Zhang F, Xu WD, Wang Y et al (2004) Static optical recording properties of super-resolution near-field structure with bismuth mask layer. Jpn J Appl Phys 43(11A):7802–7806
130. Lu YH, Dimitrov D, Liu JR et al (2001) Mask films for thermally induced superresolution readout in rewritable phase-change optical disks. Jpn J Appl Phys 40(3B):1647–1648
131. Zhang F, Wang Y, Xu WD et al (2005) Read-only memory disk with AgO_x super-resolution mask layer. Chin Opt Lett 3(2):113–115
132. Liu Q, Kim J, Fukaya T, Tominaga J (2003) Thermal-induced optical properties of a PdO_x mask layer in an optical data storage system with a supper-resolution near-field structure. Opt Express 11(21):2646–2653
133. Tsujioka T. Photochromism and its application to a high-density optical memory. Mol Cryst Liq Cryst 315:1–9
134. Hatakeyama M et al (2000) Super-resolution rewritable optical disk having a mask layer composed of thermochromic organic dye. Jpn J Appl Phys 39:752–755
135. Lu SW, Hou LS, Gan FX (1997) Structure and optical property changes of sol-gel derived VO_2 thin films. Adv Mat 9(3):244–245
136. Shintani T et al (1999) A new super-resolution film applicable to read-only and rewritable optical disks. Jpn J Appl Phys 38:1656–1660
137. Gan FX (ed) (1992) Digital optical disk and optical storage materials. Shanghai Science and Technology Press, Shanghai (in Chinese)
138. Wu YH, Khoo H, Kogure T (1994) Read-only optical disk with superresolution. Appl Phys Lett 64(24):3225–3227
139. Mori G, Yamamoto M, Tajima H et al (2005) Energy-gap-induced super-resolution (EG-SR) optical disc using ZnO interference film. Jpn J Appl Phys 44(5B):3627–3630
140. Husakou A, Herrmann J (2004) Superfocusing of light below the diffraction limit by photonic crystals with negative refraction. Opt Express 12(26):6491–6497
141. Liu L, He SL (2004) Near-field optical storage system using a solid immersion lens with a left-handed material slab. Opt Express 12(20):4835–4841
142. Chu TC, Liu W-C, Tsai DP, Yoshimasa K (2008) Readout signals enhancements of subwavelength recording marks via random nanostructures. Jpn J Appl Phys 47(7):5767–5769

143. Park S, Hahn JW (2009) Plasmonic data storage medium with metallic nano-aperture array embedded in dielectric material. Opt Express 17(22):20203
144. Challener WA, Gage Ed, Itagi A, Peng C (2006) Optical transducers for near field recording. Jpn J Appl Phys 45(8B):6632–6642
145. Matsumoto T, Shimano T, Saga H, Sukeda H, Kiguchi M (2004) Highly efficient probe with a wedge-shaped metallic plate for high density near-field optical recording. J Appl Phys 95:3901
146. Nakagawa K, Kim J, Itoh A (2006) Near-field optically assisted hybrid head for self-aligned plasmon spot with magnetic field. J Appl Phys 99:08F902
147. Miyanishi S, Iketani N, Takayama K, Innami K, Suzuki I, Kitazawa T, Ogimoto Y, Murakami Y, Kojima K, Takahashi A (2005) Near-field assisted magnetic recording. IEEE Trans Magn 41:2817
148. Hongo K, Watanabe T (2008) Lensless surface plasmon head with 1 Tbit/in.^{2} Recording density. Jpn J Appl Phys 47(7):6000–6006
149. Zakharian R, Mansuripur M, Moloney JV (2004) Transmission of light through small elliptical apertures. Opt Express 11:2631
150. Shi X, Lambertus H, Thornton RL (2003) Ultrahigh light transmission through a C-shaped nanoaperture. Opt Lett 28(15):1320–1322
151. Jin EX, Xu X (2003) Enhancement of optical transmission through planar nano-apertures in a metal film. Proceedings of IMECE'03 (Washington D.C.), p1
152. Xu X, Jin EX, Uppuluri SMV (2004) Enhancement of optical transmission through planar nano-apertures in a metal film. Proc SPIE 5515:230
153. Xu J, Xu T, Wang J, Tian Q (2005) Design tips of nanoapertures with strong field enhancement and proposal of novel L-shaped aperture. Opt Eng 44:018001
154. Jin EX, Fu X (2004) Finite-difference time-domain studies on optical transmission through planar nano-apertures in a metal film. Jpn J Appl Phys 43:407
155. Shima T, Kuwahara M, Fukaya T, et al (2004) Super-resolutional readout disk with metal-free phthalocyanine recording layer. Jpn J Appl Phys 43(1A/B):L 88–L 90
156. Gao XY, Wang SY, Li J et al (2004) Study of structure and optical properties of silver oxide films by ellipsometry, XRD and XPS methods. Thin Solid Films 455/456:438–442
157. Rausch T, Mihalcea C, Pelhos K et al (2006) Near field heat assisted magnetic recording with a planar solid immersion lens. Jpn J Appl Phys 45(2B):1314–1320
158. Kim JH, Buechel D, Nakano T et al (2000) Magneto-optical disk properties enhanced by a nonmagnetic mask layer. Appl Phys Lett 77(12):1774–1776
159. Shi LP, Chong TC, Miao XS et al (2001) A new structure of Super-resolution near-field phase-change optical disk with a Sb_2Te_3 mask layer. Jpn J Appl Phys 40(3B): 1649–1650
160. Likodimos V, Labardi M, Pardi L et al (2003) Optical nanowriting on azobenzene side-chain polymethacrylate thin films by near-field scanning optical microscopy. Appl Phys Lett 82(19):3313–3315
161. Lu Y, Wang P, Jiangying Z et al (2003) Near field optical storage based on solid immersion lens. Chin J Laser 30(2):145–148 (in Chinese)
162. Hamano M, Irie M (1996) Rewritable near-field optical recording on photochromic thin films. Jpn J Appl Phys 35(3):1764–1767
163. Kim J, Song KB, Park KH et al (2002) Near-field optical recording of photochromic materials using bent cantilever fiber probes. Jpn J Appl Phys 41(8):5222–5225
164. Kim MS, Sakata T, Kawai T et al (2003) Amorphous photochromic films for near-field optical recording. Jpn J Appl Phys 42(6A):3676–3681
165. Liu X, Pu S, Zhang F (2004) Synthesis and application of organic near field optical storage materials. Chin J Laser 31(12):1460–1465 (in Chinese)
166. Feng X, Yang W, Yongling B et al (2004) Near field optical storage characteristics of bacteriorhodopsin. Appl Laser 25(6):5–8 (in Chinese)
167. Irie M, Ishida H, Tsujioka T (1999) Rewritable near-field optical recording on photochromic perinaphthothioindigo thin films: readout by fluorescence. Jpn J Appl Phys 38(10): 6114–6117

168. Shima T, Nakano T, Kurihara K et al (2008) Super-resolution readout of 50 nm read-only-memory pits using optics based on high-definition digital versatile disc. Jpn J Appl Phys 47(7):5842–5844
169. Dvornikov S, Walker EP, Rentzepis PM (2009) Two-photon three-dimensional optical storage memory. J Phys Chem 113(49):13633
170. Göppert-Mayer M (1931) Über elementarakte mit zwei quantensprüngen. Ann Phys 9:273
171. Kaiser W, Garrett CGB (1961) Two-photon excitation in CaF2:Eu2+ Phys Rev Lett 7(6):229
172. Belfield KD, Morales AR, Andrasik S et al (2002) Novel two-photon absorbing polymers in Carraher CE and Swift GG (ed) Functional Condensation Polymer, Springer, New York, pp 135–147
173. Rumi M, Barlow S, Wang J et al (2008) Adv Polym Sci 213:1
174. Ovsianikov A, Chichkov BN (2008) Two-photon polyerization-hgih resolution 3D laser technology and its application in Korkin A, Rosei F (ed) Nanoelectronics and photonics, Springer, New York, pp 427–446
175. Kawata S, Kawata Y (2000) Three-dimensional optical data storage using photochromic materials. Chem Rev 100:1777
176. Parthenopoulos DA, Rentzepis PM (1989) Three-dimensional optical storage memory. Science 24:843
177. Dvornikov S, Malkin J, Rentzepis PM (1994) Spectroscopy and kinetics of photochromic materials for 3D optical memory devices. J Phys Chem 98:6746
178. Irie M (ed) (1994) Photoreactive materials for ultrahigh-density optical memory. Elsevier, Amsterdam
179. Hamano M, Irie M (1996) Rewritable near-field optical recording on photochromic thin films. Jpn J Appl Phys 35:1764
180. Toriumi A, Kawata S, Gu M (1998) Reflection confocal microscope readout system for three-dimensional photochromic optical data storage. Opt Lett 23:1924
181. Xia D, Wada S, Tashiro H (1998) Optical data storage in C60 doped polystyrene film by photo-oxidation. Appl Phys Lett 73:1323–1325
182. Wada S, Xia AD, Tashiro H (2002) 3D optical data storage with two-photon induced photon-oxidation in C60-doped polystyrene film. RIKEN Rev 49:11
183. Li F, Zhuang J, Jiang G, Tang H, Xia A, Jiang L, Song Y, Li Y, Zhu D (2008) A Rewritable Optical Data Storage Material System by [2 + 2] Photocycloreversion–Photocycloaddition. Chem Mater 20:1194
184. Maruo S, Nakamura O, Kawata S (1997) Three-dimensional microfabrication with two-photon-absorbed photopolymerization. Opt Lett 22:132
185. Reinhardt C, Kiyan R, Passinger S, Stepanov A, Ostendorf A, Chichkov B (2007) Rapid laser prototyping of plasmonic components. Appl Phys A: Mater Sci Process 89:321
186. Park S, Lim T, Yang D, Kim R, Lee K (2006) Improvement of spatial resolution in nano-stereolithography using radical quencher. Macromol Res 14:559–564
187. Kawata S, Sun HB, Tanaka T, Takada K (2001) Finer features for functional microdevices. Nature 412:697
188. Yanez CO, Andrade CD, Yao S, Luchita G, Bondar MV, Belfield KD (2009) Photosensitive polymeric materials for two-photon 3D WORM optical data storage systems. Appl Mater Interfaces 1:2219
189. Dvornikov S, Cokgor I, Wang M, McCormick FB, Esener SE, Rentzepis PM (1997) Materials and systems for two photon 3D ROM device. IEEE TCPMT – Part A 20:200
190. McCormick FB, Cokgor I, Esener SC, Dvornikov AS, Rentzepis PM (1996) Two-photon absorption-based 3D optical memories. Proc SPIE 2604:23
191. Dvornikov S, Walker EP, Rentzepis PM (2009) Two-photon three-dimensional optical storage memory. J Phys Chem A 113(49):13633
192. Strickler JH, Webb WW (1991) Three-dimensional optical data storage in refractive media by two-photon point excitation. Opt Lett 16:1780

193. Denk W, Strickler JH, Webb WW (1990) Two-photon laser scanning fluorescence microscopy. Science 248:73
194. Toriumi A, Kawata S, Gu M (1998) Reflection confocal microscope readout system for three-dimensional photochromic optical data storage. Opt Lett 23:1924
195. Kawata Y, Ishitobi H, Kawata S (1998) Use of two-photon absorption in a photorefractive crystal for three-dimensional optical memory. Opt Lett 23:756
196. Day D, Gu M (1998) Effects of refractive-index mismatch on three-dimensional optical data-storage density in a two-photon bleaching polymer. Appl Opt 37:6299
197. Cheng PC, Bhawalkar JD, Pan SJ, Swiatkiewicz J, Samarabandu JK, Liou WS, He GS, Ruland GE, Kumar ND, Prasad PN (1996) Three-dimensional laser scanning two-photon fluorescence confocal microscopy of polymer materials using a new efficient upconverting fluorophore. Scanning 18:129
198. Walker Ed, Dvornikov A, Coblentz K, Rentzepis P (2008) Terabyte recorded in two-photon 3D disk. Appl Opt 47:4133
199. Walker Ed, Dvornikov A, Coblentz K, Esener S, Rentzepis P(2007) Toward terabyte two-photon 3D disk. Opt Express 15:12264
200. Walker EP, Rentzepis PM (2008) Two-photon technology: A new dimension. Nat Photonics 2:406

Chapter 9
Biological Memory in Animals and in Man

Raffaele d'Isa, Nicola Solari, and Riccardo Brambilla

Abstract This chapter will deal with natural forms of learning and memory, with particular emphasis on the vertebrate and human brain function. The topographical and temporal organization of multiple interacting memory systems will be first described at the cognitive and psychological level. Recent developments in pharmacology and molecular genetics will then be discussed to provide a state-of-the-art description of the molecular and cellular mechanisms underlying memory formation, consolidation, retrieval, and reconsolidation. Finally, a comparative analysis between natural and artificial memory devices will be outlined.

Keywords Human memory · Learning · Synaptic plasticity · Hippocampus · Amygdala · Consolidation · Reconsolidation · Place cells · Grid cells

Introduction

As a general concept, memory can be defined as the persistence of previously acquired information. This definition does not refer only to biological organisms, but also considers the properties of inorganic materials. The lunar surface with the traces of hundreds of asteroids which have hit it, a memory foam mattress which keeps memory of the shape of our body, a pair of memory titanium eyeglasses that can regain their original shape if we have accidentally sit on them, or a USB flash drive are all examples of objects endowed with memory properties.

In psychology and cognitive neuroscience, memory is technically defined as the faculty of an organism to encode, store, and retrieve information. In this chapter, we will focus on biological and human memory. We will begin describing its complex organization and will then explain its neural, cellular, and molecular

R. Brambilla (✉)
San Raffaele Scientific Institute and San Raffaele University, Via Olgettina 58, 20132 Milano, Italy
e-mail: brambilla.riccardo@hsr.it

G. Campardo et al. (eds.), *Memory Mass Storage*,
DOI 10.1007/978-3-642-14752-4_9, © Springer-Verlag Berlin Heidelberg 2011

bases. Finally, in the last paragraph we will make some considerations about the comparison between biological and computer memory.

The Organization of Human Memory Systems: Short-Term Versus Long-Term Memory

One of the most important notions in the science of memory is that the ability to form memories is not a unique faculty but it can be divided into many subsystems. This idea is certainly not new. More than 200 years ago, in 1804, a French philosopher, Pierre Maine de Biran, imagined that memory could be divided into three forms, which he called representative, mechanical, and sensitive [1]. The first type, representative memory, is the recollection of a "well-circumscribed idea." The second, mechanical memory, represents a mechanism that does not generate ideas, but simply facilitates the repetition of a movement. Finally, sensitive memory is a mechanism which, again, does not recall ideas but can instead generate a feeling or vague image.

A significant aspect that should be taken in consideration throughout this chapter is that memory systems are brain machineries able to form a potentially infinite number of memory traces or "engrams." The concept of the engram was first put forward by the now almost forgotten German biologist Richard Semon early in the twentieth century [2], but it was made popular in the US by the work of Karl Lashley at the mid of the past century [3]. Essentially, an engram is a physical representation in the brain of a specific acquired memory. To make an easily understandable example, if one is asked to memorize two strings of digits, let us say two telephone numbers, the subject will actually form two distinct engrams in his/her mind and, with most probability, in the brain as well. It is interesting to note here that while in recent years we have accumulated a large body of evidence on the mechanisms of memory acquisition, consolidation, etc., we know little on how the brain keeps distinct similar engrams and is able to retrieve them in a context-specific manner. The problem of the exact encoding of each engram is a formidable one: one should only think what a cognitive failure it would be if the brain was unable to retrieve the right and only the right telephone number at the requested moment.

In any case, the most basic and widely accepted scientific classification of memory is the one based on the duration of its retention, with the classical dichotomy between short-term and long-term memory. Short-term memory (STM) relates to durations which can range from seconds to minutes. For example, when someone tells us a telephone number and we compose it soon after to make the call, we are using our STM. If we are distracted while composing it, the number seems to vanish from our memory and we have to ask it again. Long-term memory (LTM) on the other hand refers to a duration of days, years, or even decades: we remember what we did when we were children or what we have done yesterday. At the end of the nineteenth century, William James, in his *Principles of Psychology*, had already hypothesized a similar distinction between what he named as primary and secondary memory [4]. But evidences for the fact that STM and LTM are really two distinct kinds of memory came mainly from clinical neuropsychology and animal experimental studies in the late twentieth century.

The clearest evidence are patients with amnesic syndrome, the most famous and well-studied of which is without doubt the amazing case of H.M. [5]. Henry Gustav Molaison, better known as the patient H.M., was a boy suffering from epileptic seizures from the age of 16. His illness was intractable with drugs and progressively worsened though he was able to complete high school and find a job in a factory of electrical motors. In 1953, at the age of 27, his epilepsy aggravated to the point that he had to stop working. He then decided to try a new and radical treatment for epilepsy: neurosurgery. Since his symptoms indicated that the seizures had origin from the medial part of the temporal lobe of the brain, the neurologist William Scoville suggested to ablate the epileptogenic focus by surgically removing the medial temporal lobe bilaterally (a procedure known as temporal lobectomy). The operation was in itself a success: the patient survived after the complex operation and his epilepsy ameliorated. Unexpectedly though, H.M. began to develop some strange memory deficits. In 1955, 1 and a half years after the operation, he was convinced of being still in 1953 and only 27 years old! He claimed that he had never seen people that were talking with him just a few minutes before and had just left the room. If they entered the room again, H.M. received them as if he had never seen them in his life. What had happened to him? H.M. was now affected by a particular form of amnesia called anterograde amnesia. The most commonly known form of amnesia, depicted in numerous novels and movies is retrograde amnesia, in which the patient loses memory of events lived before the moment of the accident. Anterograde amnesia instead is the loss of the ability to form memories after the moment of the accident. H.M.'s STM was absolutely normal. He could carry on a conversation and he had an intact digit span, as assessed by the ability to repeat strings of six or seven digits [6]. But he had completely lost LTM for the events he lived or the facts he could for instance read in books or listen on television. As the neuroscientist Suzanne Corkin, who worked with him for decades, has written, H.M. basically lived in the past [7].

The Organization of Human Memory Systems: Explicit Versus Implicit Memory

In the last paragraph, we saw the division of memory based on the duration of its retention: short-term versus long-term memory. But this is not the only criterion on the basis of which memory can be classified. Another fundamental distinction is the one between explicit memory and implicit memory, which is based on the conscious awareness required in recalling the memory.

Explicit memory, also called declarative memory as its content can be "declared," is the body of experiences and information we can recollect consciously and intentionally. This kind of memory is further subdivided, due to the work of Endel Tulving, into episodic and semantic memory [8]. Episodic memory refers to the events of our lives and to our personal experiences. It is our autobiographical memory, centered on our person and specific to a particular context, localized in time and space. Remembering our 18th birthday party, having visited a foreign city during a holiday, or having met a friend last week are all examples of episodic memory. Semantic memory instead does not concern events but facts, considered independently of the context. It represents all our abstract knowledge about the world and is not related to specific experiences of our existence. We are for instance able to say who is the 16th president of the United States, even if we do not remember the exact circumstances in which we learned this fact. Examples of semantic memory comprise encyclopedic notions, like "Rome is the capital of Italy" or "the Battle of Hastings was fought in 1066," but also, more simply, remembering what a dog is or how our wife looks like.

Implicit (also known as nondeclarative) memory, on the other side, refers to all the knowledge that does not require intentional (conscious) recall of information. It can be revealed when previous experiences aid in the performance of a task without conscious awareness of these previous experiences [9, 10].

To make clear the distinction between explicit and implicit memory, we will return to the already-mentioned case of H.M. Initially, it seemed that H.M. had completely lost his long-term memory and could not learn anything new. However, in the 1960s, neuropsychologist Brenda Milner, who was examining H.M. in a series of various and accurate tests, made an exceptional discovery when she subjected him to a mirror drawing task, demonstrating that actually he could acquire some kind of information in a long-lasting manner [11]. The task required to trace on a paper the outline of a complex geometrical figure that was not exposed to direct view, but was hidden and could be seen only using a mirror. It is a test which depends on a hand–eye coordination skill. Initially, most subjects consider this task quite difficult and show a bad performance, but within a few trials of training they become proficient. H.M. revealed a normal learning curve in the task: given ten trials he became skilled at drawing the shape and he progressively reduced the number of errors. Even more surprisingly, he maintained this skill over the next 2 days, even if he claimed he did not remember having exercised himself in the task before nor having ever seen the apparatus! He said he had no idea of what to do, but when he tried he was actually able to complete the test capably. He had learned the skill implicitly, in complete absence of explicit or conscious memory.

Mirror drawing task represents a form of implicit memory called procedural memory, which is defined as the long-term memory for skills. Skills are procedures for operating in the world. They can be motor (for example, riding a bike), perceptual (like visual anticipation in tennis players who can anticipate which shot will be played by their opponent), or cognitive (like reading a book). Procedural memory is not the only form of implicit memory. Other important forms are priming, conditioning, and nonassociative learning.

Priming refers to facilitative changes in the ability to identify, generate, or process an item due to a specific prior encounter with the item. Psychologists have extensively studied priming in healthy human subjects. In typical experiments, subjects are presented with a list of words, pictures of objects, or nonverbal material like line drawings. In a second phase, they are tested with some of the old items alternated with new items and they are required to name words or objects, to complete a figure from its fragments or to make rapid decisions about new and old items. Subjects generally result having a better performance for the old (primed) items than for the new (nonprimed) ones. An example of test in which priming effects can be observed is the word stem completion task. If a subject reads a list of words which includes the word "sugar" and is then required, in the testing phase, to complete a series of word stems as he prefers, his probability to complete the stem "sug" with the word "sugar" is higher than for nonprimed subjects. Another test is picture completion. In this task, subjects are shown an incomplete sketch of a figure that cannot be identified. They are then shown more of the sketch, until they are able to recognize the figure. Subsequently, due to the priming effect, they will identify the picture with a smaller number of steps than the first time. Word stem completion and picture completion are two forms of perceptual priming, which is based on the physical aspect of the stimulus and derives from the match between the old and the new item: altering the format of the stimuli or the sensory modality by which they are presented affects the strength of perceptual priming. Another kind of priming is conceptual priming, which is instead based on the semantic relation between the stimuli. For instance, "spoon" can show priming effects on "fork" because they are both pieces of cutlery.

Anyway, besides psychological tests, priming can easily be observed also in everyday life. If we are talking with a friend and we casually hear a low-frequency word, such as "ephemeral," then, in a subsequent appropriate occasion, this word might have a greater probability to just spring to our mind and be used than if we had not been previously exposed to the word during the conversation with our friend. Priming can also be of legal interest since it can lead to cryptomnesia, or inadvertent plagiarism, in which a person falsely recalls having generated a thought, when this was actually generated by someone else [12]. The person is not deliberately plagiarizing, but really fails to credit the original source of the information. A famous case is the one George Harrison, the guitarist of the English pop band The Beatles, who was sued over royalties for his first solo song "My Sweet Lord," for the reason that it resembled too much "He's so fine" by The Chiffons. Harrison lost the case and was obliged to pay $587,000 to Bright Tunes Music, who owned the copyright of The Chiffons' song. The judge regarded him guilty because, although he had not voluntarily copied the song, he had been influenced in the composition of the song by his implicit memory of "He's so fine" and had hence "subconsciously plagiarized" it.

A key finding on priming has been made by Elizabeth Warrington and Lawrence Weiskrantz when they studied amnesic patients (like H.M.) starting from the end of the 1960s [13–15]. Although their performance in recall of words from a previously read list was severely impaired, these patients demonstrated to be normal in the word completion task, even though they had no idea of why they had completed the stem in such a way nor they remembered having even read a list of words previously. This indicated that priming was spared in these patients, representing a further evidence in favor for the fact that conscious explicit memory and nonconscious implicit memory are actually two distinct forms of memory.

Conditioning is a form of associative learning, that is the process by which an element is learned through association with a different, preoccurring one. It is technically defined as a learning process in which either (a) a given stimulus becomes increasingly effective in evoking a response or (b) a response occurs with increasing regularity in a well-specified and stable environment. These two cases correspond to two distinct types of conditioning, which are called, respectively, classical and operant.

Classical conditioning, also called Pavlovian conditioning, was first demonstrated by the Russian physiologist Ivan Petrovich Pavlov, during a series of experiments conducted on dogs [16]. These dogs, to whom Pavlov gave picturesque names like Zygan (gypsy), Arlekin (jester), Baikal (the Russian lake), or Gengis Khan [17], became famous all over the world for having led him to the discovery of one of the most basic and universal forms of learning. Still today, in common speaking, the phrase "Pavlov's dog" is often used to depict someone who simply reacts automatically to a situation instead of thinking in a critical manner.

In the 1890s, Pavlov was studying the secretory activity of digestion. By externalizing a salivary gland he could collect the saliva produced in response to food, in order to quantify it and analyze its composition under different conditions. While performing these studies, he casually observed that, unexpectedly, the dogs had the tendency to salivate even before food was actually delivered to their mouths, merely at the sight of the laboratory technician who used to feed them. He thought that the phenomenon, which he called "psychic secretion," deserved a particular attention and began a new series of experiments to investigate the nature of this effect, shifting his interest from digestive to psychological function. These experiments led him to formulate the laws that regulate what he named "conditional reflexes," that is, reflex responses, like salivation, that occur only conditionally upon specific previous experiences of the animal. When a dog received meat powder in his mouth, he spontaneously began salivating. Pavlov called this reflex the unconditioned response and the stimulus which elicits it unconditioned stimulus (US). If a neutral stimulus, for instance, the sound of a ringing bell, is systematically presented before the delivery of the food, after a few trials the dog will begin to salivate in response to the bell alone (even if no food will be delivered). The neutral stimulus (the bell) has become a conditioned stimulus (CS) due to the repeated pairing with the US. This association represents the conditioned response. It is worth noticing that Pavlovian conditioning implies associations between two stimuli, for example, the gustatory stimulus of meat and the auditory stimulus of the bell. For this reason it also known as stimulus–stimulus (S–S) learning. A different kind of conditioning, which does not relate two stimuli, but a stimulus and a response, is operant conditioning.

Operant or Skinnerian conditioning (from the name of the American psychologist Burrhus Frederic Skinner who studied it) is a type of learning in which the occurrence of a behavior "operating" on the environment is modified by its consequences. These consequences can be mainly two: a reinforcement or a punishment. A reinforcement is a stimulus that increases the probability of occurrence of a behavior. A punishment instead is a stimulus that decreases the probability of a behavior to occur. The laws of operant conditioning can be well exemplified by describing a typical Skinner box, an experimental apparatus realized by Skinner in the late 1920s. This box is a cage containing an "operandum," for example, a lever, which can be manipulated by an animal. The consequences of the lever pressing can be, for instance, food release (a reinforcement). In this case, if a rat is put in the cage, initially he will probably just explore the cage ignoring the lever, which he has never seen. At some point the animal might accidentally press it with his fore-paws. He would then receive the food. If he presses the lever again he will obtain other food. In a short time the rat will learn and associate the lever pressing with the food reinforcement. On the other hand, if a punishment (like an electrical shock) is released after the lever pressing, the animal will learn not to touch the lever.

Till here we have always spoken of conditioning referring to animals, but actually this kind of learning is not only valid as a mechanism of animal memory, but can also be observed in humans. John Broadus Watson, the American psychologist who founded the psychological school of Behaviorism, performed a famous study known as the "Little Albert experiment" [18]. This experiment would not be approved by any ethical committee nowadays, but remains an effective demonstration of the validity of conditioning procedures on humans. Watson showed an 11-month-old infant, Albert, a white rat. The animal induced no fear in the infant. After that Watson began to pair the appearance of the rat with a loud noise, which he produced by striking with strength a hammer on a steel bar. Albert startled violently and broke into tears. Finally, when Watson exposed Albert to the rat alone, the infant cried and tried to crawl away from the animal. The boy had been conditioned to fear (notice that this is an example of Pavlovian conditioning: the white rat is the neutral stimulus, the sound of the hammer is the US, and the final reaction to the rat, now become CS, is the conditioned response).

In the end, we will consider a last kind of implicit memory: nonassociative learning. This is probably the most elementary form of learning, which can be found even in very simple invertebrates. Its name derives from the fact that there is no association being formed (nor between two stimuli, nor between a stimulus and a response), but just a modulation of a spinal reflex evoked by a stimulus. Repeated exposure to the same stimulus allows organisms to learn about the properties of that stimulus. There are two types of nonassociative learning: habituation and sensitization. The first one, habituation, is the reduction or suppression of a response following repeated administration of a stimulus. An example is the suppression of the startle reflex. If we hear a sudden loud noise, we startle. But if this sound is repeated many times, at some point we will certainly stop to show any startle reflex. The latter one, sensitization, is instead the amplification of a response after repeated stimulation. An example is the increased response to a mild tactile stimulus after a painful one. Neuroscientist Eric Richard Kandel has won the Nobel Prize in 2000 for his pioneering studies on nonassociative memory, performed on the sea slug *Aplysia californica* between the 1960s and the 1970s. In particular, he examined the gill withdrawal reflex. If the siphon of this mollusk is touched, it will automatically retract its gill. After touching the siphon many times, the gill withdrawal reflex is attenuated (habituation). If a noxious electrical shock is then applied to the tail while touching the siphon, the sea slug will not only show the reflex again, but it will withdraw the gill at the slightest stimulation (sensitization).

In this paragraph, we have considered the differences between explicit and implicit memory and described the most important forms of both. Figure 9.1 summarizes the relation between all these different types of memory.

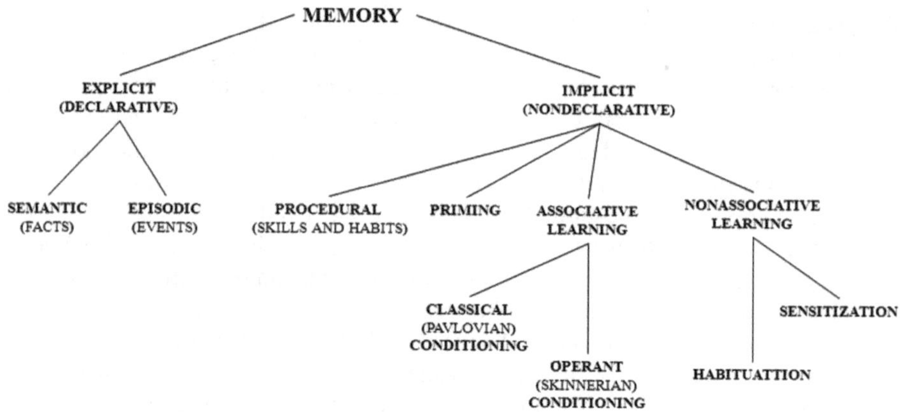

Fig. 9.1 Classical psychological partition of the memory

Memory Systems in the Brain

In the previous sections, you have been introduced with the common psychological approach to memory and its categorization; now it is time to change our point of view and see what practically all this means. How is the brain involved in learning and memory formation? Is there a correspondence between our schematic representation of the memory and the way our gray matter deals with it?

Throughout years of clinical observation, brain imaging studies, and experiments, it has emerged that memory is not a unitary faculty of the mind but is composed of multiple systems that have different operating principles and different neuroanatomy, that nonetheless mirror the classical partitions presented before (Fig. 9.2).

Here, we will focus just on a few of these structures, in order to offer a simple idea of the arrangement of the device underlining the memory: the hippocampus and the amygdala (Fig. 9.3).

Behavioral work, together with neuroanatomical studies, has finally identified the anatomical components of the medial temporal lobe memory system that support declarative memory: the hippocampus, together with the adjacent entorhinal, perirhinal, and parahippocampal cortices that make up much of the parahippocampal gyrus. The hippocampus has the shape of a curved tube, similar to a seahorse (from which the name) or, more prosaically, a banana. Its general layout holds across the wide range of mammalian species, although the details vary.

The adjacent entorhinal cortex (EC) is the greatest source of hippocampal input and target of hippocampal output; it is strongly and reciprocally connected with many other parts of the cerebral cortex, functioning in this manner as the main "interface" between the hippocampus and other parts of the brain. The signal flow forms a loop mainly unidirectional, with signals propagating through a series of tightly packed cell layers, first to the dentate gyrus, then to the CA3 layer, then to the CA1 layer, then to the subiculum, and then out of the hippocampus to the EC.

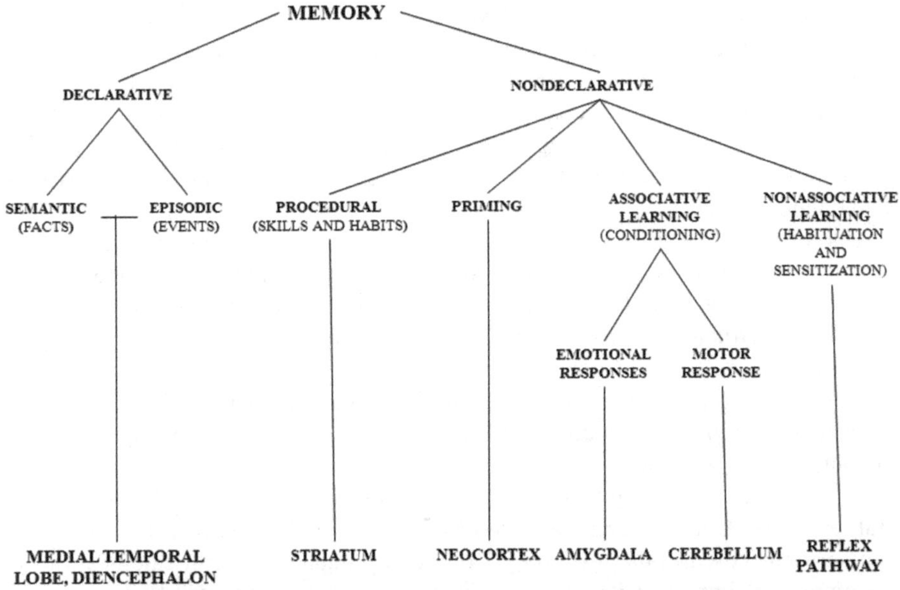

Fig. 9.2 The canonical subdivision of the memory in psychology reflects an anatomical partition of the involved brain areas

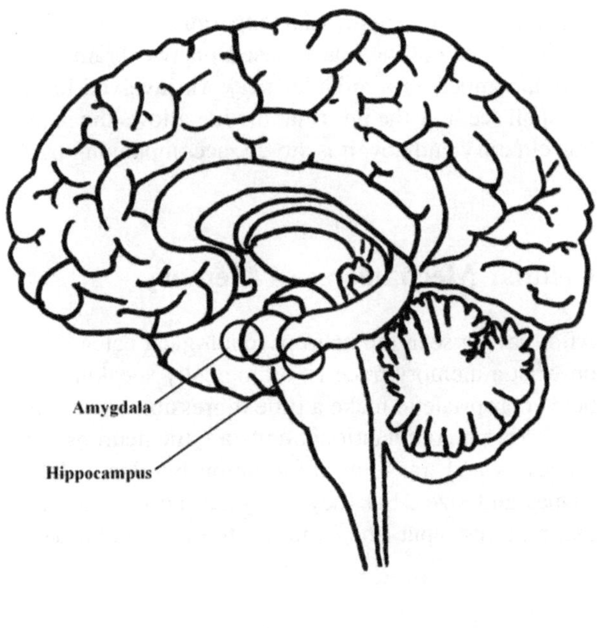

Fig. 9.3 Amygdala and hippocampus location in the human brain

Recalling the peculiar case of H.M., you remember that he was submitted to a bilateral surgical procedure involving the medial part of the temporal lobe, the area where the hippocampal circuitry is located. That explains the explicit memory impairment afflicting the patient; in fact the hippocampus plays a crucial role in coding the memories, merging their different aspects. In the next paragraph, by getting more deeply in the biological nature of the memory, these features will be studied more properly.

One of the afferent structures of the EC is the amygdala, another relevant gear of the memory machinery. In complex vertebrates, including humans, the amygdala performs primary roles in the formation of memories associated with emotional events. This structure exerts its function both in explicit and implicit memory by processing the emotional valence of an experience.

Despite the importance of the amygdala in modulating memory, however, learning can occur without it, though such learning appears to be emotionally "flat" and, in many cases, less vivid.

The relevance of amygdala not only in memory, but also in behavior in general, emerges clear in patients afflicted by the Urbach–Wiethe (UW) syndrome. It is a very rare disease, that in 50–75% of cases produces bilateral calcifications of the amygdala, impeding its functioning: those patients, submitted to several neuropsychological tests, demonstrated impaired episodic memory for visual and verbal emotionally arousing stimuli [19]. It is also interesting to mention the brief biography reported by the Portuguese neurologist Antonio Damasio of the patient RB: this woman was incapable not only of feeling fear and recognizing it in others' face, but also of a correct evaluation of danger signals, that normally evokes alarm in healthy subjects. She did not distrust anyone; in fact her behavior was inappropriately unabashed and warm with everyone, also with suspect or dangerous persons, that she could not recognize as such neither from appearance nor from behavior, causing her to live unpleasant experiences from "wrong" love to fraud [20].

This brief outline has focused on two brain structures which are reciprocally wired, composing a modular network involved in memory. In the next paragraph, we shall see how the information that allows the memory to be formed travels along this circuitry and how it is stored, accompanying us all life long.

Cellular Mechanisms of Memory

After having seen the neuropsychological categorization of memory, we can focus on what a memory trace is, biologically speaking. However, in order to do so, it is here appropriate to make a little digression about how the brain signaling works.

The brain computational units are the neurons: a neuron is an excitable cell that processes and transmits information by electrochemical signal. Neurons vary for shapes and sizes, but they all share the general organization: there is a structure that receives input (the dendritic tree), a central part that elaborates it (soma), and

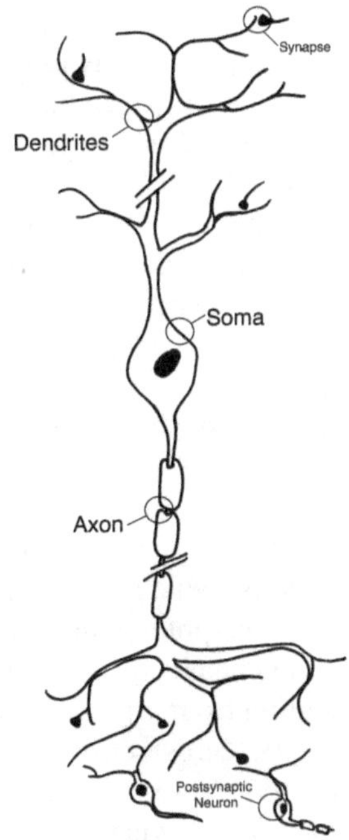

Fig. 9.4 Classical representation of a neuron and its three major subdivisions (dendrites, soma, and axon), connected by synapses to other neurons

an output branch (axon). Between the axon terminal and the dendrites of the next neuron there is a gap called synaptic cleft: the information runs along the neuron as an action potential, i.e., a propagating electrical signal that is generated by exploiting the electrically excitable membrane of the neuron, and when it reaches the end of the axon, it causes the release of certain molecules, called neurotransmitters. Those molecules, after crossing the synaptic cleft and reaching the dendrite portion of it, can bind to specific proteins called receptors. The formation of these complexes between neurotransmitters and their receptors results in cellular changes dependent on the nature of the signal, ranging from a new action potential generation to a structural modification of the neuron itself (Fig. 9.4).

Nowadays, it is quite accepted that our learning, as well as the memories formed, are coded in our brain by a modification of preexisting synapses or by a creation of brand new ones. Primitive forms of a synaptic theory of memory had already been proposed at the end of the nineteenth century, by prominent scientific personalities such as Santiago Ramon y Cajal, William James, or Sigmund Freud, but they were lacking a rigorous definition based on experimental facts. The big leap forward was taken by the psychologist Donald Hebb in the middle of the twentieth century, with the simple and elegant statement that "When an axon of cell A is near enough to excite cell B and repeatedly or persistently takes part in firing it, some growth process or metabolic change takes place in one or both cells such that A's efficiency, as one of the cells firing B, is increased," or, more simply, neurons that fire together wire together [21]. So, for example, if I dare to eat a rotten cheesecake and then I become sick the day after, the neurons that were coding the dangerous tasty morsel and those of the embarrassing following experience strengthen their connections, so the next time at the first sight of any cake the sickness memory will be coactivated and I will not be tempted anymore. Synaptic modifications underlying these learning processes are collectively termed as long-term synaptic plasticity, or stable alterations in the ability of specific neurons to fire.

In the past few decades, plenty of data has been obtained using different experimental approaches, ranging from electrophysiology to molecular genetics, contributing to shed light on the various forms of synaptic plasticity. You have been already introduced to the distinction between STM and LTM. The former requires only transient modification recruiting preexisting proteins, while the latter, that concerns stable and enduring memories, involves a more complex machinery.

The hippocampal neurons are characterized by two main forms of long-term plasticity, which are considered as valid cellular correlates of memory formation: long-term potentiation (LTP) and long-term depression (LTD). LTP is a long-lasting enhancement the efficacy of the synaptic transmission between two neurons resulting from a synchronous stimulation and leading to an improvement of the post-synaptic cell's sensitivity to neurotransmitter released from the presynaptic cell. LTD, on the other hand, is an opposite phenomenon associated with a prolonged reduction of synaptic transmission. Both LTP and LTD exhibit several important properties, including input specificity (no spreading to other synapses), associativity, and persistence (lasting from several minutes to months).

LTP, the most studied of the two phenomena, develops in two phases: an early phase (E-LTP), that takes place in the first hour or so and involves reversible modification of preexisting proteins, and a later phase (L-LTP), that endures hours to days and implies gene transcription and protein synthesis. The reader here should be struck by the analogy between E-LTP and STM, as well as between L-LTP and LTM. But what does this mean? The information of our experience travels as an electric impulse along the neural net, but then it needs to stabilize and take a physical shape: this is the pool of new and specific synapses, as a final result of extremely complex molecular events that occur in our neurons.

Synaptic transmission and long-term plasticity rely on the engagement of several neurotransmitters in the brain but the two most common ones are glutamate (GLU) and gamma-aminobutyric acid (GABA) which act as main excitatory (increasing neuronal activity) and inhibitory (reducing neuronal activity) molecules, respectively. For instance, in the hippocampal excitatory cells called pyramidal neurons, release of glutamate leads to the activation of several receptors including the NMDA (*N*-methyl-D-aspartate) receptor (NMDAr), that is the predominant molecular device for controlling synaptic plasticity and memory function. In fact, NMDAr acts as a coincidence detector: only if both the pre- and postsynaptic cells are simultaneously active, the NMDAr will be active, functioning as an AND door requiring an input both presynaptic (the glutamate) and postsynaptic (the activated state of the neuron itself, signaled by its voltage change). Thus, the NMDA receptor allows the entrance of calcium ions that can trigger long-term plasticity. Conversely, GABA

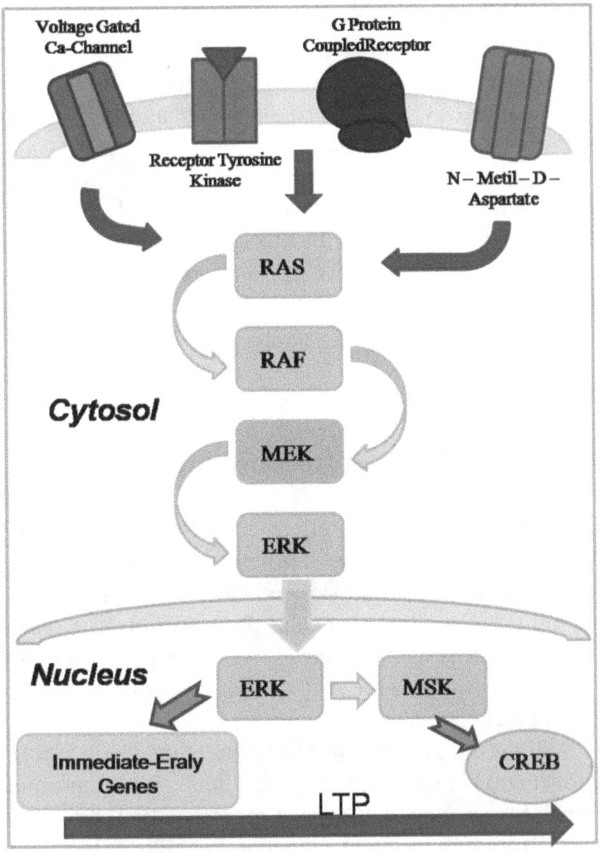

Fig. 9.5 The Ras-ERK pathway: the activation of several receptors begins the signal cascade that ends in the nucleus with the activation of several genes and transcription factor leading to LTP

receptor can act in opposition, reducing calcium influx, for instance, by attenuating glutamate release.

Despite the complexity of the cellular mechanisms occurring in the dendrites after receptor activation, few molecular pathways have directly been implicated in the process of transferring ("transducing") the information from the synapse to the soma and possibly to the nucleus where, during sustained neuronal activity occurring in the process of LTM formation, gene expression programs take place. Among these intracellular pathways lies the Ras-ERK cascade, which is indispensable for the LTP formation and LTM consolidation (Fig. 9.5).

The Ras-ERK pathway is a signal transduction pathway that couples intracellular responses to extracellular stimuli: in fact, through a chain of activations, the information is transmitted from molecule to molecule, starting from a receptor, until the activation of transcription factors that allows gene expression in the nucleus. This is normally achieved through a sequential molecular activation in which calcium ions activate first Ras proteins, thus transmitting the information to downstream molecules like Raf, MEK, and finally ERK. These events all happen in the dendrites, but once ERK proteins are activated, they can move into the nucleus where they recruit transcription factors like CREB which can activate many specific genes. These cellular modifications take then place in order to cause synaptic plasticity events like LTP, leading to stable changes in the strength of synapses and thus stabilizing memory traces.

So, changes of our hippocampal synapses connections are essential to code and store new memories, but that is not the end of the process. Getting back to H.M., he could not learn new memories, but anyhow his old memories were intact: so there should be a further passage after the hippocampus, an additional step that "transfers" the memory traces somewhere else. And this is what we will see in the next paragraph.

Consolidation and Reconsolidation

Now that the processed information has been acquired, and the neural connection has been shaped by de novo protein synthesis, this condition needs to be maintained, otherwise the whole process of memory acquisition will not lead to a stabilization of the trace. At this point the phenomenon of Consolidation (Latin for "to make firm") takes place, i.e., the progressive postacquisition stabilization of long-term memory, as well as the memory phase(s) during which such presumed stabilization takes place.

Within the first minutes to hours after the encoding has occurred, the processed information is maintained in local nodes in the neuronal circuit(s) that encode(s) the experience-dependent internal representation, i.e., the memory (synaptic consolidation). Those changes take place in the hippocampus but, if the memory will be kept just in this structure, we could only accumulate a limited number of memories, due to the storage limit.

The fact that patients suffering of amnesia after hippocampal damage could retrieve remote memories acquired before the trauma as a temporal gradient going from the most recent to the oldest ones (the Théodule Ribot's law, hypothesized in 1881) suggests that the hippocampus is essential in the first stage of memory consolidation, but after that the engram is stored somewhere else [22, 23]. This observation has led to the speculation that another level of consolidation may exist at a system level, implying the relocation of the memory traces to cortical regions. This has been confirmed by the brain imaging studies showing a progressive increased activation of the Pre-Frontal Cortex (PFC) during memory retrieval of remote traces concomitant with a progressive reduction of hippocampal activation, similar to what may be observed in the presence of retrograde amnesia in patients lesioned in this area [24, 25].

The process is called "system consolidation," and it is a gradual and extremely slow event that takes weeks, months, or even years to be accomplished. The biological processes underlying these phenomena are still widely unknown, but recently research started to shed light upon those events.

For example, structural analysis of the cortex revealed that the passage from the hippocampus to the PFC is accompanied by functional changes (by the creation of new synapses) and reorganization of these structures [26]. The aforementioned NMDAr has been shown to play a crucial role also in this case: rats that received an antagonist (i.e., a substance that prevents the functioning of its target) for those receptors in the PFC 1 or 2 weeks after a learning task showed an impaired performance, and this was correlated to a reduced LTP in the connection between hippocampus and PFC [27].

Now we have all our memories safely stored in the cortex, but what if some of them need to be modified or removed? By common experience we know that we can modify or update our memories, but how is it possible if until now we have spoken about a consolidation process as something stable and enduring?

In 1968, Misanin and colleagues reported that electroconvulsive shock (ECS), a known amnestic agent, leads to memory loss if administered immediately after retrieval, 24 h posttraining, of a seemingly consolidated, long-term conditioned memory. The observation was repeated with other memory tasks or using other amnestic agents, and the basic concept that emerged from these experiments was that an active (retrieved) trace returns to a labile, fragile state, while an inactive (stored) trace is immune to such manipulations; so the idea of a reconsolidation process emerged, i.e., that when a memory trace is recalled, it deconsolidates and then becomes solid again [28].

Unfortunately, the inconsistency of several studies and a growing focus on other aspects of neuroscientific research in the cellular and molecular mechanism of the memory in the following years confined this phenomenon to oblivion.

A new phase of the reconsolidation hypothesis started only a few years ago. Karim Nader and Joe LeDoux reported the phenomenon in fear conditioning: they trained rats on Pavlovian fear conditioning to associate a tone with foot shock. If the animal has learned, the hearing of the tone should induce an immediate freezing. The consolidation of this type of memory can be blocked by microinfusion of the protein synthesis inhibitor anisomycin into the lateral and basal nuclei of the amygdala (BLA), immediately after training. The authors reported that the consolidated fear memory returns to a labile state, during which local microinfusion of anisomycin into the BLA immediately after the retrieval of the fear memory produces amnesia on subsequent tests. This happened regardless of whether retrieval was 1 or 14 days after conditioning. The same treatment with anisomycin in absence of memory reactivation left memory intact. The conclusion was that consolidated fear memory, when reactivated, returns to a labile state that requires de novo protein synthesis for new consolidation, i.e., reconsolidation [29].

This article brought to new life the reconsolidation studies, and additional studies soon followed, showing that this mechanism is evolutionarily conserved in all species, from worms to human, but its mechanism and its function are still unclear. From a biological point of view, it has emerged that reconsolidation must undergo determined boundaries, regarding factors as the age of the memory or its strength, sharpening the scope of the experiment [30]; and that even if it requires de novo protein synthesis such as consolidation, their ways are not the same: some transcription factors such as the aforementioned CREB are required in both processes [30], but some others are engaged only by one of them, even if the memory and the structure involved are the same [31].

Also, the nature of the reconsolidation itself, as well as its function, is still subjected to a hot debate: the two most common positions see either this process as a mechanism that makes the memory trace always stronger and stable [32, 33] or as a tool to update or correct a memory in the light of new experiences [34].

Apart from the opposing theories and the different sets of data available, one crucial idea still stands clear: is it possible to erase memories in a targeted way? This is a new extremely powerful research tool for the search of therapies of cognitive disorders: what about removing a traumatic war experience from young soldiers, or craving for the drug from addicted people? As a matter of fact a lot of researchers are now focusing on deletion of memories related to addiction [35, 36], in which the Ras-ERK pathway plays a crucial role [37, 38], or connected to Post Traumatic Stress Disorder (PTSD) [39].

To conclude, we would like to tell a notorious case that took place at the time of the Misanin findings, and only nowadays can be fully comprehended in all its meaning and potential which should convince those of you who remained skeptic about reconsolidation. The clinician R.D. Rubin hypothesized that if individuals focused on the subject of their psychopathologies, this would return it in a labile state. Patients suffering from obsessive–compulsive disorder (OCD) or hallucinations were given ECS after being prompted to act out their desires or after their hallucination had begun: all of them improved dramatically for periods ranging from 3 months to 10 years later. In a case study, a 30-year-old woman with OCD was made to act out her compulsion of killing her mother with a butcher's knife and was then administered a single session of ECS while still awake. The next day, she went home and spoke kindly to her mother for the first time in years. She asked her mother "Do you love me?" and then kissed her. When the author asked if she still felt like stabbing her mother, she laughed and said, "Oh, she doesn't deserve anything like that." She returned home and to work and remained free of symptoms for the 2 years up to the publication of the study [40, 41].

Spatial Representation and Memory

In the previous paragraphs, we outlined and overviewed the neurobiological approach to memory function. Before concluding, in this paragraph, we will investigate two types of neurons, the place cells and the grid cells, which are very relevant in both everyday life and memory formation.

In every moment, while awake, we can locate ourselves in the surrounding space: I know that by sitting on the chair with the window on the right, 3 m on the left there is my console, or that, when biking to the lab along a new road, if I recognize a familiar building I can establish where I am. In the simple orienteering, in the creation of a new mental map, as well in the recall of a known scenario, we employ a spatial representation system, apparently simple on a first sight, but terribly complicated at a more accurate view.

In 1781, the philosopher Kant argued that some ideas exist as a priori intuitions, i.e., they exist before and independently of specific experience. One of these ideas is the concept of space, an innate organizing principle of the mind, essential to our perception of the world: Kant himself would be surprised knowing how unexpectedly his idea has found a scientific manifestation two centuries after, and is still nowadays one of the most fascinating areas of research in Cognitive Neuroscience [42].

In 1971, O'Keefe and Dostrovsky found spatial receptive fields in complex-spiking neurons in the rat hippocampus. These cells fired whenever the rat was in a certain location in the local environment, while neighbor neurons fired at different locations such that, throughout the hippocampus, the entire environment was represented: these cells are called place cells. In a few years it was also found that the same place cells could work together to represent different environments, but the relationship of the firing fields differed from one setting to the next, leading O'Keefe and Nadel in 1978 to hypothesize that place cells are the basic elements of a distributed noncentered map-like representation: those cells were suggested to provide the animal with a dynamic, continuously updated representation of allocentric space and the animal's own position in that space [43, 44].

In the following years of investigation, it became clear that this type of cellular behavior was not unique to the hippocampal circuitry: spatial signals were delivered to CA1 directly from some projection neurons of the entorhinal cortex (EC). Those cells exhibited sharply tuned spatial firing, much like place cells in the hippocampus, except that each cell had multiple firing fields [45]. The many fields of each neuron formed a periodic triangular array, or grid, that lined the entire environment explored by the animal [46]. Such grid cells collectively signaled the rat's changing position with a precision similar to that of place cells in the hippocampus. Moreover, experimental evidence supports a role of synaptic plasticity in the dentate gyrus (DG) and CA3 in the fast acquisition of context conditioning, suggesting that there may be specialized mechanisms for rapidly storing memories in these parts of the hippocampal circuitry.

We now have accumulated compelling evidence from a number of mammalian species that the hippocampus plays a crucial role in spatial representation and spatial memory: place cells and grid cells may determine how we perceive and remember our position in the environment as well as the events we experience in that environment.

But how do those maps work? The brain needs algorithms for linking the places in metric terms. When we move from the start position of our mental map, we must keep track of their changing positions by integrating linear and angular self-motion: this process is the path integration.

The self-motion information is probably the primary source for maintaining and updating grid representation that offers a general metric navigational system to the place cell: the current model of grid field formation suggests that EC neurons path-integrate speed and direction signals provided by specialized cells, whereas sensory information related to the environment is used for setting the initial parameters of the grid or adjusting it. Although still too little is known, it seems obvious that the spatial representation engages a wider brain circuit, which probably carries out complex computational tasks.

What about things that we know regard a place but that are not merely spatial information? Nonspatial variables are not represented primarily by a dedicated subset of neurons or a nonspatial variant of the place cells: they are represented in place cells by continuous rate modulation within the field. Support for this idea comes from the observation that, in hippocampal cell assemblies, spatial and nonspatial variables (place and color) are represented independently by variation in firing location and firing rate, respectively [47]. These studies, along with others, indicate a conjunctive spatial/nonspatial code for representation of experience in the hippocampus.

This system works efficiently and rapidly: hippocampus rapidly discriminates between similar boxes [48], and this suggests that the new neuronal codes should not only emerge rapidly, but also become quickly associated with the detailed configuration of all available sensory cues. Moreover, when animals are transferred from a known environment to a new one, the place code in CA1 appears to undergo immediate global remapping [49, 50]. Although distinct place cells appear rapidly in CA1, their firing patterns in the beginning do not discriminate between more detailed differences between recording arenas that are presented at the same location [51–53] and, along with new CA1 firing after the animal begins to explore an arena, also CA3 cells begin to fire with spatial selectivity.

At this point a question should arise: if those are the cells that, given their peculiar trait, provide us a given map time by time, does that mean that we can learn just a determined amount of maps? In fact all the maps rely on the very same population of cells. So, how are memories stored in the place cell system?

On the basis of the extensive intrinsic connectivity and modifiability of the CA3 network, theoretical work from the artificial intelligence field has indicated a possibility. In networks with discrete attractor states (a Hopfield network), associative connections would allow stored memories to be recalled from degraded versions of the original input (pattern completion) without mixing up the memory with other events stored in the network (pattern separation).

The formation of different spatial memories should thus depend on a pattern separation process: theoretically, new representations could be distinguished as a result of attractor dynamics in the associative CA3 network. The formation of novel discrete representations would benefit strongly from the existence of a control, upstream of CA3, where small differences between incoming signals could be augmented before the input is presented to the associative network. The DG serves this function: while in the CA areas pattern separation is expressed in place cells as a substantial reorganization of the collective firing pattern ("remapping") induced when inputs exceed a certain difference threshold, those changes have place in the DG at a lower threshold under conditions where sensory input patterns were made progressively more different. So, when the environment is only slightly modified, pattern separation is expressed in DG and CA3 as a change in the pattern of correlated activity, but when the changes in input to the hippocampus are more substantial, pattern separation is accomplished also by recruitment of a statistically independent cell population in CA3. These mechanisms provide a potential neuronal substrate for disambiguation of overlapping memories in the hippocampus [54]. Meanwhile, pattern completion is apparent from the fact that CA3 cells maintain their location specificity after removing many of the landmarks that originally defined the environment.

Although those processes of tie formation are still not fully characterized, not surprisingly, NMDA receptors seem to be involved [55, 56].

Many points remain to be explained, but we are beginning to see the contours of a modularly organized network of which grid cells and place cells are just a component, and we cannot be not fascinated by the existence of such a complex and elegant machinery to store memories.

Biological Versus Computer Memory

One of the most famous analogies in psychology is the one between the human brain and the computer. The view that the brain can be studied as a calculator was introduced by cognitivism in the middle of the twentieth century, with the cognitive revolution movement. In experimental psychology, the predominant school in the former decades was behaviorism, which considered that psychology should focus only on observable behaviors that can be studied and measured through adequate tests. Since internal mental states are not observable, behaviorists regarded that references to mental processes and mind should be avoided. The cognitive revolution was a response to this view and proposed that also mental processes deserved attention and could be studied scientifically. For the cognitivism, which became the dominant psychological school in the 1970–1980s, internal mental states can be considered as symbols and the manipulations operated on them can be illustrated as algorithms. In analogy with computers, the brain is the hardware and the mind is the software. The computational processes of the program (the software) are abstract so they can be described independently from the hardware, no matter if this is a collection of brain cells or of computer silicon chips. Since the cognitive revolution, many psychological faculties have been compared between humans and computers, for example, artificial versus human intelligence or the speculations on the possibility of artificial consciousness. In this paragraph, we will concentrate on the comparison between human and computer memory, highlighting the main similarities and differences.

Psychologists divide memory into three basic processes: encoding (by which physical sensory stimuli are processed and introduced into memory), storage (by which encoded information is maintained in memory), and retrieval (the recall of stored information). As this subdivision can also be applied to computers, we will compare human and computer memory following the order of these processes. Beginning with encoding, in humans, inputs like light or sound are detected through sense organs, for example, eyes or ears, and the physical energy of the stimulus is transduced into neural activity; this information is then sent to the brain through nerves. In computers, on the other hand, keystroke inputs are translated into an electronic language and transmitted to the central processing unit or CPU (the processor), through the wires which form the data buses.

For what concerns storage, it is interesting to notice that both humans and computers are endowed with a temporary memory and a long-lasting one. The first one is called short-term memory in humans and has its analog in computer's random access memory (RAM), so called because it can access any piece of the stored information in an equally quick manner, regardless of its physical location. RAM and short-term memory are both described as being volatile and have limited capacity. Like short-term memory, RAM is employed to keep in memory information on which some kind of operation is being performed at the moment, in this specific case data that is being processed by the CPU. For instance, when our computer has an open document on which we are working, this is stored in the RAM. The computer equivalent of human long-term memory is instead the read-only memory (ROM), like the hard disk drive (HDD). HDD is a nonvolatile storage device (information can also be retained if it is not powered) in which a rigid platter with magnetic surfaces rotates and an arm moves on the different positions of the platter to write or read data. Similarly to what happens in humans (in which some older memories are accessed more slowly than recent ones), data are read more slowly from the HDD than from the RAM, since the platter must spin and the arm has to reach the stored information, but a huge amount of data can be stored in this kind of memory and it can be kept for many years. A notable difference in storage is that humans forget, while computers do not. In normal conditions, a computer maintains information stored in its HDD without memory losses. Nonetheless, although computers do not have mechanisms of passive forgetting, they allow operations to actively delete data, while in humans this function is not physiologically viable (even if, as we saw in the Section "Consolidation and Reconsolidation," recent researches suggest the possibility to selectively erase target memories, by reactivating them and then interfering with the natural reconsolidation process, opening in such a way new options for possible therapies of psychiatric disorders deriving from the formation of maladaptive memories).

Finally, taking into consideration the last memory process, retrieval, we can observe that computers recall information faithfully, without altering them. Human memories, on the contrary, can be subject to distortions when retrieved and, most importantly, can be subjected to reconsolidation. Currently, we still do not know whether a reconsolidated memory in the brain is a distinct trace from the original one or an updated version. The problem of the noncomplete reliability of eyewitness testimonies is a clear example of humans' tendency to distort memories. A famous study performed by Loften and Palmer in the mid-1970s showed for instance how false facts can be introduced into the memory during recall [57]. A group of experimental subjects were presented with a series of slides which depicted a multiple car accident. A part of the group was asked "About how fast were the cars going when they *smashed* into each other?", whereas another part was questioned "About how fast were the cars going when they *hit* each other?". One week later the subjects were asked to refer if they had seen broken glass in the photos. Although there was actually no broken glass in the images shown in the slides, the 32% of the group asked if the cars had "*smashed*" replied they had seen it, while only 14% of the other group (for which the question was if the cars had "*hit*" each other) reported so. Moreover, humans can retrieve more easily a list of data if this was presented in a story form or had aroused emotions. For example, if we listen to a list of expressions and some are particularly funny, afterward we will probably recall with less effort those which made us laugh compared to the ones that we considered emotionally neutral. Computers instead are not influenced from meaning when retrieving data. But a similarity can be seen if we think at pathologies that affect retrieval. Humans can develop neurodegenerative disorders, like Alzheimer disease, which lead to deficits in memory retrieval. Analogously, computers can be attacked by viruses that can interfere with the retrieval of data from the HDD.

Glossary

***Anterograde amnesia*:** Loss of the ability to form memories after the moment of the accident.

***Associative learning*:** The process by which an element is learned through association with a different, preoccurring one.

***Classical conditioning*:** A kind of associative learning in which a given stimulus becomes increasingly effective in evoking a response.

***Consolidation*:** A process that stabilizes a memory trace after the initial acquisition. It is distinguished into two specific processes: synaptic consolidation, which occurs within the first few hours after learning, and system consolidation, where hippocampus-dependent memories become independent of the hippocampus over a period of weeks to years.

***Cryptomnesia*:** Also known as inadvertent plagiarism, it is the phenomenon whereby a person falsely recalls having generated a thought, when this was actually generated by someone else.

***Episodic memory*:** A subtype of explicit memory which refers to the events and personal experiences of our lives.

***Explicit memory*:** Also called declarative memory, as its content can be "declared"; it is the body of experiences and information we can recollect consciously and intentionally.

***Grid cells*:** Cells from the entorhinal cortex with sharply tuned spatial firing and multiple firing fields that form a periodic triangular array, or grid.

***Habituation*:** A form of nonassociative learning in which a response is reduced or suppresses following repeated administration of a stimulus.

***Hard disk drive (HDD)*:** A nonvolatile data storage device (information can be retained also if it is not powered), in which a rigid platter with magnetic surface rotates and an arm moves on the different positions of the platter to write or read data.

***Hebb's law*:** Describes the basic mechanism for synaptic plasticity wherein an increase in synaptic efficacy arises from the presynaptic cell's repeated and persistent stimulation of the postsynaptic cell: "fire together, wire together."

***Implicit memory*:** Also known as nondeclarative memory, it is constituted by all the knowledge that does not require intentional, conscious, recall of information. It can be revealed when previous experiences aid in the performance of a task without conscious awareness of these previous experiences.

***Long-term memory (LTM)*:** A long-lasting kind of memory which has a duration of days, years, or even decades.

***Long-term potentiation (LTP)*:** Long-lasting enhancement in signal transmission between two neurons that results from stimulating them synchronously. It is one of several phenomena underlying synaptic plasticity, widely considered one of the major cellular mechanisms that underlies learning and memory.

***Memory*:** As a general concept memory can be viewed as the persistence of previously acquired information. In psychology, it is defined as the faculty of an organism to encode, store, and retrieve information.

***Operant conditioning*:** A kind of associative learning in which a response occurs with increasing regularity in a well-specified and stable environment.

***Pattern completion*:** The phenomenon whereby a memory can be recalled by presentation of only a subset of the cues that was available during the learning episode. There is evidence that the CA3 subregion of the hippocampus is necessary for animals to achieve pattern completion.

***Pattern separation*:** The phenomenon whereby two similar contexts can be discriminated on the basis of subtle differences in the constituent cues. Such pattern separation allows the recall of only those memories that are relevant to one context or the other. There is evidence that the dentate gyrus is necessary for pattern separation.

***Pavlovian conditioning*:** See → Classical conditioning.

***Place cells*:** Neurons in the hippocampus that exhibit a high rate of firing whenever an animal is in a specific location in an environment corresponding to the cell's area of competence.

***Priming*:** An implicit memory represented by facilitative changes in the ability to identify, generate, or process an item due to a specific prior encounter with the item.

***Procedural memory*:** A form of implicit memory referring to skills and procedures ("how to" knowledge).

***Punishment*:** A stimulus that decreases the probability of a behavior to occur.

***Random access memory (RAM)*:** A computer memory that can be read from and written to in arbitrary sequence in a very high speed.

***Read-only memory (ROM)*:** A class of data storage media used in computers and other electronic devices.

***Reconsolidation*:** A phenomenon by which previously consolidated memories can be made labile again through reactivation of the memory trace, allowing their manipulation or erasure.

***Reinforcement*:** A stimulus that increases the probability of occurrence of a behavior.

***Retrograde amnesia*:** Condition in which the patient loses memory of events lived before the moment of the accident.

***Semantic memory*:** A subtype of explicit memory which refers to facts, considered independently of the context, and represents all our abstract knowledge about the world.

***Sensitization*:** A form of nonassociative learning in which a response is amplified after repeated stimulation.

***Short-term memory (STM)*:** A volatile type of memory which has a duration that can range from seconds to minutes.

***Synaptic plasticity*:** It is the ability of the synapse to change in strength. Since memories are postulated to be represented by vastly interconnected networks of synapses in the brain, synaptic plasticity is one of the important neurochemical foundations of learning and memory.

References

1. Maine de Biran FPG (1929) The influence of habit on the faculty of thinking. Williams & Wilkins, Baltimore, MD
2. Semon R (1921) The Mneme. George Allen & Unwin, London
3. Lashley KS (1950) In search of the engram. Symp Soc Exp Biol 4:454–482
4. James H (1890) Principles of psychology. Holt, New York, NY
5. Scoville WB, Milner B (1957) Loss of recent memory after bilateral hippocampal lesions. J Neurol Neurosurg Psychiatry 20:11–21
6. Milner B (1966) Amnesia following operation on the temporal lobes. In: Whitty CWM, Zangwill OL (eds) Amnesia. Butterworths, London, pp 109–133
7. Corkin S (2002) What's new with the amnesic patient H.M.? Nat Rev Neurosci 3:153–160
8. Tulving E (1972) Episodic and semantic memory. In: Tulving E, Donaldson W (eds) Organization of memory. Academic Press, New York, NY, pp 381–403
9. Graf P, Schacter DL (1985) Implicit and explicit memory for new associations in normal and amnesic subjects. J Exp Psychol Learn Mem Cogn 11:501–518
10. Schacter DL (1987) Implicit expressions of memory in organic amnesia: learning of new facts and associations. Hum Neurobiol 6:107–118
11. Milner B (1965) Memory disturbance after bilateral hippocampal lesions. In: Milner PM, Glickman S (eds) Cognitive processes and the brain. Van Nostrand, Princeton, pp 97–111

12. Taylor FK (1965) Cryptomnesia and plagiarism. Br J Psychiatry 111:1111–1118
13. Warrington EK, Weiskrantz L (1968) New method of testing long-term retention with special reference to amnesic patients. Nature 217:972–974
14. Warrington K, Weiskrantz L (1970) Amnesic syndrome: consolidation or retrieval? Nature 228:628–630
15. Warrington K, Weiskrantz L (1974) The effect of prior learning on subsequent retention in amnesic patients. Neuropsychologia 12:419–428
16. Pavlov IP (1927) Conditioned reflexes: an investigation of the physiological activity of the cerebral cortex. Oxford University Press, London
17. Tully T (2003) Pavlov's dog. Curr Biol 13:R117–R119
18. Watson JB, Rayner R (1920) Conditioned emotional reaction. J Exp Psychol 3:1–14
19. Siebert M, Markowitsch HJ, Bartel P (2003) Amygdala, affect and cognition: evidence from 10 patients with Urbach-Wiethe disease. Brain 126:2627–2637
20. Damasio AR (1994) Descartes' error: emotion, reason, and the human brain. Putnam, New York, NY
21. Hebb DO (1949) The organization of behavior: a neuropsychological theory. Wiley, New York, NY
22. Scoville WB, Milner B (2000) Loss of recent memory after bilateral hippocampal lesions, 1957. J Neuropsychiatry Clin Neurosci 12:103–113
23. Teng E, Squire LR (1999) Memory for places learned long ago is intact after hippocampal damage. Nature 400:675–677
24. Wiltgen BJ, Brown RA, Talton LE, Silva AJ (2004) New circuits for old memories: the role of the neocortex in consolidation. Neuron 44:101–108
25. Takashima A, Petersson KM, Rutters F, Tendolkar I, Jensen O, Zwarts MJ, McNaughton BL, Fernandez G (2006) Declarative memory consolidation in humans: a prospective functional magnetic resonance imaging study. Proc Natl Acad Sci USA 103:756–761
26. Maviel T, Durkin TP, Menzaghi F, Bontempi B (2004) Sites of neocortical reorganization critical for remote spatial memory. Science 305:96–99
27. Takehara-Nishiuchi K, Nakao K, Kawahara S, Matsuki N, Kirino Y (2006) Systems consolidation requires postlearning activation of NMDA receptors in the medial prefrontal cortex in trace eyeblink conditioning. J Neurosci 26:5049–5058
28. Misanin JR, Miller RR, Lewis DJ (1968) Retrograde amnesia produced by electroconvulsive shock after reactivation of a consolidated memory trace. Science 160:554–555
29. Nader K, Schafe GE, Le Doux JE (2000) Fear memories require protein synthesis in the amygdala for reconsolidation after retrieval. Nature 406:722–726
30. Tronson NC, Taylor JR (2007) Molecular mechanisms of memory reconsolidation. Nat Rev Neurosci 8:262–275
31. Lee JL, Everitt BJ, Thomas KL (2004) Independent cellular processes for hippocampal memory consolidation and reconsolidation. Science 304:839–843
32. Alberini CM (2005) Mechanisms of memory stabilization: are consolidation and reconsolidation similar or distinct processes? Trends Neurosci 28:51–56
33. Dudai Y, Eisenberg M (2004) Rites of passage of the engram: reconsolidation and the lingering consolidation hypothesis. Neuron 44:93–100
34. Lee JL (2009) Reconsolidation: maintaining memory relevance. Trends Neurosci 32:413–420
35. Lee JL, Di Ciano P, Thomas KL, Everitt BJ (2005) Disrupting reconsolidation of drug memories reduces cocaine-seeking behavior. Neuron 47:795–801
36. Lee JL, Milton AL, Everitt BJ (2006) Cue-induced cocaine seeking and relapse are reduced by disruption of drug memory reconsolidation. J Neurosci 26:5881–5887
37. Valjent E, Corbille AG, Bertran-Gonzalez J, Herve D, Girault JA (2006) Inhibition of ERK pathway or protein synthesis during reexposure to drugs of abuse erases previously learned place preference. Proc Natl Acad Sci USA 103:2932–2937
38. Miller CA, Marshall JF (2005) Molecular substrates for retrieval and reconsolidation of cocaine-associated contextual memory. Neuron 47:873–884

39. Taubenfeld SM, Riceberg JS, New AS, Alberini CM (2009) Preclinical assessment for selectively disrupting a traumatic memory via postretrieval inhibition of glucocorticoid receptors. Biol Psychiatry 65:249–257
40. Rubin RD, Franks C (1967) New application of ECT. In: Rubin RD (ed) Advances in behavior therapy. Academic Press, New York, NY, pp 37–44
41. Rubin RD (1976) Clinical use of retrograde amnesia produced by electroconvulsive shock. A conditioning hypothesis. Can Psychiatr Assoc J 21:87–90
42. Kant I (1855) Critique of the pure reason (translated by Meiklejohn JMD), Henry G. Bohon, London
43. O'Keefe J, Dostrovsky J (1971) The hippocampus as a spatial map. Preliminary evidence from unit activity in the freely-moving rat. Brain Res 34:171–175
44. O'Keefe J, Nadel L (1978) The hippocampus as a cognitive map. Oxford University Press, Oxford
45. Fyhn M, Molden S, Witter MP, Moser EI, Moser MB (2004) Spatial representation in the entorhinal cortex. Science 305:1258–1264
46. Hafting T, Fyhn M, Molden S, Moser MB, Moser EI (2005) Microstructure of a spatial map in the entorhinal cortex. Nature 436:801–806
47. Leutgeb S et al (2005) Independent codes for spatial and episodic memory in hippocampal neuronal ensembles. Science 309:619–623
48. Frankland PW, Cestari V, Filipkowski RK, McDonald RJ, Silva AJ (1998) The dorsal hippocampus is essential for context discrimination but not for contextual conditioning. Behav Neurosci 112:863–874
49. Muller RU, Kubie JL (1987) The effects of changes in the environment on the spatial firing of hippocampal complex-spike cells. J Neurosci 7:1951–1968
50. Wills TJ, Lever C, Cacucci F, Burgess N, O'Keefe J (2005) Attractor dynamics in the hippocampal representation of the local environment. Science 308:873–876
51. Bostock E, Muller RU, Kubie JL (1991) Experience-dependent modifications of hippocampal place cell firing. Hippocampus 1:193–205
52. Lever C, Wills T, Cacucci F, Burgess N, O'Keefe J (2002) Long-term plasticity in hippocampal place-cell representation of environmental geometry. Nature 416:90–94
53. Leutgeb S, Leutgeb JK, Moser EI, Moser MB (2006) Fast rate coding in hippocampal CA3 cell ensembles. Hippocampus 16:765–774
54. Leutgeb S, Leutgeb JK (2007) Pattern separation, pattern completion, and new neuronal codes within a continuous CA3 map. Learn Mem 14:745–757
55. Kentros C et al (1998) Abolition of long-term stability of new hippocampal place cell maps by NMDA receptor blockade. Science 280:2121–2126
56. Nakazawa K et al (2002) Requirement for hippocampal CA3 NMDA receptors in associative memory recall. Science 297:211–218
57. Loftus EF, Palmer JC (1974) Reconstruction of automobile destruction: an example of the interaction between language and memory. J Verb Learn Verb Behav 13:585–589

Chapter 10
Memories for Everybody[1]

Giovanni Campardo

Abstract This chapter discusses about semiconductor memories. We will see what does it means, today, to "remember" using solid-state electronics. Volatile memory and non-volatile memory will be analyzed, starting from the concept of the electrical parameter modification, capacitance, resistance, threshold, etc., to storage information. RAM and floating gate memory architecture and their working operations will be discussed.

Keywords Solid state memory · Floating gate · Flash · RAM

Prologue

During the last few years, both as a book publisher and editor, I wrote several books and articles for technical journals and conferences with memories as the main subject and, to be more precise, non-volatile memories. I've spent a lot of my working life on the subject of non-volatile memories and I've explored a few of many different possibilities. For this book, written by experts in various fields of mass storage, I had thought about writing a chapter relating to solid-state memories by telling, always as an expert, how to make a non-volatile memory, but then I thought about writing, just for once, for a wider audience.

So what I've done is imagined three friends who get together and talk about themselves and, during their conversations, one of them explains to the others simply, how both volatile and non-volatile solid-state memories work. Then, to make sure I've obtained my goal, I asked a few people to read the chapters and I put their

G. Campardo (✉)
Numonyx Agrate, Agrate Brianza, Italy
e-mail: giovanni.campardo@numonyx.com

[1] This chapter is dedicated to my teacher in the Integrated Circuit field, Dave Novosel, who died in 2009. Dave will be always in my heart, in my mind and in my pencil when I work on Electronics Field.
Translated by Marcello Campardo.

G. Campardo et al. (eds.), *Memory Mass Storage*,
DOI 10.1007/978-3-642-14752-4_10,

remarks at the ends of the chapters. I hope that various subjects in the same volume and a chapter with an unusual style for a scientific book, will make reading more pleasant and attractive to all.

Gianni is an integrated circuits designer, **Roberto** is an architect, and **Luca** is a business consultant.

Day One

Luca: Listen to this joke; it's wonderful:
In a foreign country, a priest, a lawyer and an engineer are about to be guillotined. The priest puts his head into the guillotine, but after pulling the rope, nothing happens – he claims that he was saved by divine intervention – so they let him go. The lawyer puts his head into the guillotine, but the rope again doesn't release the blade – he claims that you can't be executed twice for the same crime – so they let him go. Then they catch the engineer and they lock him on the guillotine; he looks at the release mechanism and says, "Wait a minute. I understood what the problem is."

Gianni: Yes, it's very good and I think it truly describes the mentality of engineers; we want to show everybody that we're able to make things work.

Roberto: Of course we architects are on good terms with engineers. Listen to this:
An architect is a man who has a bit of knowledge on a wide variety of arguments and, over time, knows less and less about a greater number of subjects until he knows pratically nothing about any thing. An engineer is a man with vast knowledge about a limited number of arguments, so that he knows practically everything about nothing.
Fortunately there are people that are graduate in economy, as they don't know nothing as from the beginning, they don't have to be worry about have any kind of information.

Luca: So now you've gotten me involved, too. My dear engineer, let's talk about this cell phone that loses its phone numbers instead!

Gianni: What do you mean by "loses"?

Luca: I left it on the heater for a day. I forgot about it, and the day after there were some wrong phone numbers.

Gianni: Well, anyway it's strange because electronic components are checked many times, so it could be the theory that is broken. It will be spoiled the memory.

Roberto: I'm also interested in understanding a little about how those memories work. What can you tell us?

Gianni: Okay, come here and listen. First we've got to understand the meaning of the word "remember."

Luca: As a matter of fact, I remember very few things.

Paolo: We're off to a fine start!

Gianni: "Remember" means taking data, organized in the most advantageous way, and putting it where we're sure it will remain for a long time, until we need it. Then you take it and use it.

Paolo: Like when you write phone numbers in your phone book, organizing them alphabetically, in order to find them easily without having to read the entire phone book to find them.

Gianni: I think that what you said is let us first differentiate between so-called serial access memories and the newest ones, called random access memories. Serial access ones are, for example, those contained on magnetic tape; to reach the information, we have to wind or rewind the entire tape. As the addresses in the phone book previously mentioned were not written in alphabetical order, to find the one that we're looking for we need to read them all. Most recent memories can assign, one address, so it can be easiest to find the same one.

Paolo: Obviously we must know where the information is written in order to find it.

Gianni: Sure. Information is arranged in the easiest way, just as you arrange a library or a warehouse, putting similar things in "easy" places to remember, or creating an index. But we'll see this later. Let's get back to us now. We were talking about the storage of data for retrieving it. For millennia paper was the basis for the storage of activities, then it was the photograph that allowed for an incredible technological leap; it was able not only to record events, like a daily scene or a happening, that "on paper" could be described only with a picture, but anyway can not imaging to carrying on canvas the same amount of detail that can be present in a photograph.

Paolo: Actually, the amount of information that a photo can record is really impressive. So explain to me how digital cameras work.

Gianni: Okay. We were saying that photography allows us not only to record a lot of data, but quickly, too – as long as it takes to click the shutter. Data recovery is very simple to do, you have only to take back a photo in your hands.

Paolo: Yes, actually photos are mixed in all together in the same drawer, and maybe the funny thing is that random research actually lets you discover other hidden pictures.

Gianni: All true. I see you're eager for it. To sum it up, when we look back on something, we alter some parameters that are on our hand; in the case of the phone book's addresses, we alter it by writing over it, whereas for the photo we change the silver grains' state.

Luca: Do you know how photographic plates work? The grains of silver halide are the detectors; they're suspended in a layer of gelatin (a huge organic molecule) forming the photographic emulsion, which is deposited on a glass plate, which provides mechanical stability and planarity to the system. The physical process at the base of the photograph is a hypothesis known as "The Hypothesis of Gurney-Mott" – a photon incident on a grain of silver halide excites an atom by sending an electron in a conduction band. All the electrons released in this way stop on centers of entrapment (impurities or defects in the crystal lattice), which purchase a negative charge in this way. This attracts free silver ions, which combine with electrons to form silver atoms. The trap becomes increasingly efficient and it captures more and more free electrons. In this way, a cluster containing from a few atoms up to a few hundred atoms of pure silver is created inside the halide crystal. At the end of the exposure the so-called latent image will be created: crystals that were in the areas enlightened by the photons will have clusters of silver, while those that were in the dark areas will remain simple crystals of silver halide.

At this point, the plate is developed, plunging it into a solution that converts the crystals of silver halide into silver. The reaction is very slow. But if the crystal already has a cluster of silver in it, this acts like a catalyst and the reaction is much faster. In most sensitive emulsions there are enough for three to six silver atoms per grain to catalyze the conversion of grain in silver, while in normal emulsions more than ten atoms are needed (this contributes to producing the characteristic low quantum efficiency of photographic plates). Then the development will be done by holding the plate in the solution for the necessary time to convert the grains exposed to the photons in the silver grains (opaque), but short enough not to turn the other grains into silver, too (which will therefore remain transparent). The development process is stopped by immersing the plate in a solution (fixing) which dissolves and washes away all the residual silver halide. Gelatin is a good support for grains because it allows a good penetration of solutions used for development and fixing.

Gianni: That's wonderful. Do you know how many "grains" there can be on a photo plate?

Paolo: I think that on the larger plates there may be up to 10^{10} grains.

Gianni: This means that each plate can contain up to ~ 8 bit $\times 10^{10}$ pixel $=$ 10,000 Mbytes of information.

Really impressive. But let's get back to us now. We arrived at the point where we realized that to remember something we have to change a parameter. For electronic devices variable parameters all in all, they're only a few. Even though it seems strange, I think that it could be easier to understand things if we start talking about non- volatile memories. To understand the operation a little bit, we must first understand the MOS transistor.

Luca: In order not to stray from the topic, could you tell me something about transistors? It has been talked about often but nobody ever talks about it in a comprehensible manner.

Gianni: But so we will not go far.

Roberto: What do you have to do? Don't tell me that you have to paint?

Gianni: Later, if you're good boys, I will let you see one of my paintings.

Luca and **Roberto** (together): Yes, you've been promising that for months, but you haven't shown us anything.

Gianni: So what were you saying about transistors? The story is quite complicated and a whole book wouldn't be enough to tell it all. However, in short, in the years after World War II, the goal was to look for a chance to build a solid-state element that could replace the valves, which consumed too much power and were cumbersome. In the end, a group of Americans, who even received the Nobel Prize for it – Bardeen, Brattain and Shockley – invented and built the first transistor. Like all the discoveries that have changed the world, this is full of anecdotes. My favorite one relates to Bardeen. In 1956, along with Brattain and Shockley, he received the Nobel Prize in Physics for his "Studies on Semiconductors and the Discovery of the Transistor." He had taken only one of his children with him to the ceremony in Stockholm and King Gustav VI had scolded him for it, saying that this was an event that the whole family should not miss. Bardeen assured him that "the next time" he would take the whole family! In 1972, along with Cooper and Schrieffer, he again received the Nobel Prize in Physics for the explanation of superconductivity. This time, as he had promised, he took all three of his children with him!

Roberto: He was a great fellow!

Gianni: Yes, truly.

Paolo: And what does transistor mean?

Gianni: Transistor means "transfer-variator" – that is transfer resistor, that this component is able to modify the strength of the line on which it is inserted, or rather, the impedance. But I don't expect you to understand. I said that there are too many new concepts together.

Roberto: Well, we'll pretend to have understood.

Gianni: Before moving on to the operation, I've got still a gem that I want to tell you.

I read on the Web that someone says that the transistor was brought here by aliens, area 51 strikes again!

Roberto: It could be true that they only make the operation start, and then we are good enough to destroy them by our own. With bows and arrows it would take too long. With modern weapons, those that are considered to be intelligent, we certainly can do it faster.

Gianni: Well, I never considered it that way. Were they extremely intelligent or are we not far enough advanced? However, you can imagine that a transistor is just a component that works like a lock on a river. Have you ever seen those rivers in the countryside that have metal plates that are raised and lowered by hand to drain the water in the channel? You should think that the transistor works like that (see Fig. 10.1).

Fig. 10.1 A channel with a lock placed perpendicular to the flow of water

The truly remarkable thing is that the flow of the water exerts great pressure on the lock's walls, but since the lock is placed perpendicular to the flow of water, the force used to lift it isn't much. So by using just a little force, you can control the flow of a lot of water. In other words, a little force controls much force.

Roberto: I understand, You act on a control terminal, where, by using little energy, you can decide how much water can pass.

Gianni: That's right! A transistor has three terminals – a control terminal and two that we can call of passage. Acting on the control terminal we can decide how much electric current is passed on the other two terminals. This way, we can call it "to modulate" how much current flows from one terminal to the other by acting on the control terminal.

Luca: Very beautiful, very elegant.

Gianni: Yes, the concept is very powerful and has been extended to a large class of devices. Transistors work like this: The terminal control keeps a check on the current flowing through the transistor, coming from the second terminal and flowing to the third. I will draw it for you (see Fig. 10.2).

Luca: I understand, maybe.

Gianni: There are different types of transistors, the first that see the light (was the first transistor produced) is known as the BJT, Bipolar Juncion Transistor, or simply a bipolar transistor. But the most commonly used one is called MOS, for Metal Oxide Semiconductor, indicating the elements that make up the transistor. I would say that this component is made just like a lock on the river. There is an element of control that is called "gate," which, by applying a few volts, puts the two terminals called "source" and "drain" together.

Luca: Are these names related to the direction of the current flow?

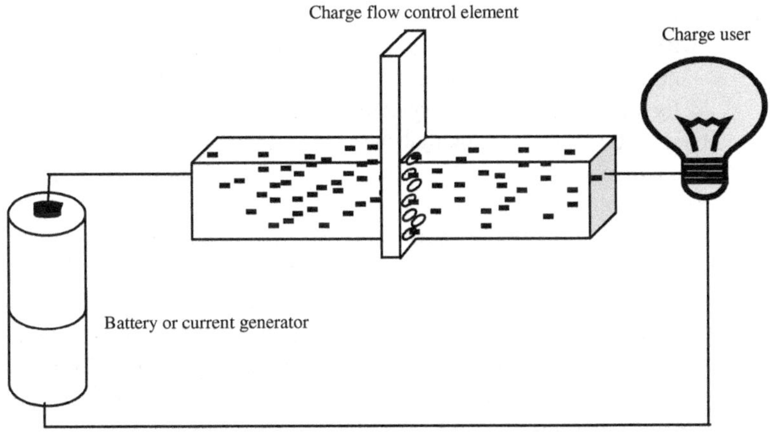

Fig. 10.2 The transistor works like a tube where the electric charges flow. The quantity of the charges can be modified by the control element. Control elements work like a sieve; it decides how many charges can pass from the source to the drain. Reality is more complicated

Gianni: Exactly. From the source come the electric charges which, thanks to the opening, even partial, of the gate, which means "door", moving from source to gate. At the end it is a bit like a light switch. You push it and two wires get in touch with themselves. The connection allows the current to reach the element which brings to fruition the function in question. In the instance of the switch, the element is typically the bulb, which lights up.

Luca: On what scale does this thing work today ?

Gianni: Sizes have been gradually reduced, but by following what is known as "Moore's Law." In the years just after the bipolar transistor patent filed by Jack Kilby, in 1965 Gordon Moore, co-founder of Intel, noticed that the number of transistors per square inch in an integrated circuit doubled every year. Moore predicted that this trend could be observed even in the future; the following years showed a doubling of the active components' density in an integrated circuit every 18 months, and today this value is one for the current definition of Moore's Law. For example, in the 18 months spent by a Pentium-1.3 processor to a Pentium-4, the number of transistors passed from 28 to 55 million.

Luca: I don't understand.

Gianni: It means that the number of transistors that were in Pentium 4 was 55 million. Fifty-five million components on a few square millimeter silicon chip. Some forecasts, which are already dated, predicted that in 2008 a typical desktop personal computer would have had between 4 and 8 CPUs, operating at 40 GHz, with a RAM between 4 and 12 GB and 1.5 TB of mass storage, hard disk, with a 100 Gbit LAN, while by 2011 the clock system could reach 150 GHz and 6 TB of mass storage.

Roberto: I'm lost.

Gianni: Moore's Law applies to all parts of microelectronics. It means that the number of CPUs is the number of processors, that is to say the "intelligent's parties" of the system. 40 GHz is the speed system. When I say 40 GHz, it means that the system can commute 40 billions times each second and, at least theoretically, it corresponds to the number of practicable operations in 1 s.

Then we'll examine RAM Memory closely but at the moment you've got to think about the size of memory. I talked about some GigaByte, GB, or TeraBytes (TB). Just as you do the Morse code, so can you encode fonts of the typewriter's keyboard, letters, numbers and special fonts, using a binary encoding, namely with "ones and zeros".

Luca: Yes, please explain something that I've never really understood: How are fonts translated into numbers?

Gianni: Okay. Suppose you have 2 available fonts and that you can use only digit number 1 or the number-digit 0, which are called bits, from *bi*nary digi*t*. The chances which you can have are four: 00, 11, 01 and 10. If we had 3 available fonts, but if we can use only 1 or 0, we would have 8 combinations: 000, 001, 010, 011, 100, 101, 110 and 111. If you take a careful look at the first case, there are 4 combinations, in the second one 8. The numbers in the first case correspond to 2 squared, and in the second case, to 2 cubed. The general rule is that the number of possible combinations is even to the number of available figures – in this case two, that is 1 and 0 fonts, raised to the power of fonts' number of, in the first case 2, 3 in the second one. If we keep a tally of the number of letters, plus the number of digits, plus the number of special fonts (periods, commas, parentheses, plus signs, minus signs, etc.) we don't come to 200. Thinking in reverse, with 2 available digits we need 8 fonts to represent all that we see on the keyboard. This gives the number of combinations equal to 256, since each binary value, from 00000000 to 11111111, is associated with one keyboard's font; the set of 8 bits is called byte. To understand each other, think about the Morse code. With two symbols, the point and the line (and the area of separation), you can represent all the characters used.

Roberto: I understand, maybe; it's just like the Morse code, but it uses only one and zero instead of dots and dashes!

Gianni: Yes, that's right. Now, the number of fonts that are in a line of this book is 85; obviously we also consider spaces. Space is a character like everyone else, and there are 46 rows. A total of 85×46 means 3,910 characters. For each page we need 3,910 bytes, where we recall that a byte comprises 8 bits. A book of 1,000 pages, and there are many, uses just under 4 Mbyte, that is 4 million bytes. Home computers' memories today are on the order of hundreds of GigaBytes, that is billions of Bytes. My computer has a 500 GB memory that is

500 billion bytes and accordingly I can record on it 125,000 books of 1,000 pages each!

Luca: But it's not true... it's all a story. It looks impossibile.

Gianni: Sometimes I think the same thing too, too complex, it look incredible. It is true instead, and when you will know how the "recording" takes place physically, you will think that it's really amazing. But you got me off track.

Let's go back to the transistor; we were talking about the MOS transistor. Look. What do you see this photograph [Fig. 10.3]? I have marked the areas that we named source, drain and gate. The part that is under the gate is called the channel and is the part that connects the source to the drain. This part must be insulated from the gate, so that current flows only between the source and drain. The insulation is made of silicon oxide, such as quartz.

Roberto: I found the crystal on the mountain near to my town.

Gianni: I let you see them, I pick up them in that neighbourhood. Look how beautiful they are [Fig. 10.4]. So I was saying that the gate is insulated from the channel by means of an oxide that is grown specifically on the channel. This oxide has the property not to be crossed by the current, which is the control electrode and the source and drain, where the current will flow. With regard to memories, the idea was to put another gate, called flottante, that is not attached to anything and is submerged in the oxide under the gate [Fig. 10.5]. This piece of material, completely surrounded by oxidation, results in a point as if it were particularly welcome to electric charges, the electrons. If you can put it over some electrons, they behave like sheep in fold. The oxide's barrier is a big obstacle to possible leakages.

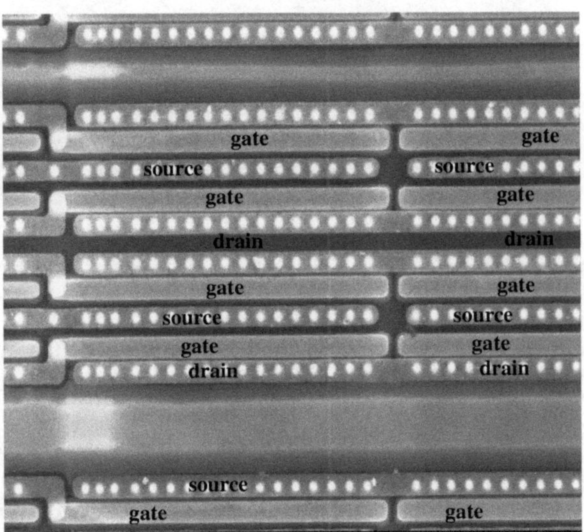

Fig. 10.3 The photograph shows some MOS transistors from an old technological process, with sources and drains identified. The gates are in between and the channel is under the gate. The "little dots" are the contacts to put the different layers in communication. The dimension of the ruler is 5 micron

Fig. 10.4 Crystal quartz

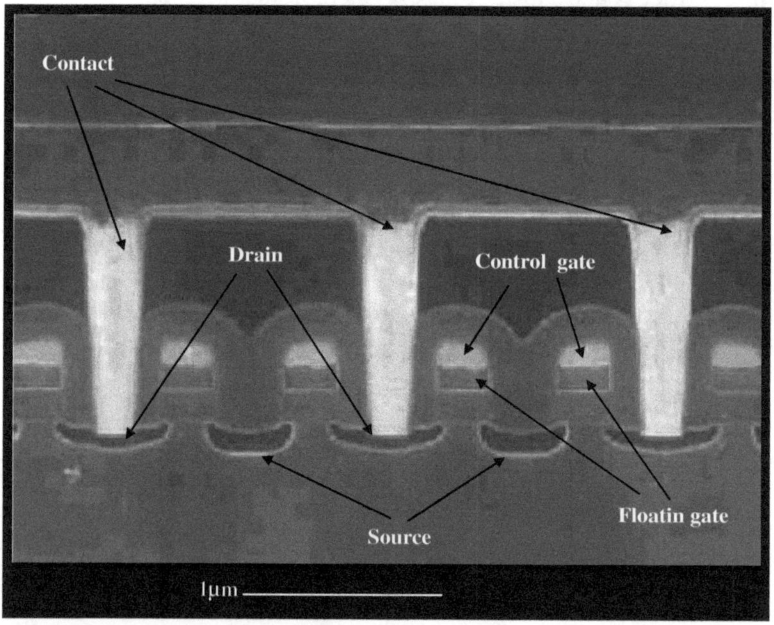

Fig. 10.5 Flash cells with floating gates, cell contacts to the circuitry, control gates, sources and drains zones highlighted. The dimension of the ruler is 1 micron

Luca: But how do you put electrons into the floating gate? With tweezers?

Gianni: I think that it is better to rely on an act of faith, because the discussion would lead us too far. To put in the electrons and to remove them, you use very high potential that is applied to only three terminals that we now know: the source, the drain and the gate. Then the electrons end up on the gate or are removed from it through quantum-type physical phenomena that can be explained only by particle physics. If we have time we'll talk about it; meanwhile let's come to a first conclusion. What we do is put electrons into the floating gate to change a parameter, called a threshold. Follow me carefully. If we think of a transistor, acting on the gate with a specific potential, which is about 1 volt, it causes a phenomenon that is called channel generation, which means that thanks to this potential, the charges that are in the transistor are recalled to the surface so that the source and drain are put in contact. The potential's value that we apply to allow the creation of the channel and therefore the passage of current is called "threshold."

Roberto: This thing is not at all easy to understand. So the two ends of the transistor are normally separate.

Gianni: Right, as in the example of the lock already mentioned, the lock separates the two sections of the channel.

Roberto: Then you begin to lift the lock and water starts to flow.

Gianni: Yes, but in transistors, the lock lifts up after a certain voltage value, which is the threshold. As if we say that not all the people have the force to lift the lock, if a child tries to do it, he can't. So the minimum strength that is needed to lift the lock is also in this case definable as "threshold".

Luca: Yes, it is a little clearer now. The analogies are very helpful.

Gianni: Good. Now we must take another step, and then we're at the end. The threshold's value depends on many things: the type of material, the geometry and the concentrations of these materials and much more. The floating gate that is included in the oxide, below the control gate, also figures into the definition of the threshold.

Luca: I'll say it myself. The electrons in the floating gate act as a shield for the channel's generation. Is that correct?

Gianni: Absolutely right! The presence or absence of electrons in the floating gate changes the threshold value. To sum it up: the insulated gate is an excellent trap for the electrons that ensure a charge's retention equal to some tens of years. Operations that allow carrying electrons to the isolated gate, and then removing them when necessary, are called programming and erasing. These operations allow you to change a macroscopic electrical parameter that corresponds to the threshold voltage of memory cells and are used for discriminating the logical value that must be stored. It is conventionally called erasing the operation performed collectively on a set of cells. We said that one cell corresponds to a logical zero when the floating gate is full of electrons; on the contrary, we said that it corresponds to a logical one when it has no electrons in excess.

Luca: I understand. You put in and take out the electrons changing the state of the cell and assigning a logical value to this state.

Gianni: It's like having a little box where you can find or not find some beans inside. If there are beans in it, you say it's a "1," and if there are not, it's a "0." Then you gather up 8 boxes together, making a byte, and use the decoding that we have seen before.

Roberto: Okay, so you add electrons or not and this is the memory, i.e., in order to put or not put electric charges at a point where they find it difficult to leave, and this alters the macroscopic parameter called threshold. Correct?

Gianni: Yes, perfect. Now to give you a more general and complex idea, you need to know that in the world market of semiconductors, a significant portion is represented by memories, which are the key components of all electronic systems. From the system's perspective, semiconductor memory can be divided into two main categories: RAM (Random Access Memories), whose content can be changed very quickly, i.e., in a few nanoseconds, which means a billionth of a second and for a virtually unlimited number of times, and ROM (Read Only Memories), whose content may not be changeable or, if it is, it's in milliseconds, which means thousandths of a second.

A second feature that differentiates the two families: RAM loses its contents when power is turned off, and ROM keeps them virtually forever. Ideal memory should combine these two features, rapid modifiability and retention. The category of non-volatile memory includes all memory whose contents can be electrically changed but is preserved when the power is removed. These are more flexible than the original ROM, whose content is defined during the manufacturing process and cannot be changed. The history of non-volatile memory begins in the 1970s with the introduction of the first EPROM (Erasable Programmable Read Only Memory). Since then, non-volatile memories have always been considered one of the most important families of semiconductors and, until the 1990s their interest was more linked to their role in product development for new technologies rather than their economic value. Since 1995, with the introduction of Flash memory and mobile products like cell phones, PDAs, digital cameras, etc., the non-volatile market had a dizzying increase, coinciding with the development of technology for floating gate nonvolatile memories that we just talked about. From 1984 to the present, the area of the single cell, i.e., the equivalent of a bit, has increased from 30 to 0.08 μm^2, climbing to 32 times in 20 years! The set of cells on a chip is called the matrix; matrices of this type reach densities of billions of cells in a few square millimeters [Fig. 10.6].

Roberto Luca together: That's enough for today; we're tired. See you tomorrow and we'll continue the story. Come on – show us one of your paintings!

Gianni: Tomorrow. Today it's too late.

End of Day One

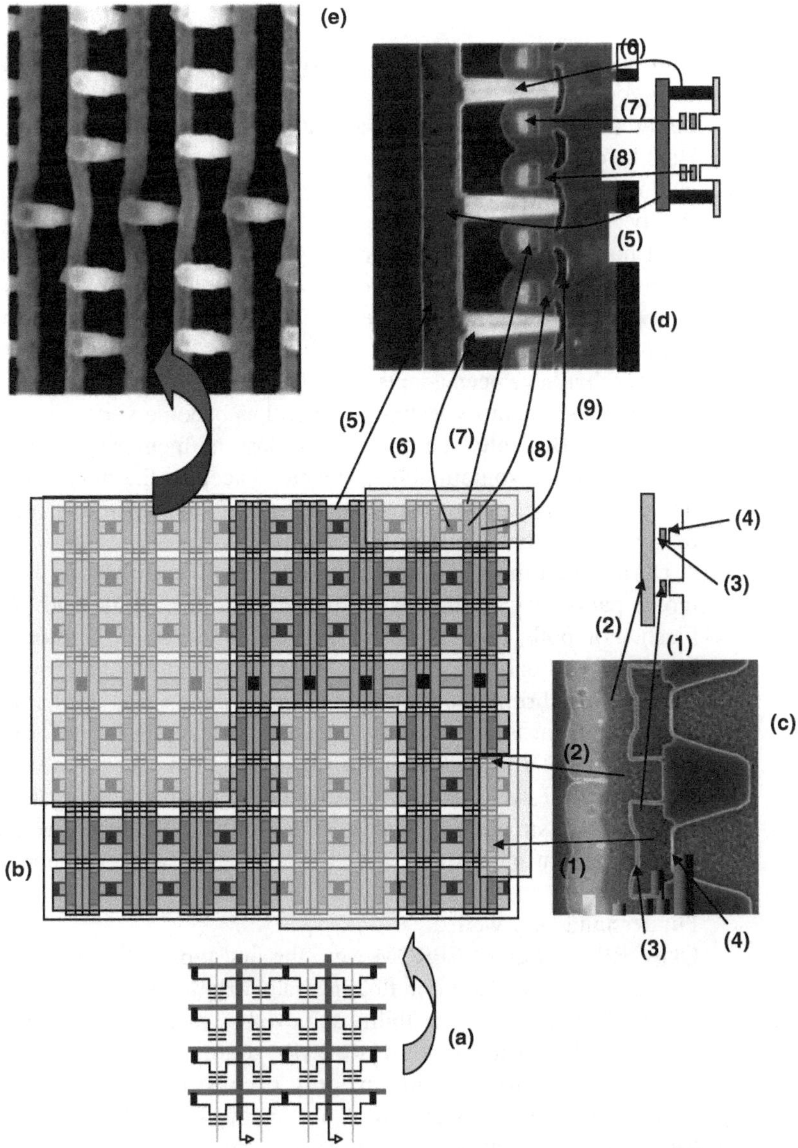

Fig. 10.6 A piece of matrix Flash NOR cell, the schematic (**a**), the layout, the drawing used to built the chip (**b**). Different parts of layout are analyzed compared versus the real chip photographs

Day Two

Luca: And here we are again. How are you? Certainly the holidays are a very good thing.

Roberto: I am really happy. Has Gianni already arrived?

Luca: Here he comes.

Gianni: Hello everyone. How are you? Are you ready for the second round?

Roberto: Ready. We're all ears!

Gianni: So, I hope you remember everything we said about the non-volatile memory. Let's summarize in order to get to the applications in use today. Non-volatile memory technology evolution has been fast and has led to a variety of technologies and architectures with peculiar features that have been adapted to different application fields. Among them is Flash memory technology that has become very popular and successful. Basically it represents a class of memories electrically erasable and re-programmable, capable of keeping the stored information even if a power loss occurs or the power supply itself is switched off by the external controller.

During the time, Flash technology has evolved following two different paths, originating two different families: NOR and NAND Flash. For both, the atomic information is stored in memory cells based on the floating gate technology evolving through a progressive lithographic shrink which determines continuous area reduction and cost effectiveness. Each cell is capable of retaining the information represented by a single bit (Single Level Cell = SLC) or combinations of bits (Multi Level Cell = MLC). Cells are grouped and organized in arrays whose structure differs in NOR and NAND Flash. Each array is organized in blocks. For NAND Flash, each block is made up of pages.

Luca: I understand very well.

Gianni: Okay. EPROM and EEPROM were the first type of solid- state non-volatile memories based on floating gate technology. EPROM's main characteristic was the erasability achieved by removing the chip from the motherboard to expose it to a UV source; 20 min of exposure were necessary to bring the memory back to the charge neutrality. Of course the process erased all the memory. EEPROM was electrically erasable but the extreme granularity of the information was obtained by means of a selection transistor, one for each memory cell increasing the memory silicon size.

Roberto – AHA! So the first non-volatile memories were not so easy; you had to disassemble everything to erase them.

Gianni: Yes, just like that, we had to remove the EPROM from the board and they had a transparent window so that UV rays could reach the chip (Fig. 10.7). A new paradigm coming from the new wireless market necessities, alterability became the name of the game. In the 1990s,

Fig. 10.7 An EPROM memory chip. Note the quartz window, transparent to the UV rays, which was used to erase the chip

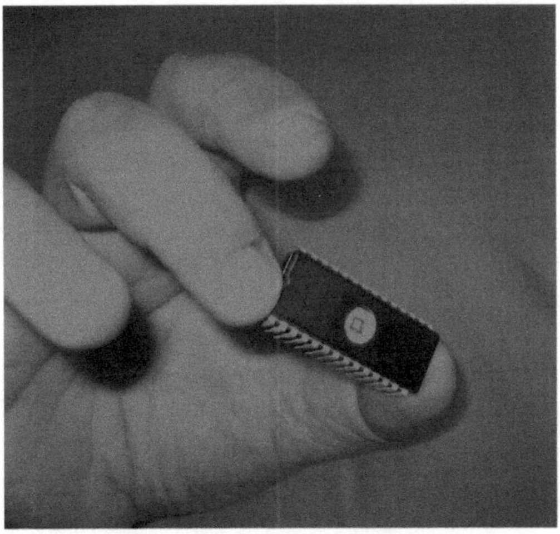

Flash memories with the possibility of modified memory content with an electrical erasing, no UV needed, decreased the silicon size with the introduction of the sector concept. Sector, a collection of a certain number of cells, is the minimum granularity for erasing. This new approach opened the door to all wireless applications – such as the first cellular phone. Running the EPROM legacy, the first Flash chip used two external voltages, 5 V for the reading operation and 12 V for the memory content modification: program and erase. To simplify the customer operation for removing one voltage on board, a single voltage device was designed, with only 5 V at the first and then 3 V and afterwards down to 1.8 V to save energy and increase battery life. This entire path implied deep changing on technology, design and new concepts introduction. Memory at that time was the "samurais" of the market.

Roberto: "Samurai" means servant, right?

Gianni: Yes, that's exactly what I meant.

Luca: So over the years, development has been driven by the needs of new applications?

Gianni: Sure, especially with the introduction of mobile technology. Over the years, multilevel emerged as another opportunity to increase memory capacity. Counting the number of the electrons that a program puts inside the floating gate, we can discriminate not only if we have or do not have electrons inside the memory, but also how many, to generate different meaning. After a real fight to subdue the whims and sometime the freaks of the technology today, the 2-bit/cell is a reality and the 3- and more bit/cell glance up on the market for NOR and NAND. And then with the new millennium, a new way to exchange music, videos, photos, etc., again changes the perspective for the memories.

Roberto: Wait a minute. We understood the history of the electrons in the floating gate, but now that you talk about counting the electrons, maybe I didn't understand everything.

Gianni: Yes, okay. Let's go back. You understand that the threshold is the parameter that is modified and that we achieved it by placing the electrons in the floating gate. The interesting and impressive thing is that the electrons are very small. To get an idea, the mass of an electron is estimated to be about 10×10^{-31} kg, i.e., zero point thirty zeroes and then a one!

Well, we put a few thousand of them in a floating gate and we must be able to understand whether or not they are present, so that we can say if we had put in the cell a logical "1" or a logical "0". As if that were not enough, during the last few years we invented techniques that allow, in a certain way, "counting" the electrons, allowing you to put in one cell, not only "1" or "0", recognizable by the presence or absence of electrons, but also more bits in the same cell, somehow counting how many electrons there are. For example, if we assume that we can put up to 3,000 electrons, we can say, in a binary logic with one bit per cell, that if there are 0 electrons, cell represents a logical "1". If there are 3,000, a logical "0". In a multibit logic per cell, if we are able to count the electrons, we can say that if there are 0, electrons cell represents a "11", if there are 1,000 a "10", 2,000 a "01" and 3,000 a "00".

Luca: Then it must be very complicated to make the whole thing work.

Gianni: Yes, you've got to consider that we demand that the information remain for years and that it not be altered by external events, such as temperature that tends to let electrons fall away from the floating gate, changing the meaning of the logical value recorded.

Roberto (turning to Luca): So when you tell us that your phone loses its memory when it is heated, the reason is obvious!

Gianni: The main causes of failure are related to electrical stress resulting from cycles. We can define a cycle as the set composed of an erasing and programming operation; the number of cycles executable on a memory cell is a key quality parameter for a device type of Flash non-volatile memory. The cycling induces mechanisms of failure due, for example, to charge trapping or spurious programming and erasing that induce erased cells to be written and vice versa.

The reality of a Flash device is that the operations of programming and erasure generate distributions of thresholds of cells that are different for each cycle. The erase operation, performed simultaneously on many cells, produces distributions of the electrical characteristics of cells that can lead to non-functioning of the device. The project must carefully consider all these problems and work to reduce any possible marginality. So the difficulty of designing a non-volatile memory device should be clear. It is a living thing that changes over time and sometimes, like all organisms, is sick in some of his parts; some cells may, over time, not function well, but are "cured" by applying algorithms of temporary recovery. On the contrary, a device which is only logical, such as a microprocessor, is much more complex in terms of system architecture, but frozen in time after working and therefore more easily reproducible on an industrial scale.

Consider then that in a 100 Watt light bulb in your home, something like 10 billion electrons are moved while we are playing with a few hundred of them now.

Luca: Very impressive. And what are the most frequent applications for these things?

Gianni: Well, among the Flash memory applications which are most used today, there are the Flash Memory Cards (FMC or Card) now widely used in portable applications such as MP3 players, digital cameras, GPS systems, etc., where they play the role of major systems for information storing. The FMC is a system composed of at least two integrated circuits – memory and controller; the controller comes with software to handle communications between memory and the outside world. The storage element consists of one or more NAND flash memory. The use of Flash memory is motivated by the benefits this technology offers in terms of integration, power consumption and durability, compared to other types of non-volatile memories. In particular, the NANDs are distinguished by the speed of data programming of up to 10 MB/s: This parameter affects directly, for example, the time that we have to wait between one shot and another one when we use a digital camera. Seen from the outside, the card is a non-volatile device that communicates with the surrounding world (usually known as the host) through a number of interface signals, depending on the type of protocol that implements the card; protocol is the "language" by which host and card can communicate with each other.

Communication between host and cards is generally designed to save data in the card (writing operations) or to retrieve data from a card reading operation), so it is error free. A first classification of cards, in terms of protocol, is based, on the one hand, on the width of the data bus shared between the card and host (in particular, it differentiates itself from serial and parallel protocols) and, on the other hand, on the presence or absence of a synchronization signal between the card and host. For example, the MultiMediaCard(tm) protocol is a serial synchronous one, while the CompactFlash(tm) protocol is a parallel asynchronous one (Fig. 10.8).

Roberto: Beautiful. Those are the ones that I use in my camera. From now on when I take a picture, I'll think of the quantum effects that make it possible!

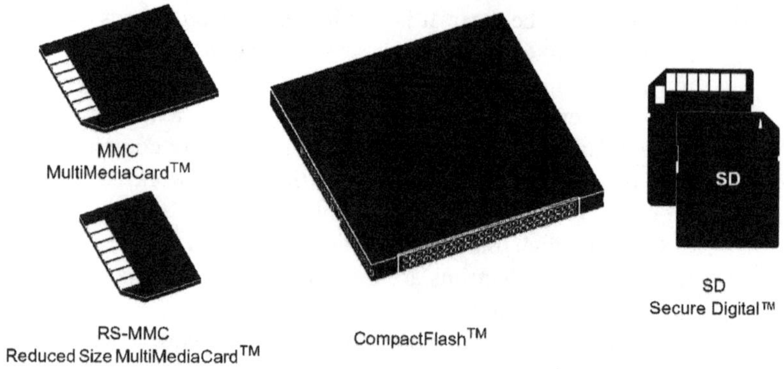

Fig. 10.8 CARD typology

Gianni: And I'm not finished. In order to increase the density of available memory, the implementation technology on silicon has made great strides, but the demand for more and more complex devices and, for example, for increasing the available memory capacity, has led to the development of an assembly technology that also allows the use of the dimension "z," putting more chips in the same package, In this way, stacking the chips in a package to form a "stack" of devices, called just a stacked one, or stacking them side by side, it was possible to obtain devices with huge capacities and different types of memory in the same product. To achieve this, the silicon chips were thinned, from an 800 μm initial thickness of a slice at the end of the processing factory of spreading, up to 50um, to be assembled. In the same package, non-volatile memory such as Flash memory with volatile-type RAM, both static (SRAM) and dynamic (DRAM or LPSDRAM) or pseudo-static (PSRAM) can be inserted. Today, in a package thickness of 1 mm, one can easily find space for 8 devices, from 16 GB of memory each, overcoming Moores Law (Fig. 10.9).

Luca: I would say that you've destroyed me. I have an incredible amount of data in my head.

Roberto: Let's talk about volatile memory now.

Gianni: As you wish. Let's begin. Well, the first thing to say is that these memories are called volatile because, unlike non-volatile, which we have just seen, they lose information when the power is removed. This is obviously very limiting.

Luca: I don't understand, then, why you don't use only the non-volatile ones that we have just seen?

Gianni: Each type of memory has different characteristics and, depending on their "skills", they're used for different purposes. Until a few years ago, high-density memories were just the RAM and non-volatile ones reached lower densities. Today it is no longer like that, but what still

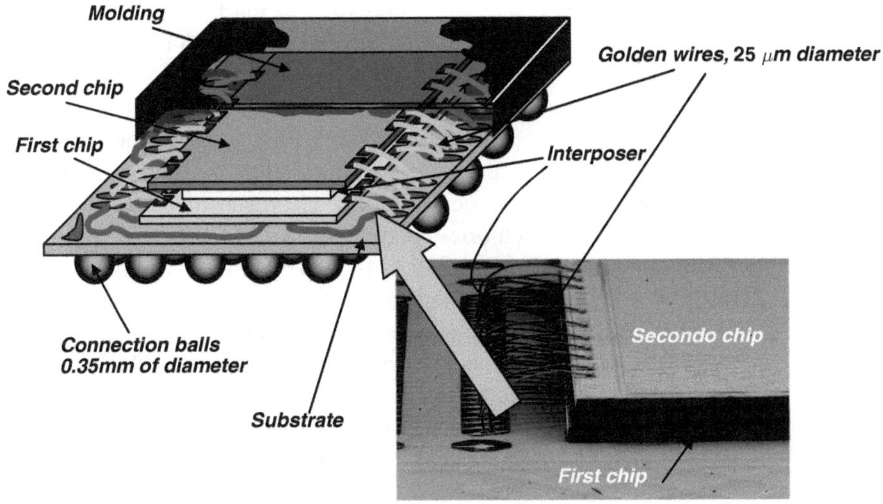

Fig. 10.9 A "stacked" with two assembled devices. Today the final total thickness dimension is less than one millimeter

makes RAM interesting is their speed in reading and in modification of information – especially the modification. In non-volatile memory the information is modified for a very long time, in scale of time of electronic systems, even up to seconds, while the information in RAM is written, erased or read in nanoseconds – that is, billionths of a second. When you perform calculations or data processing, much intermediate data is read and written and would not be possible to think of as always working with non-volatile memories.

Roberto: Then if I understand correctly, the number of times you can read and write non-volatile is limited.

Gianni: The main limitation is for the write mode, more than the read mode; the permitted maximum write number is around of hundred of thousand. Maximum reading numbers are in the millions but, for volatile memories, numbers of reading and writing are virtually unlimited.

Luca: And why do non-volatile ones have such short lives?

Gianni: It's a bit like exhausting jobs. To move charges into the floating gate, we need very high potentials and the use of these tends to subject the material to wear, and the voltages used in the RAM are much lower. So, let's start with the RAM. First, a few words on the meaning. RAM stands for Random Access Memory.

Luca: You mean that you've access in it as you want?

Gianni: Not as you want, but where you want. The first memories have been those on tape, as I said yesterday, and to find the desired information, you must go through the whole tape. They have the information related to an address instead, a bit like for the homes of a city. The postman would have a lot to do if the houses were all in a row without addresses!

Roberto: But how do you communicate the addresses to memory?

Gianni: Each writing cycle is coded in two steps. First you send the address to which you want to send or retrieve information and then you send the information, always on device's pins. However, there are two large families of RAM memories, static ones, SRAM, where S stands for Static, and dynamic ones, DRAMs, where D stands for Dynamics. DRAMs are much more complex than SRAMs but to explain how the memory cell works, which is based on two different principles in the two cases, I think it's better to start from DRAM. To understand how the SRAM memory cell works we have to understand what feedback is, and it's not so easy. We will try to understand later. So let's start from the DRAM. The element that we need to know is the capacitor.

Luca: that one of steam?

Gianni: No, of course not. The capacitor is an electronic component, but before you get there, it is better, as usual, to think of an analogy. It's things that everybody know, I hope, the existence of electrical charges.

Roberto: Without beating around the bush, could you get to the point? Well, we know that you see us as cavemen, but perhaps this is too much!

Gianni: Okay, I exaggerated. However, the electric charges are of two types, positive and negative.

Luca: Yes, yin and yang and then we do a lu of Tai ji Quan? Perhaps the chen style.

Gianni: I will, when the Master will teach me. So, back to electric charges. When you isolate an electrical charge, it tries in every way to draw opposite charges to itself. I think a better example is the lightning one (Fig. 10.10).

In various ways, the surface of the clouds becomes electrically charged and consequently the ground becomes charged with opposite

Fig. 10.10 The lightning directions could be different, as in the figure. Lightning forms the negative cloud center to the ground; other discharges can be verified between different parts of the cloud. Sometimes it is possible to have lightning between the ground and the positive top of the cloud

polarity. All proceeds until the amount of the charge has reached its permissible maximum and the air between the cloud and the ground begins to be crossed by the electrical charges, giving rise to lightning.

Roberto: Therefore, there is a limit to the number of electric charges that may be on surfaces?

Gianni: Yes, exceeded (or come through, go over, to cross, to overstep, please choose the best) this value (or amount) we "break" what is called the dielectric, that is the element that separates the two surfaces on which the charge has accumulated. This structure, i.e., the two surfaces and the element that separates them, is called the capacitor. A capacitor is an element that is able to "contain" electrical charges. There are two problems. The first is how do they get their positions and the second is why they stay there. Let's look at the first problem. Here we show a capacitor, in the simplest way, i.e., with plane parallel faces. Then one end of the capacitor is attached to the ground, as the soil of which we have said before for the lightning, then, we attached the other end, as if they were the clouds, to a switch. The switch plays the role of "*deus ex machina*".

Luca: Sure! Then what?

Roberto: Come on, you're really rough; *deus ex machina* is a Latin phrase borrowed from the Greek; it comes from Greek theater. In this context, when it was necessary to call in a god (or gods) on the scene, the actor who played the god took place on a rudimentary wooden crane, moved from a system of ropes and winches, called *mechanè*. In this way, the actor was sent down from above, thus simulating the intervention of a god descending from heaven. Indeed the term *deus ex machina* means "god who comes out of the machine". The *ex machina* intervention of the gods was often used to resolve a complicated circumstance which had seemingly no other way out. Then the meaning of this expression was expanded over time, used for any story where the internal logic of the story itself appeared quite unlikely and was used so the author could finish the story as intended. In current usage, the term is used to indicate the sudden and unexpected intervention of someone who intervened decisively on intricate and complex matters.

Luca: Wow, what a culture!

Gianni: Thanks for the explanation – really beautiful and erudite. So this capacitor, attached to the switch, forms the memory cell. Information coming from the fact that, inside the capacitor, we can have or not electrical charges, electrons.

Roberto: Therefore, in this case, too, the electric charge is put on a point as in the case of non-volatile ones.

Gianni: It has actually something similar, but this time, the point at which the charge is placed is one of the plates of the capacitor, not an isolated point; on the contrary, it should be attached to a switch that serves to bring it the charge.

Luca: But the switch is insulated by definition, meaning that when it's in the "open" position, it disconnects the two ends. More isolate than that!

Gianni: This is true, but the fact is that you cannot build a switch that is truly isolated. It could be done, only on a small plate of a few square millimeters of silicon, but building several billion of them is a little complicated.

We use a MOS transistor instead, which is a good approximation of a switch, but without being it truly. The end of the condenser attached to the MOS transistor isn't isolated but it's connected to a material that forms the drain of the transistor and so it's capable of dispersing the charge.

Luca: What sizes are we talking about? Still very small, I suppose, as in the previous case of non-volatile ones?

Gianni: Yes, the dimensions are now sub-micrometrical. Then we were saying that electric charges are carried on a capacitor plate through a switch – in our case, a MOS transistor. So it seems done; with a switch I add the charge and then you take it off, making it always pass through the same switch and so I made the cell. If there is a charge, it is a logical "1"; if the charge is absent, it's a logical "0". Any questions?

Roberto: But in this case it's like a non-volatile one? I don't understand!

Gianni: Right. The problem is that the plate of our condenser, thus produced, is not a floating node. On the contrary, since we use an MOS, without having to spell things out, every point of contact is achieved with a piece of silicon so we don't have a switch like the light switches in our houses that contain plates that physically separate the two poles; everything here is always attached but we are able to "turn off" the MOS transistor, but there are other possible paths for the electric charges. Ultimately, in our cell, made up by the capacitor, we have very few charges, on the same order of magnitude as those that we put on the floating gates of non-volatile ones and so it's enough that they lose only a few of them to change the information status in the cell, to discharge the capacitor.

Roberto: It's not easy to imagine this not-ideal switch that interrupts the charge's flow, without being insulating.

Gianni: I was actually trying to think of an analogy but I can't think of anything in particular; while I think, let's finish what I wanted to say to help you understand the core of the operation. Now, we put the charges on the condenser, then we disconnect the capacitor, but this is not a floating node as good as that of a non-volatile one and always emits a bit of charge that wants to leave. Do not forget that we have put some electrical charges in a point and electric charges of the same sign tend to reject themselves, so after putting them inside, they reject each other, trying to move away from each other. Ultimately, because of the way in which the cell is built, the few electric charges that are put into the capacitor tend to go away within a few thousandths of a second.

Luca: So what?

Gianni: And then it built a refreshments mechanism. There's an internal counter that scans, with the right frequency, all the memory addresses and before the cells lose their information they are read and rewritten, so the information is always preserved.

Luca: Beautiful, and how to reconcile this with the fact that the memory is also used from the outside? The two mechanisms do not overlap?

Gianni: No, limits are set to the maximum number of queries that can be made from the outside on the memory, then the memory takes the time it needs for the refreshments. Let's say she makes the toilet and that, as a beautiful woman, when she's occupied to make-up herself, there's no way to dissuade her from her employment.

These are the RAM memories, those that define the execution speed of your PC. Now we can briefly explain what they are and what they are for, those ones that are in your PC and which you discuss when you buy it. Most memories at stake are of three types – the hard disk drive, non-volatile memories, which contain the BIOS, and the dynamic RAMs used to run. The hard drive is really like an old record; only it is not vinyl, but glass, with around 7,000 revolutions per minute. The information is recorded on it with magnetic tracks; the advantages are the low cost. I recently bought a disk from a terabyte (one trillion bytes) at a ridiculous price. The disadvantages are the low speed of writing and reading, in milliseconds, and the high-power consumption. You write on it all the information, programs, etc. The hard disk is a non-volatile memory; it's not solid state. Then there are the non-volatile memories, which we have already seen, where you write programs that the computer runs upon waking.

Roberto: A bit like when you wake up in the morning and you automatically do what you need to do, without really thinking about what you are doing and where you are.

Luca: In your case even to who are you!

Gianni: Yes, just like that; in this case it uses the reading speed and retention of data. Finally, the RAM memories on which we've first downloaded programs and all data from your hard drive, and then make them run, by using its high speed of writing and reading. The problem is that the data does not remain when you turn off your PC.

Luca: This overview is very nice, letting us understand that there is a precise sense for everything and how the choices are always present to optimize even minimal activity.

Gianni: Few people think about the complexity of the things we use every day that are the result of the work of great minds who have put incredible imagination into the understanding of nature.

Roberto: I'm more and more disconcerted by those who claim that science and technology are dry. What does it mean? Understanding how the world works is always an incredible job. The world works in mysterious ways and it's stranger than the most amazing science fiction novel.

Gianni: I absolutely agree with you. Changing the subject, I was thinking about the analogy for the MOS transistor used as switch. Imagine two small lakes as two holes in the ground; the first one is connected to a source, while the second is connected to the first through a lock. Opening[2] the lock, we connect the first lake to the second one. However, when the lock is closed, the water is present in the first lake because this is connected to the source, whereas the water of the second lake tries to go in all directions – for example, by penetrating into the ground. This corresponds exactly to the getaway of electric charges from the plate of the condenser, which is built on the same piece of silicon as the switch.

Luca: Perhaps it's a little clearer now.

End of Day Two

Day Three

Gianni: And now let's talk about SRAM.

Luca: It looks like a dirty word.

Roberto: Or medicine for a stomachache.

Gianni: I'm getting a stomachache, because to understand how the elementary cell of an SRAM memory or static RAM works, we must understand the concept of feedback. Although it seems strange, because few people know and even fewer use it, the concept of feedback is one of those modalities that exists almost everywhere in electronics, biology, economics, etc. To summarize, it's the control of the input on the output.

Luca: Hold it! You mean it's like the control at the entrance to a parking lot, to ensure that there is a parking space for everyone. You watch how many cars go out so you know how many you can let go in.

Gianni: Yes, in some ways even this is feedback but it's a little more complex. Think about what happens when we grasp an object with one hand – trying not to drop it, or if it is fragile, like an egg, not to break it; our brains continuously monitor the pressure of fingers, and the posture of our whole body, to ensure the result. That way it seems simple, but in reality the study of feedback systems is a very important part of modern engineering; we apply it everywhere and to do it properly we have developed mathematical techniques that allow us to study them.

---

[2]In the lock's case, we say that you open it to let the water pass; in the case of the electrical current switch, the opening corresponds to the non-functioning, i.e. of absence of current passage. Analogies are always very limited.

Roberto: Try to give us some easier examples.

Gianni: Okay. Let's start with an example by borrowing what a friend of mine told me about what her son does at home. She told me that the other day she had to scold him because he was hammering on the table. The only way to stop him was to slap him. So, in response to an output – pounding the table, and an input – the slaps – the hammering stopped. In this case we are talking about negative reaction. The intervention of the reaction either stops or stabilizes the output. Here's another example: Imagine driving a car and you want to maintain a constant speed. The gas pedal is steady, and you're in the same gear. At this point, if we encounter a hill, we shift into a lower gear and we increase the pressure on the accelerator. If at the end of the climb we come to a descent, depending on its slope, we can do various things – for example, we can stay in the same gear and reduce pressure on the accelerator, maybe initially, and then begin to brake. In this way we can constantly maintain the speed of our system, until the system itself allows it. We are getting negative feedback, defined in this way when a variation of the inputs is reflected in the system by changing the parameters so that the output remains constant.

Luca: In fact, if I think about it, there are many situations such as the one that you've described.

Gianni: Yes, but the fun is yet to come. We've described the situation called negative feedback; output increasing or decreasing will "work" on the input to bring it to the fixed conditions. Now consider the case where my friend's son, after the first hammer blow to the table is slapped, and in response to that, he hammers at the table again. What does the mother do? Let's suppose that she slaps him again, but harder, and in response to that, the child hits the table even harder – until the table is destroyed. Here's a less bloody example. Have you ever been at a concert where suddenly the loudspeakers begin to whistle?

Roberto: It's what's called the "Larsen Effect".

Gianni: That's right! And do you know what causes it?

Roberto: I think that it depends on the fact that if you put the microphone too close to the loudspeakers, you generate an exaggerated amplification effect, and finally, even if I did not understand why, everything whistles.

Gianni: The equipment that is used by those who play and sing is mainly composed of three elements – the microphone, which has the job of gathering the sound and turning it into electrical information that is usable through equipment; the amplifier that produces the amplification of the electrical signal, without changing, as much as possible, the content; and, finally, the loudspeakers that do the opposite of the microphone, i.e., transform the electrical signal into audible sound. The chain works well if the sound goes from the microphone to the loudspeakers and nowhere else. If it happens that the output sound from the loudspeakers, already amplified, is picked up by the microphone, and then later sent to the amplifier that amplifies it and sends it back over to the loudspeakers and so on, you arrive very quickly at the maximum possible amplification that is translated into a noise that we perceive as an uncontrolled hissing. This is a good example of positive feedback.

Luca: is it over here?

Gianni: No, there is a third possibility, in addition to the positive and negative feedback, and that is the situation where the reaction produces neither a damping nor an amplification but an oscillation.

Luca: Wait. Before we continue, let's try to understand the concept up to this point although I do not understand what you're driving at. I think that you have take it from a in a roundabout way.

Gianni: Yes, it's true but I could not do it without it. If you want to understand, even superficially, how a static RAM works, you must learn a new concept, particularly that of the feedback, and then I'll let myself go a little. Anyway, do you understand up to this point?

Roberto: I'll try to summarize: Then, the systems of any kind, are made so that the results of the outputs are somehow carried back to the inputs, which is where the outputs are then generated. In this way you create a circle that allows the automatic control of the system. You've given us two examples of feedback – the negative one, where the signal is used on the output, carrying it back to the input, to stabilize the output to the value decided. The positive one, instead, gives rise to a kind of closed circle so that eventually the output value continues to increase, or decrease, up to the maximum, or minimum, possible.

Gianni: That's all correct. For completeness, I was saying that there is also the possibility of having a reaction that is neither good nor bad but that results in the controlled oscillation of the output. But what interests me now is the positive reaction. An analogy that explains the operation is really difficult to find. Let's think about a system such as in Fig. 10.11, where the two women are on two platforms that can go up, depending on the weights that are on the counterweight platforms.

The other two little men who are on the counterweight platforms can take the weights and move them onto the platform of the opponent. Initially, the platforms are all the same height, half-way up. When you give the starting signal, our little men on the platforms start to throw their weights onto the platforms of the other one. Imagine that one of them is able to start earlier, thereby unbalancing one side of the women's platform. The little man who is higher up makes a minor effort to throw its weight onto the opponent's platform while the little man who is lower down will make more of an effort to throw his weight on the platform of the opponent who's higher. Ultimately, if you can start in a certain direction, it becomes easier to confirm the result, rather than turn the situation around. The departure in some

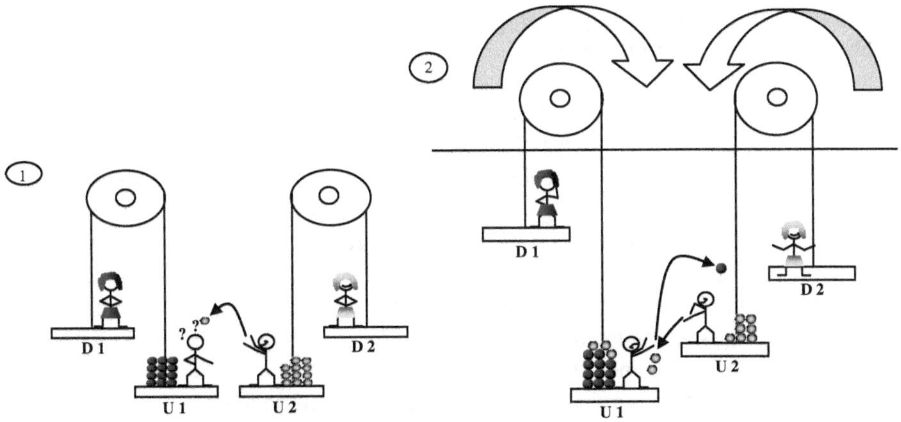

Fig. 10.11 The analogy shows a positive feedback system. The man able to start first increases his advantage thanks to the system itself. In this example win the girl that reach the ground. In effect, no one wins!

way confirms the result until reaching the equilibrium – that is, one of the two women has gone up to the point of safety and the other fell to minimum limits instead. As an example, this isn't really the best, but analogies are difficult. The difficulty in understanding the reaction is due to the fact that understanding goes through to a dynamic process that brings the system into a state of equilibrium. So we must think of the development of the whole event.

Luca: I think I've understood, the result of a moment goes in to confirm the result itself, until it reaches a point of equilibrium.

Gianni: It's just like that. And now with the transistors. The SRAM cell consists of four transistors attached in a way we can say have "crossed" each other, so that the two output nodes, which here are the two women, after a writing operation, are one down and one up. The difference, in the case of an SRAM memory cell is that the information is not up or down in space but a voltage value. So you assign the logical value "0" to the situation, looking at the example in Fig. 10.11. D1 is at the top and D2 is at the bottom, while for the logic, "1" is the reverse.

Roberto: Genius! And what does the reaction have to do with it?

Gianni: The problems consist of ensuring the electronic switching within the time that you want and then the staying of the data that, when the power is shut off, they go away; both women go towards the ground, if we can think of the weights as the electrons of the current. The power supply is removed and the electrons are no longer available to be moved to. All these things need to be prepared with ample time and hindsight to make sure that the system works. The study of this feedback system is made by applying the mathematical theory that describes it and that is called "automatic controls".

Roberto: Give me another example of feedback or one of automated systems.

Gianni: Yes, there is one that I really like. Have you ever noticed those small "handlebars" on high-voltage poles on high-voltage pylons?

Luca: I don't know what you mean by "handlebars?" Are you talking about bicycle dumbbells?

Gianni: Oh, no, I can't believe it! Let's go to the window. Do you see that high-voltage pylon? (Fig. 10.12). Do you see those things on the wires?

Roberto: I never noticed them before. What are they? What are they for?

Gianni: They're small masses that are attached to high-voltage lines to dampen any oscillations of the wires themselves. The high-voltage lines in many places may be subject to the oscillations caused by the wind and, if it's too strong, this can move the wires too much and cause damage. These dampers are calculated so that if the wires oscillate too dangerously, *they* wobble instead of the wires.

Practically, the masses that make up the dampers and linkages between themselves and the wires are calculated so that the movement of the wire is transmitted to the damper. Again, the calculations are performed at a desk, studying the vibration frequencies that most annoy high-voltage cables.

Fig. 10.12 A high voltage pylon, with the mechanical dampers indicated

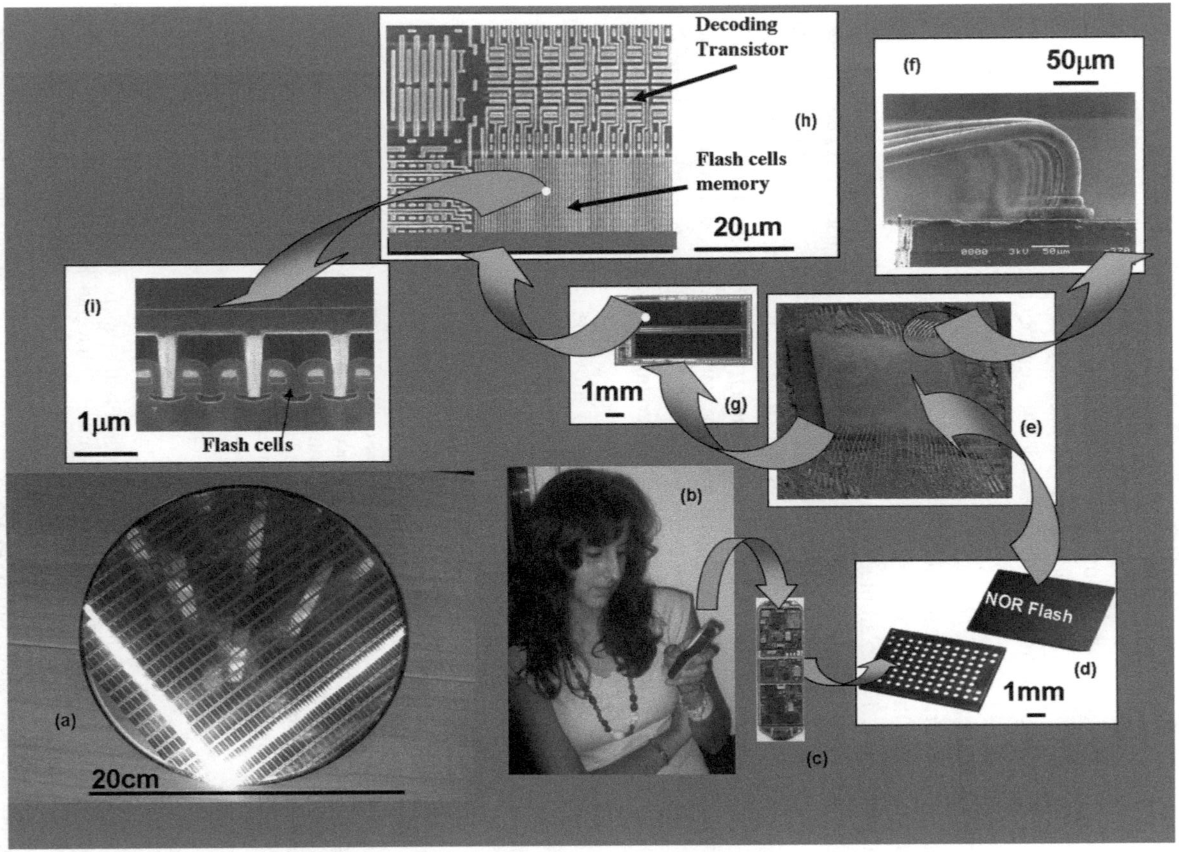

Fig. 10.13 From (**a**) the silicon wafer, (**b**) the cellular phone, (**c**) the board (**d**) package closed, (**e**) package opened, (**f**) the bonding wires, (**g**) chip photo, (**h**) a closeup of the chip and (**i**) a closeup of the matrix cells. A rules for each image

Roberto: Very interesting, really.

Gianni: We can say that I've explained the three types of memory that are most used today – the Flash, from the non-volatile family, and SRAM and DRAM, belonging to the volatile family.

I would like you to look at this summarizing panel (Fig. 10.13), where we see the story that we talked about – (a) the silicon wafer, (b) he cellular phone that contains (c) the board, (d) where we found the chip assembled inside the package, (e) the package opened, (f) a detail of the bonding wires used to connect the chip with the package, (g) the chip photography, (h) a closeup of a portion of the chip and (i) a closeup of the memory cells inside the matrix. Near each photograph is a ruler to estimate the real dimension.

Luca: But there are other kinds of them?

Gianni: Yes. For example, we have not talked about the type of memory used by your computer's hard drive, one that uses magnetic properties, whereas for solid state ones there are many others which try to exploit other properties but always with the idea of having an electrical parameter to be modified in order to store the information.

Roberto: Thanks, you've given us a good overview. When will you tell us the difference between these memories and the brain?

Gianni: To tell this, one should first understand how the brain works, but that would be another book.

Luca: Okay, gotta go. Are you going to show me a painting or not?

Roberto: Yes, come on. You promised!

Gianni: Okay. Here it is.

End of Day Three

Acknowledgments I would like to thank the following people who helped me write this book by reading the manuscript and helping me clarify many of the ideas:

My son Marcello, who is a fourth-year student of Communications; my daughter Alice, a third-year high school student concentrating in the sciences; my wife, Alessandra, who has a degree in biology and teaches high school; my brother Renato, who studied chemistry and now works on pollution issues for a government agency; my friend Giacomo Buzzetti, who works in publishing; and my friend Andrea Silvagni, who works in electronics.

References

1. (April 2003) Proc IEEE 91(4). Special Issue on the Flash Memory Technology
2. (January 2009) Proc IEEE 97(1). Special Issue on the 3D Integration Technology
3. Campardo G et al (2001) Architecture of non volatile memory with multi-bit cells. 12th Bi-annual Conference June 20–23, 2001, INFOS2001, Invited paper
4. Campardo G, Micheloni R, Novosel D (2005) VLSI-design of non-volatile memories. Springer Series in Advanced Microelectronics, 2005. ISBN: 3-540-20198-X
5. Micheloni R, Campardo G, Olivo P (2008) Memories in wireless systems. Springer, Berlin. ISBN: 978-3-540-79077-8, 2008

Index

G. Campardo et al. (eds.), *Memory Mass Storage*,
DOI 10.1007/978-3-642-14752-4, © Springer-Verlag Berlin Heidelberg 2011

Kolibri 360 GmbH
Ferdinand-Jühlke-Straße 7
99095 Erfurt, Deutschland
produktsicherheit@kolibri360.de